Lecture Notes in Computer Science 9136

More information about this series at http://www.springer.com/series/7407

Arnold Beckmann · Victor Mitrana
Mariya Soskova (Eds.)

Evolving Computability

11th Conference on Computability in Europe, CiE 2015
Bucharest, Romania, June 29 – July 3, 2015
Proceedings

Springer

Editors
Arnold Beckmann
Department of Computer Science
Swansea University
Swansea
UK

Mariya Soskova
Sofia University
Sofia
Bulgaria

Victor Mitrana
University of Bucharest
Bucharest
Romania

ISSN 0302-9743 ISSN 1611-3349 (electronic)
Lecture Notes in Computer Science
ISBN 978-3-319-20027-9 ISBN 978-3-319-20028-6 (eBook)
DOI 10.1007/978-3-319-20028-6

Library of Congress Control Number: 2015940742

LNCS Sublibrary: SL1 – Theoretical Computer Science and General Issues

Springer Cham Heidelberg New York Dordrecht London

Printed on acid-free paper

Springer International Publishing AG Switzerland is part of Springer Science+Business Media
(www.springer.com)

Preface

CiE 2015: Evolving Computability
Bucharest, Romania, June 29 – July 3, 2015

The evolution of the universe, and us within it, invites a parallel evolution in understanding. The CiE agenda – fundamental and engaged – targets the extraction and development of computational models basic to current challenges. From the origins of life, to the understanding of human mentality, to the characterizing of quantum randomness – computability theoretic questions arise in many guises. The CiE community this year met for the first time in Bucharest, to carry forward the search for coherence, depth, and new thinking across this rich and vital field of research. In line with other conferences in this series, CiE 2015 had a broad scope and provided a forum for the discussion of theoretical and practical issues in computability with an emphasis on new paradigms of computation and the development of their mathematical theory.

The conference series Computability in Europe is organized by the Association CiE. The association promotes the development of computability-related science, ranging from mathematics, computer science, and applications in various natural and engineering sciences, such as physics and biology, as well as the promotion of related fields, such as philosophy and history of computing. In particular, the conference series successfully brings together the mathematical, logical, and computer sciences communities that are interested in developing computability-related topics.

The host of CiE 2015 was the Faculty of Mathematics and Computer Science of the University of Bucharest.

The ten previous CiE conferences were held in Amsterdam (The Netherlands) in 2005, Swansea (Wales) in 2006, Siena (Italy) in 2007, Athens (Greece) in 2008, Heidelberg (Germany) in 2009, Ponta Delgada (Portugal) in 2010, Sofia (Bulgaria) in 2011, Cambridge (UK) in 2012, Milan (Italy) in 2013, and Budapest (Hungary) in 2014. The proceedings of all these meetings were published in the Springer series *Lecture Notes in Computer Science*. The annual CiE conference has become a major event and is the largest international meeting focused on computability theoretic issues. CiE 2016 will be held in Paris, France.

The series is coordinated by the CiE Conference Series Steering Committee consisting of Arnold Beckmann (Swansea, chair), Laurent Bienvenu (Paris), Alessandra Carbone (Paris), Barry Cooper (Leeds), Natasha Jonoska (Tampa FL), Benedikt Löwe (Amsterdam and Hamburg), Florin Manea (Kiel), Dag Normann (Oslo), Mariya Soskova (Sofia), and Susan Stepney (York).

The Program Committee of CiE 2015 was chaired by Victor Mitrana (Bucharest) and Mariya Soskova (Sofia). It was responsible for the selection of the invited speakers and the special session organizers and for running the reviewing process of all submitted regular contributions.

The conference had two tutorials by John Reif (Duke Unversity) and by Stephen Simpson (Pennsylvania State University), and one public lecture by Mircea Dumitru (University of Bucharest and Romanian Academy).

In addition, the Program Committee invited seven speakers to give plenary lectures: Ann Copestake (University of Cambridge), Pawel Gawrychowski (University of Warsaw), Julia Knight (University of Notre Dame), Anca Muscholl (Université Bordeaux), Gheorghe Paun (Romanian Academy), Alexander Razborov (University of Chicago and Steklov Mathematical Institute), and Vlatko Vedral (University of Oxford).

Springer generously funded a Best Student Paper Award. For the second year in a row the winner was Ludovic Patey. His contribution to this year's volume is entitled "Iterative Forcing and Hyperimmunity in Reverse Mathematics."

The conference CiE 2015 has six special sessions: two sessions, Representing Streams and Reverse Mathematics, were introduced for the first time in the conference series. In addition to this, new developments in areas frequently covered in the CiE conference series were addressed in the further special sessions on Automata, Logic and Infinite Games, Bio-inspired Computation, Classical Computability Theory, and History and Philosophy of Computing. Speakers in these special sessions were selected by the special session organizers, and were invited to contribute a paper to this volume:

Automata, Logic, and Infinite Games

Organizers. Dietmar Berwanger and Ioana Leustean
Speakers. Christian Georg Fermüller (Wien), Slawomir Lasota (Warsaw), Paulo Oliva (London), Michael Vanden Boom (Oxford)

Bio-inspired Computation

Organizers. Andrei Paun, Petr Sosik
Speakers. Erzsébet Csuhaj-Varjú (Budapest), Ion Petre (Turku), Alexandru Tomescu (Helsinki), Sergey Verlan (Paris)

Classical Computability Theory

Organizers. Marat Arslanov, Steffen Lempp
Speakers. Sergey Goncharov (Novosibirsk), Wei Li (Vienna), Frank Stephan (Singapore), Dan Turetsky (Vienna)

History and Philosophy of Computing

Organizers. Christine Proust, Marco Benini
Speakers. Felice Cardone (Turin), Laura Crosilla (Leeds), Baptiste Mélès (Nancy), Eric Vandendriessche (Paris)

Representing Streams

Organizers. Jörg Endrullis, Dimitri Hendriks
Speakers. Juhani Karhumäki (Turku), Jean-Eric Pin (Paris), Narad Rampersad (Winnipeg), Luke Schaeffer (Waterloo)

Reverse Mathematics

Organizers. Damir Dzhafarov, Alberto Marcone
Speakers. David Belanger (Ithaca, NY), Takako Nemoto (Ishikawa), Ludovic Patey (Paris), Paul Shafer (Ghent)

We received 64 non-invited contributed paper submissions, which were reviewed by the Program Committee and many expert reviewers. In the end, 42 % of the submitted papers were accepted for publication in this volume. In addition, this volume contains ten invited papers. Without the help of our expert reviewers, the production of the volume would have been impossible. We would like to thank all of them for their excellent work; their names are listed at the end of this Preface.

All authors who contributed to this conference were encouraged to submit significantly extended versions of their papers with unpublished research content to *Computability: The Journal of the Association CiE.*

The Steering Committee of the conference series CiE is concerned about the representation of female researchers in the field of computability. In order to increase female participation, the series started the Women in Computability (WiC) program in 2007, first funded by the Elsevier Foundation, then taken over by the publisher Elsevier. We are proud to continue this program with its annual WiC workshop and mentorship program for junior female researchers in 2015. Both initiatives are coordinated by Liesbeth De Mol. The workshop speakers are Johanna Franklin (University of Connecticut) Anca Muscholl (Labri, Université Bordeaux I) and Cezara Dragoi (CNRS/Inria/Equipe Antique, Ecole Normale Supérieure, Paris).

The organizers of CiE 2015 would like to acknowledge and thank the following entities for their financial support (in alphabetic order): the Association for Symbolic Logic (ASL), the European Association for Theoretical Computer Science (EATCS), Springer, the University of Bucharest, and the Asociatia Alumni Universitatii din Bucuresti. We would also like to acknowledge the support of our nonfinancial sponsor, the Association Computability in Europe (CiE).

We thank Andrej Voronkov for his EasyChair system that facilitated the work of the Program Committee and the editors considerably.

April 2015

Arnold Beckmann
Victor Mitrana
Mariya Soskova

Organization

CiE 2015 was organized by the Faculty of Mathematics and Computer Science of the University of Bucharest. The members of the Organizing Committee were Radu Gramatovici (Bucharest) and Liviu Marin (Bucharest).

Program Committee

Marat Arslanov (Kazan)
Jeremy Avigad (Pittsburgh)
Veronica Becher (Buenos Aires)
Arnold Beckmann (Swansea)
Laurent Bienvenu (Paris)
Alessandra Carbone (Paris)
S. Barry Cooper (Leeds)
Laura Crosilla (Leeds)
Liesbeth De Mol (Ghent)
Walter Dean (Warwick)
Volker Diekert (Stuttgart)
Damir Dzhafarov (Storrs, Connecticut)
Peter van Emde Boas (Amsterdam)
Rachel Epstein (Harvard)
Johanna Franklin (Hempstead, NY)
Neil Ghani (Glasgow)
Joel David Hamkins (New York)
Rosalie Iemhoff (Utrecht)
Emmanuel Jeandel (LORIA)
Natasha Jonoska (Tampa, FL)
Antonina Kolokolova (St. John's, NL)
Antonin Kucera (Prague)
Oliver Kutz (Bolzano)
Benedikt Löwe (Hamburg and Amsterdam)
Jack Lutz (Ames, IA)
Florin Manea (Kiel)
Alberto Marcone (Udine)
Radu Mardare (Aalborg)
Joe Miller (Madison, WI)
Russell Miller (Flushing, NY)
Mia Minnes (La Jolla, CA)
Victor Mitrana (Bucharest, Co-chair)
Dag Normann (Oslo)
Ian Pratt-Hartmann (Manchester)
Mehrnoosh Sadrzadeh (London)
Anne Smith (St. Andrews)
Mariya Soskova (Sofia, Co-chair)
Paul Spirakis (Patras and Liverpool)
Susan Stepney (York)
Jacobo Toran (Ulm)
Marius Zimand (Towson, MD)

Additional Reviewers

Andras, Peter
Bauwens, Bruno
Boasson, Luc
Bréard, Andréa
Bundala, Daniel
Carton, Olivier
Chailloux, André
Constantoudis, Vassilios
Dassow, Jürgen
Diener, Hannes
Dorais, François
Downey, Rod
Escardo, Martin
Freydenberger, Dominik D.
Gabbay, Murdoch
Gaspers, Serge
Gherardi, Guido
Gordeev, Lev

Goudsmit, Jeroen
Gregoriades, Vassilis
Grigorieff, Serge
Grozea, Cristian
Harizanov, Valentina
Hertrampf, Ulrich
Hetzl, Stefan
Hirst, Jeff
Hitchcock, John M.
Hoksza, David
Hu, Ting
Husfeldt, Thore
Ifrim, Georgiana
Inkpen, Diana
Istrate, Gabriel
Jervell, Herman Ruge
Kalimullin, Iskander
Kausch, Jonathan
Khoussainov, Bakhadyr
Kjos-Hanssen, Bjørn
Knowles, Joshua
Kopecki, Steffen
Kutrib, Martin
Lange, Karen
Lecroq, Thierry
Lešnik, Davorin
Lozes, Etienne
Lösch, Steffen
McNicholl, Tim
Melnikov, Alexander
Merkle, Wolfgang
Mignot, Ludovic
Mileti, Joseph
Miyabe, Kenshi
Monath, Martin
Mummert, Carl
Ng, Keng Meng Selwyn
Nies, Andre
Oger, Francis
Oitavem, Isabel
Pauly, Arno
Paun, Andrei
Paun, Gheorghe
Perifel, Sylvain
Petricek, Tomas
Petrisan, Daniela
Raffinot, Mathieu
Rao, Michael
Rettinger, Robert
Rojas, Cristóbal
Romani, Shadab
Rute, Jason
Sanders, Sam
Savchuk, Dmytro
Schlicht, Philipp
Sebastien, Labbe
Shafer, Paul
Shi, Yaoyun
Solomon, Reed
Soskova, Alexandra
Steiner, Rebecca M.
Stephan, Frank
Tadaki, Kohtaro
Terwijn, Sebastiaan
Teutch, Jason
Towsner, Henry
Tveite, Paul
Vaszil, György
Vatev, Stefan
Verlan, Sergey
Weiermann, Andreas
Westrick, Linda Brown
Woods, Damien
Zenil, Hector

Franco Montagna
1948 – 2015

Prof. Franco Montagna died on February 18, 2015, at 66. He was Professor of Mathematical Logic at the University of Siena, which he entered in 1973 working with Roberto Magari of whom he had been a student. Internationally well-known logician, Franco Montagna has authored and coauthored more than 120 papers (with an amazingly long list of national and international coauthors), which appeared in international journals of logic, algebra and computer science. Since 1973 until 1994, his main scientific interest was Provability Logic, a modal logic in which it is possible to express a counterpart of self-reference and incompleteness in arithmetic. His recent interest was the logic of uncertainty, especially manyvalued logics. In this field, his results are related to Łukasiewicz Logic with additional operators, like product and product residuation; the generalizations of Łukasiewicz logic, like Hájek's logic BL, and its further generalizations like the logic of GBL-algebras, or the monoidal t-norm logic, as well as the connections between many-valued logic and substructural logics. He had a special interest in the relationships between many-valued logics and probability, in particular the problem of coherence for probabilitic assessments. Although not a specialistic computability theorist, he cultivated and used computability theory throughout his work, with the rigour and elegance which characterized his mathematical style. He served in the Program Committee of CiE 2007. He will be remembered as a fine researcher, and as a precious teacher, beloved by his students. Colleagues and friends will remember him for his goodness, his generosity, and his gentleness.

Contents

Invited Papers

Contributed Papers

Invited Papers

Computers and the Mechanics of Communication

Outline of a Vision from the Work of Petri and Holt

Felice Cardone(✉)

Dipartimento di Informatica, Università di Torino, Turin, Italy
felice.cardone@unito.it

Abstract. Computers have become an integral part of a vast range of coordination patterns among human activities which go far beyond mere calculation. The conceptual relevance of this new field of application of computers has been advocated by Carl Adam Petri (1926–2010) and Anatol W. Holt (1927–2010), two computer scientists best known for their contributions to the subject of *Petri nets*, a graphical formalism for describing the causal dependence of events in systems distributed in space. We outline some fundamental, mainly epistemological aspects of their vision of the computer as a "communication machine."

1 Overview

One approach to the theoretical study of computation, that dates back to the early years of the discipline, has concentrated on the problem of synthesis of automata out of simple components, rather than on the closure properties of classes of computable functions. It is in this tradition that Carl Adam Petri (1926–2010) formulated a novel approach to automata theory that took seriously the physical limitations that such systems have to comply with, in particular upper bounds on the propagation speed of signals and on the density of storage of information. In Petri's 1962 PhD dissertation [26], these constraints were the guiding principles underlying the design of a class of asynchronous systems whose programming consists essentially in setting up communication rules among their parts in order to achieve coordination of behavior. The central questions here are relative to the causal relations among events: in particular, it is in this context that *concurrency* can find a general formulation as causal independence.

The ideas of Petri were pursued at an early stage, and partly in collaboration with him, by Anatol Wolf Holt (1927–2010), although he was less interested than Petri in the relations of these ideas to physics and mathematics, being primarily concerned with their use in the specification and analysis of computer systems. By setting up a theoretical approach to computing based on classes of Petri nets (a name introduced by Holt himself), and by demonstrating the expressive power of the latter in application areas ranging from the design of hardware components to the analysis of legal systems [25], Petri and Holt outlined a vision of a computer as a "general medium for strictly organized information flow" [28],

A. Beckmann et al. (Eds.): CiE 2015, LNCS 9136, pp. 3–12, 2015.
DOI: 10.1007/978-3-319-20028-6_1

a "medium for the transmission of messages between persons" [5], providing a theoretical complement to the ideas of Licklider [24] on the computer as a communication device.

An essential sketch of the scientific biography of Anatol Holt may contribute to appreciate his place in the development of computing in the United States.[1] After earning degrees in mathematics from Harvard (1950) and MIT (1953), where he was employed at the Research Laboratory for Electronics of Robert Fano as a research assistant in information theory, Holt started his reflection on computers and their role as a UNIVAC programmer in Philadelphia, where he worked for Sperry Rand Corporation from 1952 to 1960 and where, with W.J. Turansky, designed and implemented the *Generalized Programming* system [4]. In 1955, Holt became Associate Director of the Univac Applications Research Center, established by himself and J.W. Mauchly. After getting a PhD in descriptive linguistics from University of Pennsylvania under Zellig Harris (1963), while at Massachusetts Computer Associates, a subsidiary of Applied Data Research, Holt led the ARPA supported Information System Theory Project (1964–1968) [6] and subsequent projects until the end of the 1970s. These produced a large amount of theoretical and applied work on Petri nets, influencing in particular the research on asynchronous hardware and on data-flow computer architectures through relations with the Computation Structures Group led by J.B. Dennis at MIT, and contributing to spread the notions of concurrency, conflict and causality.[2]

As an example of the Petri net representation of these notions, we show the design of a two-stage asynchronous pipeline for bits, taken from an unpublished review from Holt's *Nachlass* dated November, 1976:

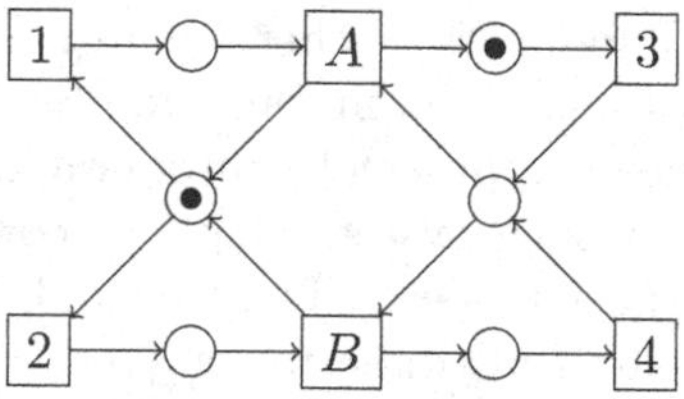

Here events 1 and 2 represent the input of a bit 0 or a bit 1, respectively, in the first stage of the pipeline, and similarly for events *A* and *B*. Conditions represent the states of each stage: empty, holding bit 0 and holding bit 1. The pairs of events 1 and 3, and 2 and 3 are concurrent: both events of each pair are enabled to fire according to the rules for the token game on nets [22]. The events 1 and 2 are in conflict: both are enabled because their common precondition is marked, but only one of them can fire, thus disabling the other. One important feature of this model of a pipeline is the synchronization of a forward flow of tokens in

1 For a biography of Petri, see the recent book by Einar Smith [34].

2 Extensive documentation on this work is now available at the page of the Defense Technical Information Center, http://www.dtic.mil/dtic/index.html, search for example for the strings "AD0704796", "ADA955303", "ADA047864".

the system, representing the bits, with a backward flow of tokens representing permits (or requests for data): this is essential to the design of safe asynchronous communication mechanisms.

As Director of the Computing Center of Boston University, Holt's interests focused more closely on the development of computer support for human organized activity and the related descriptive formalisms, like role/activity nets [12]. After an appointment at ITT as the leader of a Coordination Technology research group, in 1986 he founded and became Chief Technical Officer of Coordination Technology, working in particular on the development of new formalisms for coordination [17]. In 1991 he moved to Milano, Italy, where he continued his research on organized human activity interacting with researchers in the area of Computer Supported Cooperative Work at the Università di Milano and others interested in his ideas (like the present writer). The main outcome of this research was the book [19], soon followed by a shorter one which appeared in an Italian translation [20], and several papers on organized activity, in particular in relation to the novel approach to information suggested by his outlook [21], that he planned to investigate in a book of which several drafts survive. His last writing, as far as we know, has been an unpublished revision of the 1997 book, written in 2003.

The aim of this paper is to compose an introductory account of the vision of computer as a communication machine from the published and unpublished material by Petri and Holt, mainly focussing on the epistemological issues that underlie their endeavor rather than on the technical achievements.[3] After a short introduction to Holt's approach to systems as expression of organized human activities, we discuss its consequences on a systemic interpretation of the notion of *information* and outline two ways of looking at the interpretation of computers as communication machines: Holt's *communication mechanics* and Petri's *communication disciplines*. Finally, we comment on some mathematical aspects of the relations between discrete and continuous models of digital systems, motivated by communication mechanics and based on a topological interpretation of the primitive notions of Petri nets.

2 Computers, Systems and Communication

2.1 Communication Mechanics

Through his acquaintance with Gregory Bateson and his family[4] Holt interacted with cyberneticians and systemically oriented thinkers[5] developing a loosely

[3] Both Holt's *Nachlass*, that we are currently studying, and that of Petri, are preserved at the Deutsches Museum in Munich.

[4] Holt's mother, Claire Holt, collaborated with Bateson and his wife, Margaret Mead, as an expert of Indonesian art, especially of Balinese dance, see [2] for a biography.

[5] This happened in at least one important context, the Wenner-Gren Conference on the Effects of Conscious Purpose on Human Adaptation held in Burg Wartenstein (Austria) in 1968, as evidenced by the proceedings edited by Bateson's daughter, Mary Catherine [1]. Holt was one of the main characters of that conference which included as participants Gregory Bateson, Barry Commoner, Warren McCulloch and Gordon Pask, among others.

systemic attitude towards computing, whose main ideas can be gathered from his papers from the period 1968–1980, for example [6,7,9–11,13]. As an early illustration of Holt's interest in systemic notions, one of the concerns of the Information System Theory Project was the "analysis and description of data structures", where a data structure should not be thought of as "a static set of elements with interrelations" (ibid., p. 4), but rather should be identified with the basic operations made possible by that structure.[6] For example,

> we might consider beginning with a list of the elementary events which we suppose can occur involving a 1-dimensional array [...] Given this list of events we can now describe the constraints on their relative occurrences. [...] A set of mutually constrained events is what we call a *system* [6, pp. 4–6].

Examples of systems involve patterns of coordinated activity "such as the flow of traffic at a highway interchange, the operation of an elevator responding to calls on different floors, an iterative procedure for successively approximating the solution of a differential equation" (ibid., p. 9). In this perspective, systems are identified with the mechanical aspects of human organizations, expressed in rules whose function is to "establish certain relations of communication among a set of role players" [7]. Their investigation leads to *communication mechanics*,

> a theory about the mechanical aspects of communication – i.e., those aspects that have to do with the *rules*, insofar as these can be formalized, which define the relations among a set of communicating parts [...] We regard the words 'organization' and 'system' as referring to such bodies of rules [7].

Holt later developed communication mechanics into a theory of organized activity whose formalization instantiated the elements of the bipartite ontology of Petri nets, made generically of *conditions* and *events*, first as *roles* and *activities* [3,12] and, eventually, as *bodies* and *operations* [15,17,19]. Roles are the hallmark of organized activity: in the context of organized activities persons play *roles* and it is in such roles that they perform actions. By doing so, persons assume responsibilities that reflect organizational interests, becoming *actors*. As a machine cannot assume any responsibility for its actions, actions cannot be *performed* by machines: this remark lies at the heart of Holt's critique of the idea of artificial intelligence and, more generally, of the personification of computers [5]. In the system context, actions are units of *organizational time*: the time represented in calendars and planners and made up of lumps of human effort. Similarly, space enters this picture in the form of *organizational space*, for example files, cabinets, rooms or buildings. For all these examples,

[6] In this novel approach to the characterization of data structures we can see perhaps a first hint of techniques for data abstraction that would become a leading theme of programming language design in the next decade, culminating with the notion of (software) *object*.

> their *coordination-relevant structure* cannot be described in the language of physics. The description of this structure necessarily makes reference to *the manner of its use in the conduct of socially organized activity*, perhaps only by implication [18].

Organizational space has a topological structure whereby entities that interact directly are spatial neighbors, cfr. [12], an operational conception of topological relations which is distinctive of communication mechanics, as we shall see later.

2.2 Communication Disciplines

Holt was concerned with a notion of information subsuming the multitude of ways informational phenomena enter the picture of human organizations as supported by computers:

> Information processing and information flow within a machine cannot be adequately engineered without the power to see it as integral to information processing and flow within an organization [13] "the user" is a community with a diversity of roles and interests all of which *concurrently* affect the use of the machine (or complex of machines) [...] At any one time the machine is not serving one man but is participating in the establishment of a relation between several men [...] a very different matter than the standard concept of a "computational problem" [8].

An initial attempt at a technical notion of information supporting the study of the causal relations among decisions in systems, was explored by Holt and Fred Commoner in their contribution to the Project Mac Woods Hole conference in June 1970, in the restricted setting where a system is represented by a state machine [22]. Information is input at states where a decision has to be taken as to which of several possible actions should be performed next; dually, information is output when several actions lead to the same state: namely the information needed to backtrack from that state.

Holt's systemic notion of information matches a notion of communication which does not reduce, as in Shannon's seminal work, to "reproducing at one point either exactly or approximately a message selected at another point" [33]. The reception via fax of a perfect reproduction of a $100 bill does not count as a successful money transfer. Holt's first attempt at a criticism of traditional communication theory in a technical context, [9], reveals the hidden *organizational* assumptions underlying Shannon's theory: in the systemic treatment, information is studied as a correlate of coordination or, more generally, of organized activity.

According to the new theory of communication, in any organized human activity the material aspects, including representation, are essential: this is a recurring theme through Holt's reflection, from the early formulations of communication mechanics witnessed by Mary Catherine Bateson at the Burg Wartenstein conference:

> Tolly went up to the board and wrote the number two in several ways, 2, II, ii, 2, a tiny two and a monstrous one [...] — There is a profound illusion that it is possible in a systematic sense to separate the representation from what is represented [...] it is a fundamental error [...] The symbol is nothing apart from its uses [1, pp. 156–157]

to a late list of popular assumptions running counter to the proper foundations of a theory of organized activity, quoted from an unpublished draft translated into Italian as [20]:

> The logic of a formal procedure (and therefore organized activity) is clearly separable from the physical means by which it is performed (and therefore separable from time/spatial considerations) [...] in the study of organized activity the opposite is assumed.

The following table, extracted from [10, p. 166], summarizes the basic tenets of the systemic view of information by constrasting it with the conventional view:

Conventional	Systemic
Information is an imperishable good	Information has validity only within a given context
Information content and form are, in principle, factorable from one another	Form and content are, in principle, inseparable
Information can, in principle, flow from sender S to receiver R without information flowing from R to S	Information flow requires, in principle, circuits over which to flow (like electric current)

We have a shift from a theory of signal transmission like classical communication theory to a *formal pragmatics*[7] where computers are message processors:

> the systems point of view would [...] principally see computers as communication machines and principally see the material that flows through a computer as documents bearing messages [10, Lecture 3].

While applying his ideas to the design of electronic coordination environments [8], Holt made several examples of disciplines to be imposed on message-handling

[7] It is formal because its rules are entirely formulated in terms of roles, and it is a pragmatics in the etymological sense, because it concerns communication among actors as action performers. Pragmatics has an obvious bearing on the communication-oriented uses of information technology, and in fact we find ideas from the speech-act theories of Austin, Searle and Habermas at the basis of the 'language/action perspective' of Winograd and Flores and the related coordination programs, and also in the foundational work on information systems by the Scandinavian school of Langefors, Goldkuhl and Lyytinen, among others.

capabilities within a computer-based information system, like delegation of authority, addressing of messages and their identification and cancellation [16]. Petri [28] compiled a list of such *communication disciplines* classifying the functions of computer as a general medium for strictly organized information flow,

> disciplines of a science of communication yet to be created, and disciplines in the sense of keeping to a set of rules to be followed if communication is to be successful [30].

Understanding these rules becomes essential when the ordinary context of communication is replaced by a computer-based system, and the material carriers of messages are replaced accordingly. For example, an unproblematic notion of 'original' applies to paper based documents that can be transferred only with difficulty to electronic environments. A discipline of copying in this case is clearly relevant (after all, a copy is defined in opposition to an original), like a discipline of composition, "concerned with determining the structure of documents relative to a material or conceptual carrier" [28], relevant to the legal value of writing.

3 Continuous Discrete Behavior

The representation of communication relations in a digital context needs a way of representing motion:

> To communicate one must move [...] We will need a theory of motion suited to our ends: the analysis of communication relations. In this connection, the analysis of motion based on the mathematical continuum is not serviceable. It gives us no *systematic* way of relating communication intentions to the mechanics which implement them [7].

A sense of continuity of motion can be recovered by exploiting the topological aspects of the state/transition structure of digital systems. Petri had observed, circa 1972, that a Petri net with states S, transitions T and flow relation $F \subseteq (S \times T) \cup (T \times S)$ can be regarded as a topological space $X = S \cup T$ by forgetting the direction of flow taking a new relation $A = (F \cup F^{-1}) \cap (S \times T)$ and by defining $U \subseteq X$ to be *open* when $A^{-1}[U] \subseteq U$ [27]. In this topology $\{s\}$ is open for every $s \in S$ and $\{t\}$ is closed for every $t \in T$. In addition, this space has the properties that characterize the *net topologies* [31]:

- Arbitrary unions of closed sets are closed (equivalently, arbitrary intersections of open sets are open);
- Every singleton is either open or closed (the resulting topological space is $T_{\frac{1}{2}}$).[8]

Net topologies were intended originally to be used to define continuous mappings on nets as describing views of a system at different levels of abstractions. The basic insight at the basis of their definition has however a much wider import:

[8] In passing, we remark that closely related topologies have recently been exploited in the definition of the digital line for the purposes of digital image processing [23].

by describing the motion of bodies in space (in communication mechanics, the communication between parts) by means of nets, the events and interactions in which the moving bodies are involved are made part of the topological structure of the space:

> one ordinarily imagines that space and time can somehow be structured – i.e., subdivided into nameable entities with topological relations among them *before* describing the actions, interactions, movements, dispositions, etc. of the various distinguishable entities in which one is ultimately interested. We, on the other hand, will see temporal and spatial organization as logically related to the "drama" (or class of dramas) which are to take place in that frame [7, Sect. C].

Instead of spatial regions and their boundaries construed as point-sets, in the topological interpretation of nets we have (open) atomic regions of a state space whose (closed) transitions express the crossing of boundaries, in topological spaces whose points are better expressed by verbs (with different aspectual features) than by nouns. Net topologies allow to introduce notions like continuity, connectedness and boundary in the foundations of a new approach to models of digital phenomena.

> Though the notions 'state' and 'event' are "digital" concepts, ours will not rest on the fiction that events have no real duration – i.e., can be represented by time points in mathematical continuum – nor on the fiction that states have no real extension – i.e., can be associated with space points in a continuum [7].
>
> The coming about of an objectively verifiable change in anything takes time, and necessarily entails passage through a region of uncertainty. Our theory should also not presuppose the fiction of perfect classification schemes [...] Theories based on these fictions cannot account for the effort that must go into achieving reliability in systems [10, pp. 139–40].

On the one hand, these remarks lead to a theory of intransitive indifference relations, of which concurrency is an example, axiomatized by Petri [29,32] in relation to a finitistic view of continuity as a foundation for measurement. On the other hand, they set the stage for Holt's investigation of motion complying with the needs of communication mechanics, culminating in his unpublished contribution to the May 1981 MIT-IBM conference on Physics of Computation [14]. The basic picture behind his views, recurring through his prolonged meditation on the foundations of state/transition models, is that of a hiker who alternates the crossing of mountains (transitions) with being in valleys (states). One can set up a language describing his motion consisting of elementary statements involving the relations of the hiker with atomic regions of state space and their boundaries (like his being *in* a valley or his being *away* from a boundary). The logic of these statements permits then a purely linguistic interpretation of topological relations between regions of the state space of the hiker, their interiors and boundaries, and ultimately of net topologies. While this work has been left unfinished by

Holt, who was aware of its embryonic stage, it offers new technical and philosophical insights into the relation between continuity and discreteness and the possibility of building finite, small models of continuous phenomena that arise at the border between computing and physics. As a conclusion of our outline, we point at this as an interesting direction for future research arising from the work of Petri and Holt.

Acknowledgements. I am indebted to Anastasia Pagnoni for encouragement, and help with Holt's *Nachlass*, and to Marco Benini for his interest in this work. The financial support of Project LINTEL is gratefully acknowledged.

References

1. Bateson, M.C.: Our Own Metaphor. A Personal Account of a Conference on the Effects of Conscious Purpose on Human Adaptation. Smithsonian Institution, Washington, DC (1972)
2. Burton, D.: Sitting at the Feet of Gurus: The Life and Dance Ethnography of Claire Holt. Xlibris Corporation, Web Mounted (2009)
3. Grimes, J.D., Holt, A.W., Ramsey, H.R.: Coordination system technology as the basis for a programming environment. Electr. Commun. **57**(4), 301–314 (1983)
4. Holt, A.W.: General purpose programming systems. Commun. ACM **1**(5), 7–9 (1958)
5. Holt, A.W.: The personification of computers. Datamation **13**(3), 137–138 (1967)
6. Holt, A.W.: Information system theory project: final report. Technical report, RADC-TR-68-305, NTIS AD 676972, Applied Data Research, Inc., Princeton, NJ, September 1968
7. Holt, A.W.: Communication mechanics. Advanced Course on Operating Systems Principles, Istituto di Elaborazione dell'Informazione, Pisa, 20–31 August 1973 (course material)
8. Holt, A.W.: The design of a computer-based communication system. Technical Proposal P-7-002, Massachusetts Computer Associates Inc., Massachusetts, Wakefield, 26 February 1974
9. Holt, A.W.: Information as a system-relative concept. Technical Report CA-7409-3011, Massachusetts Computer Associates Inc., Wakefield, Massachusetts, 30 September 1974, published in Krippendorff, K. (ed.) Communication and Control in Society, pp. 279–285. Gordon and Breach Science Publishers (1979)
10. Holt, A.W.: Formal methods in system analysis. In: Shaw, B. (ed.) Computers and the Educated Individual, pp. 135–179. University of Newcastle upon Tyne (1975). http://www.ncl.ac.uk/computing/about/history/seminars/
11. Holt, A.W.: Petri nets and systems analysis. In: The MIT Conference on Petri-nets and Related Methods, July 1975 (reproduced in [10])
12. Holt, A.W.: Roles and activities. A system for describing systems, p. 58, 30 March 1979 (unpublished typescript)
13. Holt, A.W.: Computer-based information systems: the views of a quasi-wholist. IFIPS TC-9, Number 9, September 1980
14. Holt, A.W.: A mathematical model of continuous discrete beahvior. Technical report, Massachusetts Computer Associates Inc., Wakefield, Massachusetts, 11 November 1980

15. Holt, A.W.: Coordination technology and Petri nets. In: Rozenberg, G. (ed.) APN 1985. LNCS, vol. 222, pp. 278–296. Springer, Heidelberg (1985)
16. Holt, A.W.: Identification: Generally and in ICECT, 12 February 1986 (unpublished draft)
17. Holt, A.W.: Diplans: a new language for the study and implementation of coordination. ACM Trans. Off. Inf. Syst. **6**(2), 109–125 (1988)
18. Holt, A.W.: The mechanics of organized human activity, p. 100 (1988), book draft, including book overviews
19. Holt, A.W.: Organized Activity and its Support by Computer. Kluwer, Norwell (1997)
20. Holt, A.W.: Ripensare il mondo. Il computer e i vincoli del sociale. Masson, Milano (1998)
21. Holt, A.W., Cardone, F.: An organisational theory of information. In: Falkenberg, E.D., Lyytinen, K., Verrijn-Stuart, A.A. (eds.) Information System Concepts: An Integrated Discipline Emerging. IFIP Conference Proceedings, vol. 164, pp. 77–91. Kluwer, Netherlands (2000)
22. Holt, A.W., Commoner, F.: Events and conditions. In: Dennis, J.B. (ed.) Record of the Project MAC conference on concurrent systems and parallel computation, pp. 3–52. ACM, New York (1970)
23. Khalimsky, E., Kopperman, R., Meyer, P.: Computer graphics and connected topologies on finite ordered sets. Topology Appl. **36**, 1–17 (1990)
24. Licklider, J.C.R., Taylor, R.W.: The computer as a communication device. Sci. Technol. **76**, 21–31 (1968)
25. Meldman, J.A., Holt, A.W.: Petri nets and legal systems. Jurimetr. J. **12**(2), 65–75 (1971)
26. Petri, C.A.: Kommunikation mit Automaten. Ph.D. thesis, Darmstadt Technical University (1962), English Translation as Technical report RADC-TR-65-377, vol. 1, Supplement 1, Griffiss Air Force Base (1966)
27. Petri, C.A.: Concepts of net theory. In: Mathematical Foundations of Computer Science: Proceedings of Symposium and Summer School, 3–8 September, pp. 137–146. Mathematical Institute of the Slovak Academy of Sciences, Strbské Pleso (1973)
28. Petri, C.A.: Communication disciplines. In: Shaw, B. (ed.) Computing System Design, pp. 171–183. University of Newcastle upon Tyne (1976). http://www.ncl.ac.uk/computing/about/history/seminars/
29. Petri, C.A.: Modelling as a communication discipline. In: Beilner, H., Gelenbe, E. (eds.) Measuring, Modelling and Evaluating Computer Systems. North-Holland, Amsterdam (1977)
30. Petri, C.A.: Cultural aspects of net theory. Soft. Comput. **5**, 141–145 (2001)
31. Petri, C.A.: Mathematical aspects of net theory. Soft. Comput. **5**, 146–151 (2001)
32. Petri, C.A., Smith, E.: Concurrency and continuity. In: Rozenberg, G. (ed.) Advances in Petri Nets. LNCS, vol. 266, pp. 273–292. Springer, Berlin (1987)
33. Shannon, C.E.: A mathematical theory of communication. Bell Syst. Tech. J. **27**, 379–423 (1948)
34. Smith, E.: Carl Adam Petri. Eine Biographie. Springer, Berlin (2014)

Error and Predicativity

Laura Crosilla(✉)

School of Philosophy, Religion and History of Science
University of Leeds, Leeds LS2 9JT, UK
matmlc@leeds.ac.uk

Abstract. The article surveys ideas emerging within the predicative tradition in the foundations of mathematics, and attempts a reading of predicativity constraints as highlighting different levels of understanding in mathematics. A connection is made with two kinds of error which appear in mathematics: local and foundational errors. The suggestion is that ideas originating in the predicativity debate as a reply to foundational errors are now having profound influence to the way we try to address the issue of local errors. Here fundamental new interactions between computer science and mathematics emerge.

1 Certainty and Certification

Mathematics is often considered the most exact of all sciences, but error is not unusual even in print. Errors are also costly and unwelcome in computer science, where program verification is increasingly appealed to in order to minimize failure in hardware and software.

I would like to suggest a distinction between two types of error which can appear in mathematics. First of all there are what I should like to call *"local"* errors. These are errors which plague individual proofs, or possibly a relatively small group of proofs which share a similarity in structure. These mistakes are somehow confined to a small portion of the mathematical enterprise and, if corrigible, they can be amended without introducing any substantial revision of the underlying mathematical principles one appeals to when devising the given proof. Secondly, there are *"foundational"* mistakes, which instead relate to the very principles of proofs and the axioms; these occur when inconsistencies arise within our foundational systems.

Nowadays emphasis is on local errors. The ever growing specialization of mathematics has made proofs much harder not only to obtain, but also to verify. In addition to the complexity, the mere size of some proofs demands new strategies for their verification. A substantial debate on the role of computers for both the discovery and the verification processes in mathematics is presently ongoing within the mathematical community, with an increasing number of

A. Beckmann et al. (Eds.): CiE 2015, LNCS 9136, pp. 13–22, 2015.
DOI: 10.1007/978-3-319-20028-6_2

mathematicians hoping that a fruitful interaction between mathematics and computer science will produce substantial benefits for today's mathematics.[1]

As to the second kind of error, a large number of mathematicians seem by now quite confident that we have obtained inductive corroboration of our mathematical practice and also of our main mathematical systems; for example, theories like ZFC seem to have undergone sufficient scrutiny over the years to be considered reliable by most. However, a genuine concern regarding the trustworthiness of our mathematical methodology and of our foundational systems was voiced at the turn of the 20th century, as a direct reply to the deep methodological changes that mathematics was undergoing at the time, as well as the rise of the set-theoretic paradoxes.

The main topic of this article is *predicative mathematics*, a form of mathematics which originated at the beginning of the 20th century in attempts to address the threat arising from potential foundational errors, by proposing methodological constraints. Predicativity constraints are motivated by a varied family of concepts, and, as further hinted at below, inspire a number of rather different forms of mathematics. This complexity makes it very difficult to discuss predicativity within the limits of a short article and venture to draw some general conclusions. However, I would like to highlight some common themes which run through the debate on predicativity since its earliest times. I would also like to recall the difficulties encountered in understanding the demarcation between the notions of predicative and impredicative. Notwithstanding these difficulties, the hope is to be able to suggest a reading of predicativity constraints as an instrument for singling out (substantial) portions of classical mathematics which are amenable to a less abstract, and a more concrete treatment. A predicative treatment of those portions of mathematics then is often seen by its proponents as providing full conviction for the correctness of the results. That is, by restricting the methodology to more stringent canons, one ought to gain more detailed insight of the constructions carried out within a proof, and fuller grasp of the results. In the case of constructive predicativity as represented by the tradition arising within Martin-Löf type theory, the ensuing mathematics obtains a distinctive *direct* computational content. Here predicativity constraints enable an identification of mathematics with programming which has inspired fundamental research at the intersection between mathematics and computer science. In fact, recent years have seen the rise of attempts to proof-checking portions of mathematics with the support of computer systems, opening up new paths for the verification of mathematics. The suggestion I would like to make, then, is that ideas originating in the predicativity debate as a reply to foundational errors are now having profound influence to the way we try to address the issue of local errors.

[1] Dana Scott in his opening talk at the The Vienna Summer of Logic (9th–24th July 2014) suggested that we are now witnessing a paradigm change in logic and mathematics. At least in certain areas of mathematics, there is an urgent need to solve complex and large proofs, and this requires computers and logic to work together to make progress. See Dana Scott's e-mail to the Foundations of Mathematics mailing list of 28–07–14 (http://www.cs.nyu.edu/mailman/listinfo/fom).

2 Predicativity

Predicativity has its origins in the writings of Poincaré and Russell, and is only one of a number of influential programmes which arose at the beginning of the past century in an attempt to bring clarity to a fast changing mathematics. Mathematics, in fact, had undergone deep methodological alterations during the 19th century which soon prompted a lively foundational debate. The paradoxes that were discovered in Cantor's and Frege's set theories in the early 20th century were one of the principal motivations for the very rich discussions between Poincaré and Russell, within which the concept of predicativity was forged (see for example [19,22,23]). These saw impredicativity as the main source of the paradoxes, and attempted to clarify a notion of predicativity, adherence to which would hinder inconsistencies. According to one rendering of this notion, a definition is impredicative if it defines an object by quantifying on a totality which includes the object to be defined. Through Russell and Poincaré's confrontation a number of ways of capturing impredicativity and explaining its perceived problematic character emerged. One influential thought (originating in Richard and, via Poincaré, adopted and particularly pressed further by Russell) saw impredicativity as engendering from a vicious circularity, or self-reference.[2] According to this view, a vicious circle arises if we suppose that a collection of objects may contain members which can *only* be defined by means of the collection as a whole, thus bearing reference to the definiendum. As a response to these difficulties Russell introduced his well-known vicious circle principle, which in one formulation states that: "whatever in any way concerns all or any or some of a class must not be itself one of the members of a class." [24, p.198]

Perhaps an example could help clarify the issue of impredicativity. The most paradigmatic instance of antinomy is Russell's paradox, which was discovered by Russell in Frege's Grundgesetze in 1901. A modern rendering of the paradox amounts to forming Russell's set, $R = \{x \mid x \notin x\}$, by unrestricted comprehension. One then obtains: $R \in R$ if and only if $R \notin R$. A circularity arises here from the fact that R is defined by reference to (i.e. quantification on) the whole universe of sets, to which R itself would belong. Russell's vicious circle principle, then, endeavours to prevent R from selecting a collection. The perceived difficulty arising from this kind of circularity can be elucidated from a number of perspectives. For example, according to one view, we ought to have access to a well-determined meaning for the condition appearing in the above instance of the comprehension principle (i.e. $x \notin x$). The difficulty is then realated to the fact that we seem to be unable to grant this independently of whether or not there exists a set R as specified above [4].

The analysis of the paradoxes turned out to be extremely fruitful for the development of mathematical logic[3], starting from Russell's own implementation of the vicious circle principle through his type theory. In the mature version

[2] Another analysis proposed by Poincaré [20] stressed a form of "invariance" as characteristic of predicativity: a predicative set cannot be "disturbed" by the introduction of new elements, contrary to an impredicative set [3,11].

[3] See [3] for a rich discussion of the impact of the paradoxes on mathematical logic.

of [23] two crucial ideas are interwoven: that of a type restriction and of ramification. By combining these two aspects ramified type theory seems to block all vicious circularity, and thus paradoxes of both set-theoretic and semantic nature.

Russell's type theory is a first fundamental contribution to the clarification of the complex question of what is predicativity in precise, logico-mathematical terms. However, as a way of developing a predicative form of mathematics Russell's type theory encountered substantial difficulties; it eventually surrendered to the assumption, in Principia Mathematica [29], of the axiom of reducibility, whose effect (in that context) was to restore full impredicativity. However, another attempt to develop analysis from a predicative point of view was proposed by Weyl [28], who showed how to carry out (a portion of) analysis on the basis of the bare assumption of the natural number structure. Weyl's crucial idea was to take the natural number structure with mathematical induction as given, as an ultimate foundation of mathematical thought, which can not be further reduced. Restrictions motivated by predicativity concerns were then imposed at the next level of idealization: the continuum. Weyl, in fact, introduced restrictions on how we form *subsets* of the natural numbers; in today's terminology, he saw as justified only those subsets of the natural numbers of the form $\{x : \varphi(x)\}$ if the formula φ is arithmetical, that is, it does not quantify over sets (but may quantify over natural numbers). The idea was that the natural numbers with full mathematical induction constitute an intuitively given category of mathematical objects; we can then use this and some immediately exhibited properties of and relations between the objects of this category (as obtained by arithmetical comprehension) to ascend to sets of natural numbers. In this way one also avoids vicious circularity in defining subsets of the natural numbers, as the restriction to number quantifiers in the comprehension principle does not allow for the definition of a new set by quantifying over a totality of sets to which the definiendum belongs.

I wish to highlight two aspects of Weyl's contribution. First, his approach to the question of the limit of predicativity went directly at the core of the mathematical practice, to show that large parts of 19th century analysis could be recovered on the basis of this restricted methodology. He thus succeeded in reducing to predicative methodology a conspicuous segment of mathematics, including portions which prima facie required impredicativity. Second, Weyl saw only this part of classical mathematics as fully justified; as he quickly became aware that not all of classical mathematics could be so recovered, he was ready to give up the rest, as (so far) not fully justified.

After Poincaré, Russell's and Weyl's fundamental contributions, predicativity lost momentum until the 1950's, when fresh attempts were made to obtain a clearer demarcation of the boundary between predicative and impredicative mathematics. The literature from the period shows the complexity of the task, but also witnesses the fruitfulness of the mathematical methodology for the philosophy of mathematics. The celebrated upshot of that research is the logical analysis of predicativity[4] by Feferman and Schütte (independently) following

[4] According to a notion of predicativity given the natural numbers which is discussed in the next section.

lines indicated by Kreisel [4,10,25,26]. Here Russell's original idea of ramification had a crucial role, as a transfinite progression of systems of ramified second order arithmetic indexed by ordinals was used to determine a precise limit for predicativity. This turned out to be expressed in terms of an ordinal, called Γ_0, which was the least non-predicatively provable ordinal. A formal system was then considered predicatively justifiable if it is proof-theoretically reducible to a system of ramified second order arithmetic indexed by an ordinal less than Γ_0.[5]

Another crucial contribution to the clarification of the extent of predicativity was the mathematical analysis of predicativity, aiming at elucidating which parts of mathematics can be expressed in predicative terms [5,27]. Work by Feferman, as well as results obtained within Friedman and Simpson's programme of Reverse Mathematics have shown that large parts of contemporary mathematics can be framed within (weak) predicative systems. Ensuing these results, Feferman has put forth the working hypothesis that all of scientifically applicable analysis can be developed in the system W of [5], which codifies in mordern terms Weyl's system in Das Kontinuum. These recent developments help better understand the reach of predicative mathematics, and reveal that predicativity goes much further than previously thought.

3 Plurality of Predicativity

As clarified by Feferman [4,6], the logical analysis of predicativity aimed at determining the limits of a notion of predicativity *given the natural numbers*. That is, one here takes an approach to predicativity similar to Weyl's, in assuming for given the structure of the natural numbers with full induction, and then imposing appropriate predicativity constraints on the formation of subsets of the natural numbers.[6] With Kreisel and Feferman the study of predicativity becomes thus an attempt to clarify what is implicit in the acceptance of the natural number structure (with full induction).[7]

Different incarnations of predicativity have however appeared in the literature, giving rise to very different forms of mathematics. For example, predicativity constraints have motivated Nelson's predicative arithmetic [16] and Parsons' criticism of the impredicativity of standard explanations of the notion of natural number [18]. According to Nelson already the whole system of the natural numbers equipped with full mathematical induction is predicatively problematic on grounds of circularity [16]: "The induction principle assumes that the natural number system is given. A number is conceived to be an object satisfying

[5] See [6] for an informal account of this notion of predicativity and for further references.

[6] The resulting notion of predicativity is, in fact, more generous than in Weyl's original proposal. The proof theoretic strength of a modern version of Weyl's system, like, for example, Feferman's system W from [5], equates that of Peano Arithmetic, and thus lays well below Γ_0.

[7] This line of research has been brought forward with Feferman's notion of unfolding, as analysed further by Feferman and Strahm e.g. in [7].

every inductive formula; for a particular inductive formula, therefore, the bound variables are conceived to range over objects satisfying every inductive formula, including the one in question." [16, p.1] From this point of view, then, already the theory of Peano Arithmetic, with its unrestricted induction, lies well beyond predicativity. Therefore Nelson's rejection of circularity leads him to justify only systems which are interpretable in a weak fragment of primitive recursive arithmetic, Robinson's system Q.[8]

Themes stemming from the original predicativity debates also play a prominent role within constructive mathematics, for example in the work of Lorenzen and Myhill [12], and in Martin-Löf type theory [14]. For constructive foundational theories, a more 'liberal' approach to predicativity, compared with that by Kreisel-Feferman-Schütte, has been suggested. Here the driving idea is that so-called *generalised inductive definitions* ought to be allowed in the realm of constructive mathematics. The intuitive justification of inductive definitions is related to the fact that they can be expressed by means of *finite rules*, and allow for a specification of sets which proceeds from the 'bottom up'. The underlying idea is to start from a well understood structure, say the natural numbers, and then use finite rules to extend this, by a process of successive iterations. We thus build a first subset of the set of natural numbers according to the rule, then use this to build a new one, and so on. The predicativity of this process is granted provided that we can ensure that at no stage in the built up of the new set, we need to presuppose a totality "outside" the set under construction. If this were the case, then, we would rely exclusively on increasingly more complex fragments of the very set under definition, and no vicious circularity would occur.[9] An important point to make is that the proof-theoretic strength of so-called theories of inductive definitions goes well beyond Feferman and Schütte's bound (and thus also very much beyond Peano Arithmetic), as shown in [1]. Following this line of reasoning, relatively strong theories are considered predicative in today's foundations of constructive mathematics [17,21].

A remarkable fact which emerges starting from the detailed logical analysis initiated in the 1950's, is that we now witness a number of different versions of predicativity, that appear to relate to very different forms of mathematics. Thus predicativity constraints motivate Nelson's strictly finitary subsystems of Peano Arithmetic, but also the much more generous predicativity given the natural numbers, which, under the analysis by Kreisel, Feferman and Schütte, extends well beyond Peano Arithmetic. Further up in the proof theoretic scale, we have constructive predicativity, which (on the basis of intuitionistic logic) reaches the strength of rather substantial subsystems of second order arithmetic [21]. In fact, the use of intuitionistic logic and its interaction with predicativity

[8] As such, Nelson's ideas have proved extremely fruitful, as they have paved the way for substantial contributions to the area of computational complexity [2].

[9] Theories of inductive definitions are discussed in [4], where they are considered unacceptable from a predicative point of view on grounds of circularity. See also [18] for an alternative view which sees inductive definitions as justified from a constructive perspective.

makes it more difficult to assess the relation between this kind of predicativity and the others. But it would seem that in all cases predicatively motivated constraints can be "applied" to different initial "bases", different mathematical structures which are taken as accepted, or granted. A possible understanding of predicativity would then see it as a (series of) methodological constraints, often motivated by the desire to avoid vicious circularity, which can be implemented on top of a previously given base, considered secure and granted. Predicativity constraints then impose methodological restrictions on the mathematical constructions which populate the next higher level of abstraction. For example, predicativity given the natural numbers takes the natural number structure with full induction as unquestionable and builds predicatively motivated restrictions on top of it, thus constraining the notion of arbitrary set.

A very significant aspect which emerges here is the crucial role of the principle of induction for debates on predicativity. In fixing the conceptual framework which we take as basis, we have to explicitly clarify how much induction we are prepared to accept. That is, it would seem that induction (possibly appropriately restricted) is a crucial component of the structure one takes as base, and, as highlighted by Nelson and Parsons, plays a crucial role in discussions of impredicativity. In less neutral terms, it would seem that when looking at the conceptual framework of reference, we need to include not only the relevant objects, for example the natural numbers, but also the way we are to reason about them. The example of constructive predicativity also seems to support similar conclusions, suggesting to include even the logic within the base one takes for granted.

There is here some complex philosophical work which is required to justify the choice of the privileged base as well as the methodological restrictions to be put on place. It is not unusual within the literature on predicativity to find reference to the time-honoured distinction between potential and actual infinity in mathematics. Often then predicativity constraints are seen as ways of avoiding full commitment to actual infinity; this, in turn, is frequently linked to the philosophical debate on realism versus anti-realism in mathematics. From a perspective of this kind, for example, one might be prompted to accept predicativity given the natural numbers, from the desire to subscribe to some form of realism with respect to the natural number structure, while maintaining an anti-realist (e.g. a definitionist) position on arbitrary sets [6]. Here I would like to suggest another possible reading of predicativity, which cashes it out in terms of our understanding of mathematical concepts.[10] Predicativity now becomes a crucial instrument in arguing for differences in levels of understanding, and conceptual clarity. Predicativity given the natural numbers, for example, would now represent a way of vindicating a commonly preceived difference in understanding between the concept of natural number and that of real number or, more generally, of arbitrary set [6]. That is, one here attempts to capture a distinction between forms of understanding, rather than ontological status, claiming that some concepts are more fundamental, or clearer, or more evident than others.

[10] A view along similar lines is also hinted at by Feferman in [6].

Predicativity constraints then could be seen as ways of extending beyond those more fundamental concepts (the conceptual basis) in ways which are somehow already implicit in the basis itself, that is, withouth extending the very conceptual apparatus in substantial ways. Here again a difficult philosophical task lays ahead in attempting to further explicate the distinction between different forms of understanding, especially in light of the logical analysis briefly discussed above, which brings to the fore a plurality of versions of predicativity.[11] A crucial aspect of this view is that predicativity becomes a tool for clarifying different forms of mathematics and various ways of understanding, but it does not entail a claim that only predicative mathematics of some kind is justified. In fact, predicativity, like other restrictions to standard methodology, in the hands of the logician become a tool for exploring in precise terms which parts of standard mathematics are amenable to be reframed in terms of more elementary assumptions or ways of reasoning.

Predicativity is an essential component of constructive type theory [14,15]. In fact, predicativity made a very dramatic appearance within Martin-Löf type theory, which bears surprising similarities to how it entered the mathematical landscape at the beginning of the 20th century. The appeal to an impredicative type of all types in the first formulation of intuitionistic type theory, in fact, gave rise to Girard's paradox [8]. Martin-Löf promptly corrected his type theory by eliminating the all-encompassing type of all types, and introduced in its place a hierarchy of type universes, each "reflecting" on previously constructed sets and universes [14]. Type theoretic universes are indeed at the centre of the generous notion of predicativity which arises in intuitionistic type theory [21].

Martin-Löf type theory embodies the Curry–Howard isomorphism, and thus identifies propositions with types (and their proofs with the elements of the corresponding types). As a consequence, type theory is simultaneously a very general programming language and a mathematical formalism. Girard's paradox is usually read as implying that in this context impredicativity (in the form of arbitrary quantification on types) is inconsistent with the Curry–Howard isomorphism. In a sense, predicativity signs the limit of the strong identification of mathematics with programming which is at the heart of constructive type theory.[12]

An observation naturally comes to mind: recent years have seen the flourishing of research on formalization of mathematics, with the purpose of verification. Here a new interplay between computer science and mathematics emerges. For

[11] Further challenges are also posed by technical developments in proof theory which have brought Gerhard Jäger to introduce a notion of *metapredicative* [9]. A thorough analysis of predicativity also ought to clarify its relation with metapredicativity.

[12] Predicativity is also at the centre of Martin-Löf's meaning explanations for type theory, which explain the type theoretic constructions of this theory "from the bottom up". A key concept here is that of *evidence*: constructive type theory represents a form of mathematics which is, according to its proponents, intuitively evident, amenable to contentual and computational understanding. This contentual understanding is then seen as supporting the belief in the consistency of this form of mathematics [13].

example, as observed by Georges Gonthier, one strategy which proved useful in proof checking is to turn mathematical concepts into data structures or programs, thus converting proof checking into program verification. Here constructive type theory has played a pivotal role, and inspired the development of other systems, like the (impredicative) calculus of constructions which underlines the Coq system.[13] One would then be tempted to conclude that ideas which originated through the fear of foundational errors are now having profound impact on new ways of addressing the ever pressing issue of local errors.

Acknowledgements. The author would like to thank Andrea Cantini and Robbie Williams for reading a draft of this article. She also gratefully acknowledges funding from the School of Philosophy, Religion and History of Science, University of Leeds.

References

1. Buchholz, W., Feferman, S., Pohlers, W., Sieg, W.: Iterated Inductive Definitions and Subsystems of Analysis. Springer, Berlin (1981)
2. Buss, S.: Bounded Arithmetic. Studies in Proof Theory Lecture Notes. Bibliopolis, Naples (1981)
3. Cantini, A.: Paradoxes, self-reference and truth in the 20th century. In: Gabbay, D. (ed.) The Handbook of the History of Logic, pp. 5–875. Elsevier, UK (2009)
4. Feferman, S.: Systems of predicative analysis. J. Symb. Log. **29**, 1–30 (1964)
5. Feferman, S.: Weyl vindicated: Das Kontinuum seventy years later. In: Temi e prospettive della logica e della scienza contemporanee, pp. 59–93 (1988)
6. Feferman, S.: Predicativity. In: Shapiro, S. (ed.) Handbook of the Philosophy of Mathematics and Logic. Oxford University Press, Oxford (2005)
7. Feferman, S., Strahm, T.: The unfolding of non-finitist arithmetic. Ann. Pure Appl. Log. **104**(1–3), 75–96 (2000)
8. Girard, J.Y.: Interprétation fonctionnelle et élimination des coupures de l'arithmetique d'ordre supérieur (1972)
9. Jäger, G.: Metapredicative and explicit mahlo: a proof-theoretic perspective. In: Cori, R., et al. (eds.) Proceedings of Logic Colloquium '00. Association of Symbolic Logic Lecture Notes in Logic, vol. 19, pp. 272–293. AK Peters, AK Peters (2005)
10. Kreisel, G.: Ordinal logics and the characterization of informal concepts of proof. In: Proceedings of the International Congress of Mathematicians (August 1958), pp. 289–299. Gauthier-Villars, Paris (1958)
11. Kreisel, G.: La prédicativité. Bulletin de la Societé Mathématique de France **88**, 371–391 (1960)
12. Lorenzen, P., Myhill, J.: Constructive definition of certain analytic sets of numbers. J. Symb. Log. **24**, 37–49 (1959)

[13] The calculus of constructions takes an opposite route compared with Martin-Löf type theory to the impasse given by Girard's paradox: it relinquishes the Curry–Howard hisomorphism in favour of impredicative type constructions. Although the Coq sytem was originally developed on the impredicative calculus of constructions, recent versions are based on a predicative core, although they also allow for impredicative extensions.

13. Martin-Löf, P.: The Hilbert-Brouwer controversy resolved? In: van Atten, M. (ed.) One Hundred Years of Intuitionism (1907–2007). Birkhäuser, Basel (2008)
14. Martin-Löf, P.: An intuitionistic theory of types: predicative part. In: Rose, H.E., Shepherdson, J.C. (eds.) Logic Colloquium 1973. North-Holland, Amsterdam (1975)
15. Martin-Löf, P.: Constructive mathematics and computer programming. In: Choen, L.J. (ed.) Logic, Methodology, and Philosophy of Science VI. North-Holland, Amsterdam (1982)
16. Nelson, E.: Predicative Arithmetic. Princeton University Press, Princeton (1986)
17. Palmgren, E.: On universes in type theory. In: Sambin, G., Smith, J. (eds.) Twenty-Five Years of Type Theory. Oxford University Press, Oxford (1998)
18. Parsons, C.: The impredicativity of induction. In: Detlefsen, M. (ed.) Proof, Logic, and Formalization, pp. 139–161. Routledge, London (1992)
19. Poincaré, H.: Les mathématiques et la logique. Revue de métaphysique et de morale **14**, 294–317 (1906)
20. Poincaré, H.: La logique de linfini. Revue de Métaphysique et Morale **17**, 461–482 (1909)
21. Rathjen, M.: The constructive Hilbert program and the limits of Martin-Löf type theory. Synthese **147**, 81–120 (2005)
22. Russell, B.: Les paradoxes de la logique. Revue de métaphysique et dc morale **14**, 627–650 (1906)
23. Russell, B.: Mathematical logic as based on the theory of types. Am. J. Math. **30**, 222–262 (1908)
24. Russell, B.: Essays in Analysis. George Braziller, New York (1973)
25. Schütte, K.: Eine Grenze für die Beweisbarkeit der Transfiniten Induktion in der verzweigten Typenlogik. Archiv für mathematische Logik und Grundlagenforschung **7**, 45–60 (1965)
26. Schütte, K.: Predicative well-orderings. In: Crossley, J., Dummett, M. (eds.) Formal Systems and Recursive Functions. North-Holland, Amsterdam (1965)
27. Simpson, S.G.: Subsystems of Second Order Arithmetic. Perspectives in Logic, 2nd edn. Cambridge University Press, Cambridge (2009)
28. Weyl, H.: Das Kontinuum Kritischen Untersuchungen über die Grundlagen der Analysis. Veit, Leipzig (1918)
29. Whitehead, A.N., Russell, B.: Principia Mathematica, vol. 1. Cambridge University Press, Cambridge (1925)

Is Human Mind Fully Algorithmic? Remarks on Kurt Gödel's Incompleteness Theorems

Mircea Dumitru(✉)

University of Bucharest and Romanian Academy,
Bdul M. Kogalniceanu nr. 36-46, 050107 Bucharest, Romania
mircea.dumitru@unibuc.ro

Abstract. In this paper I shall address an issue in philosophy of mind related to philosophy of mathematics, or more specifically to the nature of mathematical knowledge and reasoning. The issue concerns whether the human mind is fully algorithmic. I shall develop my answer against the background which is created by Kurt Gödel's celebrated incompleteness theorems. In what follows: (i) I shall first sketch the main programs and responses to the mind-body problem in philosophy of mind; (ii) then, I shall provide an informal overview of the two Gödelian incompleteness theorems; (iii) finally, I shall present and comment upon some of the main views advocated by Gödel about minds and machines, mind and matter, and the contrast between Turing machines and the so-called Gödel minds. In the process, Gödel's very unorthodox and unfashionable views against computabilism, neuralism, physicalism, psychoneural parallelism, and even against the underlying philosophical presuppositions of the Turing machines will emerge. Shocking as they, understandably, are, as compared to the standard psychological and philosophical orthodoxy underlying the received computabilistic views on mind, Gödel's own views are worth exploring and they fully deserve our undivided philosophical attention. Gödel is, after all, the founding father and one of the essential inspiring sources for the whole domain and range of topics that I address in my paper.

Philosophy of mind is thriving nowadays. The field has been developed extensively and intensively receiving all sorts of input from other connected fields, notably from computer science and cognitive science. The complexity and vitality of the domain is reflected by the vast literature which ramifies in various sub-fields and directions of research in which one tackles a batch of interrelated topics: the ontological problem (the so-called mind-body problem), the semantic problem, the epistemological problem, the methodological problem, artificial intelligence, and problems of neuroscience.

In my paper I shall deal with an issue in philosophy of mind related to philosophy of mathematics, or more specifically to the nature of mathematical knowledge and reasoning. The issue is whether or not human mind and intelligent consciousness is fully algorithmic. I shall develop my answer against the background which is created by Kurt Gödel's celebrated incompleteness theorems. In what follows: (i) I shall first sketch the main programs and responses

A. Beckmann et al. (Eds.): CiE 2015, LNCS 9136, pp. 23–33, 2015.
DOI: 10.1007/978-3-319-20028-6_3

to the mind-body problem in philosophy of mind; (ii) then, I shall provide an informal overview of the two Gödelian incompleteness theorems; (iii) finally, I shall present and comment upon some of the main views advocated by Gödel about minds and machines, mind and matter, and the contrast between Turing machines and the so-called Gödel minds. In the process, Gödel's very unorthodox and unfashionable views against computabilism, neuralism, physicalism, psychoneural parallelism, and even against the underlying philosophical presuppositions of the Turing machines will emerge. Shocking as they, understandably, are as compared to the standard psychological and philosophical orthodoxy underlying the received computabilistic views on mind, Gödel's own views are worth exploring and they fully deserve our undivided philosophical attention. Gödel is, after all, the founding father and one of the essential inspiring sources for the whole domain and range of topics that I address in my paper. And even if we do not, and perhaps cannot, take everything that he thought on those issues on board, one can still have a lot to learn from how he framed the questions and what he had to say about those fascinating issues concerning the nature and the functioning of our (mathematical) mind.

1 Sketch of the Main Programs and Responses to the Mind-Body Problem in Philosophy of Mind

So, let us first canvass the metaphysics of mind. A very useful resource for this topic is [1], whom I basically follow for the systematization of the main philosophical responses to the mind-body problem What we aim at clarifying here is the problem of the nature of the mind's states and processes. More specifically, the questions that we raise are: where do mental states and processes occur, and how are they correlated to the physical world? Is my consciousness going to survive my physical decay after I am dead? Is it possible that a purely physical system (a computer) be built in such a way that it can have conscious experience with qualia? Where do minds come from? What are they?

The reasoned answers to those difficult issues are theory and methodological driven. They are dependent upon the particular theory of mind that we may favor, which is based on its explicative and predictive power, and also on its coherence and simplicity. The main theories that have been advocated in the philosophy of mind are:

Dualism. The essence of all forms of dualism, such as substance dualism and property dualism, is that the nature of the mental resides in a nonphysical entity, which escapes the domain of physics, neurophysiology and computer science. Dualism, nowadays, undergoes a sort of paradoxical fate. By far the most popular and traditional philosophical perspective on mind, akin to the position advocated by various major religions on the relation between mind, soul, and body, dualism is almost completely rejected by professional philosophers today.

Philosophical behaviorism. This has been a tremendously influential conception in the metaphysics of mind for several decades in the XX-th century.

The rise of cognitivism in linguistics and psychology led to the demise of this once powerful position and critical tool against traditional speculative metaphysics. As such, philosophical behaviorism is not a theory about the essence of mental states per se; it is, rather, a kind of analysis of the language in which we talk about our mental states. Thus, sentences about various mental episodes, such as emotions, sensations, beliefs, desires, wants, etc. are not about would-be inner occurrences of mental events, but instead, they are abbreviated ways of speaking about actual and possible behavior. Therefore, any sentence about a mental state or process can be rephrased in a longer sentence about behavior.

Reductive materialism (Identity theory). The main claim of this form of materialist theory is that mental states are (identical with) physical states of the brain. More specifically, each type (or token, in weaker versions of this theory) of mental states or processes is numerically identical to some type (token) of physical state or process which takes place in the brain or in the central nervous system.

Functionalism. This doctrine, which is the prevalent view on mind today, says that the characteristic feature of any type of mental state is the set of causal relations it bears to the input coming from the environment, to other types of mental states, and to the output of our behavior. A mental state plays a causal role, and that mental state is defined through its network of causal roles. According to functionalism, as opposed to philosophical behaviorism, reference to mental states cannot be eliminated; and in order to define such a type of mental state, one has to refer to a number of other mental states with which that state is causally connected. Functionalism acknowledges the reality of mental states which should be studied systematically. It follows that psychology should be an autonomous science from, and not reducible to, the physical sciences (physics, biology, neurophysiology). Psychology is a science in its own right with its own irreducible laws, and its own domain.

Eliminative materialism. This is a profoundly skeptical view on the mind. It casts doubts upon the concepts and explanations of folk psychology (which explains the intelligent actions of human beings in terms of the causal powers of propositional attitude ascriptions, such as belief, desire, hope, etc.). Eliminative materialism also goes against reductive materialism, since part of that reduction program is to achieve a one-to-one correspondence between the mental states and processes acknowledged by folk psychology and some neuro-physiological processes that occur in the brain. This intended reduction cannot be done. And the reason is not lack of ingenuity from the part of the theorists. The reason is the non-existence of such things as mental states, processes or attitudes which are posited by this common-sense psychological framework which, in its turn, is hypothesized by folk psychology through an inference to the best explanation. Therefore, one key reason for this reduction being impossible resides in the fact that the common-sense psychological framework is fraught with some fatal problems: it is literally false, and consequently, it is also a misleading conception of what determines causally our behavior and mental activity. Through scientific education, it is expected that, and it is hoped that, gradually one can get rid of this false representation about our own psychology, based on propositional attitude ascriptions. This framework will be eliminated by future neuroscientific discovery. This is the motivation for the name of the conception, viz. "eliminative materialism".

A general idea that emerges from the various responses to the mind-body problem, which is essentially a leit-motif of the ongoing dialectics running through those questions and answers, is that what best explains mental states and processes is a computational paradigm of the mind. Most theorists argue that the mind is a sort of computation on symbols and representational mental contents. This computational paradigm will offer a coherent answer to the hard problem of integrating two distinct views on human beings: the causal view, which underlies the explanation of the bio-chemical complex structures in which human beings qua biological entities consist, with the intentionality of the mental representation view of human beings qua rational and socio-cultural decision-making agents. The hope is that functionalism will solve this integration problem. The most influential version of current functionalism considers a computational theory of mind to be the best available explanation of human behavior via the causal role of mental states to mediate, explicitly in computational terms, between the environmental input and the behavioral output. We shall see in a moment that Gödel rejects both this view on mind and its presuppositions.

2 A Short Informal Overview of the Two Gödelian Incompleteness Theorems

What Gödel's First Incompleteness Theorem shows is that any consistent formal axiom system or deductive system T, which is sound (i.e. proves only true sentences) and powerful enough to express elementary arithmetic, is bound to be incomplete because a sentence, that we shall call G_T, can be true according to the interpretation of that formal system T, but cannot be derived as a theorem in that system.

Thus, Gödel shows that the common idea, according to which arithmetical truth equals proof within a formal deductive system, is wrong. Gödel was able to prove this following a series of ingenious steps of ([4], p. 1–7). First, he constructed a sentence G_T, in the language of arithmetic (via the technique of Gödel-numbering), which represents the meta-mathematical sentence: "The sentence G_T is not provable in the system T". That is, G_T says of itself that it is unprovable in T. It follows that G_T is true if and only if (iff) G_T cannot be proved in T. Let's suppose further that T is sound. If G_T were provable in T then G_T would be false, and hence unprovable in T, since T is sound and it can only prove true sentences. So, up to this point, if G_T were provable, then it could not be proven. Therefore G_T is not provable after all in T meaning that G_T is true. Suppose now that G_T were not provable. Then G_T is true and, of course, its negation, $\sim G_T$, is false. But T is sound and it proves only true sentences. Thus, T cannot prove $\sim G_T$ either. So, there is a true sentence, G_T, which says of itself that it is not provable in a system T, and neither that sentence G_T, nor its negation $\sim G_T$ is provable in T. Hence the sentence G_T is undecidable by the means of the system T and, assuming that T is sound, the system T is incomplete. Adding G_T to the system T does not solve the issue because, according to the same method, a new sentence G'_T can be constructed in such a way as to be able to say of itself that it is not provable in $T + G_T$, while being true, and while neither G'_T nor its negation $\sim G'_T$ being provable in $T + G_T$.

Thus far Gödel has shown that, since G_T is true and unprovable in T, the axioms of the system T are incomplete. Summing up this part of the proof, that culminates in Gödel's First Incompleteness Theorem, Nagel & Newman in ([3], p. 67) cogently argue that "we cannot deduce all arithmetical truths from the axioms. Moreover, Gödel established that arithmetic is essentially incomplete: even if additional axioms were assumed so that the true formula G_T could be formally derived from the augmented set, another true but formally undecidable formula could be constructed."

In the Second Incompleteness Theorem, Gödel shows how to construct an arithmetical statement A that has the meta-mathematical content: "Arithmetic is consistent". He goes one to prove that the sentence "$A \longrightarrow G_T$" is formally provable; however, since G_T itself is not provable, Gödel shows that A is not provable either. What follows from this is the Second Incompleteness Theorem which establishes the fact "that the consistency of arithmetic cannot be established by an argument that can be represented in the [very same] formal arithmetical calculus" ([3], p. 67).

Do all these Gödelian ground-braking meta-mathematical results have any philosophical significance? And if so, what would that significance be? From among many reactions and comments that Gödel's Incompleteness Theorems have prompted[1], I shall take a look at Gödel's own philosophical views correlated with his own results, and make some comments on three issues, concerning (a) Gödel's view on minds, machines and computabilism, (b) Gödel's view on mind, matter, physicalism, and psycho-physical parallelism, and (c) Gödel vs. Turing, i.e. Gödel's view on Turing Machines and on Gödel Minds.

3 Comments upon Some of the Main Views Advocated by Gödel

This section of the paper is based on [6], which is an extremely reach source for Gödel's philosophical views.

3.1 Gödel About Minds and Machines

Gödel had a strong conviction that neither computabilism, i.e. the view that the brain and the mind work essentially like a computer, nor neuralism, i.e. the view that the brain is a sufficient explanans for mental phenomena is right, and

[1] For some of the reactions see [2, 5, 6].

consequently he argued vigorously against and rejected both views. Hao Wang [6] tells us that Gödel was preoccupied with the problem of whether computabilism was a complete explanation of mental processes, "that is, the issue of whether all thinking is computational - with special emphasis on mathematical thinking. Gödel's main concern was to demonstrate that not all mathematical thinking is computational" ([6], p. 183).

In one if its several formulations, Gödel's Second Incompleteness Theorem states something that is relevant to the mathematical capacity of the human mind, namely that if a reasonably strong theorem-proving computer or program is sound and consistent, then it cannot prove the truth that expresses its own consistency. According to [6], Gödel drew a relevant conclusion from this concerning the human mind: "6.1.1 The human mind is incapable of formulating (or mechanizing) all its mathematical intuitions. That is, if it has succeeded in formulating some of them, this very fact yields new intuitive knowledge, for example the consistency of this formalism. This fact may be called the "incompletability" of mathematics. On the other hand, on the basis of what has been proved so far, it remains possible that there may exist (and even be empirically discoverable) a theorem-proving machine which in fact is equivalent to mathematical intuition, but cannot be proved to be so, nor even be proved to yield only correct theorems of finitary number theory" ([6], p. 184–185).

Hao Wang, again, tells us that Gödel was very attached to some ideas about creation in mathematics and the algorithmic nature of human mind and mathematical thought ([6], p. 186). Those ideas are relevant for the implications of his theorem, as one can see from the following remark made by Gödel: "6.1.8 My incompleteness theorem makes it likely that mind is not mechanical, or else mind cannot understand its own mechanism. If my result is taken together with the rationalistic attitude which Hilbert had and which was not refuted by my results, then [we can infer] the sharp result that mind is not mechanical. This is so, because, if the mind were a machine, there would, contrary to this rationalistic attitude, exist number-theoretic questions undecidable for the human mind" ([6], p. 186–187).

The upshot of all those remarks is Gödel's strong conviction that human mind, through its intuitive powers and creativity, is superior over computers, and that the partaking of individual minds to the collective experience of the human species gives a whole new range of possibilities, which allows the human mind and spirit to surpass the power of computing machines. Here are some of Gödel's thoughts in this regard:

"6.1.19 The brain is a computing machine connected with a spirit."
"6.1.21 Consciousness is connected with one unity. A machine is composed of parts."
"6.1.23 By *mind* I mean an individual mind of unlimited life span[2]. This is still different from the collective mind of the species. Imagine a person engaged in solving a whole set of problems: this is close to reality; people constantly introduce new axioms." ([6], p. 189).

---

[2] Gödel believed the human soul is immortal, that science will prove that fact one day. His philosophical hero was Leibniz.

This fragment tellingly shows Gödel's trust that eventually we can prove mind's superiority over computers, because of its creativity and power to give new forceful ideas and insights:
"6.1.24 It would be a result of great interest to prove that the shortest decision procedure requires a long time to decide comparatively short propositions. More specifically, it may be possible to prove: For every decidable system and every decision procedure for it, there exists some proposition of length less than 200 whose shortest proof is longer that 10^{20}. Such a result would actually mean that computers cannot replace the human mind, which can give short proofs by giving a new idea." ([6], p. 189).

3.2 Gödel About Mind and Matter

The making of the distinction between mind and matter imposes upon us the metaphysical idea that they are distinct from each other, a doctrine which commits us to some form of dualism (see above). The difficulty to tie the two together (causally or otherwise), once we separated them essentially, has been notorious since Descartes' work. Gödel has his own way of framing this celebrated metaphysical issue, namely by asking whether "the brain suffices for the explanation of all mental phenomena". Gödel rephrases the question in a more precise, quantitative fashion, raising the issue of whether there are enough brain operations that represent the mental operations in such a manner that the correspondence between physical brain and mental operations is one-to-one or even many-to-one.

Scientific and philosophical orthodoxy argues that such a correlation exists, a view which is known as *psychoneural parallelism.* If, further on, one makes the physicalist assumption, which is quite common today, that all neural operations are physical operations of a special kind, the view turns into *psychophisical parallelism.*

Gödel's own argument is that there is mind which is separate from brain (matter). Gödel, contrary to the whole scientific establishment and current orthodoxy, refutes both psychoneural and psychophysical parallelism. His remarks with regard to this topic are very daring, and surely shocking for many of us. Thus, says Gödel: "6.2.1 Parallelism is a prejudice of our time. 6.2.2 Parallelism will be disproved scientifically (perhaps by the fact that there aren't enough nerve cells to perform the observable operations of the mind)" ([6], p. 190).

Gödel is quick to recognize that not all prejudices are necessarily false. A prejudice is a widely shared belief whose strength is not backed by solid pieces of evidence. Why do we hold so strongly to the parallelism prejudice? We do it, because we are impressed by the power of science and technology, often leading us to uncritically accepting scientism. Gödel makes the further extraordinary remark that the philosophical point of the parallelism in the aforementioned 6.2.2 is not only a philosophical prejudice, but also a scientific and empirical stance that will be disproved. Gödel emphasizes this idea whenever he feels that it is important to make more room for it in conceptual space. He refers to this notion in the following passages: "6.2.3 It is a logical possibility that the existence of mind [separated from matter] is an empirically decidable question. This possibility is not a conjecture. [...] there is an empirical question behind it. 6.2.4 Logic deals with more general concepts; monadology, which contains general laws of biology, is more specific. The limits of science: Is it possible that all mind activities [...] are brain activities? There can be a factual answer to this question. Saying no to thinking as a property of a specific nature calls for saying no also to elementary particles. Matter and mind are two different things. 6.2.5 The mere possibility that there may not be enough nerve cells to perform the function of the mind introduces an empirical component into the problem of mind and matter" ([6], p. 191).

Gödel puts a lot of emphasis on the important and difficult metaphysical issue of the relation between mind and matter, considering it central to philosophical inquiry and critical to understanding philosophy's importance to science. Thus, one can read the following remark made by Gödel in conversation with Hao Wang: "6.2.6 Many so-called philosophical problems are scientific problems, only not yet treated by scientists. One example is whether mind is separate from matter. Such problems should be discussed by philosophers before scientists are ready to discuss them, so that philosophy has as one of its functions to guide scientific research. Another function of philosophy is to study what the meaning of the world is" ([6], p. 191).

Gödel is very interested in clarifying the issue of the parallelism between mind and matter, clearly stating his stance: "6.2.9 Mind is separate from matter: it is a separate object ..." Moreover, he boldly conjectures, completely going against the grain of the scientific establishment, that science itself will eventually refute this prejudice of the psychoneural parallelism: "6.2.11 [...] I believe that mechanism in biology[3] is a prejudice of our time which will be disproved. In this case, one disproof, in my opinion, will consist in a mathematical theorem to the effect that the formation within geological times of a human body by the laws of physics (or any other laws of a similar nature), starting from a random distribution of the elementary particles and the field, is as unlikely as the separation by chance of the atmosphere into its components" ([6], p. 192).

And a last remark in this regard. The remark shows Gödel's conviction that: (a) the brain is a physical object, (b) the mind (or the spirit) is a separate entity from the brain, and (c) the brain, as a normal physical object, functions the way it does just because it is connected to a mind: "6.2.14 Even if the finite brain cannot store an infinite amount of information, the spirit may be able to. The brain is a computing machine connected with a spirit. If the brain is taken to be

[3] By 'mechanism in biology' Hao Wang says that Gödel meant Darwinism, "which he apparently sees as a set of algorithmic laws (of evolution). Even though he seems to believe that the brain - and presumably also the human body - functions like a computer [...], he appears to be saying here that the human body is so complex that the laws of physics and evolution are insufficient to account for its formation within the commonly estimated period of time" ([6], p. 192).

physical and as [to be] a digital computer, from quantum mechanics [it follows that] there are then only a finite number of states. Only by connecting it [the brain] to a spirit might it work in some other way" ([6], p. 193).

3.3 Turing Machines vs. the So-Called Gödel Minds

Gödel thought profoundly of the nature of algorithms, and of the formalization of logical systems. Consequently, he was very interested in the ground-braking work of Alan Turing, holding Turing's work in very high esteem. Gödel came to believe that his own incompleteness theorems hit upon an important aspect of the limits of formalization only after Turing developed his analysis, which Gödel fully endorsed, of the concept of mechanical (or computational) procedures, the so-called Turing machine. Moreover, Gödel was satisfied with the fact that Turing machines provide evidence for the thesis that sharp concepts really exist, and that human minds can perceive them clearly. Nevertheless, Gödel spotted a problem in Turing's argument of the adequacy of his analysis of algorithms, namely a fallacious proof of the conclusion that minds and machines are equivalent ([6], p. 194).

Gödel's position is made clear through the following remark: "6.3.5 Attempted proofs for the equivalence of minds and machines are fallacious. One example is Turing's alleged proof that every mental procedure for producing an infinite series of integers is equivalent to a mechanical procedure" ([6], p. 197). Gödel explains why he considers the proof attempted by Turing to be fallacious: "6.3.6 Turing gives an argument which is supposed to show that mental procedures cannot carry farther than mechanical procedures. However, this argument is inconclusive, because it depends on the supposition that a finite mind is capable of only a finite number of distinguishable states" ([6], p. 197).

Gödel rejects the supposition that mind (spirit) is matter; he says: "6.3.7 It is a prejudice of our time that (1) there is no mind separate from matter; indeed, (1) will be disproved scientifically" ([6], p. 198). He, then, continues to interpret and reconstruct Turing's argument, and finds it valid, only after certain presuppositions are guaranteed and accepted: "6.3.8 It is very likely that (2) the brain functions basically like a digital computer. 6.3.9 It is practically certain that (2') the physical laws, in their observable consequences, have a finite limit of precision. 6.3.10 If we accept (1), together with either (2) or (2'), then Turing's argument becomes valid" ([6], p. 198).

It is hard for us today not to accept all these presuppositions just because we are accustomed, or indeed perhaps prejudiced, to thinking of the brain and the mind as being two aspects of the same thing. However, Gödel did not consider the matter to be so: "6.3.11 If (i) a finite mind is capable only of a finite number of distinguishable states, then (ii) mental procedures cannot carry any farther than mechanical procedures. 6.3.12 Turing's argument (iii) for the condition (i) is his idea which centers on the following sentence: *We will also suppose that the number of states of mind which need be taken into account is finite. The reasons for this are of the same character as those which restricted the number of symbols.*

If we admit an infinity of states of mind, some of them will be 'arbitrarily close' and will be confused." ([6], p. 198).

Gödel is happy with the inference from (i) to (ii). He believes, though, that (i) can be inferred from (iii) only if some additional assumptions are forthcoming. And since Gödel does not accept that brain is equivalent to mind, he goes on to reject both (i) and (ii).

At the end of the day, Gödel's refutation of mental computerism, his deeply held conviction that mind can carry farther than machines, is based on the idea that "*6.3.13 Mind, in its use, is not static, but constantly developing.*" When we focus, introspectively, on the stream of our consciousness, we are struck by the fact that the mental states and their succession do not enjoy the sharpness and clarity of the states of Turing machines. Wang comments the following concerning Gödel's idea: "... we develop over time, both individually and collectively; and so, for instance, what appeared to be complex becomes simple, and we understand things we did not understand before. Here again, we feel that the process of development is somewhat indefinite and not mechanical" ([6], p. 200).

Do we have a proof that minds can carry farther than computers, and that they are not fully mechanical? We do not. However, Gödel promotes a dynamic and developing kind of vision of mind that is both telling and credible: "6.3.14 Although at each stage of the mind's development the number of its possible states is finite, there is no reason why this number should not converge to infinity in the course of its development" ([6], p. 200).

A mechanical brain connected to a creative, ever-evolving, developing, and non-mechanical mind (spirit) is a set-up that goes beyond the individualism of the atomic, and isolated minds. Thus, brains and minds can create thoughts in a manner which reflects that minds can carry farther than brains and computers, eventually indicating mind's superiority over computers.

Acknowledgements. I want to thank Dr. Victor Mitrana, University of Bucharest, for commenting upon an earlier version of the paper which contributed to the improvement of the arguments, and for introducing the text in LaTeX. I am grateful to Dr. Daniela Dumitru, Bucharest University of Economic Studies, for stimulating discussions about cognitivism and computerism. I also want to thank Miss Ioana Andrada Dumitru, PhD student at Johns Hopkins University, for making comments and stylistic suggestions which I happily accepted and which improved the clarity and the readability of the paper.

References

1. Churchland, P.M.: Matter and Consciousness. The MIT Press, Cambridge (1992)
2. Hintikka, J.: On Gödel. Wadsworth, Belmont (2000)
3. Nagel, E., Newman, J.R.: Gödel's Proof. Routledge, London (1958)

4. Smith, P.: An Introduction to Gödel's Theorems, 2nd edn. Cambridge, New York (2013)
5. Tieszen, R.: After Gödel. Platonism and Rationalism in Mathematics and Logic. Oxford, UK (2011)
6. Wang, H.: A Logical Journey. From Gödel to Philosophy. The MIT Press, Cambridge (1996)

A New Approach to the Paperfolding Sequences

Daniel Goč[1,2], Hamoon Mousavi[1,3], Luke Schaeffer[1,4], and Jeffrey Shallit[1(✉)]

[1] School of Computer Science, University of Waterloo,
Waterloo, ON N2L 3G1, Canada
{dgoc,hamoon.mousavihaji,l3schaeffer,shallit}@cs.uwaterloo.ca

[2] Queen's University, Kingston, Ontario

[3] McAfee Software, Waterloo, Ontario

[4] MIT, Cambridge, MA

Abstract. In this paper we show how to re-derive known results about the paperfolding sequences, and obtain new ones, using a new approach using a decision method and some machine computation. We also obtain exact expressions for the recurrence and appearance function of the paperfolding sequences, and solve an open problem of Rampersad about factors shared in common between two different paperfolding sequences.

1 Introduction

To simplify notation, we often write -1 as $\overline{1}$ in this paper. The (regular) paperfolding sequence $\mathbf{R} = (R_n)_{n\geq 1} = 1 1 \overline{1}\, 1 1 \overline{1}\, \overline{1} \cdots$ is defined as the limit, as $n \to \infty$, of the following sequence of finite words over the alphabet $\{1, \overline{1}\}$:

$$F_0 = 1 \qquad\qquad F_{n+1} = F_n\ 1\ (-F_n^R) \qquad (n \geq 0).$$

Here by w^R we mean the reversal of the word w, by $-x$ we mean the negation of each symbol in x, and juxtaposition means concatenation. The sequence arises from the following process, which explains the name: take a piece of paper, fold it lengthwise over and over n times, then unfold it so all the angles are right angles. Read left to right, the sequence of "hills" (right turns) $(+1)$ and "valleys" (left turns) (-1) so obtained is a prefix of length $2^n - 1$ of $\mathbf{R}$. The sequence $\mathbf{R}$ is 2-automatic in the sense of Cobham [6], and has been extensively studied.

Davis and Knuth [7] generalized the regular paperfolding sequence by allowing the folds to be right-hand over left, or vice-versa. During unfolding, we can read off the sequence of fold choices that we made, and encode it by letting 1 denote a hill and $\overline{1}$ a valley. For every infinite sequence of folds $\mathbf{f} = (f_n)_{n\geq 0}$ we get a corresponding infinite paperfolding sequence $P_\mathbf{f} = p_1 p_2 p_3 \cdots$, given as the limit of the following finite words:

$$F_0 = f_0 \qquad\qquad F_{n+1} = F_n\ f_{n+1}\ (-F_n^R) \qquad (n \geq 0).$$

For example, if we choose the sequence of unfolding instructions $1 \overline{1}\, 1 \overline{1} \cdots$, then the resulting sequence is $1 \overline{1}\, \overline{1}\, 1 1 1 \overline{1}\, \overline{1}\, 1 \overline{1}\, \overline{1}\, \overline{1}\, 1 1 \overline{1} \cdots$. This gives uncountably many distinct paperfolding sequences.

A. Beckmann et al. (Eds.): CiE 2015, LNCS 9136, pp. 34–43, 2015.
DOI: 10.1007/978-3-319-20028-6_4

Dekking, Mendès France and van der Poorten [8] observed that we can easily determine the paperfolding sequence $(p_n)_{n\geq 1}$ from the sequence of *unfolding instructions* $(f_i)_{i\geq 0}$, as follows: write $n = 2^s \cdot r$, where r is odd. Then

$$p_n = \begin{cases} f_s, & \text{if } r \equiv 1 \pmod 4; \\ -f_s, & \text{if } r \equiv 3 \pmod 4). \end{cases} \tag{1}$$

One of the earliest themes in the study of the paperfolding sequences is the properties of their finite factors. By a finite factor of an infinite sequence $(a_i)_{i\geq 0}$ we mean a word of the form $a_i a_{i+1} \cdots a_j$ for $0 \leq i \leq j+1$. For example, Prodinger and Urbanek [11] showed that the regular paperfolding sequence contains no square factors xx with $|x| > 5$. Allouche [1] proved that the subword complexity (aka "factor complexity" or just "complexity" — the number of distinct factors of length n) of every paperfolding sequence is $4n$ for $n \geq 7$. Allouche and Bousquet-Mélou [2] gave an upper bound of $44n$ on the recurrence function of every paperfolding sequence. In another paper, Allouche and Bousquet-Mélou [3] showed that there are no 3^+ powers in any paperfolding sequence, and that the only cubes are 111 or $\overline{1}\,\overline{1}\,\overline{1}$. They also studied the "almost-squares" in the paperfolding sequences; these are words of the form wcw, where w is a nonempty word and c is a single-letter. They claimed that any such almost-square satisfies $|w| \in \{1, 2, 3, 4, 7\}$, but this is not quite correct. The correct statement, as proved in [10, Proposition. 3], is that either $|w| \in \{2, 4\}$ or $|w| = 2^k - 1$ for some $k \geq 1$. Furthermore, there are almost-squares corresponding to all these orders. In this paper we reconfirm the corrected version of their results.

All the proofs of the results above are essentially based on case analysis, some of it rather intricate. It is natural to wonder if there is some more general approach to obtaining these results, and perhaps more. In this paper, we give such a general approach. We provide a technique whereby, most of these results can be proved purely mechanically, using a decision procedure and a machine computation — at least in principle. (In some cases the decision procedure may result in extremely long running times or extremely large space requirements.) We also obtain some new results.

Parts of this paper have already appeared in the master's thesis of the third author [12].

2 The Main Idea

The main idea is to observe that Eq. (1) implies that the n'th term of a paperfolding sequence specified by unfolding instructions $f_0, f_1, \ldots, f_{t-1}$ can be computed by the 5-state deterministic finite automaton that takes, as input, the base-2 expansion of n *in parallel* with the unfolding instructions, *provided* that $1 \leq n < 2^t$ — in other words, provided enough unfolding instructions have been furnished. The first components of the inputs are the unfolding instructions, starting with f_0, and the second components form the base-2 expansion of n, starting with the least-significant digit (lsd) first, and padded with 0's on the right, if necessary. The symbol "$*$" is a wildcard and matches anything in the input. The output associated with the input is contained in the last state reached. The automaton is depicted below in Fig. 1.

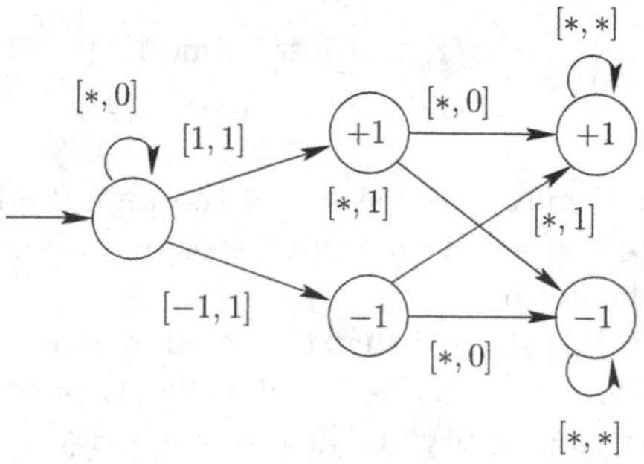

Fig. 1. The paperfolding automaton for arbitrary unfolding instructions, lsd first

Alternatively we could feed in n with the most significant digit first. This requires reversing the order of the paperfolding instructions, so that f_{t-1} is fed in first and f_0 last. The resulting automaton has 7 states and is omitted from this version. Thus, although there are *uncountably* many distinct paperfolding sequences, they can all be encoded by a *single* finite automaton.

Now we adapt the technique discussed previously in [4,5,9] to the paperfolding sequences. Briefly, this technique allows us to translate certain predicates $P(n)$ concerning the factors of a k-automatic sequence into an automaton M accepting the base-k representations of those n for which $P(n)$ holds. Given M we can easily determine if there exists such an n, or if there exist infinitely many n.

Since the sequence of unfolding instructions is infinite, an appropriate decision procedure can be based on the theory of ω-words (i.e., infinite words) [12]. However, for our purposes, it is simpler to implement a version where we only consider finite lists of unfolding instructions. Thus, we need to rephrase assertions about infinite paperfolding sequences in terms of finite prefixes of paperfolding sequences. This adds a bit of complication: when we evaluate a predicate involving terms of the paperfolding sequence, the result makes sense only if we have been provided sufficiently many terms of the unfolding instructions.

During the course of our computation, we will have to query the automaton in Fig. 1 on various inputs consisting of a finite list of unfolding instructions f and an index n into the paperfolding sequence whose unfolding instructions begin with f. We can only guarantee the correctness of the computation if $1 \leq n \leq 2^{|f|} - 1$.

All computations reported here were performed on a Macintosh 2.6 GHz Intel Core i5 machine.

3 Our Results

In what follows, the unfolding instructions are over $\{1, \overline{1}\}$ and by convention are indexed, starting at 0, as follows: $\mathbf{f} = f_0, f_1, \ldots$. The resulting paperfolding sequence $P_\mathbf{f}$ is, by convention, indexed starting at 1 and is given by $P_\mathbf{f}[1]P_\mathbf{f}[2]\cdots$.

We start by recovering a known result about the orders of squares occurring in paperfolding sequences. By the *order* of a square xx we mean $|x|$.

3.1 Orders of Squares

Theorem 1. *If xx is a nonempty factor of any paperfolding sequence, then $|x| \in \{1, 3, 5\}$. Furthermore, every paperfolding sequence contains squares of orders* $1, 3,$ *and* 5.

Proof. We want to determine those positive integers n such that there exists a sequence of unfolding instructions $\mathbf{f}$ and an integer $i \geq 1$ such that

$$P_{\mathbf{f}}[i..i+n-1] = P_{\mathbf{f}}[i+n..i+2n-1].$$

However, since the number of unfolding sequences is uncountably infinite, we cannot check this as stated. Instead, we reason as follows: if there is some paperfolding sequence $P_{\mathbf{f}}$ satisfying $P_{\mathbf{f}}[i..i+n-1] = P_{\mathbf{f}}[i+n..i+2n-1]$, this equality will also hold for the finite prefix of P_f of length $2^t - 1$ specified by the prefix of f of length t, provided $2^t - 1 \geq i + 2n - 1$. So we create an automaton M accepting all those words w over $\{1, \overline{1}\} \times \Sigma_2 \times \Sigma_2$ where the projection $\pi_1(w)$ onto the first coordinate specifies the unfolding instructions, and the projections $\pi_2(w)$ (resp., $\pi_3(w)$) onto the second (resp., third coordinates) gives the base-2 representation of i (resp., n) (starting with the least significant digit), such that $i + 2n - 1 \leq 2^{|w|} - 1$, and $P_{\mathbf{f}}[i..i+n-1] = P_{\mathbf{f}}[i+n..i+2n-1]$. (This last condition is checked by nondeterministically guessing an index where it fails, creating the appropriate NFA, determinizing it, and then interchanging the role of final and nonfinal states.) Once we have M, we project onto the third coordinate to obtain an automaton accepting all n corresponding to orders of squares in all paperfolding sequences.

We implemented this idea using two independent programs, created by the first and second authors, respectively. Hamoon Mousavi's Java program, called `Walnut`, uses the following command:

```
eval paperfolding_square_orders
"?lsd_2 (n > 0) & (Ef Ei (i >= 1) & (Ak k < n => PF[f][i+k] = PF[f][i+k+n]))":
```

Here the "E" is an abbreviation for "there exists" (often written $\exists$) and the "A" is an abbreviation for "for all" (often written $\forall$).

When we run this, the result is given by the automaton depicted below in Fig. 2.

By inspection of Fig. 2, we can see that the only orders of squares that can possibly occur in paperfolding sequences are 1, 3, and 5. This completes the proof. The total computation time was 109 ms.

Next, we demonstrate that every infinite paperfolding sequence contains squares of orders 1, 3, and 5. We use the following command in Mousavi's `Walnut` prover:

```
eval paper1
 "?lsd_2 An ((n=1) | (n=3) | (n=5)) =>
 (Ei (i > 0) &  (Ak k < n => PF[f][i+k] = PF[f][i+k+n]))":
```

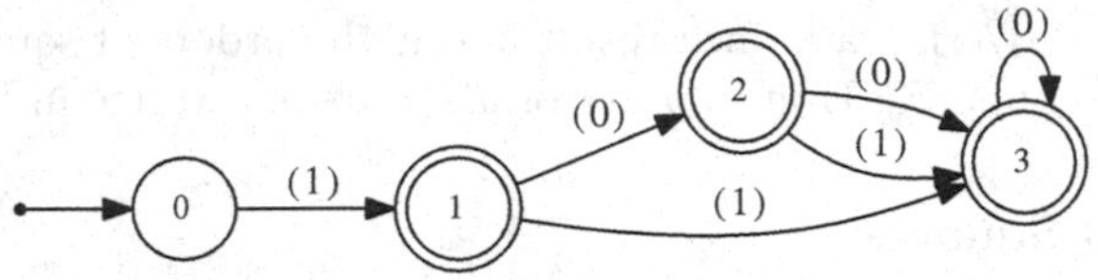

Fig. 2. Possible orders of squares in paperfolding sequences, lsd first

In other words, we are asking for those finite sequences of unfolding instructions having squares of order 1, 3, and 5. These are given below in Fig. 3. By inspection of Fig. 3, we easily see that every sufficiently long sequence of unfolding instructions is accepted. The total computation time was 129 ms.

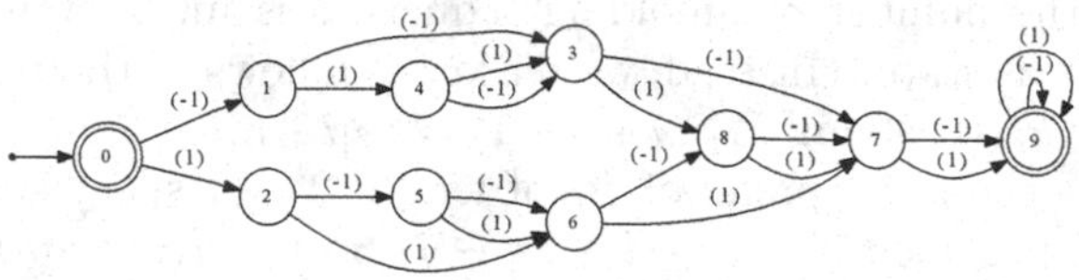

Fig. 3. Finite unfolding sequences having squares of length 1, 3, and 5

3.2 Cubes

Theorem 2. *The only cubes contained in any paperfolding sequence are* 111 *and* $\overline{1}\,\overline{1}\,\overline{1}$. *Furthermore, every infinite paperfolding sequence contains these cubes.*

Proof. For the first part, it suffices to determine the lengths of cubes xxx that can appear in any paperfolding sequence. We use the technique described in the preceding section, modulo the change of the condition:

$$P_{\mathbf{f}}[i..i+2n-1] = P_{\mathbf{f}}[i+n..i+3n-1].$$

We implemented this using

```
eval paperfolding_cube_orders
"?lsd_2 (n > 0) & (Ef Ei (i >= 1) & (Ak k < 2*n => PF[f][i+k] = PF[f][i+k+n]))":
```

The output is an automaton of two states accepting only the order 1. The computation took 400 ms. The second part is proved as in the case of squares.

3.3 Higher Powers

Theorem 3. *No paperfolding sequence contains a* 3^+*-power.*

Proof. Using the ideas above, it suffices to determine those lengths n for which $P_{\mathbf{f}}[i..i+2n] = P_{\mathbf{f}}[i+n..i+3n]$.
We implemented this using

```
eval higher_power_orders
 "?lsd_2 (n > 0) & Ef Ei ((i >= 1) & (Ak k <= 2*n => PF[f][i+k] = PF[f][i+k+n]))":
```

The resulting automaton accepts no n at all. The computation took 189 ms.

3.4 Orders of Almost-Squares

Theorem 4. *If wcw is a factor of a paperfolding sequence, where w is a word and c is a single letter, then either $|w| \in \{2, 4\}$ or $|w| = 2^k - 1$ for some $k \geq 1$.*

Furthermore, for all $k \geq 1$, and all paperfolding sequences, and every $n \in \{2, 4\} \cup \{2^k - 1 \ : \ k \geq 1\}$. there is a factor of the form wcw with $|w| = n$.

Proof. We can verify the first claim with the methods above, computing those n for which

$$P_{\mathbf{f}}[i..i+n-1] = P_{\mathbf{f}}[i+n+1..i+2n].$$

We implemented this using

```
eval almost_square_orders
 "?lsd_2 (n > 0) & Ef Ei ((i >= 1) & (Ak (k < n) => PF[f][i+k] = PF[f][i+k+n+1]))":
```

and found the following automaton (Fig. 4):

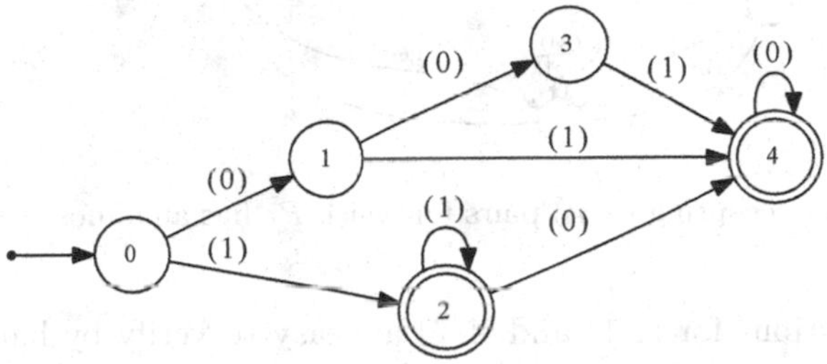

Fig. 4. Orders of almost-squares in all paperfolding sequences, lsd first

Note that the n accepted are precisely those with base-2 expansion 10, 100, and $11 \cdots 1$ (in reversed form). The computation took 196 ms.

For the second statement, we make an automaton accepting those pairs (f, n) for which P_f has an almost-square of order n.

```
eval paper2
 "?lsd_2 (n > 0) & (Ei (i > 0) &  (Ak k < n => PF[f][i+k] = PF[f][i+k+n+1]))":
```

The resulting automaton (Fig. 5) has 9 states and was computed in 72 ms.

By inspection of the automaton, one can verify that every path, no matter what the first component is, having second component labeled with a member of 0100 or 00100 or 11*00, reaches the accepting state 8.

3.5 Appearance Function for Paperfolding Sequences

In what follows, we need a bound on the so-called "appearance" function $A_{\mathbf{x}}(n)$, which is the least integer i such that every factor of length n of an infinite word $\mathbf{x}$ appears somewhere in the prefix $\mathbf{x}[1..i]$.

The appearance function is not identical for every paperfolding sequence. For example, $A_{P_{111\cdots}}(3) = 24$, while $A_{P_{11\overline{1}\overline{1}\overline{1}\cdots}}(3) = 26$. However, we have the following theorem.

Theorem 5. *Let $A(n)$ denote the supremum of the appearance function for length n factors, taken over all paperfolding sequences. Then*

$$A(1) = 3 \qquad A(2) = 7 \qquad A(n) = 6 \cdot 2^h + n - 1, \quad \textit{where } h = \lceil \log_2 n \rceil.$$

Furthermore, the upper bound is best possible and is achieved by the unfolding instructions $\overline{1}\,\overline{1}\,\overline{1}\,1^{\omega}$.

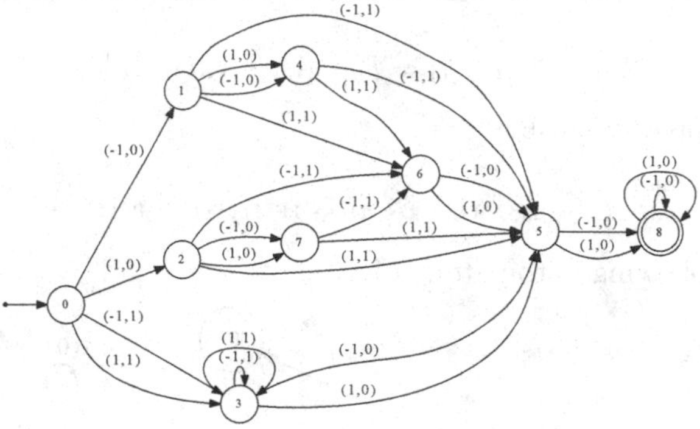

Fig. 5. Automaton accepting (f, n) pairs for which P_f has an almost-square of order n

Proof. The assertions for $A(1)$ and $A(2)$ are easy to verify by hand. For $n \geq 3$, our proof has two parts. First, we show that in every paperfolding sequence P_f, every factor of length n appears in the prefix $P_f[1..A(n)]$. To see this, we attempt to find a prefix of a paperfolding sequence for which this assertion *fails*, that is, we write a predicate asserting the existence of some $n \geq 3$ and $i \geq 1$ for which $P_f[i..i+n-1] \neq P_f[j..j+n-1]$ for all $j \leq 6 \cdot 2^{\lceil \log_2 n \rceil}$.

```
reg power2 lsd_2 "0*10*":
 def twoceilinglog "?lsd_2 $power2(x) & (x >= y) & x < 2*y":
 eval appearance "?lsd_2 En Ei (n >= 3) & (i >= 1) &
 (Aj (Ex x >= 1 & j <= 6*x & $twoceilinglog(x,n)) =>
 (Ek k < n & PF[f][i+k] != PF[f][j+k]))":
```

Here `power2` is a regular expression stating that its argument is a power of 2 (in lsd form, allowing trailing zeroes) and `twoceilinglog` applied to (x, y) asserts that $x = 2^{\lceil \log_2 y \rceil}$. When we run this, we discover that the resulting automaton has 1 state and accepts nothing. Hence there are no counterexamples for $n \geq 3$, and hence $A(n) \leq 6 \cdot 2^h + n - 1$, as desired.

In the next step, we verify that $A(n)$ is the best possible bound. To do so, we reduce our bound on j in the expression above by 1, and ask for the n corresponding to the particular f in the statement. In this case all $n \geq 3$ are accepted.

Corollary 6. *Let x be a factor of length n of any paperfolding sequence. Then x appears for the first time beginning at a position i with $1 \leq i \leq 12n$.*

3.6 The Minimum of the Appearance Function

The minimum value of the appearance function, taken over all paperfolding sequences, is given by

$$m(1) = 2 \qquad m(2) = 5 \qquad m(3) = 16$$
$$m(4) = 17 \qquad m(5) = 32 \qquad m(6) = 36$$
$$m(n) = 4 \cdot 2^h + n - 1, \quad \text{where } h = \lceil \log_2 n \rceil, \text{ for } n \geq 7.$$

No single paperfolding sequence seems to achieve these bounds for all n. However $m(n)$ for $n \geq 7$ is achieved by $11111\overline{1}\,1\overline{1}\,1\overline{1}\,1\overline{1} \cdots$. This can be proved by an approach similar to that in the previous section, and we omit the details.

3.7 Recurrence and the Recurrence Function

A sequence is recurrent if every factor that occurs, occurs infinitely often. As is well known, this is equivalent to the statement that every factor that occurs, occurs at least twice.

Theorem 7. *Every paperfolding sequence is recurrent.*

Trying to prove this directly illustrates a deficiency in our method. We would like to show for all paperfolding sequences **f** for all $i \geq 1, n \geq 1$ there exists $j > i$ such that

$$P_{\mathbf{f}}[i..i+n-1] = P_{\mathbf{f}}[j..j+n-1].$$

During the course of our computation, we can only deal with a prefix of the folding instructions of the same length as the representations of i, j, and n. Since j is unbounded in this formulation, we cannot implement this predicate as stated (unless we use ω-automata).

The solution is to come up with a plausible bound on j. In fact, we can find a conjectured exact bound empirically, and then use our method to verify our conjecture. More precisely, for every paperfolding sequence, we would like to find the recurrence function $r(n)$; this is the smallest integer m such that every block of size m contains every factor of length n.

Theorem 8. *Define*

$$r(1) = 4 \quad r(2) = 9 \quad r(n) = 8 \cdot 2^h + n - 1, \quad \textit{where } h = \lceil \log_2 n \rceil, \textit{form} \geq 3.$$

Then $r(n)$ is the recurrence function for every paperfolding sequence.

This improves the bound given in Allouche and Bousquet-Mélou [2].

Proof. The proof is quite similar to that of Theorem 5. Again, we look for counterexamples. In Mousavi's `Walnut` prover we use

```
reg power2 lsd_2 "0*10*":
def twoceilinglog "?lsd_2 $power2(x) & (x >= y) & x < 2*y":
eval recur4 "?lsd_2 En Ei (n >= 3) & (i >= 1) &
(Aj (Ex x >= 1 & j <= 8*x & j >= 1 & $twoceilinglog(x,n)) =>
(Ek k < n & PF[f][i+k] != PF[f][i+j+k]))":
```

and when we run this we discover there are none (in 276 ms).

As before we can prove the bounds are optimal.

The *recurrence quotient* of a sequence $\mathbf{x}$ is defined to be $\limsup_{n\geq 1} r(n)/n$.

Corollary 9. *Every paperfolding sequence has recurrence quotient equal to* 17.

3.8 Intersection of Sets of Paperfolding Factors

Narad Rampersad asked (personal communication) if it is true that for any two distinct paperfolding sequences, there is some length l such that the sequences have no factors of length l in common. The answer is yes:

Theorem 10. *If* $\mathbf{f}$ *and* $\mathbf{g}$ *are two different unfolding sequences and* l *the smallest index for which* $f_l \neq g_l$, *then* $P_\mathbf{f}$ *and* $P_\mathbf{g}$ *have no factors of length* $\geq 14 \cdot 2^l$ *in common.*

Proof. It suffices to show that $P_\mathbf{f}$ and $P_\mathbf{g}$ have no factors of length exactly $14 \cdot 2^l$ in common.

To do this, we build an automaton that accepts words w where

- $\pi_1(w) = f$, a prefix of $\mathbf{f}$
- $\pi_2(w) = g$, a prefix of $\mathbf{g}$
- $[\pi_3(w)]_2 = i$, a starting position of a factor in P_f
- $[\pi_4(w)]_2 = j$, a starting position of a factor in P_g

such that f and g differ at some position, say position l (starting indexing at position 0) and $P_f[i..i + 14 \cdot 2^l - 1] = P_g[j..j + 14 \cdot 2^l - 1]$. In order to calculate these factors we know from Theorem 5 that any such factor can be found with $i, j \leq 12 \cdot 14 \cdot 2^l$. It therefore suffices to assume that i and j end with at least 7 zeroes.

A program by the first author has verified this assertion.

4 Remarks on Correctness

One referee asked whether we have proved our program correct. The answer is no. In the vast literature of combinatorics on words, machine computations commonly form part of the arguments presented, but to our knowledge nobody attempts to prove their programs correct. For one thing, programs are frequently hundreds or thousands of lines long, and proving even short programs correct is a nontrivial task. For another, typically one proves programs correct with the aid of a prover, but then who has verified the prover? In our opinion, the only reasonable prescription is to (a) provide enough details that a reader could duplicate the computations and (b) provide access to the code used. We hope we have done (a). As for (b), the software can be downloaded from https://cs.uwaterloo.ca/~shallit/papers.html.

References

1. Allouche, J.P.: The number of factors in a paperfolding sequence. Bull. Aust. Math. Soc. **46**, 23–32 (1992)
2. Allouche, J.P., Bousquet-Mélou, M.: Canonical positions for the factors in the paperfolding sequences. Theor. Comput. Sci. **129**, 263–278 (1994)
3. Allouche, J.P., Bousquet-Mélou, M.: Facteurs des suites de Rudin-Shapiro généralisées. Bull. Belg. Math. Soc. **1**, 145–164 (1994)
4. Allouche, J.P., Rampersad, N., Shallit, J.: Periodicity, repetitions, and orbits of an automatic sequence. Theor. Comput. Sci. **410**, 2795–2803 (2009)
5. Charlier, E., Rampersad, N., Shallit, J.: Enumeration and decidable properties of automatic sequences. Int. J. Found. Comp. Sci. **23**, 1035–1066 (2012)
6. Cobham, A.: Uniform tag sequences. Math. Syst. Theory **6**, 164–192 (1972)
7. Davis, C., Knuth, D.E.: Number representations and dragon curves-I, II. J. Recreat. Math. **3**(66–81), 133–149 (1970)
8. Dekking, F.M., Mendès France, M., Poorten, A.J.v.d.: Folds! Math. Intell. **4**, 130–138, 173–181, 190–195 (1982). Erratum **5**, 5 (1983)
9. Goč, D., Henshall, D., Shallit, J.: Automatic theorem-proving in combinatorics on words. In: Moreira, N., Reis, R. (eds.) CIAA 2012. LNCS, vol. 7381, pp. 180–191. Springer, Heidelberg (2012)
10. Kao, J.Y., Rampersad, N., Shallit, J., Silva, M.: Words avoiding repetitions in arithmetic progressions. Theor. Comput. Sci. **391**, 126–137 (2008)
11. Prodinger, H., Urbanek, F.J.: Infinite 0-1-sequences without long adjacent identical blocks. Discrete Math. **28**, 277–289 (1979)
12. Schaeffer, L.: Deciding properties of automatic sequences. Master's thesis, University of Waterloo, School of Computer Science (2013)

Covering the Recursive Sets

Bjørn Kjos-Hanssen[1], Frank Stephan[2](✉), and Sebastiaan A. Terwijn[3]

[1] Department of Mathematics,
University of Hawaii at Manoa, Honolulu, HI 96822, USA
bjoernkh@hawaii.edu

[2] Department of Mathematics and Department of Computer Science,
National University of Singapore, Singapore 119076, Republic of Singapore
fstephan@comp.nus.edu.sg

[3] Department of Mathematics, Radboud University Nijmegen,
P.O. Box 9010, 6500 GL Nijmegen, The Netherlands
terwijn@math.ru.nl

Abstract. We give solutions to two of the questions in a paper by Brendle, Brooke-Taylor, Ng and Nies. Our examples derive from a 2014 construction by Khan and Miller as well as new direct constructions using martingales.

At the same time, we introduce the concept of i.o. subuniformity and relate this concept to recursive measure theory. We prove that there are classes closed downwards under Turing reducibility that have recursive measure zero and that are not i.o. subuniform. This shows that there are examples of classes that cannot be covered with methods other than probabilistic ones. It is easily seen that every set of hyperimmune degree can cover the recursive sets. We prove that there are both examples of hyperimmune-free degree that can and that cannot compute such a cover.

1 Introduction

An important theme in set theory has been the study of cardinal characteristics. As it turns out, in the study of these there are certain analogies with recursion theory, where the recursive sets correspond to sets in the ground model. In a recent paper by Brendle, Brooke-Taylor, Ng and Nies [1], the authors point out analogies between cardinal characteristics and the study of algorithmic randomness. We address two questions raised in this paper that are connected to computing covers for the recursive sets.

In the following, we will assume that the reader is familiar with various notions from computable measure theory, in particular, with the notions of

This work was partially supported by a grant from the Simons Foundation (#315188 to Bjørn Kjos-Hanssen) and by a grant from the NUS (R146-000-181-112 to F. Stephan). A substantial part of the work was performed while the first and third authors were supported by the Institute for Mathematical Sciences of the National University of Singapore during the workshop on *Algorithmic Randomness* during 2–30 June 2014.

A. Beckmann et al. (Eds.): CiE 2015, LNCS 9136, pp. 44–53, 2015.
DOI: 10.1007/978-3-319-20028-6_5

Martin-Löf null, Schnorr null and Kurtz null set. For background on these notions we refer the reader to the books of Downey and Hirschfeldt [4], Li and Vitányi [12] and Nies [14].

Our notation from recursion theory is mostly standard. The natural numbers are denoted by ω, and 2^ω denotes the Cantor space and $2^{<\omega}$ the set of all finite binary sequences. We denote the concatenation of strings σ and τ by $\sigma\tau$. The notation $\sigma \sqsubseteq \tau$ denotes that the finite string σ is an initial segment of the (finite or infinite) string τ. We identify sets $A \subseteq \omega$ with their characteristic sequences, and $A{\upharpoonright}n$ denotes the initial segment $A(0)\dots A(n-1)$. We use λ to denote the empty string. Throughout, μ denotes the Lebesgue measure on 2^ω.

Definition 1. A function $M : 2^{<\omega} \to \mathbb{R}^{\geqslant 0}$ is a *martingale* if for every $x \in 2^{<\omega}$, M satisfies the averaging condition

$$2M(\sigma) = M(\sigma 0) + M(\sigma 1), \tag{1}$$

A martingale M *succeeds on* a set A if

$$\limsup_{n\to\infty} M(A{\upharpoonright}n) = \infty.$$

The class of all sets on which M succeeds is denoted by $S[M]$.

The following definition is taken from Rupprecht [17].

Definition 2. An oracle A is *Schnorr covering* if the union of all Schnorr null sets is Schnorr null relative to A. An oracle A is *weakly Schnorr covering* if the set of recursive reals is Schnorr null relative to A.

Definition 3. A *Kurtz test* relative to A is an A-recursive sequence of closed-open sets G_i such that each G_i has measure at most 2^{-i}; these closed-open sets are given by explicit finite lists of strings and they consist of all members of $\{0,1\}^\omega$ extending one of the strings. Note that $i \to \mu(G_i)$ can be computed relative to A. The intersection of a Kurtz test (relative to A) is called a Kurtz null set (relative to A). An oracle A is *Kurtz covering* if there is an A-recursive array $G_{i,j}$ of closed-open sets such that each i-th component is a Kurtz test relative to A and every unrelativized Kurtz test describes a null-set contained in $\cap_j G_{i,j}$ for some i; A is *weakly Kurtz covering* if there is such an array and each recursive sequence is contained in some A-recursive Kurtz null set $\cap_j G_{i,j}$.

Brendle, Brook-Taylor, Ng and Nies [1] called the notion of (weakly) Schnorr covering in their paper (weakly) Schnorr *engulfing*. In this paper, we will use the original terminology of Rupprecht [17]. We have analogous notions for the other notions of effective null sets. For example, a set A is weakly Kurtz covering if the set of recursive reals is Kurtz null relative to A. We also have Baire category analogues of these notions of covering: A set A is *weakly meager covering* if it computes a meager set that contains all recursive reals. Recall that a set A is diagonally nonrecursive (DNR) if there is a function $f \leqslant_T A$ such that, for all x, if $\varphi_x(x)$ is defined then $\varphi_x(x) \neq f(x)$. A set A has hyperimmune-free Turing degree if for every $f \leqslant_T A$ there is a recursive function g with $\forall x\,[f(x) \leqslant g(x)]$.

2 Solutions to Open Problems

In [1, Question 4.1], Brendle, Brooke-Taylor, Ng and Nies posed three questions, (7), (8) and (9). In this section, we will provide the answers to the questions (7) and (9). For this we note that by [1, Theorem 3] and [9, Theorem 5.1] we have the following result.

Theorem 4. *A set A is weakly meager covering iff it is high or of* DNR *degree.*

We recall the following well-known definitions and results.

Definition 5. A function ψ, written $e \mapsto (n \mapsto \psi_e(n))$, is a recursive numbering if the function $(e, n) \mapsto \psi_e(n)$ is partial recursive. For a given recursive numbering ψ and a function h, we say that f is DNR_h^ψ if for all n, $f(n) \neq \psi_n(n)$ and $f(n) \leq h(n)$. An *order function* is a recursive, nondecreasing, unbounded function.

Theorem 6 (Khan and Miller [8, Theorem 4.3]). *For each recursive numbering ψ and for each order function h, there is an $f \in \text{DNR}_h^\psi$ such that f computes no Kurtz random real.*

Wang (cf. [4, Theorem 7.2.13]) gave a martingale characterization of Kurtz randomness. While it is obvious that weakly Kurtz covering implies weakly Schnorr covering for the martingale notions, some proof is needed in the case that one uses tests (as done here).

Proposition 7. *If A is weakly Kurtz covering then A is weakly Schnorr covering.*

Proof. Suppose A is weakly Kurtz covering, as witnessed by the A-recursive array of closed-open sets $G_{i,j}$. Then the sets $F_j = \cup_i G_{i,i+j+1}$ form an A-recursive Schnorr test, as each F_j has at most the measure $\sum_j 2^{-i-j-2} = 2^{-i-1}$ and the measures of the F_j is uniformly A-recursive as one can relative to A compute the measure of each $G_{i,i+j+1}$ and their sum is fast converging. As for each recursive set there is an i such that all $G_{i,i+j+1}$ contain the set, each recursive set is covered by the Schnorr test. □

Theorem 8. *There is a recursive numbering ψ and an order function h such that for each set A, if A computes a function f that is DNR_h^ψ then A is weakly Kurtz covering.*

Proof. Fix a correspondence between strings and natural numbers num : $2^{<\omega} \to \omega$ such that

$$2^{|\sigma|} - 1 \leq \text{num}(\sigma) \leq 2^{|\sigma|+1} - 2.$$

For instance, num(σ) could be the position of σ in the length-lexicographically lexicographic ordering of all strings as proposed by Li and Vitányi [12]. Let $\text{str}(n) = \text{num}^{-1}(n)$ be the string representation of the number n. Thus

$$2^{|\text{str}(n)|} - 1 \leq \text{num}(\text{str}(n)) = n \leq 2^{|\text{str}(n)|+1} - 2.$$

Let φ be any fixed recursive numbering, let

$$\langle a, b\rangle = \text{num}(1^{|\text{str}(a)|}0\text{str}(a)\text{str}(b))$$

in concatenative notation. Let $\psi_{2\langle e,n\rangle}(x) = \varphi_e(n)$ for any x and $\psi_{2y+1} = \varphi_y$. Note that ψ is an acceptable numbering. Let $s(e, n) = 2\langle e, n\rangle$. Then if f is DNR with respect to ψ then f has the following property with respect to φ:

$$f(s(e, n)) \neq \varphi_e(n).$$

Indeed,

$$f(s(e, n)) = f(2\langle e, n\rangle) \neq \psi_{2\langle e,n\rangle}(2\langle e, n\rangle) = \varphi_e(n).$$

Moreover,

$$\begin{aligned} s(a, b) = 2\langle a, b\rangle &\leq 2(2^{|1^{|\text{str}(a)|}0\text{str}(a)\text{str}(b)|}) = 4(2^{|1^{|\text{str}(a)|}|}2^{|\text{str}(a)|}2^{|\text{str}(b)|}) \\ &= 4(2^{|\text{str}(a)|}2^{|\text{str}(a)|}2^{|\text{str}(b)|}) \leq 4(a+1)^2(b+1). \end{aligned}$$

Consider a partition of ω into intervals I_m such that $|I_m|$ is $2 + \log(m + 1)$ rounded down, and let $h(m) = |I_m|$. If f is DNR_h^ψ then we have

$$\forall \varphi_e \forall n\, (f(s(e, n)) \in \{0, 1\}^{I_{s(e,n)}} \text{ and } f(s(e, n)) \neq \varphi_e(n)).$$

Given a recursive set R, there is, by the fixed point theorem, an index e such that, for all n, $\varphi_e(n) = R \upharpoonright I_{s(e,n)}$ and $f(s(e, n)) \neq R \upharpoonright I_{s(e,n)}$. Note that for every fixed e,

$$\prod_{n=0}^{\infty}(1 - 2^{-|I_{s(e,n)}|}) \leqslant \prod_{n=e+2}^{\infty}(1 - 2^{-(2+\log(4(e+1)^2(n+1)+1))}) \leqslant$$

$$\prod_{n=e+2}^{\infty}(1 - 2^{-(3+\log(4(e+1)^2(n+1)))}) = \prod_{n=e+2}^{\infty}(1 - 2^{-(5+2\log(e+1)+\log(n+1))}).$$

The last product in this formula is 0, as the sum

$$\sum_{n=e+2}^{\infty} 2^{-(5+2\log(e+1)+\log(n+1))} = 1/32 \cdot (e+1)^{-2} \cdot \sum_{n=e+2}^{\infty} 1/(n+1)$$

diverges. Thus

$$\mu(\{B : \exists e \forall n\, [B \upharpoonright I_{s(e,n)} \neq f(s(e, n))]\}) \leqslant \sum_e \prod_{n=0}^{\infty}(1 - 2^{-|I_{s(e,n)}|}) = 0$$

So if f is computable from A then we have a $\Sigma_2^0(A)$ null set that contains all recursive sets, as desired. □

Theorem 9 (Affirmative answer to Brendle, Brooke-Taylor, Ng and Nies [1, Question 4.1(7)]). *There exists a set A such that*

1. *A is weakly meager covering,*
2. *A does not compute any Schnorr random set,*
3. *A is of hyperimmune-free degree,*
4. *A is weakly Schnorr covering.*

Proof. Let h and ψ as in Theorem 8. By Theorem 6, there is an $f \in \mathrm{DNR}_h^\psi$ such that f computes no Kurtz random real. Let A be a set Turing equivalent to f.

1. By Theorem 4, A is weakly meager covering. Alternatively, one could use the fact that every weakly Kurtz covering oracle is also weakly meager covering and derive the item 1 from the proof of item 4.
2. Since each Schnorr random real is Kurtz random, A does not compute any Schnorr random real.
3. Since A does not compute any Kurtz random real, A is of hyperimmune-free degree.
4. By Theorem 8, A is weakly Kurtz covering. In particular, by Proposition 7, A is weakly Schnorr covering.

This completes the proof. □

Franklin and Stephan [6] characterised that a set A is *Schnorr trivial* iff for every $f \leqslant_{tt} A$ there is a recursive function g such that, for all n, $f(n) \in \{g(n,0), g(n,1), \ldots, g(n,n)\}$; this characterisation serves here as a definition.

Theorem 10 (Affirmative answer to Brendle, Brooke-Taylor, Ng and Nies [1, Question 4.1 (9)]). *There is a hyperimmune-free oracle A which is not DNR (and thus low for weak 1-genericity) and which is not Schnorr trivial and which does not Schnorr cover all recursive sets.*

Remark 11. The reader may object that the original question in [1] asked for a set that was *not low for Schnorr tests* rather than *not Schnorr trivial.* However, we can recall the following facts:

- Kjos-Hanssen, Nies and Stephan [10] showed that if A is low for Schnorr tests then A is low for Schnorr randomness;
- Franklin [5] showed that if A is low for Schnorr randomness then A is Schnorr trivial.

3 Infinitely Often Subuniformity and Covering

Let $\langle .,. \rangle$ denote a standard recursive bijection from $\omega \times \omega$ to ω. For a function $P : \omega \to \omega$ define

$$P_n(m) = P(\langle n, m \rangle)$$

and say that P *parametrizes* the class of functions $\{P_n : n \in \omega\}$. We identify sets of natural numbers with their characteristic functions. A class $\mathcal{A}$ is *(recursively) uniform* if there is a recursive function P such that $\mathcal{A} = \{P_n : n \in \omega\}$, and *(recursively) subuniform* if $\mathcal{A} \subseteq \{P_n : n \in \omega\}$. These notions relativize to any oracle A to yield the notions of *A-uniform* and *A-subuniform.*

It is an elementary fact of recursion theory that the recursive sets are not uniformly recursive. The following theorem, as cited in Soare's book [18, p. 255], quantifies exactly how difficult it is to do this:

Theorem 12 (Jockusch). *The following conditions are equivalent:*

(i) *A is high, that is, $A' \geqslant_T \emptyset''$,*
(ii) *the recursive functions are A-uniform,*
(iii) *the recursive functions are A-subuniform,*
(iv) *the recursive sets are A-uniform.*
If A has r.e. degree then (i)–(iv) *are each equivalent to:*
(v) *the recursive sets are A-subuniform.*

In the following we study infinitely often parametrizations and the relation to computing covers for the recursive sets.

3.1 Infinitely Often Subuniformity

Definition 13. We say that a set X *covers* a class $\mathcal{A}$ if there is an X-recursive martingale M such that $\mathcal{A} \subseteq S[M]$.

Note that for X recursive this is just the definition of recursive measure zero. For basics about computable martingales see [4, p. 207].

Definition 14. A class $\mathcal{A} \subset 2^\omega$ is called *infinitely often subuniform* (i.o. subuniform for short) if there is a recursive function $P \in \{0,1,2\}^\omega$ such that

$$\forall A \in \mathcal{A}\, \exists n \left[\exists^\infty x\big(P_n(x) \neq 2\big) \wedge \forall x\big(P_n(x) \neq 2 \rightarrow P_n(x) = A(x)\big)\right]. \qquad (2)$$

That is, for every $A \in \mathcal{A}$ there is a row of P that computes infinitely many elements of A without making mistakes. Again, we can relativize this definition to an arbitrary set X: A class $\mathcal{A}$ is i.o. X-subuniform if P as above is X-recursive.

Let REC denote the class of recursive sets. Recall that A is a PA-complete set if A can compute a total extension of every $\{0,1\}$-valued partial recursive function. Note that if a set A is PA-complete then REC is A-subuniform (cf. Proposition 15 below).

For every recursive set A there is a recursive set $\hat{A}$ such that A can be reconstructed from any infinite subset of $\hat{A}$. Namely, let $\hat{A}(x) = 1$ precisely when x codes an initial segment of A. So it might seem that any i.o. sub-parametrization of REC can be converted into a subparametrization in which every recursive set is completely represented. However, we cannot do this uniformly (since we cannot get rid of the rows that have $P_n(x) = 2$ a.e.) and indeed the implication does not hold.

Proposition 15. *We have the following picture of implications:*

$$\begin{array}{ccccc} A \text{ is PA-complete} \Rightarrow & \text{REC is } A\text{-subuniform} \Rightarrow & & \text{REC is i.o. } A\text{-subuniform} \\ & \Uparrow & & \Uparrow \\ & A \text{ is high} & \Rightarrow & A \text{ has hyperimmune degree} \end{array}$$

No other implications hold than the ones indicated.

Proposition 16. *Every i.o. subuniform class has recursive measure zero. This relativizes to: If $\mathcal{A}$ is i.o. X-subuniform then X covers $\mathcal{A}$.*

Proof. The ability to compute infinitely many bits from a set clearly suffices to define a martingale succeeding on it. The uniformity is just what is needed to make the usual sum argument work. □

Proposition 17. *There exists a class of recursive sets that has recursive measure zero and that is not i.o. subuniform.*

Proof. The class of all recursive sets A satisfying $\forall x\,[A(2x) = A(2x+1)]$ has recursive measure 0 but is not i.o. subuniform: If P would witness this class to be i.o. subuniform then Q defined as $Q_i(x) = \min\{P_i(2x), P_i(2x+1)\}$ would witness REC to be i.o. subuniform, a contradiction. □

Above the recursive sets, the 1-generic sets are a natural example of such a class that has measure zero but that is not i.o. subuniform: It is easy to see that the 1-generic sets have recursive measure zero because for every such set A there are infinitely many n such that $A \cap [n, 2n] = \emptyset$. On the other hand, a variation of the construction in the proof of Proposition 17 shows that the 1-generic sets are not i.o. X-subuniform for any X:

Proposition 18. *The 1-generic sets are not i.o. X-subuniform for any set X.*

Proof. Let $P \subseteq \{0,1,2\}^\omega$ be an X-recursive parametrization and let A be 1-generic relative to X (so that A is in particular 1-generic). Then for every n, if $P_n(x) \neq 2$ for infinitely many x then

$$\{\sigma \in 2^{<\omega} : \exists x\,[P_n(x) \neq 2 \wedge P_n(x) \neq \sigma(x)]\}$$

is X-recursive and dense, hence A meets this set of conditions and consequently P does not i.o. parameterize A. □

Now both the example from Proposition 17 and the 1-generic sets are counterexamples to the implication "measure 0 $\Rightarrow$ i.o. subuniform" because of the *set structure* of the elements in the class. One might think that for classes closed downwards under Turing reducibility (i.e. classes defined by information content rather than set structure) the situation could be different, i.e. that for $\mathcal{A}$ closed downwards under $\leqslant_T$ the implication "X covers $\mathcal{A} \Rightarrow \mathcal{A}$ i.o. X-subuniform" would hold. Note that for X recursive this is not interesting, since any nonempty class closed downwards under Turing reducibility contains REC and REC does not have recursive measure zero. However, this is also not true: Let $\mathcal{A}$ be the class

$$\{A : A \leqslant_T G \text{ for some 1-generic} G\}.$$

Clearly $\mathcal{A}$ is closed downwards under Turing reducibility and it follows from proofs by Kurtz [11] and by Demuth and Kučera [2] (a proof is also given by Terwijn [19]), that $\mathcal{A}$ is a Martin-Löf nullset, and that in particular $\emptyset'$ covers $\mathcal{A}$. However, by Proposition 18 the 1-generic sets are not i.o. $\emptyset'$-subuniform so that in particular $\mathcal{A}$ is not i.o. $\emptyset'$-subuniform.

3.2 A Nonrecursive Set that Does Not Cover REC

It follows from Proposition 15 and Proposition 16 that if A is of hyperimmune degree then A covers REC. In particular every nonrecursive set comparable with $\emptyset'$ covers REC. We see that if A cannot cover REC then A must have hyperimmune-free degree. We now show that there are indeed nonrecursive sets that do not cover REC. Indeed, the following result establishes that there are natural examples of such sets.

Theorem 19. *If A is Martin-Löf random then there is no martingale $M \leqslant_{tt} A$ which covers* REC. *In particular if A is Martin-Löf random and of hyperimmune-free Turing degree then it does not cover* REC.

Proof. Let A be Martin-Löf random and M^A be truth-table reducible to A by a truth-table reduction which produces on every oracle a savings martingale, that is, a martingale which never goes down by more than 1. Without loss of generality, the martingale starts on the empty string with 1 and is never less than or equal to 0. Note that because of the truth-table property, one can easily define the martingale N given by

$$N(\sigma) = \int_{E \subseteq \omega} M^E(\sigma)\, dE.$$

As one can replace the E by the strings up to $use(|\sigma|)$ using the recursive use-function use of the truth-table reduction, one has that

$$N(\sigma) = \sum_{\tau \in \{0,1\}^{use(|\sigma|)}} 2^{-|\tau|} M^{\tau}(\sigma)$$

and N is clearly a recursive martingale. Let B be a recursive set which is adversary to N, that is, B is defined inductively such that

$$\forall n\, [N(B{\restriction}(n+1)) \leqslant N(B{\restriction}n)].$$

Define the uniformly r.e. classes S_n by

$$S_n = \{E : M^E \text{ reaches on } B \text{ a value beyond } 2^n + 1\}.$$

By the savings property, once M^E has gone beyond $2^n + 1$ on B, M^E will stay above 2^n afterwards. It follows that the measure of these E can be at most 2^{-n}. So $\mu(S_n) \leqslant 2^{-n}$ for all n and therefore the S_n form a Martin-Löf test. Since A is Martin-Löf random, there exists n such that $A \notin S_n$, and hence M^A does not succeed on B. □

We note that the set $\{A \in 2^\omega : A \text{ covers REC}\}$ has measure 1. This follows from Proposition 15 and the fact that the hyperimmune sets have measure 1 (a well-known result of Martin, cf. [4, Theorem 8.21.1]). We note that apart from the hyperimmune degrees, there are other degrees that cover REC:

Proposition 20. *There are sets of hyperimmune-free degree that cover the class* REC.

Proof. As in Proposition 15, take a PA-complete set A of hyperimmune-free degree. Then the recursive sets are A-subuniform, so by Proposition 16 A covers REC. □

3.3 Computing Covers Versus Uniform Computation

We have seen above that in general the implication "X covers $\mathcal{A} \Rightarrow \mathcal{A}$ i.o. X-subuniform" does not hold, even if $\mathcal{A}$ is closed downwards under Turing reducibility. A particular case of interest is whether there are sets that can cover REC but relative to which REC is not i.o. subuniform.

Theorem 21. *There exists a set A that covers* REC *but relative to which* REC *is not i.o. A-subuniform.*

Theorem 22. *We have the following picture of implications:*

$$
\begin{array}{ccc}
 & & A \textit{ is PA-complete} \\
 & & \Downarrow \\
A \textit{ is high} & \Rightarrow & \text{REC } \textit{is } A - \textit{subuniform} \\
\Downarrow & & \Downarrow \\
A \textit{ has hyperimmune degree} & \Rightarrow & \text{REC } \textit{is i.o. } A - \textit{subuniform} \\
 & & \Downarrow \\
A \textit{ is nonrecursive} & \Leftarrow & A \textit{ covers } \text{REC} \\
 & & \Downarrow \\
\multicolumn{3}{c}{A \textit{ has hyperimmune degree or } A \textit{ is not Martin-Löf random}}
\end{array}
$$

No other implications hold than the ones indicated.

The following interesting question is still open.

Question 23. *Are there sets A such that A covers* REC, *but not the class of recursively enumerable sets* RE*?*

Acknowledgements. The authors would like to thank George Barmpalias and Michiel van Lambalgen for discussions about Sect. 3 and André Nies and Benoit Monin for correspondence and thorough checking of Theorem 10.

References

1. Brendle, J, Brooke-Taylor, A, Ng, K.M., Nies, A.: An analogy between cardinal characteristics and highness properties of oracles. Technical report on http://arxiv.org/abs/1404.2839 (2014)
2. Demuth, O., Kučera, A.: Remarks on 1-genericity, semigenericity and related concepts. Commentationes Math. Univ. Carol. **28**(1), 85–94 (1987)
3. Downey, R.G., Hirschfeldt, D.R., Lempp, S., Solomon, R.: A Δ_2^0 set with no infinite low subset in either it or its complement. J. Symbolic Logic **66**(3), 1371–1381 (2001)
4. Downey, R.G., Hirschfeldt, D.R.: Algorithmic Randomness and Complexity. Theory and Applications of Computability. Springer, New York (2010)
5. Franklin, J.: Hyperimmune-free degrees and Schnorr triviality. J. Symbolic Logic **73**, 999–1008 (2008)
6. Franklin, J., Stephan, F.: Schnorr trivial sets and truth-table reducibility. J. Symbolic Logic **75**, 501–521 (2010)
7. Hirschfeldt, D.R., Terwijn, S.A.: Limit computability and constructive measure. In: Computational Prospects of Infinity II: Presented Talks. Lecture Notes Series, Institute for Mathematical Sciences, National University of Singapore, vol. 15, pp. 131–141. World Scientific Publishing Co., Pte. Ltd., Hackensack (2008)
8. Khan, M., Miller, J.S.: Forcing with bushy trees. Manuscript (2014)
9. Kjos-Hanssen, B., Merkle, W., Stephan, F.: Kolmogorov complexity and the recursion theorem. Trans. Am. Math. Soc. **363**(10), 5465–5480 (2011)
10. Kjos-Hanssen, B., Nies, A., Stephan, F.: Lowness for the class of Schnorr random reals. SIAM J. Comput. **35**(3), 647–657 (2005)
11. Kurtz, S.A.: Randomness and genericity in the degrees of unsolvability. Thesis (Ph.D.) - University of Illinois at Urbana-Champaign, ProQuest LLC, Ann Arbor, MI (1981)
12. Li, M., Vitányi, P.: An Introduction to Kolmogorov Complexity and Its Applications, 3rd edn. Springer, Heidelberg (2008)
13. Miller, W., Martin, D.A.: The degrees of hyperimmune sets. Zeitschrift für Math. Logik und Grundlagen der Math. **14**, 159–166 (1968)
14. Nies, A.: Computability and Randomness. Oxford Science Publications, New York (2009)
15. Odifreddi, P.G.: Classical Recursion Theory. Studies in Logic and the Foundations of Mathematics, vol. 125. North-Holland Publishing Co., Amsterdam (1989)
16. Odifreddi, P.G.: Classical Recursion Theory II. Studies in Logic and the Foundations of Mathematics, vol. 143. North-Holland Publishing Co., Amsterdam (1999)
17. Rupprecht, N.: Relativized Schnorr tests with universal behavior. Arch. Math. Logic **49**(5), 555–570 (2010)
18. Soare, R.I.: Recursively enumerable sets and degrees. In: Feferman, S., Lerman, M., Magidor, M., Scedrov, A. (eds.) Perspectives in Mathematical Logic. Springer, Berlin (1987)
19. Terwijn, S.A.: On the quantitative structure of Δ_2^0. In: Berger, U., Osswald, H., Schuster, P. (eds.) Reuniting the Antipodes–Constructive and Nonstandard Views of the Continuum (Venice. 1999). Synthese Library, pp. 271–283. Kluwer Academic Publishers, Dordrecht (2001)

On Distributed Monitoring and Synthesis

Anca Muscholl(✉)

LaBRI, University of Bordeaux, Talence, France
anca@labri.fr

1 Context

Modern computing systems are increasingly distributed and heterogeneous. Software needs to be able to exploit these advances, providing means for applications to be more performant. Traditional concurrent programming paradigms, as Java for example, are based on threads, shared-memory, and locking mechanisms that guard access to common data. More recent paradigms, such as the reactive programming model of Erlang [2] and Scala/Akka [1,3] replace shared memory by asynchronous message passing, where sending a message is non-blocking.

In all these concurrent frameworks, writing reliable software is a big challenge because programmers tend to think about code mostly in a sequential way, and have difficulties in overviewing all possible interleavings of executions by different entities. For the same reason, formal verification and analysis of concurrent programs is very challenging. Testing, which is still the main method for error detection in software, has low coverage for concurrent programs. The reason is that bugs in such programs are difficult to reproduce: they may happen under very specific thread schedules and the likelihood of taking such corner-case schedules is very low. Formal verification, such as model-checking and other traditional exploration techniques, can handle very limited instances of concurrent programs, mostly because of the very large number of possible states and of possible interleavings of executions.

Formal verification of programs requires as a pre-requisite a clear mathematical model for programs. Usually, verification of sequential programs starts with an abstraction step – reducing the value domains of variables to finite domains, viewing conditional branching as non-determinism, etc. Another major simplification consists in disallowing recursion. This leads to a very robust computational model, namely *finite-state automata* and *regular languages*. Regular languages of words (and trees) are particularly well understood notions. The deep connections between logic and automata revealed by the foundational work of Büchi, Rabin and others, are crucial pieces in automata-based verification and synthesis.

Synthesis means to translate a specification into a program that conforms with the specification, and thus can provide solutions that are correct by construction. Synthesis of *reactive systems*, that is of systems that interact with an environment, started as a problem in logics. In the sixties, A. Church asked for an algorithm to construct devices that transform sequences of input bits into sequences of output bits in a way required by a logical formula [9]. Later,

A. Beckmann et al. (Eds.): CiE 2015, LNCS 9136, pp. 54–62, 2015.
DOI: 10.1007/978-3-319-20028-6_6

Ramadge and Wonham proposed the *supervisory control* formulation [33], where a plant and a specification are given; a controller should be designed such that its product with the plant satisfies the specification. Thus, control means restricting the behavior of the plant. Synthesis is the particular case of control where the plant allows for every possible behavior. Rabin's result about the decidability of monadic second-order logic over infinite trees solved Church's question for MSO specifications [32].

When adding concurrency, the landscape of verification and automated synthesis becomes much more complicated. First, there is no canonical model for concurrent systems, simply because there can be very different kinds of interaction between processes. Compare for example multi-threaded shared memory systems and programs with asynchronous function calls. A second serious obstacle for developing automata-based verification techniques for concurrent systems is the lack of a general framework for distributed synthesis, and this even for systems without environment. The question whether a sequential specification can be turned into a distributed implementation over a given distributed architecture was first raised in the context of Petri nets. Ehrenfeucht and Rozenberg introduced the notion of regions to describe how to associate places of nets with states of a transition system [12].

Inspired by Petri nets, Mazurkiewicz proposed in the late seventies the theory of *Mazurkiewicz traces* [27], that we present in the next section. Within this theory, Zielonka's theorem [36] is a prime example for distributed synthesis. Our survey aims at introducing Mazurkiewicz traces and Zielonka's theorem, and describe how this theory can help to verify and design concurrent programs.

2 Mazurkiewicz Traces and Zielonka Automata

Mazurkiewicz traces [27] are one of the simplest formalisms able to describe concurrency. To define the model we fix an alphabet of actions Σ and a *dependence relation* $D \subseteq \Sigma \times \Sigma$ on actions, that is reflexive and symmetric. The idea behind this definition is that two dependent actions are always ordered and cannot be permuted. For instance, in a multi-threaded program all actions belonging to one thread must be ordered according to the program order. The actions of acquiring or releasing the same lock are also ordered, since a thread needs to wait that a lock is released before acquiring it. By contrast, independent actions can be permuted.

A by now classical way to express such dependencies is Lamport's *happens-before* partial order [25]. Mazurkiewicz traces capture this partial order through the dependence relation: from a linear execution $w = a_1 \dots a_n \in \Sigma^*$ a partial order $T(w) = \langle E, \preceq \rangle$ is defined, where:

- $E = \{a_1, \dots, a_n\}$ is the set of *events*, in one-to-one relation with the positions of w,
- $\preceq$ is the reflexive-transitive closure of $\{(a_i, a_j) \mid i < j,\ a_i \ D\ a_j\}$.

Partial orders $T(w)$ as above are called *Mazurkiewicz traces.*

Example 1. As an example consider a concurrent program with threads $T \in \mathcal{T}$ that have read/write access to shared variables $x \in X$. The dependence relation D over the alphabet of actions $\Sigma = \{r(T,x), w(T,x) \mid T \in \mathcal{T}, x \in X\}$ is given by $a\ D\ b$ if

- a, b are actions of the same thread T, or
- a, b access to the same variable $x \in X$ and at least one of them is a write.

This dependence relation simply describes that two actions are independent only if they belong to different threads. Moreover, if they access the same shared variable, then they must be both read actions.

From a language-theoretical viewpoint, traces are almost as attractive as words, and a rich body of results on automata and logics over finite and infinite traces exists, see the handbook [11]. One of the cornerstone results in Mazurkiewicz trace theory is based on a simple notion of finite-state distributed automata, Zielonka automata, that we present in the remaining of the section.

Informally, a Zielonka automaton [36] is a finite-state automaton with control distributed over several *processes* that synchronize on shared actions. There is no global clock, for instance between two synchronizations, two processes can do a different number of actions. Because of this, Zielonka automata are also known as *asynchronous automata.* Sharing of actions is defined through a fixed distributed action alphabet.

A *distributed action alphabet* on a finite set $\mathbb{P}$ of processes is a pair (Σ, dom), where Σ is a finite set of *actions* and $dom : \Sigma \to (2^{\mathbb{P}} \setminus \emptyset)$ is a *location function.* The location $dom(a)$ of action a comprises all processes that synchronize in order to perform this action. The location induces a natural dependence relation D over Σ by letting $a\ D\ b$ if $dom(a) \cap dom(b) \neq \emptyset$.

Example 2. As an example of distributed alphabet reconsider Example 1. A pair (Σ, dom) corresponding to the dependence relation D defined above can be obtained from the set of processes: $\mathbb{P} = \mathcal{T} \cup \{\langle T, x\rangle \mid T \in \mathcal{T}, x \in X\}$. Informally each thread represents a process, and there is a process for each pair $\langle T, x\rangle$, representing the cached value of x in thread T.

The location function defined below satisfies $a\ D\ b$ iff $\mathrm{dom}(a) \cap \mathrm{dom}(b) \neq \emptyset$:

$$\mathrm{dom}(a) = \begin{cases} \{T, \langle T, x\rangle\} & \text{if } a = r(T,x) \\ \{T, \langle T', x\rangle \mid T' \in \mathcal{T}\} & \text{if } a = w(T,x) \end{cases}$$

Formally, a *Zielonka automaton* $\mathcal{A} = \langle (S_p)_{p\in\mathbb{P}}, (s_p^{init})_{p\in\mathbb{P}}, \delta\rangle$ over (Σ, dom) consists of:

- a finite set S_p of (local) states with an initial state $s_p^{init} \in S_p$, for every process $p \in \mathbb{P}$,
- a partial transition relation $\delta \subseteq \bigcup_{a\in\Sigma} \left(\prod_{p\in dom(a)} S_p \times \{a\} \times \prod_{p\in dom(a)} S_p\right)$.

As usual, an automaton is called deterministic if the transition relation is a (partial) function. The reader may be more familiar with synchronous products of finite automata, where a joint action means that every automaton having this action in its alphabet executes it according to its transition relation. Joint transitions in Zielonka automata follow a *rendez-vous* paradigm, meaning that the processes having action a in their alphabet can exchange information via the execution of a. The following example illustrates this effect:

Example 3. The CAS operation is available as atomic operation in the JAVA package java.util.concurrent.atomic, and supported by many architectures. It takes as parameters the thread identifier T, the variable name x, and two values, `old` and `new`. The effect of the instruction `y = CAS(T,x,old,new)` is conditional: the value of x is replaced by `new` if it is equal to `old`, otherwise it does not change. The method returns `true` if the value was swapped, and `false` otherwise.

We can view the `CAS` instruction as a synchronization between two processes, P_T associated with the thread T and P_x associated with the variable x. The states of P_T are valuations of the local variables of T. The states of P_x are the values x can take. An instruction a of the form `y = CAS(T,x,old,new)` becomes a synchronization action between P_T and P_x with the following two transitions (represented for convenience as Petri net transitions; places on the left represent states of P_T, and on the right of P_x):

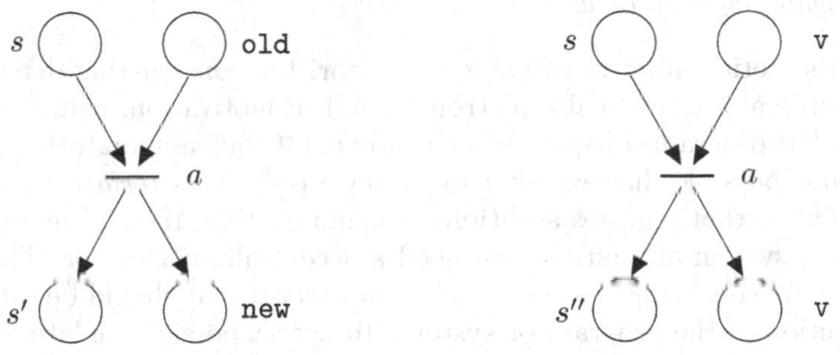

On the left side of the figure we have the case where the value of x is `old`, and on the right half when it is different from `old`. Notice that in state s' the value of y is `true`, whereas in s'', it is `false`.

For convenience, we abbreviate a tuple $(s_p)_{p \in P}$ of local states by s_P.

Notice that a Zielonka automaton can be seen as a usual finite-state automaton with the state set $S = \prod_{p \in \mathbb{P}} S_p$ given by the global states, and transitions $s \xrightarrow{a} s'$ if $(s_{dom(a)}, a, s'_{dom(a)}) \in \delta$, and $s_{\mathbb{P} \setminus dom(a)} = s'_{\mathbb{P} \setminus dom(a)}$. Thus states of this automaton are the tuples of states of the processes of the Zielonka automaton. As a language acceptor, a Zielonka automaton $\mathcal{A}$ accepts a *trace-closed language* $L(\mathcal{A})$, that is, a language closed under permutation of adjacent independent symbols.

3 Distributed Synthesis

A cornerstone result in the theory of Mazurkiewicz traces is a construction that transforms sequential automata into deterministic Zielonka automata, whenever the language is trace-closed. This important result is one of the rare examples of distributed synthesis, next to Ehrenfeucht and Rozenberg's theory of regions.

Theorem 1 ([36]). *For a given distributed alphabet (Σ, dom), and a regular trace-closed language $L \subseteq \Sigma^*$ over (Σ, dom), a deterministic Zielonka automaton $\mathcal{A}$ can be effectively constructed with $L(\mathcal{A}) = L$.*

The intricacy of Zielonka's construction is such that there has been a lot of work to simplify it and to improve its complexity, see e.g. [10,16,19,28]. The most recent construction produces deterministic Zielonka automata of size that is exponential only in the number of processes. It was shown in [16] that the construction is optimal modulo a technical assumption (that is actually required for monitoring).

Theorem 2 ([16]). *There is an algorithm that takes as input a distributed alphabet (Σ, dom) over n processes and a DFA $\mathcal{A}$ accepting a trace-closed language over (Σ, dom), and computes an equivalent deterministic Zielonka automaton $\mathcal{B}$ with at most $4^{n^4} \cdot |\mathcal{A}|^{n^2}$ states per process. Moreover, the algorithm computes the transitions of $\mathcal{B}$ on-the-fly in polynomial time, and checks whether a state is final in polynomial time as well.*

Besides a theoretical interest of having an algorithm constructing deterministic Zielonka automata, there is also a strong practical motivation, namely to monitor distributed programs or systems at runtime. Of course, monitoring a system offline is also possible, however it can be done only *a posteriori* or by a centralized monitor that requires additional communication. If we want to monitor a distributed system at runtime, we need a decentralized monitor. The idea is simple: we have some trace-closed, regular property ϕ that should be satisfied by every execution of the program or system. To detect possible violations of ϕ at runtime, we construct a monitor for ϕ and run it in parallel with the program. Assuming that we model our program P by a Zielonka automaton $\mathcal{A}_P$, running monitor M, that is also a Zielonka automaton $\mathcal{A}_M$, amounts to build the usual product automaton on each process between $\mathcal{A}_P$ and $\mathcal{A}_M$.

It is worth noting that the properties one would like to monitor on distributed programs can be often expressed in terms of the partial order between specific events. To illustrate this, consider as an example the *race detection problem* for multi-threaded programs. Informally, a race occurs whenever there are conflicting accesses to the same shared variable without proper lock synchronization. Detecting races is important since executions with races may yield non-deterministic, unexpected behaviors. Two accesses to the same variable are called *conflicting*, if at least one of them is a write. A *race* is given by two conflicting accesses that are unordered in the happens-before relation. This relation is a dependence relation in terms of Mazurkiewicz traces, that orders the events of each thread and lock access operations for each lock. So a violation of the "no-race" safety property consists in monitoring for two unordered occurrences of such conflicting accesses.

The construction of a deterministic Zielonka automata for properties asking for the partial ordering between specific events is in fact very close to the critical part of all available proofs of Zielonka's theorem. This critical part is known as the *gossip automaton* [28], and the name reflects already its rôle: it computes what a process knows about the knowledge of other processes.

In general, the gossip automaton is already responsible for the exponential complexity of the Zielonka construction. Thus, an important practical question is whether the construction of the gossip automaton can be avoided, or at least simplified. As the theorem below shows, gossiping is not needed when the communication structure is hierarchical.

A distributed alphabet (Σ, dom) is called *acyclic* if all actions have unary or binary domains, and the following graph $G(\Sigma, \text{dom})$ (called *communication graph*) is acyclic: the set of nodes of $G(\Sigma, \text{dom})$ is the set $\mathbb{P}$ of processes and the set of edges is $\{(p, q) \mid \exists a \in \Sigma : \text{dom}(a) = \{p, q\}\}$.

Theorem 3 ([22]). *Let (Σ, dom) be a distributed alphabet whose communication graph is acyclic. Then every regular, trace-closed language L over Σ can be recognized by a deterministic Zielonka automaton with $O(s^2)$ states per process, where s is the size of the minimal DFA for L.*

We need to stress that the practical use of Zielonka automata for e.g. monitoring properties does not depend exclusively on the efficiency of the constructions from the above theorems. Further properties are required for a monitoring automaton $\mathcal{A}_M$ besides determinism. A first requirement is that violations of the property to monitor should be detectable locally, i.e., by at least one thread. The reason is that local detection enables a thread to start some recovery actions, like rollback of a transaction and a new try. A Zielonka automaton $\mathcal{A}$ with this property is called *locally rejecting* [16]. More formally, each process p has a subset of states $R_p \subseteq S_p$, and an execution leads a process p into a state from R_p if and only if the causal past of p cannot be extended to a trace in $L(\mathcal{A})$. A second requirement is that the monitoring automaton should not block the monitored system $\mathcal{A}_P$. This can be achieved by asking that in every global state of $\mathcal{A}_M$ such that no process is a rejecting state, every action is enabled. A related discussion of desirable properties of Zielonka automata and on an implementation of the construction of [16] is reported in [5] (see also [34]).

4 Related Work

This brief overview aimed at presenting the motivation behind distributed synthesis and how Mazurkiewicz trace theory can be useful in this respect. In the following we point out some related results.

Synthesis. Zielonka's algorithm has been applied for solving the synthesis problem, for models that go beyond Mazurkiewicz traces. One example is synthesis of communicating automata from graphical specifications known as *message sequence charts.* Communicating automata are distributed finite-state automata communicating over point-to-point FIFO channels. As such, the model is Turing powerful. However, if the communication channels are bounded, there is a strong link between execution sequences of the communicating automaton and Mazurkiewicz traces [21]. Actually we can even handle even the case where the assumption about bounded channels is relaxed by asking that they are are bounded for *at least one* scheduling of message receptions [18]. Producer-consumer behaviors are captured by this second setting.

Multiply nested words with various bounds on stacks [23,24,31] are an attractive model for concurrent programs with recursion, because of their decidability properties and expressiveness. In [7] the model is extended to nested Mazurkiewicz traces and Zielonka's construction is lifted to this setting.

We do not survey here recent results on synthesis of open systems and control for Zielonka automata. The interested reader is referred to [14,15,17,26,29].

Verification. As we already mentioned, automated verification of concurrent systems encounters major problems due to state explosion. One particularly efficient technique able to addresses these problems is known as *partial order reduction* (POR) [20,30,35]. It consists of restricting the exploration of the state space by avoiding the execution of similar, or equivalent runs. The notion of equivalence of runs used by POR is based on the model of *Mazurkiewicz traces.* The efficiency of POR methods depends of course on the precise equivalence notion between executions. More recent methods such as dynamic POR work without storing explored states explicitly and aim at improving the precision by computing additional information about (non)-equivalent executions [4].

There are many other contexts in verification where analysis can be made more efficient using equivalences based on Mazurkiewicz traces. One such example is counter-example generation based on partial (Mazurkiewicz) traces instead of linear traces, as done in [8]. Another example is the detection of concurrency bugs such as atomicity violations [13], non-linearizability and sequential inconsistency [6].

References

1. Akka. http://akka.io/
2. Erlang programming language. http://www.erlang.org/
3. Scala programming language. http://www.scala-lang.org/
4. Abdulla, P., Aronis, S., Jonsson, B., Sagonas, K.: Optimal dynamic partial order reduction. In: POPL 2014, pp. 373–384. ACM (2014)
5. Akshay, S., Dinca, I., Genest, B., Stefanescu, A.: Implementing realistic asynchronous automata. In: FSTTCS 2013, LIPIcs, pp. 213–224. Schloss Dagstuhl - Leibniz-Zentrum fuer Informatik (2013)
6. Alur, R., McMillan, K., Peled, D.: Model-checking of correctness conditions for concurrent objects. In: LICS 1996, pp. 219–228. IEEE (1996)
7. Bollig, B., Grindei, M.-L., Habermehl, P.: Realizability of concurrent recursive programs. In: de Alfaro, L. (ed.) FOSSACS 2009. LNCS, vol. 5504, pp. 410–424. Springer, Heidelberg (2009)

8. Černý, P., Henzinger, T.A., Radhakrishna, A., Ryzhyk, L., Tarrach, T.: Efficient synthesis for concurrency by semantics-preserving transformations. In: Sharygina, N., Veith, H. (eds.) CAV 2013. LNCS, vol. 8044, pp. 951–967. Springer, Heidelberg (2013)
9. Church, A.: Logic, arithmetics, and automata. In: Proceedings of the International Congress of Mathematicians, pp. 23–35 (1962)
10. Cori, R., Métivier, Y., Zielonka, W.: Asynchronous mappings and asynchronous cellular automata. Inf. Comput. **106**, 159–202 (1993)
11. Diekert, V., Rozenberg, G. (eds.): The Book of Traces. World Scientific, Singapore (1995)
12. Ehrenfeucht, A., Rozenberg, G.: Partial (set) 2-structures: parts i and ii. Acta Informatica **27**(4), 315–368 (1989)
13. Farzan, A., Madhusudan, P.: Monitoring atomicity in concurrent programs. In: Gupta, A., Malik, S. (eds.) CAV 2008. LNCS, vol. 5123, pp. 52–65. Springer, Heidelberg (2008)
14. Gastin, P., Lerman, B., Zeitoun, M.: Distributed games with causal memory are decidable for series-parallel systems. In: Lodaya, K., Mahajan, M. (eds.) FSTTCS 2004. LNCS, vol. 3328, pp. 275–286. Springer, Heidelberg (2004)
15. Gastin, P., Sznajder, N.: Fair synthesis for asynchronous distributed systems. ACM Trans. Comput. Log. **14**(2), 9 (2013)
16. Genest, B., Gimbert, H., Muscholl, A., Walukiewicz, I.: Optimal Zielonka-Type construction of deterministic asynchronous automata. In: Abramsky, S., Gavoille, C., Kirchner, C., Meyer auf der Heide, F., Spirakis, P.G. (eds.) ICALP 2010. LNCS, vol. 6199, pp. 52–63. Springer, Heidelberg (2010)
17. Genest, B., Gimbert, H., Muscholl, A., Walukiewicz, I.: Asynchronous games over tree architectures. In: Fomin, F.V., Freivalds, R., Kwiatkowska, M., Peleg, D. (eds.) ICALP 2013, Part II. LNCS, vol. 7966, pp. 275–286. Springer, Heidelberg (2013)
18. Genest, B., Kuske, D., Muscholl, A.: A Kleene theorem and model checking algorithms for existentially bounded communicating automata. Inf. Comput. **204**(6), 920–956 (2006)
19. Genest, B., Muscholl, A.: Constructing exponential-size deterministic zielonka automata. In: Bugliesi, M., Preneel, B., Sassone, V., Wegener, I. (eds.) ICALP 2006. LNCS, vol. 4052, pp. 565–576. Springer, Heidelberg (2006)
20. Godefroid, P., Wolper, P.: Using partial orders for the efficient verification of deadlock freedom and safety properties. Form. Meth. Syst. Des. **2**(2), 149–164 (1993)
21. Henriksen, J.G., Mukund, M., Kumar, K.N., Sohoni, M., Thiagarajan, P.S.: A theory of regular MSC languages. Inf. Comput. **202**(1), 1–38 (2005)
22. Krishna, S., Muscholl, A.: A quadratic construction for Zielonka automata with acyclic communication structure. Theor. Comput. Sci. **503**, 109–114 (2013)
23. La Torre, S., Madhusudan, P., Parlato, G.: A robust class of context-sensitive languages. In: LICS 2007, pp. 161–170. IEEE (2007)
24. La Torre, S., Parlato, G.: Scope-bounded multistack pushdown systems: fixed-point, sequentialization, and tree-width. In: FSTTCS 2012, LIPIcs, pp. 173–184. Schloss Dagstuhl - Leibniz-Zentrum fuer Informatik (2012)
25. Lamport, L.: Time, clocks, and the ordering of events in a distributed system. Oper. Syst. **21**(7), 558–565 (1978)
26. Madhusudan, P., Thiagarajan, P.S., Yang, S.: The MSO theory of connectedly communicating processes. In: Sarukkai, S., Sen, S. (eds.) FSTTCS 2005. LNCS, vol. 3821, pp. 201–212. Springer, Heidelberg (2005)
27. Mazurkiewicz, A.: Concurrent program schemes and their interpretations. DAIMI report PB 78, Aarhus University, Aarhus (1977)

28. Mukund, M., Sohoni, M.A.: Keeping track of the latest gossip in a distributed system. Distrib. Comput. **10**(3), 137–148 (1997)
29. Muscholl, A., Walukiewicz, I.: Distributed synthesis for acyclic architectures. In: FSTTCS 2014, LIPIcs, pp. 639–651. Schloss Dagstuhl - Leibniz-Zentrum fuer Informatik (2014)
30. Peled, D.A.: All from one, one for all: on model checking using representatives. In: Courcoubetis, C. (ed.) CAV 1993. LNCS, vol. 697, pp. 409–423. Springer, Heidelberg (1993)
31. Qadeer, S., Rehof, J.: Context-bounded model checking of concurrent software. In: Halbwachs, N., Zuck, L.D. (eds.) TACAS 2005. LNCS, vol. 3440, pp. 93–107. Springer, Heidelberg (2005)
32. Rabin, M.O.: Automata on Infinite Objects and Church's Problem. American Mathematical Society, Providence (1972)
33. Ramadge, P.J.G., Wonham, W.M.: The control of discrete event systems. Proc. IEEE **77**(2), 81–98 (1989)
34. Stefanescu, A.: Automatic synthesis of distributed transition systems. Ph.D. thesis, Universität Stuttgart (2006)
35. Valmari, A.: Stubborn sets for reduced state space generation. In: Rozenberg, G. (ed.) APN 1990. LNCS, vol. 483, pp. 491–515. Springer, Heidelberg (1991)
36. Zielonka, W.: Notes on finite asynchronous automata. RAIRO-Theor. Inf. Appl. **21**, 99–135 (1987)

Unconventional Computing: Do We Dream Too Much?

Gheorghe Păun(✉)

Institute of Mathematics of the Romanian Academy, PO Box 1-764, 014700 Bucureşti, Romania
gpaun@us.es

Abstract. After briefly mentioning the motivation and the "dreams" of unconventional computing (with an eye on natural computing, especially on bio-inspired computing), we ask ourselves whether these "dreams" are realistic, and end with a couple of related research issues from the membrane computing area.

1 Motivation of Unconventional Computing

The motivation of unconventional computing, in particular, of natural computing (here we understand it mainly in the sense of bio-inspired computing, although in many cases – see, e.g., [22] – also quantum computing and cellular automata are included), comes from at least three directions: (i) the limits of current ("Turing-von Neumann") computers, (ii) the need for new modeling and simulating tools for sciences like biology, ecology, even physics, (iii) the intrinsic human curiosity, the need to know, to predict, to build mathematical models.

The computers are the most influential invention of the last century, but they have both theoretical limits and practical limits (the two categories do not completely overlap – but this can be the subject of a separate discussion). Among the theoretical ones, two are fundamental: (1) current computers (although they are not Turing machines in the strict, mathematical meaning of the term) cannot compute "beyond the Turing barrier", cannot compute what is Turing uncomputable, and (2) for current computers, problems of a complexity higher than the polynomial one are intractable, they cannot be solved in a feasible time. In general, problems which are **NP**-complete are (considered) intractable – although most non-trivial practical problems are of this type. (Cryptography is crucially dependent on this assumption.)

Breaking "the Turing barrier" was a constant concern of computability, and the term *hypercomputability* was coined to name research in this direction. The bibliography is large, we mention here only [6, 24].

Much larger is the literature related to complexity classes **P** and **NP** and, especially in the unconventional computing area, the bibliography of attempts to "break the **NP** barrier". Symmetrical to the hypercomputation concept, the term *fypercomputation* was proposed in [18] as a name for the research aiming to find polynomial time solutions to computationally hard problems, typically, **NP**-complete problems (the initial "f" is taken from "fast").

A. Beckmann et al. (Eds.): CiE 2015, LNCS 9136, pp. 63–70, 2015.
DOI: 10.1007/978-3-319-20028-6_7

Fypercomputability is the primary goal of unconventional computing and this is a very important issue from a practical point of view, [5], although it is believed, see, e.g., [6], that "computing the uncomputable" could have even more important consequences than proving that **P** = **NP**.

2 Dreams of Unconventional Computing

The term "unconventional" is not precisely defined, we use it here in the vague sense of "non-classic" in theory (not belonging to the "standard" automata theory, based on Turing machines and their variants, restrictions and generalizations) and "not electronic" in the form of implementation, having in mind especially bio-inspired computing (DNA computing, membrane computing, evolutionary and neural computing, the long list of algorithms abstracted from bio-processes, such as immune, ant colony, bee colony, swarm, cuckoo, strawberry algorithms), water flowing algorithms and cultural computing, as well as quantum computing and analog computation, optical computing and many others.

As said before, solving problems considered of an exponential complexity (we assume **P** ≠ **NP**) in polynomial time is the first dream of all these directions of research (even if the solution is not provably optimal, but "good enough", whatever this means – e.g., close to optimal with a known probability). In evolutionary-like computing the strategy is based on an intriguing slogan: when you do not know where to, go randomly! Actually, the random walk through the space of candidate solutions is controlled in a way which imitates the Darwinian evolution, or the way the ants and the bees look for food, and so on and so forth. Combined with the impressive brute force of the modern computers, this strategy, which is only a metaphoric imitation of biological processes, proves to work surprisingly well in surprisingly many situations.

Although usually not explicitly stated, also hypercomputability is a dream of unconventional computing, starting, for instance, from the (debatable, of course) observation that "the brain is not a Turing machine" and, similarly, from other "computations" taking place in nature which seem to be of a non-Turing type. There are many basic ideas of hypercomputing, [24], which lead to devices more powerful than Turing machines (typically, able to answer the halting problem). Some of these ideas were extended also to natural computing models. An example is the acceleration, see [3] (the first step takes one time unit, but the machine learns, so the next step takes only half of a time unit, and so on, each step taking half of the time needed for the previous step; note the important distinction between the internal clock and the external one: infinitely many internal steps are performed in two external time units).

There also are other goals/dreams of unconventional computing, which can themselves be called "unconventional" with respect to the classical computer science. We only list some of them: energy efficiency, also related to the reversibility issue; adaptability, learnability, evolvability; self-healing, robustness with respect to hardware errors.

We somewhat moved from theory to engineering, so let us go further towards practice (the three dimensions, theory, computer engineering, and applications, are intimately related, of course). The internet brought into the stage old-new concepts, such as unsynchronized/asynchronous computation, amorphous computing, cloud computing. Biology asks for models of cells (this is considered the main challenge of the bioinformatics, after the completion of the Genome Project), medicine asks for nano-robots able to scan the body and deliver medicines in the necessary places, repair genes, kill viruses (a project of such a nano-robot, built from DNA molecules, was presented in [2]; note that repairing genes, identifying and breaking in parts viruses are string editing operations, hence they pertain to DNA computing).

Then, on top of all these, there is a "meta-dream", actually, a forecast: both physics and biology will gain a lot from bringing among their central paradigms the information and the computability, new ages of these sciences are foreseen, based on these paradigms. For physics, this seems to be enhanced by the progresses in quantum computing (see, e.g., [10,27]). For biology, the issue is more urgent, taking into account that biology is not yet a mathematized science, the biologists need tools and techniques for modeling and simulating processes at all levels, from cells to eco-systems. A huge quantity of empirical data was gathered, but the tools to process these data were not correspondingly developed. In [14] the term *infobiotics* was proposed for this informational-computational biology, in [19] we propose the term *infobiology* (symmetrical to the currently used term *bioinformatics*).

Significant for our discussion, a lot of words were spent around the *systems biology* syntagma, a planned research area aiming to transform biology and medicine into a "precise engineering" (see, e.g., [12]).

3 Difficulties and Limits

And now, we come to the question in the title: are these goals and dreams realistic, or we dream too much? Of course, we do not have to underestimate the progresses in science and technology, in particular, in bioengineering, there are many funny examples of this kind in the history. However, in many cases, after a successful experiment (e.g., in bio-computing), the progresses were disappointing, not confirming the initial enthusiasm. The typical example is that of DNA computing, with Adleman's experiment, [1], opening a new research area, which, after twenty years has not confirmed yet the big hopes of the beginning.

There are many details to be discussed here. The scientists are, in general, moderately enthusiastic, but they are "forced" to be so in order to "sell" their results, to get projects, hence money. Then, mass-media is always ready to inflate the facts, to predict "scientific revolutions" (see, as an example, the media echoes of the "doctor in a cell" from [2] and of similar – science fiction at this moment – ideas). As another illustration we can mention the noise around systems biology, in many respects not too much more than an reincarnation of system theory applied to biology, [25], not very successful in its first stages, in sixties, because of various reasons, connected to both computer science and biology development at that time.

All these pertain to the politics and the sociology of science. Here we are more interested in the difficulties and, especially, the limits coming from science and technology. There are many of them.

On paper we work with idealized objects, e.g., DNA molecules; in a test tube or in a cell they behave "slightly" differently. There are three different "worlds", *in vivo, in vitro, in info*, and the notions, ideas, models cannot be directly transferred among them. The models of a cell are made of symbols, they are reductionistic, abstract, they always work as prescribed, which is not the case in reality. The goal of life is life, not computing. We, the computer scientists see computations everywhere. The biochemistry is in a large extent nondeterministic, probabilistic, context-dependent, the controls are complex and not always known or easy to find. Nature is redundant, it has "enough time", it affords to try and, if the result is not acceptable, to discard it. All these are far from what we need and what we can do in computer science. That is why the modeling and the simulation of a cell, the minimal entity which is unanimously considered alive, is so difficult. That is why it is difficult to ask the biomolecules or the cells to compute.

Going upwards from the cell, the things become still more difficult, first because the systems and processes we have to deal with are much more complex. Small is beautiful, big is necessary (but difficult to handle). At the level of "big" there appear such phenomena as emergence, synergy, system effect, which are often non-predictable (this can be related to precise mathematical results: Rice theorem tells us that no non-trivial question about a model of the cell which is known to be Turing equivalent – and most such models in membrane computing are so – can be algorithmically answered).

How to model such "things" like life and intelligence providing that we do not even have good (mathematical) definitions of them?!... This raises the question whether such definitions are possible, in general, or, more realistic, possible in the framework of the today mathematics. It was said in several places (see the references of [17], from where several ideas are recalled here) that it is perhaps necessary to wait for a new mathematics in order to face such tasks, like modeling life and intelligence.

But, also in terms of what we have now we can find intrinsic limits of unconventional computing.

Three are my favorite results of this kind. They are mentioned in the order of the time of their publication.

The first one is Gandy's approach to what is (to use Hilbert's term) *mechanically computable*, see [9]. In the attempt to free the Turing-Church thesis of any anthropic meaning, Gandy considered a general algebraic definition of a computing device, and then proved that any computing machine fulfilling the (four) conditions in this definition can be simulated by a Turing machine. The generality of Gandy's definition and his theorem can be seen as a proof that hypercomputing is a difficult task – and this fits with Martin Davis opinion [7,8] that "hypercomputing is a myth", moreover, that the computing machineries more powerful than the Turing machine which were reported so far are based on dishonest tricks (the power is introduced in the definition, in disguise, in the form of infinite processes, real numbers, other infinite ingredients, hence there is no surprise that the device is powerful).

Next, one must recall Conrad theorems, [4]. Basically, they say that the three desired characteristics of computing devices, universality (hence programmability), efficiency and learnability/evolvability, are contradictory, there cannot exist a computing device simultaneously having all these three properties. These theorems are of a kind which is famous in mathematics, *impossibility theorems*, proving that when we demand simultaneously certain properties, it happens that there is no object having all these properties. Gödel theorems and Arrow theorem (in social choice) are classic examples.

As the third limiting theorem in natural computing/optimization we can consider the (in) famous *no free lunch* theorem of Wolpert and Macready, [26], saying, in short, that in average all methods of approximate optimization are equally good over all optimization problems. "Equally good" can also be read "equally bad", which can explain the plethora of approximate optimization methods – each new one finds a niche where it is better than others, and so on....

Returning to applications in biology, one often makes lists of properties which, for instance, the mathematical models should possess: adequacy and relevance, programmability and scalability, efficiency, understandability, emergent behavior. The question which arises is obvious: can all these properties be simultaneously reached, or also in the biological modeling area there exist impossibility theorems, like Conrad theorems? I bet for the existence of such impossibility results.

4 Research Topics

The last lines of the previous section already formulated a research topic. There are many others – of course, we are interested only in those which are related to the discussion here, concerning the goals/dreams of unconventional computing.

Two are the basic directions of future research in this respect: hypercomputing and fypercomputing. We consider them only in terms of membrane computing, the area we know better, [21]. The only ideas considered so far in the direction of hypercomputing were acceleration [3] and evolutionary lineages of P systems [23]. Many further ideas (well, "tricks", in terms of Martin Davis) can be found in the literature [6, 24] – which ones can be extended also to membrane computing? Which of them can also have a biological motivation? In particular, what about spiking neural P systems [11] able of hypercomputation? No result of this type is known, not even the acceleration used in [3] was extended to SN P systems. This issue is especially relevant, taking into account the fact that the brain is considered non-Turing.

Fypercomputing is a basic research topic in membrane computing, encouraged by the fact that there are many biological processes which can be used in order to produce an exponential space in linear time, so that we can trade-off time for space, thus essentially speeding-up the computation. Membrane division, membrane creation (plus the possibility of producing exponentially many objects in linear time), membrane separation, neuron division and neuron budding, string replication are such processes. An almost systematic study of the effect of these operations on various classes of P systems was carried out – but still efforts are needed in this respect to complete the map.

A special case is that of numerical P systems from [20], proved in [13] to be efficient for a variant used in robot control, those with "enzymatic control", [16]. What about numerical P systems without the enzymatic control? In the above mentioned papers, these systems are used for computing functions and no research was reported where numerical P systems are used for solving decision problems. This remains as a research topic. To this aim, is it necessary to introduce further ingredients, such as membrane division, or these systems are intrinsically efficient?

An interesting and natural issue is to try to transform the ideas which lead to hypercomputation to tools for obtaining fypercomputation. As an example, acceleration is such a tool, in two time units (external time), any computation ends (although internally one performs an infinite number of steps). The first task is to define complexity classes for such devices, then to compare them with each other and with standard complexity classes.

Natural (bio-inspired) computing raises certain complexity problems which were not considered in the classical complexity theory, [15]. We only mention three of them, in the form already discussed – but not completely settled – in the membrane computing framework: (1) allowing non-uniform solutions (called semi-uniform when the algorithms are produced in polynomial time, starting from instances, not from the size of the problem), comparing uniform and semi-uniform complexity classes; (2) using pre-computed resources, an arbitrarily large initial workspace, without containing "too much" information, activated by introducing a problem in a well delimited portion of it; (3) allowing nondeterminism, but taking care that the device is *confluent*, either converges to a unique configuration, from where the computation continues deterministically (strong confluence), or all computations halt and provide the same result (weak, logical confluence).

Of course, a major problem concerns the implementations. Most chapters of natural computing aim at finding ways to better use the existing computers – see as a typical example the case of evolutionary computing. DNA computing came with a new promise, of using molecules as a support for computation. This goes close to analog computing, where the device is the big novelty, not the possible model behind it. At this moment, no commercial unconventional computer is known – maybe the D-wave computers are a counterexample, but the extent in which they can be considered quantum computers is still debatable.

But, as we said before, we have to be careful, not to underestimate the progress, in particular, that of unconventional computing!...

References

1. Adleman, L.M.: Molecular computation of solutions to combinatorial problems. Science **226**, 1021–1024 (1994)
2. Benenson, Y., Shapiro, E., Gill, B., Ben-Dor, U., Adar, R.: Molecular computer. A 'Smart Drug' in a test tube. In: Ferretti, C., Mauri, G., Zandron, C. (eds.) Proceedings of Tenth DNA Computing Conference, vol. 49, Milano, 2004, (abstract of invited talk), Univ. of Milano-Bicocca (2004)
3. Calude, C., Păun, G.: Bio-steps beyond turing. BioSystems **77**, 175–194 (2004)
4. Conrad, M.: The price of programmability. In: Herken, R. (ed.) The Universal Turing Machine: A Half-Century Survey, pp. 285–307. Kammerer and Unverzagt, Hamburg (1988)
5. Cook, S.: The importance of the P versus NP question. J. ACM **50**(1), 27–29 (2003)
6. Copeland, B.J.: Hypercomputation. Mind. Mach. **12**(4), 461–502 (2002)
7. Davis, M.: The myth of hypercomputation. In: Teuscher, C. (ed.) Alan Turing: The Life and Legacy of a Great Thinker, pp. 195–212. Springer, Berlin (2004)
8. Davis, M.: Why there is no such discipline as hypercomputation. Appl. Math. Comput. **178**, 4–7 (2006)
9. Gandy, R.: Church's thesis and principles for mechanisms. In: Barwise, J., et al. (eds.) The Kleene Symposium, pp. 123–148. North-Holland, Amsterdam (1980)
10. Gruska, J.: Quantum Computing. McGraw-Hill, Maidenhead (1999)
11. Ionescu, M., Păun, G., Yokomori, T.: Spiking neural P systems. Fundamenta Informaticae **71**, 279–308 (2006)
12. Kitano, H.: Systems biology. a brief overview. Science **295**, 1662–1664 (2002)
13. Leporati, A., Mauri, G., Porreca, A.E., Zandron, C.: Enzymatic numerical P systems using elementary arithmetic operations. In: Alhazov, A., Cojocaru, S., Gheorghe, M., Rogozhin, Y., Rozenberg, G., Salomaa, A. (eds.) CMC 2013. LNCS, vol. 8340, pp. 249–264. Springer, Heidelberg (2014)
14. Manca, V.: Infobiotics: Information in Biotic Systems. Springer, Berlin (2013)
15. Papadimitriou, C.H.: Computational Complexity. Addison-Wesley, Reading (1994)
16. Pavel, A.B., Vasile, C.I., Dumitrache, I.: Robot localization implemented with enzymatic numerical p systems. In: Prescott, T.J., Lepora, N.F., Mura, A., Verschure, P.F.M.J. (eds.) Living Machines 2012. LNCS, vol. 7375, pp. 204–215. Springer, Heidelberg (2012)
17. Păun, G.: From cells to (silicon) computers, and back. In: Cooper, B.S., Lowe, B., Sorbi, A. (eds.) New Computational Paradigms. Changing Conceptions of what is Computable, pp. 343–371. Springer, New York (2008)
18. Păun, G.: Towards "fypercomputations" (in membrane computing). Essays dedicated to jürgen dassow on the occasion of his 65th birthday. In: Bordihn, H., Kutrib, M., Truthe, B. (eds.) Languages Alive. LNCS, vol. 7300, pp. 207–220. Springer, Heidelberg (2012)
19. Păun, G.: Looking for Computers in the Biological Cell. After Twenty Years. The Publishing House of the Romanian Academy, Bucharest (2014). (in Romanian)
20. Păun, G., Păun, R.: Membrane computing and economics: numerical P systems. Fundamenta Informaticae **73**, 213–227 (2006)
21. Păun, G., Rozenberg, G., Salomaa, A. (eds.): The Oxford Handbook of Membrane Computing. Oxford University Press, New York (2010)
22. Rozenberg, G., Bäck, T., Kok, J.N. (eds.): Handbook of Natural Computing. Springer, Berlin (2012)

23. Sosík, P., Valik, O.: On evolutionary lineages of membrane systems. In: Freund, R., Păun, G., Rozenberg, G., Salomaa, A. (eds.) WMC 2005. LNCS, vol. 3850, pp. 67–78. Springer, Heidelberg (2006)
24. Syropoulos, A.: Hypercomputation: Computing Beyond the Church-Turing Barrier. Springer, Berlin (2008)
25. Wolkenhauer, O.: Systems biology: the reincarnation of systems theory applied in biology? Brief. Bioinform. **2**(3), 258–270 (2001)
26. Wolpert, D.H., Macready, W.G.: No free lunch theorems for optimization. IEEE Trans. Evol. Comput. **1**, 67 (1997)
27. Zenil, H. (ed.): A Computable Universe. Understanding and Exploring Nature as Computation. World Scientific, Singapore (2013)
28. The P Systems Website. http://ppage.psystems.eu

Newton's Forward Difference Equation for Functions from Words to Words

Jean-Éric Pin(✉)

LIAFA, Université Paris-Diderot and CNRS, Case 7014, 75205 Paris Cedex 13, France
jean-eric.pin@liafa.univ-paris-diderot.fr

Abstract. Newton's forward difference equation gives an expression of a function from $\mathbb{N}$ to $\mathbb{Z}$ in terms of the initial value of the function and the powers of the forward difference operator. An extension of this formula to functions from A^* to $\mathbb{Z}$ was given in 2008 by P. Silva and the author. In this paper, the formula is further extended to functions from A^* into the free group over B.

Let A be a set. In this paper, we denote by A^* the free monoid over A and by $FG(A)$ the free group over A. The empty word, which is the unit of both A^* and $FG(A)$, is denoted by 1.

Original motivation. The characterization of the regularity-preserving functions is the original motivation of this paper, but since there is a long way to go from this problem to Newton's forward difference equation, it is worth relating the story step by step.

A function f from A^* to B^* is *regularity-preserving* if, for each regular language L of B^*, the language $f^{-1}(L)$ is also regular. Several families of regularity-preserving functions have been identified in the literature [3,8,10–12,18,19], but finding a complete description of these functions seems to be currently out of reach. Following a dubious, but routine mathematical practice consisting to offer generalizations rather than solutions to open problems, I proposed a few years ago the following variation: given a class $\mathcal{C}$ of regular languages, characterize the $\mathcal{C}$-preserving functions. Of course, a function f is *$\mathcal{C}$-preserving* if $L \in \mathcal{C}$ implies $f^{-1}(L) \in \mathcal{C}$.

For instance, a description of the sequential functions preserving star-free languages (respectively group-languages) is given in [17]. A similar problem was also recently considered for formal power series [4]. The question is of special interest for varieties of languages. Recall that a *variety of languages* $\mathcal{V}$ associates with each finite alphabet A a set $\mathcal{V}(A^*)$ of regular languages closed under finite Boolean operations and quotients, with the further property that, for each morphism $\varphi : A^* \to B^*$, the condition $L \in \mathcal{V}(B^*)$ implies $\varphi^{-1}(L) \in \mathcal{V}(A^*)$.

Algebra and topology step in. It is interesting to see how algebra and topology can help characterizing $\mathcal{V}$-preserving functions. Let us start with algebra.

Eilenberg [5] proved that varieties of languages are in bijection with varieties of finite monoids. A *variety of finite monoids* is a class of finite monoids

A. Beckmann et al. (Eds.): CiE 2015, LNCS 9136, pp. 71–82, 2015.
DOI: 10.1007/978-3-319-20028-6_8

closed under taking submonoids, homomorphic images and finite products. For instance, the variety of all finite monoids corresponds to the variety of regular languages, and the variety of aperiodic finite monoids corresponds to the variety of star-free languages.

Topology is even more relevant to our problem. To each variety of finite monoids $\mathbf{V}$, one can attach a pseudometric $d_{\mathbf{V}}$, (called the *pro-*$\mathbf{V}$ *pseudometric*, see [1,14,16] for more details). Now, if $\mathcal{V}$ is the variety of languages corresponding to $\mathbf{V}$, the following property holds: a function is $\mathcal{V}$-preserving if and only if it is uniformly continuous with respect to $d_{\mathbf{V}}$. This result motivated P. Silva and the author to investigate more closely uniform continuity with respect to various varieties of monoids [14]. Simultaneously, we started to investigate a specific example, the variety $\mathbf{G}_p$ of finite p-groups, where p is a given prime [13,15]. Then the corresponding pseudometric is a metric denoted by d_p.

This case is interesting because there are relevant known results both in algebra and in topology. First, Eilenberg and Schützenberger [5, p. 238] gave a very nice description of the languages recognized by a p-group. Secondly, the free monoid over a one-letter alphabet is isomorphic to $\mathbb{N}$, and the metric d_p is the *p-adic metric*, a well known mathematical object. The completion of the metric space $(\mathbb{N}, d_p)$ is the space of *p-adic numbers*. Thirdly, the uniformly continuous functions from $(\mathbb{N}, d_p)$ to itself are characterized by Mahler's theorem, a celebrated result of number theory. This is the place where *Newton's forward difference equation* is needed.

Newton's forward difference equation. This result states that for each function $f : \mathbb{N} \to \mathbb{Z}$ and for all $n \in \mathbb{N}$, the following equality holds:

$$f(n) = \sum_{k=0}^{\infty} \binom{n}{k} (\Delta^k f)(0) \tag{1}$$

where Δ is the *difference operator*, defined by $(\Delta f)(n) = f(n+1) - f(n)$.

Mahler's theorem states that a function $f : \mathbb{N} \to \mathbb{N}$ is uniformly continuous for d_p if and only if $\lim_{k\to\infty} |\Delta^k f(0)|_p = 0$, where $|n|_p$ denotes the *p-adic norm* of n. This gives a complete characterization of the d_p-uniformly continuous functions from a^* to a^*.

An extension of Mahler's theorem to functions from A^* to $\mathbb{N}$ was given in [13,15], giving in turn a complete characterization of the d_p-uniformly continuous functions from A^* to a^*. This result relies on an extension of Newton's forward difference equation which works as follows. For each function $f : A^* \to \mathbb{Z}$ and for all $u \in A^*$, the following equality holds:

$$f(u) = \sum_{v \in A^*} \binom{u}{v} (\Delta^v f)(1) \tag{2}$$

where $\binom{u}{v}$ denotes the binomial coefficient of two words u and v (see [5, p. 253] and [9, Chap. 6]). If $v = a_1 \cdots a_n$, the binomial coefficient of u and v is defined as follows

$$\binom{u}{v} = \big|\{(u_0, \ldots, u_n) \mid u = u_0 a_1 u_1 \ldots a_n u_n\}\big|.$$

The difference operator Δ^w is now defined by induction on the length of the word w by setting $\Delta^1 f = f$ and, for each letter a,

$$\Delta^a f(u) = f(ua) - f(u)$$
$$\Delta^{aw} f(u) = (\Delta^a(\Delta^w f)(u)$$

In order to further extend Mahler's theorem to functions from A^* to B^* (for arbitrary finite alphabets A and B), one first need to find a Newton's forward difference equation for functions from A^* to $FG(B)$ and this is precisely the objective of this paper. As the reader will see, it is relatively easy to guess the right formula, but the main difficulty is to find the appropriate framework to prove it formally.

The paper is organized as follows. An intuitive approach to the forward difference equation is given in Sect. 1. The main tools to formalize this intuitive approach are the near rings, introduced in Sect. 2 and the noncommutative Magnus transformation presented in Sect. 3. The formal statement and the proof of the forward difference equation are given in Sect. 4.

1 The Difference Operator

Let $f : A^* \to FG(B)$ be a function. For each letter a, the difference operator $\Delta^a f$ is the map from A^* to $FG(B)$ defined by

$$(\Delta^a f)(u) = f(u)^{-1} f(ua) \tag{3}$$

One can now define inductively an operator $\Delta^w f : A^* \to FG(B)$ for each word $w \in A^*$ by setting $\Delta^1 f = f$, and for each letter $a \in A$ and each word $w \in A^*$,

$$\Delta^{aw} f = \Delta^a(\Delta^w f). \tag{4}$$

One could also make use of Δ^{wa} instead of Δ^{aw} in the induction step, but the result would be the same, in view of the following result:

Proposition 1.1. *The following formulas hold for all $v, w \in A^*$:*

$$\Delta^{vw} f = \Delta^v(\Delta^w f)$$

Proof. By induction on $|v|$. The result is trivial if v is the empty word. If $v = au$ for some letter a, we get $\Delta^{vw} f = \Delta^{auw} f = \Delta^a(\Delta^{uw} f)$. Now by the induction hypothesis, $\Delta^{uw} f = \Delta^u(\Delta^w f)$ and thus $\Delta^{vw} f = \Delta^a(\Delta^u(\Delta^w f)) = \Delta^{au}(\Delta^w f) = \Delta^v(\Delta^w f)$. □

For instance, we get

$$(\Delta^1 f)(u) = f(u)$$
$$(\Delta^a f)(u) = f(u)^{-1} f(ua)$$
$$(\Delta^{aa} f)(u) = f(ua)^{-1} f(u) f(ua)^{-1} f(uaa)$$

$$(\Delta^{baa} f)(u) = f(uaa)^{-1} f(ua) f(u)^{-1} f(ua) f(uba)^{-1} f(ub) f(uba)^{-1} f(ubaa)$$
$$(\Delta^{abaa} f)(u) = f(ubaa)^{-1} f(uba) f(ub)^{-1} f(uba) f(ua)^{-1} f(u) f(ua)^{-1} f(uaa)$$
$$f(uaaa)^{-1} f(uaa) f(ua)^{-1} f(uaa) f(uaba)^{-1} f(uab) f(uaba)^{-1}$$
$$f(uabaa)$$

A forward difference equation should express f in terms of the values of $(\Delta^w f)(1)$, for all words w. To simplify notation, let us set, for all $w \in A^*$:

$$\Delta^w = (\Delta^w f)(1)$$

A little bit of computation leads to the formulas

$$\begin{aligned} f(1) &= \Delta^1 \\ f(a) &= \Delta^1 \Delta^a & f(b) &= \Delta^1 \Delta^b \\ f(ab) &= \Delta^1 \Delta^a \Delta^b \Delta^{ab} & f(ba) &= \Delta^1 \Delta^b \Delta^a \Delta^{ba} \\ f(bab) &= \Delta^1 \Delta^b \Delta^a \Delta^{ba} \Delta^b \Delta^{bb} \Delta^{ab} \Delta^{bab} & f(aba) &= \Delta^1 \Delta^a \Delta^b \Delta^{ab} \Delta^a \Delta^{aa} \Delta^{ba} \Delta^{aba} \end{aligned}$$

which give indeed a forward difference equation for $f(w)$ for a few values of w. But how to find a closed formula valid for all values of w? To do so, acting as a physicist, we will generate some formulas without worrying too much about correctness. Then we will describe a rigorous formalism to justify our equations.

As a first step, our exponential notation suggests to write Δ^{u+v} for $\Delta^u \Delta^v$, which gives

$$\begin{aligned} f(1) &= \Delta^1 \\ f(a) &= \Delta^{1+a} & f(b) &= \Delta^{1+b} \\ f(ab) &= \Delta^{1+a+b+ab} & f(ba) &= \Delta^{1+b+a+ba} \\ f(bab) &= \Delta^{1+b+a+ba+b+bb+ab+bab} & f(aba) &= \Delta^{1+a+b+ab+a+aa+ba+aba} \end{aligned}$$

The next step is to observe that, in an appropriate noncommutative setting, one can write

$$\left.\begin{aligned} (1+a)(1+b) &= 1+a+b+ab \\ (1+b)(1+a) &= 1+b+a+ba \\ (1+b)(1+a)(1+b) &= 1+b+a+ba+b+bb+ab+bab \\ (1+a)(1+b)(1+a) &= 1+a+b+ab+a+aa+ba+aba \end{aligned}\right\} \quad (5)$$

which gives for instance the noncommutative difference equations

$$f(aba) = \Delta^{(1+a)(1+b)(1+a)} \quad \text{and} \quad f(bab) = \Delta^{(1+b)(1+a)(1+b)}$$

It is now easy to guess a similar equation for $f(u)$, for any word u.

But it is time to tighten the bolts and justify our adventurous notation. A little bit of algebra is in order to give grounds to the foregoing formulas. Let us start by introducing the relatively little-known notion of a near-ring.

2 Near-Rings

A (left) *near-ring* (with unit) is an algebraic structure K equipped with two binary operations, denoted additively and multiplicatively, and two elements 0 and 1, satisfying the following conditions:

(1) K is a group (not necessarily commutative) with identity 0 under addition,
(2) K is a monoid with identity 1 under multiplication,
(3) multiplication distributes on the left over addition: for all $x, y, z \in K$, $z(x + y) = zx + zy$.

An element of z of K is *distributive* if, for all $x, y \in K$, $(x + y)z = xz + yz$.

It follows from the axioms that $x0 = 0$ and $x(-y) = -xy$ for all $x, y \in K$. However, it is not necessarily true that $0x = 0$ and $(-x)y = -xy$. It is even possible that $(-1)x$ is not equal to $-x$.

A well-known example of near-ring is the set of all transformations on a group G, equipped with pointwise addition as addition and composition as product.

Let us now survey a construction first introduced by Fröhlich [6,7]. We follow the presentation of Banaschewski and Nelson [2]. Let M be a monoid. We want to construct a near-ring $FG[M]$ in which the additive group is the free group $FG(M)$ on the set M and the multiplication extends the operation on M. This leads us to denote the operation on M multiplicatively and to use an additive notation for the free group[1].

Let us consider terms of the form

$$\varepsilon_1 u_1 + \cdots + \varepsilon_k u_k$$

with $\varepsilon_1, \ldots, \varepsilon_k \in \{-1, +1\}$ and $u_1, \ldots, u_k \in M$. A term is *reduced* if it does not contain any subterms of the form $u + -u$ or $-u + u$. The *reduction* of a term is obtained by iteratively ruling out the subterms of the form $u + -u$ or $-u + u$ until the term is reduced. One can show that these operations can be done in any order and lead to the same reduced term.

The elements of $FG[M]$ can be represented by reduced terms. The sum of two elements $\varepsilon_1 u_1 + \cdots + \varepsilon_r u_r$ and $\varepsilon_1 v_1 + \cdots + \varepsilon_s v_s$ is obtained by reducing the term

$$\varepsilon_1 u_1 + \cdots + \varepsilon_r u_r + \varepsilon_1 v_1 + \cdots + \varepsilon_s v_s$$

The empty term (corresponding to the case $k = 0$) is the identity for this addition and is simply denoted by 0. The inverse of $\varepsilon_1 u_1 + \cdots + \varepsilon_k u_k$ is $-\varepsilon_k u_k + \cdots + -\varepsilon_1 u_1$.

We now define a multiplication on $FG[M]$ in two steps. First, given an element $\varepsilon_1 u_1 + \cdots + \varepsilon_r u_r$ of $FG[M]$ and $m \in M$, we set

$$(\varepsilon_1 u_1 + \cdots + \varepsilon_r u_r)m = (\varepsilon_1 u_1 m + \cdots + \varepsilon_r u_r m)$$
$$(\varepsilon_1 u_1 + \cdots + \varepsilon_r u_r)(-m) = (-\varepsilon_r u_r m + \cdots + -\varepsilon_1 u_1 m)$$

[1] Therefore, the notation $FG(M)$ and $FG[M]$ refer to the same set, but to different structures: the free group on M in the first case, the free near semiring on M in the latter case.

Now, the product of two elements $\varepsilon_1 u_1 + \cdots + \varepsilon_r u_r$ and $\varepsilon'_1 u'_1 + \cdots + \varepsilon_s u'_s$ of $FG[M]$ is defined by

$$\begin{aligned}(\varepsilon_1 u_1 + \cdots + \varepsilon_r u_r)(\varepsilon_1 u'_1 + \cdots + \varepsilon_s u'_s) = (\varepsilon_1 u_1 + \cdots + \varepsilon_r u_r)(\varepsilon'_1 u'_1) \\ + (\varepsilon_1 u_1 + \cdots + \varepsilon_r u_r)(\varepsilon'_2 u'_2) + \cdots + (\varepsilon_1 u_1 + \cdots + \varepsilon_r u_r)(\varepsilon'_s u'_s)\end{aligned} \tag{6}$$

This operation defines a multiplication on $FG[M]$. Together with the addition, $FG[M]$ is now equipped with a structure of near-ring.

Since $(u)(v) = (uv)$, the monoid M embeds into the multiplicative monoid $FG[M]$ and it is convenient to simplify the notation (u) to u. With this convention, the identity of the multiplication of $FG[M]$ is denoted by 1. Furthermore an element $(u_1, \ldots, u_r)$ can be written as $u_1 + \cdots + u_r$ and thus (6) is a consequence of the following natural formulas, where $u_1, \ldots, u_r, v_1, \ldots, v_s, v \in M$ and $w \in FG[M]$:

$$(u_1 + \cdots + u_r)v = u_1 v + \cdots + u_r v \tag{7}$$

$$w(v_1 + \cdots + v_s) = wv_1 + \cdots + wv_s \tag{8}$$

The near-ring $FG[M]$ has the further convenient property that 0 is distributive in $FG[M]$ since $0x = 0$ by definition. Moreover, the equality $(-x)y = -xy$ holds if $y \in M$ but is not necessarily true otherwise. Even the relation $(-1)y = -y$ may fail if y is not an element of M. For instance, if M is the free monoid $\{a, b\}^*$, then $(-1)(a + b) = -a - b$ but $-(a + b) = -b - a$.

Note that if M is the trivial monoid, then $FG[M]$ is isomorphic to the ring $\mathbb{Z}$ of integers. In the sequel, M will be the free monoid A^*.

3 Noncommutative Magnus Transformation

Our goal in this section is to justify and to extend the Eq. (5). As explained in Sect. 2, we view $FG[A^*]$ as a near-ring.

3.1 Definition of the Magnus Transformation

The monoid morphism μ from A^* into the multiplicative monoid $FG[A^*]$ defined, for each letter $a \in A$, by

$$\mu(a) = 1 + a$$

is called the *Magnus transformation.* It extends uniquely to a group morphism from $FG(A^*)$ to the additive group $FG[A^*]$. For instance, if $A = \{a, b\}$, we get

$$\mu(1) = 1 \qquad \mu(a) = 1 + a \qquad \mu(b) = 1 + b$$

$$\mu(ab) = 1 + a + b + ab \qquad \mu(1 + a) = 1 + 1 + a$$

$$\mu(-1 + a - ab) = -1 + 1 + a - ab - b - a - 1 = a - ab - b - a - 1$$

$$\mu(aba) = 1 + a + b + ab + a + aa + ba + aba$$

More generally, for each $u \in A^*$,

$$\mu(au) = \mu(a)\mu(u) = (1 + a)\mu(u) = \mu(u) + a\mu(u)$$

Proposition 3.1. *The following formula holds for all $u \in FG[A^*]$ and $v \in A^*$:*

$$\mu(uv) = \mu(u)\mu(v) \tag{9}$$

Proof. Since μ is a monoid morphism μ from A^* into the multiplicative monoid $FG[A^*]$, (9) holds if $u \in A^*$. Next, if $u = \varepsilon_1 u_1 + \cdots + \varepsilon_k u_k$, with $u_1, \ldots, u_k \in A^*$ and $\varepsilon_1, \ldots, \varepsilon_k \in \{-1, 1\}$, then $uv = \varepsilon_1 u_1 v + \cdots + \varepsilon_k u_k v$ and hence $\mu(uv) = \varepsilon_1 \mu(u_1)\mu(v) + \cdots + \varepsilon_k \mu(u_k)\mu(v) = \mu(u)\mu(v)$. This proves (9). □

However, μ is not a monoid morphism for the multiplicative structure of $FG[A^*]$, since, for instance, $\mu((1+a)(1+b)) \neq \mu(1+a)\mu(1+b)$.

3.2 The Inverse of the Magnus Transformation

Let π be the monoid morphism from A^* into the multiplicative monoid $FG[A^*]$ defined, for each letter $a \in A$, by

$$\pi(a) = -1 + a$$

Then π has a unique extension to a group morphism from $FG[A^*]$ into itself and enjoys properties similar to those of μ. Just like μ, π is not a monoid morphism for the multiplicative structure of $FG[A^*]$, but a result analoguous to Proposition 3.1 also holds for π.

Proposition 3.2. *The following formula holds for all $u \in FG[A^*]$ and $v \in A^*$:*

$$\pi(uv) = \pi(u)\pi(v) \tag{10}$$

For instance

$$\begin{aligned}
\pi(aba) &= -ab + b - 1 + a - aa + a - ba + aba \\
\pi(abaa) &= -aba + ba - a + aa - a + 1 - b + ab - aba + ba - a + aa \\
&\quad - aaa + aa - baa + abaa \\
\pi(abab) &= -aba + ba - a + aa - a + 1 - b + ab - abb + bb - b + ab \\
&\quad - aab + ab - bab + abab
\end{aligned}$$

Observe that, for each letter $a \in A$,

$$\mu(\pi(a)) = \mu(-1 + a) = \mu(-1) + \mu(a) = -1 + (1 + a) = a \tag{11}$$

$$\pi(\mu(a)) = \pi(1 + a) = \pi(1) + \pi(a) = 1 + (-1 + a) = a \tag{12}$$

It is tempting to conclude from these equalities that π is the inverse of μ, but the right answer is slightly more involved.

The *reversal* of a word $u = a_1 \cdots a_n$ is the word $\overline{u} = a_n \cdots a_1$. The reversal map is a permutation on A^* which extends by linearity to a group automorphism of the free group $FG[A^*]$.

Proposition 3.3. *The following relations hold for all* $u, v \in A^*$,

$$\mu(v\overline{\pi(\overline{u})}) = \mu(v)u \tag{13}$$

$$\overline{\pi(\overline{\mu(u)v})} = u\overline{\pi(\overline{v})} \tag{14}$$

Proof. We prove (13) (for all $v \in A^*$) by induction on the length of u. The result is trivial if u is the empty word. Suppose that $u = aw$ for some letter a. Observing that $\overline{u} = \overline{w}a$, we get

$$\pi(\overline{u}) = \pi(\overline{w})\pi(a) = \pi(\overline{w})(-1 + a) = -\pi(\overline{w}) + \pi(\overline{w})a$$

whence

$$\overline{\pi(\overline{u})} = -\overline{\pi(\overline{w})} + a\overline{\pi(\overline{w})}$$

and

$$v\overline{\pi(\overline{u})} = -v\overline{\pi(\overline{w})} + va\overline{\pi(\overline{w})}$$

Applying the induction hypothesis to w, we obtain

$$\begin{aligned}\mu(v\overline{\pi(\overline{u})}) &= -\mu(v\overline{\pi(\overline{w})}) + \mu(va\overline{\pi(\overline{w})}) = -\mu(v)w + \mu(va)w \\ &= -\mu(v)w + \mu(v)\mu(a)w = (-\mu(v) + \mu(v)(1 + a))w \\ &= \mu(v)aw = \mu(v)u\end{aligned}$$

which proves (13).

We also prove (14) by induction on the length of u. The result is trivial if u is the empty word. Suppose that $u = wa$ for some letter a. We get

$$\mu(u) = \mu(wa) = \mu(w)\mu(a) = \mu(w) + \mu(w)a$$

whence

$$\overline{\mu(u)v} = \overline{\mu(w)v} + \overline{\mu(w)av}$$

and

$$\overline{\pi(\overline{\mu(u)v})} = \overline{\pi(\overline{\mu(w)v})} + \overline{\pi(\overline{\mu(w)av})}$$

Applying the induction hypothesis to w, we obtain

$$\overline{\pi(\overline{\mu(u)v})} = w\overline{\pi(\overline{v})} + w\overline{\pi(\overline{av})} = w(\overline{\pi(\overline{v})} + \overline{\pi(\overline{av})})$$

Now, since $\overline{av} = \overline{v}a$, one gets $\pi(\overline{av}) = \pi(\overline{v})\pi(a)$ and hence

$$\begin{aligned}\overline{\pi(\overline{v})} + \overline{\pi(\overline{av})} &= \overline{\pi(\overline{v})} + \overline{\pi(\overline{v})\pi(a)} = \overline{\pi(\overline{v})} + \overline{\pi(\overline{v})(-1 + a)} \\ &= \overline{\pi(\overline{v})a} = a\overline{\pi(\overline{v})}\end{aligned}$$

and finally

$$\overline{\pi(\overline{\mu(u)v})} = wa\overline{\pi(\overline{v})} = u\overline{\pi(\overline{v})}$$

which proves (14). □

Corollary 3.4. *The function* $\mu : FG[A^*] \to FG[A^*]$ *is a bijection and its inverse is defined by*

$$\mu^{-1}(u) = \overline{\pi(\overline{u})} \tag{15}$$

Proof. Taking $v = 1$ in (13) and (14) shows that for all $u \in A^*$, $\mu(\overline{\pi(\overline{u})}) = u$ and $\pi(\overline{\mu(u)}) = u$. The result follows since μ, π and the maps $u \to \overline{\mu(\overline{u})}$ and $u \to \overline{\pi(\overline{u})}$ are group morphisms. □

4 Forward Difference Equation

Let $\mathcal{F}$ be the set of all functions from A^* into $FG(B)$. Then $\mathcal{F}$ is a group under pointwise multiplication defined by setting

$$(fg)(x) = f(x)g(x)$$

whose identity is the constant map onto the identity of $FG(B)$. Furthermore, the inverse of f in this group is given by the formula

$$f^{-1}(x) = (f(x))^{-1}$$

The map $(u, f) \to \Delta^u f$ from $A^* \times \mathcal{F}$ to $\mathcal{F}$ defines a left action of A^* on $\mathcal{F}$, since $\Delta^1 f = f$ and, by Proposition 1.1, $\Delta^{uv} f = \Delta^v(\Delta^u f)$ for all $u, v \in A^*$.

This action can be extended by linearity to a map from $FG[A^*] \times \mathcal{F}$ to $\mathcal{F}$ as follows: for each element $u = \varepsilon_1 u_1 + \cdots + \varepsilon_k u_k$ of $FG[A^*]$, we define the function $\Delta^u f$ by

$$(\Delta^u f) = (\Delta^{u_1} f)^{\varepsilon_1} \cdots (\Delta^{u_k} f)^{\varepsilon_k}$$

In particular, $\Delta^0 f$ is the constant map onto the identity of $FG(B)$ and $\Delta^1 f = f$.

We are interested in the coefficients $(\Delta^u f)(1)$. To simplify notation, we introduce the following short forms, for all $u, v \in FG[A^*]$:

$$\Delta^u = (\Delta^u f)(1) \qquad \Delta^u \cdot v = (\Delta^u f)(v)$$

The next proposition gives some useful relations between these coefficients.

Proposition 4.1. *The following formulas hold for all* $u, v \in A^*$ *and* $a \in A$*:*

$$(\Delta^u \cdot v)(\Delta^{au} \cdot v) = \Delta^u \cdot va \tag{16}$$

$$\Delta^{\mu(vu)} = \Delta^{\mu(u)} \cdot v \tag{17}$$

Proof. By definition, $\Delta^u \cdot v = (\Delta^u f)(v)$ and thus we get

$$\Delta^{au} \cdot v = (\Delta^{au} f)(v) = (\Delta^a(\Delta^u f))(v) =$$
$$= \big((\Delta^u f)(v)\big)^{-1}(\Delta^u f)(va) = (\Delta^u \cdot v)^{-1}\Delta^u \cdot va$$

from which (16) follows immediately.

By induction, it suffices to establish (17) for $v = a$. If $\mu(u) = u_1 + \cdots + u_k$, then by Proposition 3.2, $\mu(au) = \mu(a)\mu(u) = u_1 + au_1 + \cdots + u_k + au_k$. Now, (16) shows that for $1 \leqslant i \leqslant k$, $(\Delta^{u_i}\Delta^{au_i}) = \Delta^{u_i} \cdot a$. It follows that

$$\Delta^{\mu(au)} = (\Delta^{u_1}\Delta^{au_1})\cdots(\Delta^{u_k}\Delta^{au_k}) = (\Delta^{u_1} \cdot a)\cdots(\Delta^{u_k} \cdot a) = \Delta^{\mu(u)} \cdot a$$

which concludes the proof. □

Proposition 4.2. *The following formulas hold for all $u, v \in A^*$:*

$$f(vu) = (\Delta^{\mu(u)} f)(v) \tag{18}$$

$$f(u) = \Delta^{\mu(u)} \tag{19}$$

Proof. Applying (17) with $u = 1$, we get $\Delta^{\mu(v)} = \Delta^{\mu(1)} \cdot v = f(v)$ which gives (19). It follows that $f(vu) = \Delta^{\mu(vu)}$. Now by (17) we also have $\Delta^{\mu(vu)} = (\Delta^{\mu(u)} f)(v)$, which yields (19). □

4.1 Difference Expansion

The formula $f(u) = \Delta^{\mu(u)}$ gives a representation of $f(u)$ as a product of elements of the form Δ^v. This expression is called the *difference expansion* of f. For instance we have

$$f(abaa) = \Delta^1\Delta^a\Delta^b\Delta^{ab}\Delta^a\Delta^{aa}\Delta^{ba}\Delta^{aba}\Delta^a\Delta^{aa}\Delta^{ba}\Delta^{aba}\Delta^{aa}\Delta^{aaa}\Delta^{baa}\Delta^{abaa}$$

We now show that this decomposition is unique in a sense that we now make precise.

Let $(c_u)_{u \in A^*}$ be a family of elements of $FG(B)$. The map $u \mapsto c_u$ extends uniquely to a group morphism from $FG[A^*]$ to $FG(B)$. In particular, for each element $\varepsilon_1 u_1 + \cdots + \varepsilon_k u_k$ in $FG[A^*]$, we set

$$c_{\varepsilon_1 u_1 + \cdots + \varepsilon_k u_k} = c_{u_1}^{\varepsilon_1} \cdots c_{u_k}^{\varepsilon_k}$$

We can now state:

Theorem 4.3. *Let f be a function from A^* to $FG(B)$. There is a unique family $(c_u)_{u \in A^*}$ of elements of $FG(B)$ such that, for all $u \in A^*$, $f(u) = c_{\mu(u)}$. This family is given by $c_u = (\Delta^u f)(1)$.*

Proof. The existence follows from (19). Unicity can be proved by induction on the length of u. Necessarily, $c_1 = f(1) = \Delta^1(f)(1)$. Suppose that the coefficients c_u are known to be uniquely determined for $|u| \leqslant n$. Let u be a word of length n and let a be a letter. Then $\mu(u) = u_1 + \cdots + u_{k-1} + u$, where the words $u_1, \ldots, u_{k-1}$ are shorter than u. Furthermore

$$\mu(ua) = u_1 + \cdots + u_{k-1} + u + u_1 a + \cdots + u_{k-1} a + ua$$

where again, ua is the only word of length $n+1$. The condition $f(ua) = c_{\mu(ua)}$ now gives

$$f(ua) = c_{u_1} \cdots c_{u_{k-1}} c_u c_{u_1 a} \cdots c_{u_{k-1} a} c_{ua}$$

It follows than c_{ua} is necessarily equal to

$$f(ua)(c_{u_1} \cdots c_{u_{k-1}} c_u c_{u_1 a} \cdots c_{u_{k-1} a})^{-1}$$

which proves unicity. □

It is a well-known fact that every sequence of real numbers can appear as coefficients of the Maclaurin series of a smooth function. The following corollary can be viewed as a discrete, non-commutative analogue of this result.

Corollary 4.4. *Given, for each $u \in A^*$, an element c_u of $FG(B)$, there exists a unique function $f : A^* \to FG(B)$ such that, for all $u \in A^*$, $\Delta^u f = c_u$. This function is defined by $f(u) = c_{\mu(u)}$ for all $u \in A^*$.*

Corollary 4.4 can also be interpreted as an answer to the following interpolation problem: determine f knowing the coefficients $\Delta^u f$ for all $u \in A^*$.

4.2 The Inversion Formula

The definition of $\Delta^u f$ was given by induction on the length of u. To conclude this article, we give a close formula that allows one to compute $\Delta^u f$ directly.

Let $f : A^* \to FG(B)$ be a function. Then f can be extended by linearity into a group morphism from $FG[A^*]$ to $FG(B)$.

Proposition 4.5. *The following formula holds for all u in A^* and v in A^*:*

$$\Delta^u f(v) = f(v\overline{\pi(\overline{u})}) \tag{20}$$

Proof. Substituting $\overline{\pi(\overline{u})}$ for u in (18) and using (13) we get

$$f(v\overline{\pi(\overline{u})}) = \Delta^{\mu(\overline{\pi(\overline{u})})} f(v) = \Delta^u f(v)$$

which gives the result. □

Corollary 4.6. *The following formula holds for all $u \in FG[A^*]$*

$$\Delta^u f = f(\overline{\pi(\overline{u})}) = f(\mu^{-1}(u)) \tag{21}$$

Example 4.7. For instance, for $u = abb$, we get $\overline{u} = bba$ and

$$\begin{aligned}
\pi(\overline{u}) &= \pi(bba) = (-1+b)(-1+b)(-1+a) = (-b+1-b+bb)(-1+a) \\
&= -bb+b-1+b-ba+a-ba+bba \\
\overline{\pi(\overline{u})} &= -bb+b-1+b-ab+a-ab+abb \\
\Delta^{abb} &= -f(bb)+f(b)-f(1)+f(b)-f(ab)+f(a)-f(ab)+f(abb)
\end{aligned}$$

Acknowlegements. I would like to thank the anonymous referees for their valuable comments.

References

1. Almeida, J.: Finite semigroups and universal algebra. World Scientific Publishing Co., River Edge (1994). Translated from the 1992 Portuguese original and revised by the author
2. Banaschewski, B., Nelson, E.: On the non-existence of injective near-ring modules. Can. Math. Bull. **20**(1), 17–23 (1977)
3. Berstel, J., Boasson, L., Carton, O., Petazzoni, B., Pin, J.É.: Operations preserving recognizable languages. Theor. Comput. Sci. **354**, 405–420 (2006)
4. Droste, M., Zhang, G.Q.: On transformations of formal power series. Inf. Comput. **184**(2), 369–383 (2003)
5. Eilenberg, S.: Automata, Languages and Machines, vol. B. Academic Press, New York (1976)
6. Fröhlich, A.: On groups over a d.g. near-ring. I. Sum constructions and free *R*-groups. Q. J. Math. Oxf. Ser. (2) **11**, 193–210 (1960)
7. Fröhlich, A.: On groups over a d.g. near-ring. II. Categories and functors. Q. J. Math. Oxf. Ser. (2) **11**, 211–228 (1960)
8. Kosaraju, S.R.: Regularity preserving functions. SIGACT News **6**(2), 16–17 (1974)
9. Lothaire, M.: Combinatorics on Words. Cambridge Mathematical Library. Cambridge University Press, Cambridge (1997)
10. Pin, J.É., Sakarovitch, J.: Operations and transductions that preserve rationality. In: Cremers, A.B., Kriegel, H.-P. (eds.) Theoretical Computer Science. LNCS, vol. 145, pp. 617–628. Springer, Heidelberg (1982)
11. Pin, J.É., Sakarovitch, J.: Une application de la représentation matricielle des transductions. Theor. Comput. Sci. **35**, 271–293 (1985)
12. Pin, J.É., Silva, P.V.: A topological approach to transductions. Theor. Comput. Sci. **340**, 443–456 (2005)
13. Pin, J.É., Silva, P.V.: A Mahler's theorem for functions from words to integers. In: Albers, S., Weil, P. (eds.) 25th International Symposium on Theoretical Aspects of Computer Science (STACS 2008), pp. 585–596. Internationales Begegnungs- Und Forschungszentrum für Informatik (IBFI), Schloss Dagstuhl, Germany (2008)
14. Pin, J.É., Silva, P.V.: On profinite uniform structures defined by varieties of finite monoids. Int. J. Algebr. Comput. **21**, 295–314 (2011)
15. Pin, J.É., Silva, P.V.: A noncommutative extension of Mahler's theorem on interpolation series. Eur. J. Comb. **36**, 564–578 (2014)
16. Pin, J.É., Weil, P.: Uniformities on free semigroups. Int. J. Algebr. Comput. **9**, 431–453 (1999)
17. Reutenauer, C., Schützenberger, M.P.: Variétés et fonctions rationnelles. Theor. Comput. Sci. **145**(1–2), 229–240 (1995)
18. Seiferas, J.I., McNaughton, R.: Regularity-preserving relations. Theor. Comp. Sci. **2**, 147–154 (1976)
19. Stearns, R.E., Hartmanis, J.: Regularity preserving modifications of regular expressions. Inf. Control **6**, 55–69 (1963)

Degrees of Unsolvability: A Tutorial

Stephen G. Simpson(✉)

Department of Mathematics, Pennsylvania State University, State College, PA, USA
simpson@math.psu.edu
http://www.math.psu.edu/simpson

Abstract. Given a problem P, one associates to P a *degree of unsolvability*, i.e., a quantity which measures the amount of algorithmic unsolvability which is inherent in P. We focus on two degree structures: the semilattice of Turing degrees, $\mathcal{D}_{\rm T}$, and its completion, $\mathcal{D}_{\rm w} = \widehat{\mathcal{D}_{\rm T}}$, the lattice of Muchnik degrees. We emphasize specific, natural degrees and their relationship to reverse mathematics. We show how Muchnik degrees can be used to classify tiling problems and symbolic dynamical systems of finite type. We describe how the category of sheaves over $\mathcal{D}_{\rm w}$ forms a model of intuitionistic mathematics, known as the *Muchnik topos.* This model is a rigorous implementation of Kolmogorov's nonrigorous 1932 interpretation of intuitionism as a "calculus of problems".

Keywords: Degrees of unsolvability · Mass problems · Turing degrees · Muchnik degrees · Algorithmic randomness · Kolmogorov complexity · Tiling problems · Symbolic dynamics · Intuitionism · Sheaves · Topoi

1 Turing Degrees

The existence of unsolvable[1] mathematical problems was discovered by Turing [77]. Indeed, Turing exhibited a *specific, natural example*[2] of such a problem: the halting problem for Turing machines. Later, in the 1950 s and 1960s, it was discovered that there are specific, natural, unsolvable problems in virtually every branch of mathematics: *number theory* (Hilbert's Tenth Problem [14]), *geometry* (the homeomorphism problem for finite simplicial complexes, the diffeomorphism

MSC2010: Primary 03D28; Secondary 03D80, 03D32, 03D35, 03D55, 03F55, 03G30, 18F20, 37B10.

S.G. Simpson—This paper is a preview of a three-hour tutorial to be given at CiE in Bucharest, June 29 to July 3, 2015. The author's research is supported by the Eberly College of Science and by Simons Foundation Collaboration Grant 276282.

[1] By *unsolvable* we mean algorithmically unsolvable, i.e., not solvable by a Turing program.

[2] We are not offering a rigorous definition of what is meant by "specific and natural." However, it is well known that considerations of specificity and naturalness play an important role in mathematics. Without such considerations, it would be difficult or impossible to pursue the ideal of "exquisite taste" in mathematical research, as famously enunciated by von Neumann.

A. Beckmann et al. (Eds.): CiE 2015, LNCS 9136, pp. 83–94, 2015.
DOI: 10.1007/978-3-319-20028-6_9

problem for compact manifolds [42, Appendix]), *group theory* (the word problem [1] and the triviality problem [47] for finitely presented groups), *combinatorics* (the problem of tileability of the plane with a finite set of tiles [6,49]), *mathematical logic* (the validity problem for predicate calculus [12,77], the decision problem for first-order arithmetic [75]), and even *elementary calculus* (the problem of integrability in finite terms [48]).

A scheme for classifying unsolvable problems was developed by Post [46] and Kleene/Post [34]. Two reals[3] X and Y are said to be *Turing equivalent* if each is computable using the other as a Turing oracle. The *Turing degree* of a real is its equivalence class under this equivalence relation. Each of the specific, natural, unsolvable problems mentioned above is a *decision problem* and may therefore be straightforwardly described or "encoded"[4] as a real. It was then shown that each of these problems is of the same Turing degree as the halting problem. This Turing degree is denoted $\mathbf{0}'$. Thus the specific Turing degree $\mathbf{0}'$ is extremely useful and important.

Given a real X, the Turing degree of X is denoted $\deg_T(X)$. If $\mathbf{a} = \deg_T(X)$ and $\mathbf{b} = \deg_T(Y)$ are the Turing degrees of reals X and Y respectively, we write $X \leq_T Y$ or $\mathbf{a} \leq \mathbf{b}$ to mean that Y is "at least as unsolvable as" X in the following sense: X is computable using Y as a Turing oracle. We also write $X <_T Y$ or $\mathbf{a} < \mathbf{b}$ to mean that $X \leq_T Y$ and $Y \not\leq_T X$. Let $\mathcal{D}_T$ be the set of all Turing degrees. Clearly $\leq$ is a partial ordering of $\mathcal{D}_T$, and every pair of degrees in $\mathcal{D}_T$ has a *supremum*, i.e., a least upper bound. In other words, $\mathcal{D}_T$ is a *semilattice*. Kleene and Post proved that there are infinitely many degrees in $\mathcal{D}_T$ which are less than $\mathbf{0}'$, and there are uncountably many other degrees in $\mathcal{D}_T$ which are incomparable with $\mathbf{0}'$. Thus $\mathcal{D}_T$ has a rich algebraic structure. However, despite recent remarkable progress [59,71], no one has yet discovered a specific, natural example of an unsolvable problem of Turing degree $\not\geq \mathbf{0}'$.

Given a real X, let X' be a real which encodes the halting problem *relative to* X, i.e., with X used as a Turing oracle. If $\mathbf{a}$ is the Turing degree of X, let $\mathbf{a}'$ be the Turing degree of X'. It can be shown that $\mathbf{a}'$ is independent of the choice of X such that $\deg_T(X) = \mathbf{a}$. The operator $\mathbf{a} \mapsto \mathbf{a}' : \mathcal{D}_T \to \mathcal{D}_T$ is called the *Turing jump* operator. Generalizing Turing's proof of the unsolvability of the halting problem, one shows that $\mathbf{a} < \mathbf{a}'$. In other words, X' is "more unsolvable than" X. Inductively we write $\mathbf{a}^{(0)} = \mathbf{a}$ and $\mathbf{a}^{(n+1)} = (\mathbf{a}^{(n)})'$ for all natural numbers n. Extending this induction into the transfinite, it is possible to define $\mathbf{a}^{(\alpha)}$ where α ranges over a large initial segment of the ordinal numbers including the constructibly countable ordinal numbers. We then have $\mathbf{a}^{(\alpha)} < \mathbf{a}^{(\beta)}$ whenever $\alpha < \beta$. See [53, Part A] and [27,60].

---

3 In this paper we take *reals* to be points in the Baire space $\mathbb{N}^{\mathbb{N}}$, i.e., functions $X : \mathbb{N} \to \mathbb{N}$ where $\mathbb{N} = \{0, 1, 2, \ldots\}$ = the natural numbers.

4 More specifically, each of the mentioned problems amounts to the question of deciding whether or not a given string of symbols from a fixed finite alphabet belongs to a particular set of such strings. The problem is then identified with the characteristic function of the set of Gödel numbers of the strings which belong to the set.

Let $\mathbf{0}$ be the bottom degree in $\mathcal{D}_{\mathrm{T}}$, i.e., the Turing degree of any solvable problem. We then have a transfinite hierarchy of specific, natural Turing degrees

$$\mathbf{0} < \mathbf{0}' < \mathbf{0}'' < \cdots < \mathbf{0}^{(\alpha)} < \mathbf{0}^{(\alpha+1)} < \cdots$$

where α ranges over a large initial segment of the ordinal numbers [60]. Moreover, this hierarchy of specific, natural Turing degrees has been useful for the classification of unsolvable mathematical problems. See for instance [43] and [50, Sect. 14.8] and Sect. 4 below. However, no other specific, natural Turing degrees are known.

The semilattice $\mathcal{D}_{\mathrm{T}}$ is large and complicated, so it is reasonable to examine subsemilattices which are hopefully more manageable. One such subsemilattice has been studied in great depth. A Turing degree is said to be *recursively enumerable*[5] if it is the Turing degree of the characteristic function of a subset of $\mathbb{N}$ which is the range of a recursive function. Let $\mathcal{E}_{\mathrm{T}}$ be the subsemilattice of $\mathcal{D}_{\mathrm{T}}$ consisting of the recursively enumerable Turing degrees. The top and bottom degrees in $\mathcal{E}_{\mathrm{T}}$ are $\mathbf{0}'$ and $\mathbf{0}$. It is known that $\mathcal{E}_{\mathrm{T}}$ is structurally rich. Two key results due to Sacks [51,52] are the *Splitting Theorem*[6] and the *Density Theorem*[7], and many other results have been obtained [37,38,58,72]. For instance, the Turing degree of the first-order theory of $\mathcal{E}_{\mathrm{T}}$ is $\mathbf{0}^{(\omega)}$ [45]. However, except for $\mathbf{0}'$ and $\mathbf{0}$ no specific, natural, recursively enumerable Turing degrees are known.

2 Muchnik Degrees

There are many specific, natural, unsolvable problems to which it is impossible to assign a Turing degree.

As an example, let T be an effectively essentially undecidable theory. For instance, we could take $T = \mathsf{PA} = \mathsf{Z}_1 =$ first-order arithmetic, or $T = \mathsf{Z}_2 =$ second-order arithmetic [62], or $T = \mathsf{ZFC} =$ Zermelo/Fraenkel set theory [25], or $T = \mathsf{Q} =$ Robinson's arithmetic [75], or $T =$ any consistent recursively axiomatizable extension of one of these. Consider the problem $\mathrm{C}(T)$ of "finding" a complete and consistent theory which extends T. A solution of the problem would be any such theory. Lindenbaum's Lemma implies that such theories exist, and by [75] no such theory is algorithmically decidable.[8] In this sense the problem $\mathrm{C}(T)$ is algorithmically unsolvable. On the other hand, the problem $\mathrm{C}(T)$ cannot correspond to a Turing degree, because for any solution X of $\mathrm{C}(T)$ there exists a solution Y of $\mathrm{C}(T)$ such that $Y <_{\mathrm{T}} X$.

In order to overcome this limitation of the Turing degrees, we now extend $\mathcal{D}_{\mathrm{T}}$ to its completion, $\mathcal{D}_{\mathrm{w}}$, the lattice of Muchnik degrees.

---

5 A.k.a., computably enumerable [73].

6 The Sacks Splitting Theorem says that $\mathcal{E}_{\mathrm{T}}$ satisfies $\forall x\,(x > 0 \Rightarrow \exists u\,\exists v\,(u < x$ and $v < x$ and $\sup(u,v) = x))$.

7 The Sacks Density Theorem says that $\mathcal{E}_{\mathrm{T}}$ satisfies $\forall x\,\forall y\,(x < y \Rightarrow \exists z\,(x < z < y))$.

8 When speaking of decidable theories, we identify a theory with the characteristic function $X \in \{0,1\}^{\mathbb{N}}$ of the set of Gödel numbers of theorems of the theory.

A *mass problem* is defined to be a set of reals.[9] The idea here is that a mass problem P "represents" (i.e., is the solution set of) the problem of "finding" or "computing" some real X which belongs to P. Accordingly, a mass problem P is said to be *unsolvable* if it contains no Turing computable real, i.e., if $P \cap \mathrm{REC} = \emptyset$ where $\mathrm{REC} = \{X \mid X \text{ is computable}\}$. Following the same idea, we generalize the notion of Turing reducibility as follows. For mass problems P and Q, we say that P is *Muchnik reducible* to Q, abbreviated $P \leq_{\mathrm{w}} Q$, if every solution of Q can be used as a Turing oracle to compute some solution of P. In other words, $P \leq_{\mathrm{w}} Q$ if and only if $\forall Y\,(Y \in Q \Rightarrow \exists X\,(X \in P \text{ and } X \leq_{\mathrm{T}} Y))$.[10] We say that P is *Muchnik equivalent* to Q, abbreviated $P \equiv_{\mathrm{w}} Q$, if $P \leq_{\mathrm{w}} Q$ and $Q \leq_{\mathrm{w}} P$. The *Muchnik degree* of P, written $\deg_{\mathrm{w}}(P)$, is the equivalence class of P under $\equiv_{\mathrm{w}}$. Let $\mathcal{D}_{\mathrm{w}}$ be the set of all Muchnik degrees, partially ordered by letting $\deg_{\mathrm{w}}(P) \leq \deg_{\mathrm{w}}(Q)$ if and only if $P \leq_{\mathrm{w}} Q$. It is easy to see that $\mathcal{D}_{\mathrm{w}}$ is a complete and completely distributive lattice. Given a real X, we identify X with the mass problem $\{X\}$ = the singleton set whose only member is X. Thus $\deg_{\mathrm{T}}(X) = \deg_{\mathrm{w}}(\{X\})$ and $\mathcal{D}_{\mathrm{T}}$ is now a subset of $\mathcal{D}_{\mathrm{w}}$.

The relationship between $\mathcal{D}_{\mathrm{T}}$ and $\mathcal{D}_{\mathrm{w}}$ may be viewed as an instance of a general construction. Namely, for any partially ordered set K let $\widehat{K}$ be the set of upwardly closed subsets of K partially ordered by reverse inclusion, i.e., $U \leq V$ if and only if $U \supseteq V$. Identifying $a \in K$ with the upwardly closed set $U_a = \{x \in K \mid x \geq a\} \in \widehat{K}$, we see that K is a subordering of $\widehat{K}$, i.e., $a \leq b$ if and only if $U_a \leq U_b$. Thus $\widehat{K}$ is a complete and completely distributive lattice, the *completion* of K. There is a unique isomorphism of $\mathcal{D}_{\mathrm{w}}$ onto $\widehat{\mathcal{D}_{\mathrm{T}}}$ which extends the identity map on $\mathcal{D}_{\mathrm{T}}$, and in this sense $\mathcal{D}_{\mathrm{w}}$ is the completion of $\mathcal{D}_{\mathrm{T}}$. The upshot here is that Muchnik degrees can be identified with upwardly closed sets of Turing degrees.[11] This remark will be important in Sect. 5 below.

In the above example, let us identify $\mathrm{C}(T)$ with the mass problem $\{X \mid X$ is a complete and consistent extension of $T\}$. Under this identification, $\mathrm{C}(T)$ is Muchnik reducible to the halting problem.[12] However, the halting problem is not Muchnik reducible to $\mathrm{C}(T)$, because the halting problem has a Turing degree while $\mathrm{C}(T)$ does not. Thus, letting $\mathbf{1}$ = the Muchnik degree of $\mathrm{C}(T)$, we have $\mathbf{0} < \mathbf{1} < \mathbf{0}'$. Furthermore, the particular Muchnik degree $\mathbf{1} = \deg_{\mathrm{w}}(\mathrm{C}(T))$ can be characterized abstractly in a way which does not depend on T. We now see that $\mathbf{1}$ is a very specific, very natural, very important Muchnik degree which is not a Turing degree.

In addition to the Muchnik degree $\mathbf{1}$ and the Turing degrees $\mathbf{0}^{(\alpha)}$ for ordinal numbers $\alpha = 0, 1, 2, \ldots, \omega, \omega + 1, \ldots$, there are many other specific, natural Muchnik degrees. Here are some examples and references.

1. Let λ be the fair coin probability measure on $\{0,1\}^{\mathbb{N}}$. A set $S \subseteq \{0,1\}^{\mathbb{N}}$ is said to be *effectively null* if $S \subseteq \bigcap_n U_n$ for some uniformly effectively open sequence of sets U_n such that $\lambda(U_n) \leq 2^{-n}$ for all n. A real $Z \in \{0,1\}^{\mathbb{N}}$ is

[9] This concept is from Medvedev [39]. As in footnote 3 a *real* is a function $X \in \mathbb{N}^{\mathbb{N}}$.

[10] This is Muchnik's notion of *weak reducibility* [41, Definition 2].

[11] For a more precise statement, see [5, Theorem 5.8].

[12] This follows from a theorem of Kleene [33, p. 398]. See also [29,57].

said to be *Martin-Löf random* [16,44] if it does not belong to any effectively null set. Let $\mathbf{r}_1 = \deg_{\mathrm{w}}(\{Z \in \{0,1\}^{\mathbb{N}} \mid Z \text{ is Martin-Löf random}\})$. It is not difficult to show that $\mathbf{0} < \mathbf{r}_1 < \mathbf{1}$.

2. More generally, for any constructibly countable ordinal number α, let $\mathbf{r}_\alpha = \deg_{\mathrm{w}}(\{Z \mid (\forall \xi < \alpha)\,(Z \text{ is Martin-Löf random relative to } 0^{(\xi)})\})$. It is not difficult to show that $\mathbf{0} = \mathbf{r}_0 < \mathbf{r}_1 < \mathbf{r}_2 < \cdots < \mathbf{r}_\alpha < \mathbf{r}_{\alpha+1} < \cdots$. Moreover, each $\mathbf{r}_\alpha$ for $\alpha \geq 2$ is incomparable with $\mathbf{1}$.
3. A partial recursive function $\psi : \subseteq \mathbb{N} \to \mathbb{N}$ is said to be *universal* if for each partial recursive function $\varphi : \subseteq \mathbb{N} \to \mathbb{N}$ there exists a recursive function $p : \mathbb{N} \to \mathbb{N}$ such that $\varphi(n) \simeq \psi(p(n))$ for all n.[13] Fix such a function ψ and let $\mathbf{d} = \deg_{\mathrm{w}}(\{Z \in \mathbb{N}^{\mathbb{N}} \mid Z \cap \psi = \emptyset\})$ and $\mathbf{d}_{\mathrm{REC}} = \deg_{\mathrm{w}}(\{Z \in \mathbb{N}^{\mathbb{N}} \mid Z \cap \psi = \emptyset$ and Z is recursively bounded$\})$. Clearly $\mathbf{d}$ and $\mathbf{d}_{\mathrm{REC}}$ are independent[14] of our choice of ψ. By [3,26] we have $\mathbf{0} < \mathbf{d} < \mathbf{d}_{\mathrm{REC}} < \mathbf{r}_1$.
4. Given a recursive function $f : \mathbb{N} \to \mathbb{N}$, define $Z \in \{0,1\}^{\mathbb{N}}$ to be *f-complex* if $\exists c \forall n\, (\mathrm{K}(Z{\upharpoonright}\{1,\ldots,n\}) > f(n) - c)$ where K denotes Kolmogorov complexity. In this way each specific, natural,[15] recursive function f gives rise to a specific, natural Muchnik degree $\mathbf{k}_f = \deg_{\mathrm{w}}(\{Z \in \{0,1\}^{\mathbb{N}} \mid Z \text{ is } f\text{-complex}\})$, and there is also $\mathbf{k}_{\mathrm{REC}} = \deg_{\mathrm{w}}(\{Z \in \{0,1\}^{\mathbb{N}} \mid Z$ is f-complex for some unbounded recursive function $f\})$. By [32] we have $\mathbf{k}_{\mathrm{REC}} = \mathbf{d}_{\mathrm{REC}}$, and by [16, Theorem 6.2.3] we have $\mathbf{k}_1 = \mathbf{r}_1$ where $1 : \mathbb{N} \to \mathbb{N}$ is the identity function. Building on the methods of Miller [40], Hudelson [24] has shown that $\mathbf{d}_{\mathrm{REC}} < \mathbf{k}_f < \mathbf{k}_g \leq \mathbf{r}_1$ holds for many pairs of unbounded recursive functions f, g. In particular, this holds whenever $\forall n\,(f(n) \leq f(n+1) \leq f(n) + 1$ and $f(n) + 2\log_2 f(n) \leq g(n) \leq n)$.
5. Let $\mathrm{MLR}^X = \{Z \in \{0,1\}^{\mathbb{N}} \mid Z \text{ is Martin-Löf random relative to } X\}$. We say that X *is* LR-*reducible to* Y, abbreviated $X \leq_{\mathrm{LR}} Y$, if $\mathrm{MLR}^X \supseteq \mathrm{MLR}^Y$ [16,44]. Letting $\mathbf{b}_\alpha = \deg_{\mathrm{w}}(\{Y \mid 0^{(\alpha)} \leq_{\mathrm{LR}} Y\})$, it is not difficult to show that $\mathbf{0} = \mathbf{b}_0 < \mathbf{b}_1 < \mathbf{b}_2 < \cdots < \mathbf{b}_\alpha < \mathbf{b}_{\alpha+1} < \cdots$. On the other hand, by [68] we know that the Muchnik degrees $\mathbf{b}_\alpha$ for $\alpha \geq 1$ are incomparable with the Muchnik degrees $\mathbf{d}$, $\mathbf{1}$, and $\mathbf{r}_\alpha$ for all $\alpha \geq 1$.
6. A partial recursive function $\psi : \subseteq \mathbb{N} \to \mathbb{N}$ is said to be *linearly universal* if it is "universal via linear functions," i.e., for each partial recursive function $\varphi : \subseteq \mathbb{N} \to \mathbb{N}$ there exist $a, b \in \mathbb{N}$ such that $\varphi(n) \simeq \psi(an + b)$ for all n. Let $D = \{Z \in \mathbb{N}^{\mathbb{N}} \mid Z \cap \psi = \emptyset$ for some linearly universal partial recursive function $\psi\}$, and let $D_{\mathrm{REC}} = \{Z \in D \mid Z \text{ is recursively bounded}\})$. Clearly $\deg_{\mathrm{w}}(D) = \mathbf{d}$ and $\deg_{\mathrm{w}}(D_{\mathrm{REC}}) = \mathbf{d}_{\mathrm{REC}}$ where $\mathbf{d}$ and $\mathbf{d}_{\mathrm{REC}}$ are as above. However, letting $D_h = \{Z \in D \mid Z \text{ is } h\text{-bounded}\}$ where h is a specific

[13] Here $E_1 \simeq E_2$ means that E_1 and E_2 are both undefined or both defined and equal.

[14] Let φ_n, $n \in \mathbb{N}$ be a fixed, standard, partial recursive enumeration of the partial recursive functions. A function $Z \in \mathbb{N}^{\mathbb{N}}$ is said to be *diagonally nonrecursive* [3, 23,26,32,65] if $Z \cap \psi = \emptyset$ where ψ is the well known *diagonal function*, defined by $\psi(n) \simeq \varphi_n(n)$. Letting $\mathrm{DNR} = \{Z \in \mathbb{N}^{\mathbb{N}} \mid Z \text{ is diagonally nonrecursive}\}$ and $\mathrm{DNR}_{\mathrm{REC}} = \{Z \in \mathrm{DNR} \mid Z \text{ is recursively bounded}\}$, we have $\mathbf{d} = \deg_{\mathrm{w}}(\mathrm{DNR})$ and $\mathbf{d}_{\mathrm{REC}} = \deg_{\mathrm{w}}(\mathrm{DNR}_{\mathrm{REC}})$.

[15] For example, $f(n)$ could be $n/2$ or $n/3$ or $\sqrt{n}$ or $\sqrt[3]{n}$ or $\log_2 n$ or $\log_3 n$ or $\log_2 \log_2 n$, etc., or f could be the inverse Ackermann function.

recursive function, we get a family of Muchnik degrees $\mathbf{d}_h = \deg_{\mathrm{w}}(D_h)$ which are of considerable interest [64, Sect. 10] [31]. In particular, for any unbounded recursive function h such that $\forall n\,(1 \leq h(n) \leq h(n+1))$ we know by [3,23] and [7, Sect. 7.3] that $\mathbf{d}_{\mathrm{REC}} < \mathbf{d}_h < \mathbf{1}$, and if $\sum_n h(n)^{-1} < \infty$ then $\mathbf{d}_h < \mathbf{r}_1$, and if $\sum_n h(n)^{-1} = \infty$ then $\mathbf{d}_h$ is incomparable with $\mathbf{r}_\alpha$ for all $\alpha \geq 1$. Also of interest is the Muchnik degree $\mathbf{d}_{\mathrm{slow}} = \deg_{\mathrm{w}}(\{Z \mid Z \in D_h$ for some recursive function h such that $\forall n\,(h(n) \leq h(n+1))$ and $\sum_n h(n)^{-1} = \infty\})$.

7. There are many other examples of specific, natural Muchnik degrees. See for instance the Computability Menagerie[16] [30]. Our choice of examples in this paper is oriented toward Sect. 3 below.

3 The Lattices $\mathcal{E}_{\mathrm{w}}$ and $\mathcal{S}_{\mathrm{w}}$

The lattice $\mathcal{D}_{\mathrm{w}}$ is large and complicated, so it is desirable to consider more manageable sublattices. The smallest such sublattice which comes immediately to mind is the countable lattice $\mathcal{E}_{\mathrm{w}}$ consisting of the Muchnik degrees of nonempty, effectively closed subsets of $\{0,1\}^{\mathbb{N}}$. The explicit study of $\mathcal{E}_{\mathrm{w}}$ was undertaken only relatively recently [8,10,61,63,64] but was implicit in some much older literature [22,28,29,54,55]. By [55] the top and bottom degrees in $\mathcal{E}_{\mathrm{w}}$ are $\mathbf{1}$ and $\mathbf{0}$, and by [10] every countable distributive lattice is lattice-embeddable into $\mathcal{E}_{\mathrm{w}}$. The only Turing degree in $\mathcal{E}_{\mathrm{w}}$ is $\mathbf{0}$, but there is an obvious analogy

$$\frac{\mathcal{E}_{\mathrm{w}}}{\mathcal{D}_{\mathrm{w}}} = \frac{\mathcal{E}_{\mathrm{T}}}{\mathcal{D}_{\mathrm{T}}}$$

and indeed the Splitting Theorem and the Density Theorem hold for $\mathcal{E}_{\mathrm{w}}$ [8,9]. The Turing degree of the first-order theory of $\mathcal{E}_{\mathrm{w}}$ is known to be $\geq \mathbf{0}^{(\omega)}$ [56] and conjectured to be $= \mathbf{0}^{(\omega_1^{\mathrm{CK}}+\omega)}$ [13, p. 127] [69, Remark 3.2.3].

An advantage of $\mathcal{E}_{\mathrm{w}}$ over $\mathcal{E}_{\mathrm{T}}$ is that $\mathcal{E}_{\mathrm{w}}$ contains a great variety of specific, natural Muchnik degrees in addition to its top and bottom degrees $\mathbf{1}$ and $\mathbf{0}$. In particular, it is not difficult [65, Sect. 3] to show that the Muchnik degrees $\mathbf{d}, \mathbf{d}_{\mathrm{REC}}, \mathbf{k}_f, \mathbf{r}_1, \mathbf{d}_h$, and $\mathbf{d}_{\mathrm{slow}}$ which were discussed in Sect. 2 belong to $\mathcal{E}_{\mathrm{w}}$.

Also of interest is the countable lattice $\mathcal{S}_{\mathrm{w}}$ consisting of the Muchnik degrees of nonempty, effectively closed subsets of $\mathbb{N}^{\mathbb{N}}$. An easy argument [69, Lemma 3.3.5] shows that $\mathcal{S}_{\mathrm{w}}$ has an alternative characterization as the lattice of Muchnik degrees of nonempty, lightface Σ^0_3 subsets of $\mathbb{N}^{\mathbb{N}}$. This is important, because it implies that $\mathcal{S}_{\mathrm{w}}$ contains many specific, natural Muchnik degrees beyond those which are already in $\mathcal{E}_{\mathrm{w}}$. In particular, the Muchnik degree $\mathbf{r}_2$ which was discussed in Sect. 2 belongs to $\mathcal{S}_{\mathrm{w}}$, as do the Turing degrees $\mathbf{0}^{(\alpha)}$ and the Muchnik degrees $\mathbf{b}_\alpha$ for all recursive ordinal numbers $\alpha < \omega_1^{\mathrm{CK}}$ [68].

Trivially $\mathcal{E}_{\mathrm{w}}$ is a sublattice of $\mathcal{S}_{\mathrm{w}}$, and by [69, Theorem 3.3.1] we know that $\mathcal{E}_{\mathrm{w}}$ is an initial segment of $\mathcal{S}_{\mathrm{w}}$. This is important, because it means that we have a specific, natural, lattice homomorphism $\mathbf{s} \mapsto \inf(\mathbf{s}, \mathbf{1}) : \mathcal{S}_{\mathrm{w}} \to \mathcal{E}_{\mathrm{w}}$. With

[16] The inhabitants of this menagerie are downwardly closed sets of Turing degrees, but the complements of such sets are essentially the same thing as Muchnik degrees.

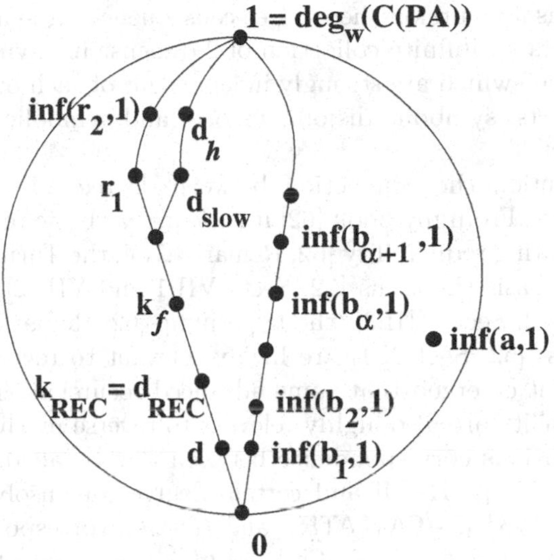

Fig. 1. A picture of $\mathcal{E}_w$.

this homomorphism, each of the specific, natural Muchnik degrees in $\mathcal{S}_w$ has a specific, natural image in $\mathcal{E}_w$. In particular, the Muchnik degrees $\inf(\mathbf{r}_2, \mathbf{1})$ [65, Sect. 3] and $\inf(\mathbf{b}_\alpha, \mathbf{1})$ for all ordinal numbers $\alpha < \omega_1^{\text{CK}}$ [66,68] belong to $\mathcal{E}_w$.

Clearly $\mathcal{E}_T$ is a subsemilattice of $\mathcal{S}_w$, and by the Arslanov Completeness Criterion [26, Theorem 1] (see also [65, Sect. 5]) our homomorphism of $\mathcal{S}_w$ onto $\mathcal{E}_w$ is one-to-one when restricted to $\mathcal{E}_T$. Thus we have a semilattice embedding $\mathbf{a} \mapsto \inf(\mathbf{a}, \mathbf{1}) : \mathcal{E}_T \hookrightarrow \mathcal{E}_w$ which carries the top and bottom degrees $\mathbf{0}', \mathbf{0} \in \mathcal{E}_T$ to the top and bottom degrees $\mathbf{1}, \mathbf{0} \in \mathcal{E}_w$. Unfortunately, the range of this embedding does not appear to contain any specific, natural Muchnik degrees other than $\mathbf{1}$ and $\mathbf{0}$. Thus the problem of finding a specific, natural, recursively enumerable Turing degree in the range $\mathbf{0} < \mathbf{a} < \mathbf{0}'$ remains open.

Figure 1 is a picture of $\mathcal{E}_w$. In this picture, $\mathbf{a}$ is any recursively enumerable Turing degree in the range $\mathbf{0} < \mathbf{a} < \mathbf{0}'$. The black dots other than $\inf(\mathbf{a}, \mathbf{1})$ denote some of the specific, natural Muchnik degrees which we have discussed.

4 Applications

We briefly mention an application of $\mathcal{E}_w$ to tiling problems. A *Wang tile* is a unit square with colored edges. Given a finite set A of Wang tiles, let P_A be the problem of tiling the plane with copies of tiles from A. More formally, P_A is the set of mappings $X : \mathbb{Z} \times \mathbb{Z} \to A$ such that for all $(i, j) \in \mathbb{Z} \times \mathbb{Z}$ the right edge of $X(i, j)$ matches the left edge of $X(i+1, j)$ and the top edge of $X(i, j)$ matches the bottom edge of $X(i, j+1)$. Clearly $\deg_w(P_A) \in \mathcal{E}_w$ provided $P_A \neq \emptyset$. It turns out [17,70] that conversely, every Muchnik degree in $\mathcal{E}_w$ is $\deg_w(P_A)$ for some finite set A of Wang tiles. This result plus the existence of an infinite independent set of degrees in $\mathcal{E}_{\mathrm{w}}$ has a recursion-theory-free consequence for symbolic dynamics. Namely, there exists an infinite collection of 2-dimensional symbolic dynamical systems of finite type which are strongly independent of each other with respect to symbolic products, symbolic disjoint unions, and symbolic morphisms. For details see [70, Sect. 3].

We briefly mention the connection between degrees of unsolvability and reverse mathematics. From my book [62] it is clear that basic recursion-theoretic concepts such as Turing reducibility [62, Remark I.7.5], the Turing jump operator [62, Remark I.3.4], basis theorems [62, Sects. VII.1 and VIII.2], the hyperarithmetical hierarchy [62, Sect. VIII.3], the hyperjump [62, Remark I.5.4], and algorithmic randomness [62, Sect. X.1] are highly relevant to reverse mathematics. More recently [68] it emerged that some advanced recursion-theoretic concepts such as LR-reducibility are also highly relevant to reverse mathematics. Beyond this, there is an obvious correspondence between the so called "Big Five" subsystems of Z_2 [62, Chaps. II–VI] and certain degrees of unsolvability. Namely, the systems RCA_0, WKL_0, ACA_0, ATR_0, and Π^1_1-CA_0 correspond to the Muchnik degrees $\mathbf{0}$, $\mathbf{1}$, $\mathbf{0}'$, $\mathbf{0}^{(\alpha)}$ for $\alpha < \omega_1^{\mathrm{CK}}$, and $\mathbf{0}^{(\omega_1^{\mathrm{CK}})}$ respectively, where ω_1^{CK} is the least nonrecursive ordinal. In addition, the system WWKL_0 [62, Sect. X.1] corresponds to the Muchnik degree $\mathbf{r}_1$.

5 The Muchnik Topos

From Medvedev's 1955 paper introducing mass problems [39] and Muchnik's 1963 paper introducing Muchnik reducibility [41][17], it is evident that both authors were motivated by Kolmogorov's nonrigorous 1932 interpretation of intuitionistic propositional calculus as a "calculus of problems" [35,36]. Kolmogorov's idea was to view intuitionistic propositions as "problems," and intuitionistic proofs of propositions as "solutions" of the corresponding "problems." Intuitionistic propositional connectives are then viewed as methods of combining "problems" to form new "problems." Two "problems" are viewed as being "equivalent" if from any solution of either of them a "solution" of the other can be "easily" or "immediately" extracted. We cannot expect the Law of the Excluded Middle to hold, because it would mean that for any proposition there should be an "easy" proof of either the proposition or its negation.

Muchnik's rigorous implementation of Kolmogorov's idea [41, Theorem 4] is based on mass problems, Muchnik reducibility, and lattice operations in $\mathcal{D}_{\mathrm{w}}$. Given two Muchnik degrees $\mathbf{p}$ and $\mathbf{q}$, we interpret $\mathbf{p} \wedge \mathbf{q}$ as $\sup(\mathbf{p},\mathbf{q})$, $\mathbf{p} \vee \mathbf{q}$ as $\inf(\mathbf{p},\mathbf{q})$, $\mathbf{p} \Rightarrow \mathbf{q}$ as $\inf(\{\mathbf{x} \mid \sup(\mathbf{p},\mathbf{x}) \geq \mathbf{q}\})$, "true" as $\mathbf{0}$, "false" as $\deg_{\mathrm{w}}(\emptyset)$, and $\mathbf{p} \vdash \mathbf{q}$ as $\mathbf{p} \geq \mathbf{q}$. For more details and references, see [69, Sect. 4] and [67,74].

Recently Muchnik's interpretation of intuitionistic propositional calculus [41] has been extended to an interpretation of intuitionistic mathematics as a whole [5]. The extension is based on a category which we call the *Muchnik topos*. The idea here is to consider $\mathcal{D}_{\mathrm{T}}$ as a topological space in which the open sets are

[17] See also the English translation in [5, Appendix].

the upwardly closed sets of Turing degrees.[18] In general, for any topological space $\mathcal{T}$, a *sheaf* over $\mathcal{T}$ consists of a topological space $\mathcal{X}$ together with a local homeomorphism $p : \mathcal{X} \to \mathcal{T}$. A *sheaf morphism* from a sheaf $p : \mathcal{X} \to \mathcal{T}$ to a sheaf $q : \mathcal{Y} \to \mathcal{T}$ is a continuous function $f : \mathcal{X} \to \mathcal{Y}$ such that $p(x) = q(f(x))$ for all $x \in \mathcal{X}$. The sheaves and sheaf morphisms over $\mathcal{T}$ form a category called $\mathrm{Sh}(\mathcal{T})$. As noted by Fourman and Scott [21], $\mathrm{Sh}(\mathcal{T})$ is a topos and provides a model of intuitionistic higher-order logic in which the truth values are the open subsets of $\mathcal{T}$. The Muchnik topos is then the special case $\mathrm{Sh}(\mathcal{D}_{\mathrm{T}})$ with truth values in $\mathcal{D}_{\mathrm{w}}$. All of this background material concerning sheaves and intuitionistic higher-order logic is explained at length in our paper [5].

Within the Muchnik topos $\mathrm{Sh}(\mathcal{D}_{\mathrm{T}})$, there are two versions of the real number system $\mathbb{R}$: the sheaf $\mathbb{R}_C = \mathbb{R} \times \mathcal{D}_{\mathrm{T}}$ of *Cauchy reals*, and the sheaf $\mathbb{R}_M = \{(r, \mathbf{a}) \in \mathbb{R}_C \mid \deg_{\mathrm{T}}(r) \leq_{\mathrm{T}} \mathbf{a}\}$ of *Muchnik reals*. Roughly speaking, the difference between $\mathbb{R}_C$ and $\mathbb{R}_M$ is that a Cauchy real can exist anywhere within the topological space $\mathcal{D}_{\mathrm{T}}$, but a Muchnik real can exist only where we have enough Turing oracle power to compute it. For precise definitions, see [5]. It turns out [5, Theorem 5.18] that the Muchnik topos satisfies a *Choice and Bounding Principle*:

$$forall x\, \exists y\, \Phi(x, y)) \Rightarrow \exists w\, \exists z\, \forall x\, (wx \leq_{\mathrm{T}} (x, z) \wedge \Phi(x, wx))$$

where x, y, z are variables ranging over Muchnik reals, w is a variable ranging over functions from Muchnik reals to Muchnik reals, and $\Phi(x, y)$ is any formula of intuitionistic higher-order logic in which w and z do not occur. Our Choice and Bounding Principle reflects a well known intuitonistic idea: if for all real numbers x there exists a real number y which bears a certain relationship to x, then there should be a function $x \mapsto y$ which computes such a y using x as a Turing oracle.

We feel that, among various interpretations of intuitionistic mathematics, our interpretation in terms of the Muchnik topos stands out because of its relationship to the ideas of Kolmogorov, Medvedev, and Muchnik.

References

1. Aanderaa, S., Cohen, D.E.: Modular machines I, II. In: [2], pp. 1–18, 19–28 (1980)
2. Adian, S.I., Boone, W.W., Higman, G. (eds.): Word Problems II: The Oxford Book. Studies in Logic and the Foundations of Mathematics, X + 578 p., North-Holland (1980)
3. Ambos-Spies, K., Kjos-Hanssen, B., Lempp, S., Slaman, T.A.: Comparing DNR and WWKL. J. Symbolic Logic **69**, 1089–1104 (2004)
4. Barwise, J., Keisler, H.J., Kunen, K. (eds.): The Kleene Symposium. Studies in Logic and the Foundations of Mathematics, XX + 425 p., North-Holland (1980)
5. Basu, S.S., Simpson, S.G.: Mass problems and intuitionistic higher-order logic, 44 p., 12 August 2014. http://arxiv.org/abs/1408.2763
6. Berger, R.: The Undecidability of the Domino Problem. Memoirs of the American Mathematical Society, vol. 66, p. 72. American Mathematical Society, Providence (1966)

[18] This topological space was considered by Muchnik [41, p. 1332] [5, p. 35].

7. Bienvenu, L., Porter, C.P.: Deep Π^0_1 classes, 37 p., 4 June 2014. http://arxiv.org/abs/1403.0450v2
8. Binns, S.: A splitting theorem for the medvedev and muchnik lattices. Math. Logic Q. **49**(4), 327–335 (2003)
9. Binns, S., Shore, R.A., Simpson, S.G.: Mass problems and density, in preparation, 5 p., 1 March 2014
10. Binns, S., Simpson, S.G.: Embeddings into the medvedev and muchnik lattices of Π^0_1 classes. Arch. Math. Logic **43**, 399–414 (2004)
11. Chong, C.T., Feng, Q., Slaman, T.A., Woodin, W.H., Yang, Y. (eds.): Computational Prospects of Infinity. In: Proceedings of the Logic Workshop at the Institute for Mathematical Sciences, Part I: Tutorials in Lecture Notes Series, Institute for Mathematical Sciences, National University of Singapore, World Scientific, 20 June–15 August, 2005, no. 14, 264 p. (2008)
12. Church, A.: A note on the Entscheidungsproblem. J. Symbolic Logic **1**, 40–41 (1936)
13. Cole, J.A., Simpson, S.G.: Mass problems and hyperarithmeticity. J. Math. Logic **7**(2), 125–143 (2008)
14. Davis, M.: Hilbert's tenth problem is unsolvable. Am. Math. Mon. **80**, 233–269 (1973)
15. Dekker, J.C.E. (ed.): Recursive function theory. In: Proceedings of Symposia in Pure Mathematics, American Mathematical Society, VII + 247 p. (1962)
16. Downey, R.G., Hirschfeldt, D.R.: Algorithmic Randomness and Complexity. Theory and Applications of Computability, XXVIII + 855 p. Springer, New York (2010)
17. Durand, B., Romashchenko, A., Shen, A.: Fixed-point tile sets and their applications. J. Comput. Syst. Sci. **78**(3), 731–764 (2012). doi:10.1016/j.jcss.2011.11.001
18. Fenstad, J.E., Frolov, I.T., Hilpinen, R. (eds.): Logic, Methodology and Philosophy of Science VIII. No. 126 in Studies in Logic and the Foundations of Mathematics, XVII + 702 p., North-Holland (1989)
19. FOM e-mail list September 1997 to the present. http://www.cs.nyu.edu/mailman/listinfo/fom/
20. Fourman, M.P., Mulvey, C.J., Scott, D.S. (eds.): Applications of Sheaves, Proceedings, Durham, 1977. No. 753 in Lecture Notes in Mathematics, XIV + 779 p., Springer (1979)
21. Fourman, M.P., Scott, D.S.: Sheaves and logic. In: [20], pp. 302–401 (1979)
22. Gandy, R.O., Kreisel, G., Tait, W.W.: Set existence. In: Bulletin de l'Académie Polonaise des Sciences, Série des Sciences Mathématiques, Astronomiques et Physiques, vol. 8, pp. 577–582 (1960)
23. Greenberg, N., Miller, J.S.: Diagonally non-recursive functions and effective Hausdorff dimension. Bull. Lond. Math. Soc. **43**(4), 636–654 (2011)
24. Hudelson, W.M.P.: Mass problems and initial segment complexity. J. Symbolic Logic **79**(1), 20–44 (2014)
25. Jech, T.: Set Theory, XI + 621 p., Academic Press (1978)
26. Jockusch Jr. C.G.: Degrees of functions with no fixed points. In: [18], pp. 191–201 (1989)
27. Jockusch Jr. C.G., Simpson, S.G.: A degree-theoretic definition of the ramified analytical hierarchy. Ann. Math. Logic **10**, 1–32 (1976)
28. Jockusch Jr. C.G., Soare, R.I.: Degrees of members of Π^0_1 classes. Pac. J. Math. **40**, 605–616 (1972)

29. Jockusch Jr. C.G., Soare, R.I.: Π_1^0 classes and degrees of theories. Trans. Am. Math. Soc. **173**, 35–56 (1972)
30. Khan, M., Kjos-Hanssen, B., Miller, J.S.: The Computability Menagerie (2015). http://www.math.wisc.edu/~jmiller/
31. Khan, M., Miller, J.S.: Forcing with bushy trees, 18 p., 30 March 2015. http://arxiv.org/abs/1503.08870v1
32. Kjos-Hanssen, B., Merkle, W., Stephan, F.: Kolmogorov complexity and the recursion theorem. Trans. Am. Math. Soc. **363**, 5465–5480 (2011)
33. Kleene, S.C.: Introduction to Metamathematics, X + 550 p., Van Nostrand (1952)
34. Kleene, S.C., Post, E.L.: The upper semi-lattice of degrees of recursive unsolvability. Ann. Math. **59**, 379–407 (1954)
35. Kolmogoroff, A.: Zur Deutung der intuitionistischen Logik. Math. Z. **35**, 58–65 (1932)
36. Kolmogorov, A.N.: On the interpretation of intuitionistic logic. In: [76], translation of [35] with commentary and additional references, pp. 151–158 and 451–466 (1991)
37. Lerman, M.: Degrees of Unsolvability. Perspectives in Mathematical Logic, XIII + 307 p., Springer, Berlin (1983)
38. Lerman, M.: A Framework for Priority Arguments. Lecture Notes in Logic, Association for Symbolic Logic, XVI + 176 p., Cambridge University Press (2010)
39. Medvedev, Y.T.: Degrees of difficulty of mass problems. Dokl. Akad. Nauk SSSR **104**, 501–504 (1955). in Russian
40. Miller, J.S.: Extracting information is hard: a turing degree of non-integral effective Hausdorff dimension. Adv. Math. **226**, 373–384 (2011)
41. Muchnik, A.A.: On strong and weak reducibilities of algorithmic problems. Sib. Mat. Zh. **4**, 1328–1341 (1963). in Russian
42. Nabutovsky, A.: Einstein structures: existence versus uniqueness. Geom. Funct. Anal. **5**, 76–91 (1995)
43. Nabutovsky, A., Weinberger, S.: Betti numbers of finitely presented groups and very rapidly growing functions. Topology **46**, 211–233 (2007)
44. Nies, A.: Computability and Randomness, XV + 433 p., Oxford University Press (2009)
45. Nies, A., Shore, R.A., Slaman, T.A.: Interpretability and definability in the recursively enumerable degrees. Proc. Lon. Math. Soc. **77**, 241–291 (1998)
46. Post, E.L.: Recursively enumerable sets of positive integers and their decision problems. Bull. Am. Math. Soc. **50**, 284–316 (1944)
47. Rabin, M.O.: Recursive unsolvability of group theoretic problems. Ann. Math. **67**, 172–194 (1958)
48. Richardson, D.: Some undecidable problems involving elementary functions of a real variable. J. Symbolic Logic **33**, 514–520 (1968)
49. Robinson, R.M.: Undecidability and nonperiodicity of tilings of the plane. Inventiones Math. **12**, 177–209 (1971)
50. Rogers Jr. H.: Theory of Recursive Functions and Effective Computability, XIX + 482 p.. MIT Press, Cambridge (1967)
51. Sacks, G.E.: Degrees of Unsolvability. No. 55 in Annals of Mathematics Studies, IX + 174 p., Princeton University Press, London (1963)
52. Sacks, G.E.: The recursively enumerable degrees are dense. Ann. Math. **80**, 300–312 (1964)
53. Sacks, G.E.: Higher Recursion Theory. Perspectives in Mathematical Logic, XV + 344 p., Springer (1990)
54. Scott, D.S.: Algebras of sets binumerable in complete extensions of arithmetic. In: [15], pp. 117–121 (1962)

55. Scott, D.S., Tennenbaum, S.: On the degrees of complete extensions of arithmetic (abstract). Not. Am. Math. Soc. **7**, 242–243 (1960)
56. Shafer, P.: Coding true arithmetic in the Medvedev and Muchnik degrees. J. Symbolic Logic **76**(1), 267–288 (2011)
57. Shoenfield, J.R.: Degrees of models. J. Symbolic Logic **25**, 233–237 (1960)
58. Shoenfield, J.R.: Degrees of Unsolvability. No. 2 in North-Holland Mathematics Studies, VIII + 111 p., North-Holland (1971)
59. Shore, R.A.: The Turing degrees: an introduction, IV + 77 p. (2012). http://www.math.cornell.edu/~shore/
60. Simpson, S.G.: The hierarchy based on the jump operator. In: [4], pp. 267–276 (1980)
61. Simpson, S.G.: FOM: natural r.e. degrees; Pi01 classes. FOM e-mail list [19], 13 August 1999
62. Simpson, S.G.: Subsystems of Second Order Arithmetic. Perspectives in Mathematical Logic, pp. XIV + 445, Second Edition, Springer (1999), Perspectives in Logic, Association for Symbolic Logic, XVI+ 444 p., Cambridge University Press (2009)
63. Simpson, S.G.: FOM: natural r.e. degrees. FOM e-mail list [19], 27 February 2005
64. Simpson, S.G.: Mass problems and randomness. Bull. Symbolic Logic **11**, 1–27 (2005)
65. Simpson, S.G.: An extension of the recursively enumerable turing degrees. J. Lond. Math. Soc. **75**(2), 287–297 (2007)
66. Simpson, S.G.: Mass problems and almost everywhere domination. Math. Logic Quarterly **53**, 483–492 (2007)
67. Simpson, S.G.: Mass problems and intuitionism. Notre Dame J. Formal Logic **49**, 127–136 (2008)
68. Simpson, S.G.: Mass problems and measure-theoretic regularity. Bull. Symbolic Logic **15**, 385–409 (2009)
69. Simpson, S.G.: Mass problems associated with effectively closed sets. Tohoku Math. J. **63**(4), 489–517 (2011)
70. Simpson, S.G.: Medvedev degrees of 2-dimensional subshifts of finite type. Ergodic Theor. Dyn. Syst. **34**(2), 665–674 (2014). doi:10.1017/etds.2012.152
71. Slaman, T.A.: Global properties of the turing degrees and the turing jump. In: [11], pp. 83–101 (2008)
72. Soare, R.I.: Recursively Enumerable Sets and Degrees. Perspectives in Mathematical Logic, XVIII + 437 p., Springer (1987)
73. Soare, R.I.: Computability and recursion. Bull. Symbolic Logic **2**, 284–321 (1996)
74. Sorbi, A., Terwijn, S.A.: Intuitionistic logic and Muchnik degrees. Algebra Univers. **67**, 175–188 (2012). doi:10.1007/s00012-012-0176-1
75. Tarski, A., Mostowski, A., Robinson, R.M.: Undecidable Theories. Studies in Logic and the Foundations of Mathematics, IX + 98 p., North-Holland (1953)
76. Tikhomirov, V.M. (ed.): Selected Works of A.N. Kolmogorov, vol. I, Mathematics and Mechanics. Mathematics and its Applications, Soviet Series, XIX + 551 p., Kluwer Academic Publishers (1991)
77. Turing, A.M.: On computable numbers, with an application to the Entscheidungsproblem. Proc. Lond. Math. Soc. **42**, 230–265 (1936)

Universality in Molecular and Cellular Computing

Sergey Verlan[1,2](✉)

[1] Laboratoire d'Algorithmique, Complexité et Logique, Université Paris Est – Créteil Val de Marne, 61, av. gén. de Gaulle, 94010 Créteil, France
[2] Institute of Mathematics and Computer Science, Academy of Sciences of Moldova, Academiei 5, 2028 Chisinau, Moldova
verlan@u-pec.fr

Abstract. In this article we present an overview of the study of the universality problem in the area of molecular and cellular computing. We consider the results that deal explicitly with this problem and that aim to optimize the obtained construction. A particular attention is given to models based on the splicing operation as well as to multiset-rewriting based models.

1 Introduction

The concept of universality was first formulated by A. Turing in [56]. He constructed a universal (Turing) machine capable of simulating the computation of any other (Turing) machine. This universal machine takes as input a description of the machine to simulate, the contents of its input tape, and computes the result of its execution on the given input.

More generally, the universality problem for a class of computing devices (or functions) $\mathfrak{C}$ consists in finding a fixed element $\mathcal{M}$ of $\mathfrak{C}$ able to simulate the computation of any element $\mathcal{M}'$ of $\mathfrak{C}$ using an appropriate fixed encoding. More precisely, if $\mathcal{M}'$ computes y on an input x (we will write this as $\mathcal{M}'(x) = y$), then $\mathcal{M}'(x) = f(\mathcal{M}(\langle g(\mathcal{M}'), h(x)\rangle))$, where h and f are the encoding and decoding functions, respectively, g is the function retrieving the number of $\mathcal{M}'$ in some fixed enumeration of $\mathfrak{C}$ and $\langle\rangle$ is a coding function for couples, e.g. the Cantor coding. We note that in the case of generating devices, e.g. formal grammars, the universality concept is slightly different as the input is empty. However, the universal element $\mathcal{M}$ has an input, which is just the code of the element to be simulated. Another solution is to define the computation in the generating framework by requiring the generated result to be singleton for any input.

We will use the terminology considered by Korec [26] and call a construction *strongly* universal if the encoding and decoding functions are identities, otherwise the corresponding construction will be called (*weakly*) universal. Some authors [26,30] implicitly consider only the strong notion of universality as the encoding and decoding functions can perform quite complicated transformations, which are not necessarily doable in the original devices [5,52]. We refer to [26] for

A. Beckmann et al. (Eds.): CiE 2015, LNCS 9136, pp. 95–104, 2015.
DOI: 10.1007/978-3-319-20028-6_10

a more detailed discussion of different variants of the universality and to [33,39] for a survey on this topic.

Let us stress here an important distinction between *computational completeness* and *universality* as in the literature the latter term is often wrongly used to denote the computational completeness. Given a class $\mathfrak{C}$ of computability models, we say that $\mathfrak{C}$ is computationally complete if the devices in $\mathfrak{C}$ can characterize the power of Turing machines (or of any other type of equivalent devices). This means that given a Turing machine M one can find an element C in $\mathfrak{C}$ such that C is equivalent with M. Thus, completeness refers to the capacity of covering the level of computability – in grammatical terms, this means to generate all recursively enumerable languages (RE). Universality is an internal property of $\mathfrak{C}$ and it means the existence of a fixed element $\mathcal{M}$ of $\mathfrak{C}$ which is able to simulate any given element $\mathcal{M}'$ (of $\mathfrak{C}$), providing that an appropriate encoding of $\mathcal{M}'$ and of the input is given. If $\mathfrak{C}$ does not have a universal element, then there is a family $\mathfrak{C}' \supset \mathfrak{C}$ that contains an element $\mathcal{M}$ universal for $\mathfrak{C}$. We note that $\mathcal{M}$ is not necessarily universal for RE.

It is clear that if a family of computability models $\mathfrak{C}$ is computationally complete, then it has an infinite number of universal elements. For example, such an element U can be the equivalent in $\mathfrak{C}$ of some universal Turing machine $\mathcal{M}_U$. Then in order to simulate an element $C \in \mathfrak{C}$ it is sufficient to simulate by $\mathcal{M}_U$ (and hence by U) the Turing machine $\mathcal{M}_C$ that is equivalent to C.

We would like to highlight another notion of universality, the *intrinsic* universality, very common in the field of cellular automata. It arises as an observation of the fact that it is not trivial to properly define the above (Turing) universality for devices working without halting, input and output on infinite configurations. A more natural notion in this case is to construct a cellular automata able to simulate the work of any other automata in a shift-invariant and time-invariant way. This is a stronger notion than the (Turing) universality as it corresponds to a bisimulation with a fixed number of steps (and within a fixed space) [50]. For the class of computationally complete devices intrinsic universality implies (Turing) universality, while the converse is not true [40].

It is interesting to find a universal element that has a small descriptional complexity giving in this way an upper bound on the required complexity of the universality construction. In 1956 Shannon [54] considered the question of finding the smallest possible universal Turing machine where the size is calculated as the number of states and symbols. In the early sixties, Minsky and Watanabe had a running competition to see who could find the smallest universal Turing machine [37,59]. Later, Rogozhin showed the construction of several small universal Turing machines [49]. An overview of recent results on this topic can be found in [39]. Other computational models were also considered, e.g. cellular automata [58] with a construction of universal cellular automata of rather small size; see [41,61] for an overview.

Small universal devices have mostly theoretical importance as they demonstrate the minimal ingredients needed to achieve a complex (universal) computation. Their construction is a long-standing and fascinating challenge involving a lot of interconnections between different models, constructions, and encodings.

In this paper we overview the results related to the universality problem for different models in molecular and cellular computing areas. The area of molecular (DNA) computing is a recent and quickly growing field of computer science which centers on the study of the operations and processes issued from the biochemical reactions involving DNA molecules [46,50]. The cellular computing (also called membrane computing or P systems) gets its inspiration from the functioning and the structure of a living cell [44,45,47]. We consider only the results explicitly building universal devices, because as we mentioned above the computational completeness already implies the existence of universal devices, which can then be trivially constructed. Hence, we suppose that explicit constructions are somehow optimized in terms of the descriptional complexity and in number of rules, in particular. We mainly focus on two models: (controlled) splicing (from DNA computing) and (distributed) multiset rewriting (from cellular computing). These models have undergone heavy theoretical investigations and present most of the small universality constructions from the the two areas.

2 Models in Molecular Computing

In this section we shall give a brief discussion on the origin of models in the area of molecular computing.

The first direction explores the representation of DNA or RNA sequences as strings and the corresponding biological operations as operations on strings, eventually making abstraction of some technical details. This brings the possibility of a theoretical study of these operations in the framework of the formal language theory. Hence, the studied questions and methods are intrinsically related to that area. As examples of this approach we can cite splicing-based models, related to the operation of DNA recombination, and insertion-deletion based models, related to the mismatched annealing of DNA and RNA sequences [46]. In most of the cases closure properties are considered, however, like in the case of regulated rewriting [51], it is possible to consider additional external control mechanisms like graph, matrix, random-context time-varying and distributed controls. We note that, traditionally, most of the models consider the generation of a language, like formal grammars. This implies that in order to consider universality constructions proper notions of input and output should be defined. We will consider that the input is finite and is given as part of the axiom(s) of the system, while in order to obtain the output we require that the corresponding systems generate a singleton language. The universality constructions discussed below fit this scheme, however it should be clear that there are other ways to define the input and the output.

The second modeling possibility relies on the annealing of some specially designed DNA structures (tiles) that produce 2-D and 3-D shapes by self-assembly. In this case the computation is performed by self-assembly and its result is a shape that encodes the result, in a way similar to the cellular automata computations. The well known example of this approach is the tile assembly model (TAM) developed by Winfree [60]. In this framework only the intrinsic universality was explicitly investigated [13]. We also note that it was shown that one tile is sufficient to obtain this result [12], so in some sense this closes the quest for "small" universal systems with respect to the number of tiles. Since the TAM is computationally complete [60], it contains universal elements. Many papers on self-assembly, e.g. [28,43,55], give constructions for Turing machine simulation by different variants of TAM, however since the universality problem is not directly addressed it is extremely difficult to compute the parameters of the universal elements obtained by given constructions.

In the third case the modeling is performed at higher functional levels: DNA strings are considered as signals that can trigger more complex actions using more sophisticated machinery, naturally leading to the construction of boolean gates and boolean networks. Because of the natural restrictions circuits using dual-rail encoding are considered. A typical example of this approach are the models from [25,53]. As far as we know, the universality problem was not considered in this framework yet, although there exist models able to simulate the work of a Turing machine [48].

3 Splicing Based Models

The splicing operation was first considered by T. Head [18,19] and it is conceptually different from the (string) rewriting: its main difference is being a binary operation. Splicing considers some specified context in each of two strings entering the operation and performs a crossover of the two strings at the context location [24,46,62]. Hence, the methods used for universality constructions are substantially different as a parallel evolution of a set of strings (instead of a single string) should be considered as well as other specificities related to the fact that splicing systems are a language generating device. This operation is very powerful allowing the construction of extremely small universal devices competing with small universal Turing machines. Since the language generated by any H system is regular [45], additional control mechanisms should be used to achieve the computational completeness.

Double Splicing. The double splicing operation is in some sense a counterpart of matrix grammars [51] for splicing systems. However, instead of sequences of prescribed rules the double splicing operation is composed of two splicings and requires that the result of the first splicing to be the input of the second one. A universal construction using only 5 splicing rules is given in [3].

(E)TVDH Systems. This model is the adaptation of the idea of time-varying grammars [51] to the area of splicing. We recall that time-varying grammars can be seen as graph-controlled grammars where the corresponding control graph has the form of a ring. In the splicing case the definition is slightly different as one allows using words that are simultaneously obtained to perform a single (or multiple in the case of ETVDH systems) splicing at each step.

In the case of TVDH systems of degree 1 the set of rules is applied to the result of the previous application and this corresponds to a simple iteration of the splicing operation. Comparing to ordinary splicing systems this provides a powerful feature that permits to eliminate all strings that are not produced by a splicing operation at the corresponding step. It is somehow surprising that this modification by itself suffices to achieve the computational completeness [34,35]. Article [1] presents constructions for universal TVDH systems of degree 1 and 2 and having 17 and 15 splicing rules, respectively.

Test Tube Systems. One of the particularities of previously discussed models is that the system starts from a single initial set of axioms and at each step the current set of words is replaced by a new one, computed according to the control of the system. It is possible to consider splicing systems based on a different idea. Namely, instead of evolving a single set of words, a fixed number of such sets (a vector) is evolved. This corresponds to a distributed system containing some units that we call *components*. The computation is then divided in two different steps: a *computation* step and a *communication* step. During the computation step splicing rules are applied in each component, independently from each other according to the underlying control. During the communication step the contents of components is redistributed in the system according to a predefined algorithm.

The model considered in this subsection is the splicing counterpart of the parallel communicating grammar systems with communication by command [51] and in the literature it is also known under the name of *splicing test tube* systems. The idea is to use the (closure of the) splicing operation for the computational step in each component and input permitting filters for the communication according to a communication graph [8].

In [3] a construction of a universal test tube system having 8 rules distributed in 3 components is given. Universal constructions with 2 tubes and different filter restrictions are given in [1], but in this case 10 splicing rules are necessary.

Splicing P Systems. Another distribution possibility is to apply a single splicing operation in each component during the computation step and redistribute the contents during the communication step based on the splicing rule that has been applied. With some reserve (because splicing is a binary operation) the model can be considered as graph-controlled splicing. The obtained model, called splicing P systems, was introduced in [44] and more details can be found in [45,47].

It is worth to note that splicing P systems allow one of the smallest universal constructions for the splicing area as well as for string-based universal devices. The construction from [3] has 5 splicing rules, 6 axioms, 3 components, an alphabet of 7 symbols and the diameter (size of rules' contexts) (3,2,2,1).

4 Networks of Evolutionary Processors

Networks of evolutionary processors (NEPs) [7] are an example of distributed systems based on three evolutionary operations: *point mutation* that can replace a symbol by another symbol randomly in the string, *insertion* of a symbol anywhere in the string and *deletion* of a specific symbol. During the computational step the corresponding operation is applied in the nodes once. During the communication step, the contents of a node is checked for an output condition given as a regular language and any words passing this condition are removed from that node. Next, the words that exit a node can enter another connected node if they satisfy its input condition also given by a regular language. The language determined by the system is defined as the set of words which appear at some distinguished node during the computation.

The regular filtering is quite a powerful mechanism that allows to finely control the computation. In [23] the following constructions are given: strongly universal NEPs with 5 rules and weakly universal NEPs with 4 rules (using a unary encoding of integers in NEPs) as well as a universal NEP with 7 rules that is able to efficiently simulate any Turing machine in polynomial time.

It is known that the expressivity of NEPs is depending on the power of the filter, on the communication graph structure and on the allowed set of node operations [2,10,11]. An interesting restriction of NEPs are Accepting Hybrid Networks of Evolutionary Processors (AHNEP) which limit the filters to random-context conditions [32]. Within this model there are obtained several universality results; we cite only [9,31] as well as [29] for the variant of the model using splicing operation. However, from the proofs of these results it is not easy to compute all the parameters of the system, in particular, the number of used rules. We refer to [32] for a more comprehensive survey on this topic.

5 Cellular Computing Based Models

We note that membrane systems with a static structure (that does not change in time) are equivalent to distributed multiset rewriting [15], which in turn can be reduced (using some standard flattening techniques) to the parallel multiset rewriting. In what follows we will use the term and the definitions related to the multiset rewriting, but we would like to note that there is one-to-one correspondence between many models based on this operation, e.g. vector addition systems, Petri nets, population protocols and P systems. So, the constructions presented below can be easily interpreted in terms of any of these models. Moreover, we would like to note that models using multiset rewriting and models using numbers directly (like register machines) are extremely similar, because multiset rewriting permits to represent any numerical mapping. The difference between two representations mostly relies in the encoding and decoding concepts as well as in the descriptional complexity issues.

Multiset Rewriting with Inhibitors. We consider the variant corresponding to forbidding grammars [51] and we note that this model is identical to P systems with symport/antiport and inhibitors [45], to Petri nets with inhibitor arcs [42] and to vector addition systems with zero checks [6,20].

In [21] several universal multiset rewriting systems are constructed exhibiting trade-offs between the cardinality of the alphabet, the number of rules, the total number of used inhibitors and the maximal size of the rule. Deterministic strong and weak universal constructions with following parameters (in the above order) are presented: $(30, 34, 13, 3)$, $(14, 31, 51, 8)$, $(11, 31, 79, 11)$, $(21, 25, 13, 5)$, $(67, 64, 8, 3)$, $(58, 55, 8, 5)$. In [22] the following parameters are obtained for the non-deterministic case: (5, 877, 1022, 729), (5, 1024, 1316, 379), (4, 668, 778, 555), (4, 780, 1002, 299).

Register Machines. Register Machines [38] can be viewed as a particular case of (finite-state) multiset rewriting with inhibitors, where (at most) one state symbol is present in reachable configurations, and the rules have a special reduced form.

Register machines with the smallest number of instructions are reported by Korec [26], where he gives a construction for a strongly universal register machine having 22 instructions and using 8 registers and a weakly universal register machine having 20 instructions, as well as several small machines with other type of instructions. It is known that machines with two registers can only be weakly universal [5,52], while with 3 registers the strong universality can be achieved. The article [22] mentions such constructions having 278 instructions for the case of the 2-register machine and 365 instructions for the case of the 3-register machine.

One can even consider generalized register machines that allow several register decrements and increments by more than one. This is a useful tool for obtaining some small universal systems. In the case of register machines such a generalization permits to have only 13 (decrement with zero test and with add) instructions for the strong universality [14]

Maximally Parallel Multiset Rewriting. The application of rules in the case of the maximally parallel multiset rewriting follows the maximality principle – a set of rules is maximally parallel if the rules are all applicable in parallel and no other rule can be added to this set maintaining this property.

A register machine viewed as a finite-state multiset rewriting system behaves sequentially even under maximal parallelism, because (at most one) state symbol is present in any configuration, and any rule contains a state symbol in its left side. In [4] a maximally parallel multiset rewriting system with 23 rules is given. The construction is based on the simulation of the universal register machine U_{32} from [26] and takes profit of the parallelism to encode efficiently the representation of states.

6 Conclusions

In this paper we gave an overview of the study of the universality problem for several models originated from molecular and cellular computing. As it can be seen, in the field of molecular computing there are almost no results exhibiting small universality constructions, except for splicing-based models. So it is natural to consider the construction of "small" universal devices for other models, like insertion-deletion systems, sticker systems [46] and (AH)NEPs [32].

Another interesting research direction is to give intrinsic universality constructions for aforementioned models. The first attempt in the case of splicing was done in [16,17], but the obtained system is not time-invariant. In the area of insertion-deletion systems there are time invariant bisimulation constructions between several classes of models [27,36,57].

Finally, we would like to note that in the case of genetic circuits the computational completeness is achieved by a simulation of boolean networks with memory. It could be interesting to encode other type of models in order to achieve smaller universal constructions.

References

1. Alhazov, A., Kogler, M., Margenstern, M., Rogozhin, Y., Verlan, S.: Small universal TVDH and test tube systems. Int. J. Found. Comput. Sci. **22**(1), 143–154 (2011)
2. Alhazov, A., Martín-Vide, C., Truthe, B., Dassow, J., Rogozhin, Y.: On networks of evolutionary processors with nodes of two types. Fundamenta Informaticae **91**(1), 1–15 (2009)
3. Alhazov, A., Rogozhin, Y., Verlan, S.: On small universal splicing systems. Int. J. Found. Comput. Sci. **23**(07), 1423–1438 (2012)
4. Alhazov, A., Verlan, S.: Minimization strategies for maximally parallel multiset rewriting systems. Theor. Comput. Sci. **412**(17), 1581–1591 (2011)
5. Barzdin, I.M.: On a class of turing machines (Minsky machines). Algebra i Logika **1**, 42–51 (1963). (in Russian)
6. Bonnet, R.: The reachability problem for vector addition system with one zero-test. In: Murlak, F., Sankowski, P. (eds.) MFCS 2011. LNCS, vol. 6907, pp. 145–157. Springer, Heidelberg (2011)
7. Castellanos, J., Martín-Vide, C., Mitrana, V., Sempere, J.M.: Solving NP-complete problems with networks of evolutionary processors. In: Mira, J., Prieto, A.G. (eds.) IWANN 2001. LNCS, vol. 2084, pp. 621–628. Springer, Heidelberg (2001)
8. Csuhaj-Varjú, E., Kari, L., Păun, G.: Test tube distributed systems based on splicing. Comput. Artif. Intell. **15**(2–3), 211–232 (1996)
9. Dassow, J., Manea, F.: Accepting hybrid networks of evolutionary processors with special topologies and small communication. In: Proceedings of DCFS 2010, of EPTCS, vol. 31, pp. 68–77 (2010)
10. Dassow, J., Manea, F., Truthe, B.: On the power of accepting networks of evolutionary processors with special topologies and random context filters. Fundamenta Informaticae **136**(1–2), 1–35 (2015)
11. Dassow, J., Mitrana, V.: Accepting networks of non-inserting evolutionary processors. In: Priami, C., Back, R.-J., Petre, I. (eds.) Transactions on Computational Systems Biology XI. LNCS, vol. 5750, pp. 187–199. Springer, Heidelberg (2009)
12. Demaine, E.D., Demaine, M.L., Fekete, S.P., Patitz, M.J., Schweller, R.T., Winslow, A., Woods, D.: One tile to rule them all: simulating any tile assembly system with a single universal tile. In: Esparza, J., Fraigniaud, P., Husfeldt, T., Koutsoupias, E. (eds.) ICALP 2014. LNCS, vol. 8572, pp. 368–379. Springer, Heidelberg (2014)
13. Doty, D., Lutz, J.H., Patitz, M.J., Schweller, R.T., Summers, S.M., Woods, D.: The tile assembly model is intrinsically universal. In: Proceedings of the 53rd Annual IEEE Symposium on Foundations of Computer Science, pp. 302–310 (2012)
14. Freund, R., Oswald, M.: A small universal antiport P system with forbidden context. In: Leung, H., Pighizzini, G. (eds.) 8th International Workshop on Descriptional Complexity of Formal Systems, pp. 259–266. Proceedings, New Mexico (2006)
15. Freund, R., Verlan, S.: A formal framework for static (Tissue) P systems. In: Eleftherakis, G., Kefalas, P., Păun, G., Rozenberg, G., Salomaa, A. (eds.) WMC 2007. LNCS, vol. 4860, pp. 271–284. Springer, Heidelberg (2007)

16. Frisco, P.: Direct constructions of universal extended H systems. Theor. Comput. Sci. **296**(2), 269–293 (2003)
17. Frisco, P., Hoogeboom, H.J., Sant, P.: A direct construction of a universal P system. Fundamenta Informaticae **49**(1–3), 103–122 (2002)
18. Head, T.: Formal language theory and DNA: an analysis of the generative capacity of specific recombinant behaviors. Bull. Math. Biol. **49**(6), 737–759 (1987)
19. Head, T.: Splicing languages generated with one sided context. In: Paun, G. (ed.) Computing with Bio-Molecules. Theory and Experiments, pp. 158–181. Springer, Singapore (1998)
20. Hopcroft, J., Pansiot, J.-J.: On the reachability problem for 5-dimensional vector addition systems. Theor. Comput. Sci. **8**(2), 135–159 (1979)
21. Ivanov, S., Pelz, E., Verlan, S.: Small universal Petri nets with inhibitor arcs (2013). arXiv, CoRR. abs/1312.4414
22. Ivanov, S., Pelz, E., Verlan, S.: Small universal non-deterministic Petri nets with inhibitor arcs. In: Jürgensen, H., Karhumäki, J., Okhotin, A. (eds.) DCFS 2014. LNCS, vol. 8614, pp. 186–197. Springer, Heidelberg (2014)
23. Ivanov, S., Rogozhin, Y., Verlan, S.: Small universal networks of evolutionary processors. J. Autom. Lang. Comb. **19**(1—-4), 133–144 (2014)
24. Kari, L.: DNA computing: arrival of biological mathematics. Math. Intell. **19**(2), 9–22 (1997)
25. Kim, J., White, K.S., Winfree, E.: Construction of an in vitro bistable circuit from synthetic transcriptional switches. Mol. Syst. Biol. **2**, 68 (2006)
26. Korec, I.: Small universal register machines. Theor. Comput. Sci. **168**(2), 267–301 (1996)
27. Krassovitskiy, A., Rogozhin, Y., Verlan, S.: Further results on insertion-deletion systems with one-sided contexts. In: Martín-Vide, C., Otto, F., Fernau, H. (eds.) LATA 2008. LNCS, vol. 5196, pp. 333–344. Springer, Heidelberg (2008)
28. Lathrop, J.I., Lutz, J.H., Patitz, M.J., Summers, S.M.: Computability and complexity in self-assembly. Theory Comput. Syst. **48**(3), 617–647 (2011)
29. Loos, R., Manea, F., Mitrana, V.: On small, reduced, and fast universal accepting networks of splicing processors. Theor. Comp. Sci. **410**(45), 406–416 (2009)
30. Malcev, A.I.: Algorithms and Recursive Functions. Wolters-Noordhoff, Groningen (1970)
31. Manea, F., Martín-Vide, C., Mitrana, V.: On the size complexity of universal accepting hybrid networks of evolutionary processors. Math. Struct. Comput. Sci. **8**, 17:753–771 (2007)
32. Manea, F., Martín-Vide, C., Mitrana, V.: Accepting Networks of Evolutionary Word and Picture Processors: A Survey, pp. 525–561. Imperial College Press, London (2010). Chap. 10
33. Margenstern, M.: Frontier between decidability and undecidability: a survey. Theor. Comput. Sci. **231**(2), 217–251 (2000)
34. Margenstern, M., Rogozhin, Y.: Time-varying distributed H systems of degree 1 generate all recursively enumerable languages. In: Ito, M., Păun, G., Yu, S. (eds.) Words, Semigroups, and Transductions, pp. 329–339. World Scientific, Singapore (2001)
35. Margenstern, M., Rogozhin, Y., Verlan, S.: Time-varying distributed H systems with parallel computations: the problem is solved. In: Chen, J., Reif, J.H. (eds.) DNA 2003. LNCS, vol. 2943, pp. 48–53. Springer, Heidelberg (2004)
36. Matveevici, A., Rogozhin, Y., Verlan, S.: Insertion-deletion systems with one-sided contexts. In: Durand-Lose, J., Margenstern, M. (eds.) MCU 2007. LNCS, vol. 4664, pp. 205–217. Springer, Heidelberg (2007)

37. Minsky, M.: Size and structure of universal Turing machines using tag systems. In: Recursive Function Theory: Proceedings, Symposium in Pure Mathematics, vo. 5, pp. 229–238. Provelence (1962)
38. Minsky, M.: Computations: Finite and Infinite Machines. Prentice Hall, USA (1967)
39. Neary, T., Woods, D.: The complexity of small universal turing machines: a survey. In: Bieliková, M., Friedrich, G., Gottlob, G., Katzenbeisser, S., Turán, G. (eds.) SOFSEM 2012. LNCS, vol. 7147, pp. 385–405. Springer, Heidelberg (2012)
40. Ollinger, N.: Automates cellulaires: structures. Ph.D. thesis, ENS Lyon (2002)
41. Ollinger, N.: The quest for small universal cellular automata. In: Widmayer, P., Triguero, F., Morales, R., Hennessy, M., Eidenbenz, S., Conejo, R. (eds.) ICALP 2002. LNCS, vol. 2380, pp. 318–329. Springer, Heidelberg (2002)
42. Patil, S.S.: Coordination of asynchronous events. Ph.D. thesis, MIT (1970)
43. Patitz, M.J., Summers, S.M.: Self-assembly of decidable sets. In: Calude, C.S., Costa, J.F., Freund, R., Oswald, M., Rozenberg, G. (eds.) UC 2008. LNCS, vol. 5204, pp. 206–219. Springer, Heidelberg (2008)
44. Păun, G.: Computing with membranes. J. Comput. Syst. Sci. **1**(61), 108–143 (2000). Also TUCS Report No. 208, 1998
45. Păun, G.: Membrane Computing: An Introduction. Springer, Heidelberg (2002)
46. Păun, G., Rozenberg, G., Salomaa, A.: DNA Computing: New Computing Paradigms. Springer, Heidelberg (1998)
47. Păun, G., Rozenberg, G., Salomaa, A.: The Oxford Handbook Of Membrane Computing. Oxford University Press, New York (2009)
48. Qian, L., Soloveichik, D., Winfree, E.: Efficient turing-universal computation with DNA polymers. In: Sakakibara, Y., Mi, Y. (eds.) DNA 16 2010. LNCS, vol. 6518, pp. 123–140. Springer, Heidelberg (2011)
49. Rogozhin, Y.: Small universal turing machines. Theor. Comput. Sci. **168**(2), 215–240 (1996)
50. Rozenberg, G., Bäck, T., Kok, J.N. (eds.): Handbook of Natural Computing. Springer, Heidelberg (2012)
51. Rozenberg, G., Salomaa, A.: Handbook of Formal Languages. Springer, Heidelberg (1997)
52. Schroeppel, R.: A two counter machine cannot calculate 2N. AI Memos, Cambridge (1972)
53. Seelig, G., Soloveichik, D., Zhang, D.Y., Winfree, E.: Enzyme-free nucleic acid logic circuits. Science **314**(5805), 1585–1588 (2006)
54. Shannon, C.E.: A universal Turing machine with two internal states. Autom. Stud. Ann. Math. Stud. **34**, 157–165 (1956)
55. Soloveichik, D., Winfree, E.: Complexity of self-assembled shapes. SIAM J. Comput. **36**(6), 1544–1569 (2007)
56. Turing, A.M.: On computable numbers, with an application to the Entscheidungsproblem. Proc. Lond. Math. Soc. **42**(2), 230–265 (1936)
57. Verlan, S.: Study of language-theoretic computational paradigms inspired by biology. Habilitation thesis, Université Paris Est (2010)
58. von Neumann, J.: Theory of self-reproducing automata. University of Illinois (1966)
59. Watanabe, S.: 5-symbol 8-state and 5-symbol 6-state universal turing machines. J. ACM **8**(4), 476–483 (1961)
60. Winfree, E.: Algorithmic self-assembly of DNA. Ph.D. thesis, Caltech (1998)
61. Wolfram, S.: A New Kind of Science. Wolfram Media Inc., UK (2002)
62. Zizza, R.: Splicing systems. Scholarpedia **5**(7), 9397 (2010)

Contributed Papers

Some Results on Interactive Proofs for Real Computations

Martijn Baartse and Klaus Meer(✉)

Computer Science Institute, BTU Cottbus-Senftenberg,
Platz der Deutschen Einheit 1, 03046 Cottbus, Germany
baartse@tu-cottbus.de, meer@b-tu.de

Abstract. We study interactive proofs in the framework of real number complexity theory as introduced by Blum, Shub, and Smale. Shamir's famous result characterizes the class IP as PSPACE or, equivalently, as PAT and PAR in the Turing model. Since space resources alone are known not to make much sense in real number computations the question arises whether IP can be similarly characterized by one of the latter classes. Ivanov and de Rougemont [9] started this line of research showing that an analogue of Shamir's result holds in the additive Blum-Shub-Smale model of computation when only Boolean messages can be exchanged. Here, we introduce interactive proofs in the full BSS model. As main result we prove an upper bound for the class $IP_{\mathbb{R}}$. It gives rise to the conjecture that a characterization of $IP_{\mathbb{R}}$ will not be given via one of the real complexity classes $PAR_{\mathbb{R}}$ or $PAT_{\mathbb{R}}$. We report on ongoing approaches to prove as well interesting lower bounds for $IP_{\mathbb{R}}$.

1 Introduction

One inspiring source of research problems to deal with in the framework of real number computations [4] is the question, whether important results from Turing complexity theory hold as well over the reals as underlying structure. And in case they do what does it need to prove them. Beside the importance for the respective computational model this might as well shed light on a better understanding what is the intrinsic reason for a result to hold. Examples are provided by the huge amount of research on quantifier elimination algorithms yielding decidability of all problems in $NP_{\mathbb{R}}$, a recent analogue of Toda's theorem [3], or the proof of a real version of the classical PCP theorem [2], to mention only a few.

Along the same lines in [9] the authors introduced interactive proofs in the framework of real number complexity theory. More precisely, they considered the additive version of the BSS model, see [4] and interaction is restricted to exchange boolean messages only. In this setting, IP again can be characterized via parallel polynomial time.

K. Meer—Both authors were supported under projects ME 1424/7-1 and ME 1424/7-2 by the Deutsche Forschungsgemeinschaft DFG. We gratefully acknowledge the support.

A. Beckmann et al. (Eds.): CiE 2015, LNCS 9136, pp. 107–116, 2015.
DOI: 10.1007/978-3-319-20028-6_11

It is thus natural to extend the definition of interactive proofs to the full BSS model by allowing real messages to be exchanged and the verifier to use multiplications as well. By a folklore result in real number complexity [11] each real decision problem can be decided in linear space. Thus the class $IP_{\mathbb{R}}$ of real decision problems acceptable by such an interactive protocol cannot meaningfully be characterized using space resources alone. As with other real complexity classes that classically correspond to space classes the question then is whether and which other characterizations hold. Note that the two classes PAR and PAT of problems decidable in parallel and in alternating polynomial time, respectively, equal PSPACE in the Turing model. However, for their real counterparts $PAR_{\mathbb{R}}, PAT_{\mathbb{R}}$ it is known that the first is a proper subset of the latter [7]. So the question is whether one of them (and which) equals $IP_{\mathbb{R}}$?

It has been shown in [9] that $PAR_{\mathbb{R}}$ is different from $IP_{\mathbb{R}}$; there are problems in $IP_{\mathbb{R}}$ that cannot be solved in parallel polynomial time. A reason for this is that $PAR_{\mathbb{R}}$ is too weak to capture certain real quantifier elimination tasks. One might then expect that the larger class $PAT_{\mathbb{R}}$ is the correct one to capture $IP_{\mathbb{R}}$. However, as our main result shows, this seems unlikely. We establish as an upper bound for $IP_{\mathbb{R}}$ the class $MA\exists_{\mathbb{R}}$. It was introduced in [8], is strictly larger than $PAR_{\mathbb{R}}$ but conjectured to be strictly included in $PAT_{\mathbb{R}}$. After introducing interactive protocols and the real complexity classes important for this paper we prove the upper bound result in Sect. 3. The remaining part of the paper then discusses approaches to achieve good lower bounds for $IP_{\mathbb{R}}$ as well. This, however, seems to be much harder and we only report on some problems for which Shamir's classical technique can be extended to design real protocols as well. We try to substantiate why the lower bound problem in the full model seems harder by re-analyzing the result of Ivanov and de Rougemont using counting problems. This viewpoint leads to a couple of interesting open problems which seem important on the way to a full real analogue of Shamir's theorem.

2 Basic Notions

As underlying algorithm model we work in the Blum-Shub-Smale BSS model over $\mathbb{R}$ [4]. Decision problems considered in this model are subsets of $\mathbb{R}^\infty := \bigsqcup_{i\geq 1} \mathbb{R}^i$. The model allows to perform the basic arithmetic operations $+, -, \times$ and test instructions of the form 'is $x \geq 0$?' at unit cost. Below, in addition we allow both the verifier and the prover to exchange real numbers at unit cost.

The prover P is a BSS machine unlimited in computational power. The verifier V is a randomized polynomial time algorithm. It is important to point out that randomization (still) is discrete, i.e., V generates a sequence of random bits $r = (r_1, r_2, \ldots)$ during its computation. Now, the computation proceeds as follows:

- Given an input $x \in \mathbb{R}^n$ of size $|x| = n$ and (some of) the random bits of r the verifier V computes a real $V(x, r) =: w_1 \in \mathbb{R}$ and sends it to P;
- using x and w_1 the prover P sends a real $P(x, w_1) =: p_1 \in \mathbb{R}$ back to V;

- in general, if after i rounds of communication $(w_1, p_1, w_2, \ldots, p_i)$ denotes the information sent forth and back, in round $i+1$ V computes a real $V(x, r, w_1, p_1, \ldots, p_i) =: w_{i+1}$ and sends it to P; P then computes a real $P(x, r, w_1, p_1, \ldots, p_i, w_{i+1}) =: p_{i+1}$ and sends it to V;
- the communication halts after a polynomial number $m = poly(|x|)$ of rounds. Then V computes its final result $V(x, r, w_1, \ldots, p_{m-1}) =: w_m \in \{0,1\}$ representing its decision to reject or accept the input, respectively.

We denote the result of an interaction between V and P on input x and using r as random string by $(P,V)(x,r)$. All computations by V have to be finished in (real) polynomial time; thus, in particular the number of random bits generated as well as the number of rounds is polynomially bounded in the algebraic size $|x|$ of x.

Definition 1. *(a) A language $L \subseteq \mathbb{R}^\infty$ has an interactive protocol if there exists a polynomial time randomized verifier V such that*
(i) if $x \in L$ there exists a prover P such that $\Pr_{r \in \{0,1\}^*} \{(P,V)(x,r) = 1\} = 1$
and
(ii) if $x \notin L$, then for all provers P it holds $\Pr_{r \in \{0,1\}^*} \{(P,V)(x,r) = 1\} \leq \frac{1}{4}$.
Above, the length of r can be polynomially bounded in the length of x.
(b) The class $\mathrm{IP}_\mathbb{R}$ is the class of all decision problems $L \subseteq \mathbb{R}^\infty$ which have an interactive protocol.

The real number complexity class most important for our considerations here was introduced and studied by Cucker and Briquel in [8] and is denoted by $\mathrm{MA}\exists_\mathbb{R}$. Starting point of defining it is the fact that over the reals classes which are classically defined or characterized using space resources turn out to have a more subtle relation among each other than they do classically. Taken alone, space resources have no meaning at all; each decision problem can be decided in linear space using an elementary coding trick [11]. As consequence, for many equivalent characterizations especially of the class PSPACE in classical complexity it is unclear what they should become in the real number framework. Recall that PAR, PSPACE, PAT, and IP, denoting the classes of languages acceptable by parallel polynomial time with exponentially many processors, in polynomial space, in polynomial alternating time, and by interactive proofs, respectively, all are the same in Turing complexity (see the textbook [1] for references and proofs). In contrast, over $\mathbb{R}$ it is known that the real counterparts of the first three classes mentioned above satisfy $\mathrm{PAR}_\mathbb{R} \subsetneq \mathrm{PSPACE}_\mathbb{R} \subseteq \mathrm{PAT}_\mathbb{R}$, where $\mathrm{PSPACE}_\mathbb{R}$ denotes the class of real decision problems decidable by an algorithm using both exponential time and polynomial space[1] and the other two classes are defined by extending the classical definitions straightforwardly, see [4,7]. As a consequence, if a new class like $\mathrm{IP}_\mathbb{R}$ is studied which classically gives yet another characterization of PSPACE via Shamir's famous result [12], it is not at all obvious where

[1] The simultaneous requirement of exponential time and polynomial space excludes the above mentioned coding trick from [11] and makes the definition meaningful.

it has to be located over the reals. The above chain of inclusions also gives the option to define new classes which do not make much sense over finite alphabets. This is precisely where [8] starts by defining classes which can be located between $\mathrm{PSPACE}_{\mathbb{R}}$ and $\mathrm{PAT}_{\mathbb{R}}$. The class $\mathrm{MA}\exists_{\mathbb{R}}$ is one such and turns out to be important for interactive protocols. It at least gives a non-trivial upper bound for $\mathrm{IP}_{\mathbb{R}}$, just indicating the latter to be likely weaker than $\mathrm{PAT}_{\mathbb{R}}$.

Definition 2 ([8]).

(a) *The class* $\mathrm{MA}\exists_{\mathbb{R}}$ *consists of all decision problems* $L \subseteq \mathbb{R}^\infty$ *for which there exists a problem* $B \in \mathrm{P}_{\mathbb{R}}$ *together with a polynomial* p *such that an* $x \in \mathbb{R}^\infty$ *belongs to* L *if and only if the following formula holds:* $\forall_B z_1 \exists_{\mathbb{R}} y_1 \ldots \forall_B z_{p(|x|)} \exists_{\mathbb{R}} y_{p(|x|)} (x, y, z) \in B$. *The subscripts* $B, \mathbb{R}$ *for quantifiers indicate whether a quantified variable ranges over* $B := \{0, 1\}$ *or* $\mathbb{R}$*, respectively.*

(b) *The class* $\mathrm{MA}\forall_{\mathbb{R}}$ *consists of the complements of problems in* $\mathrm{MA}\exists_{\mathbb{R}}$*. Thus, a language* $L \in \mathrm{MA}\forall_{\mathbb{R}}$ *contains precisely the points* x *for which a formula of the following form holds:* $\exists_B z_1 \forall_{\mathbb{R}} y_1 \ldots \exists_B z_{p(|x|)} \forall_{\mathbb{R}} y_{p(|x|)} (x, y, z) \in B$.

The following easy lemma is used later on.

Lemma 1. *Let* $\tilde{B} \in \mathrm{MA}\exists_{\mathbb{R}}$ *and* p *be a polynomial. Then the set* $L \subseteq \mathbb{R}^\infty$ *of all* x *such that* $\forall_B y_1 \exists_{\mathbb{R}} z_1 \ldots \forall_B y_{p(|x|)} \exists_{\mathbb{R}} z_{p(|x|)} (x, y, z) \in \tilde{B}$ *belongs to* $\mathrm{MA}\exists_{\mathbb{R}}$*, i.e., if in the definition of* $\mathrm{MA}\exists_{\mathbb{R}}$ *condition* $B \in \mathrm{P}_{\mathbb{R}}$ *is replaced by* $B \in \mathrm{MA}\exists_{\mathbb{R}}$ *we stay within* $\mathrm{MA}\exists_{\mathbb{R}}$[2].

Proof. The quantifier structure related to the definition of $\mathrm{MA}\exists_{\mathbb{R}}$ guarantees the existence of a polynomial q and a problem $B \in \mathrm{P}_{\mathbb{R}}$ such that $(x, y, z) \in \tilde{B} \iff \forall_B s_1 \exists_{\mathbb{R}} t_1 \forall_B s_2 \ldots \forall_B s_{q(|(x,y,z)|)} \exists_{\mathbb{R}} t_{q(|(x,y,z)|)} (x, y, z, s, t) \in B$. So $x \in L$ iff $\forall_B y_1 \exists_{\mathbb{R}} z_1 \ldots \forall_B y_{p(|x|)} \exists_{\mathbb{R}} z_{p(|x|)} \forall_B s_1 \exists_{\mathbb{R}} t_1 \forall_B s_2 \ldots \forall_B s_{q(|(x,y,z)|)} \exists_{\mathbb{R}} t_{q(|(x,y,z)|)} (x, y, z, s, t) \in B$. Since the lengths of both y and z are polynomially bounded in $|x|$ this holds as well for the lengths of s and t. Thus, L has the required representation. □

One of the main results in [8] is proving the inclusion of $\mathrm{PSPACE}_{\mathbb{R}}$ both in $\mathrm{MA}\exists_{\mathbb{R}}$ and $\mathrm{MA}\forall_{\mathbb{R}}$, which in turn implies the strict inclusion of $\mathrm{PAR}_{\mathbb{R}}$ in both latter classes.

Theorem 1 ([8]). $\mathrm{PAR}_{\mathbb{R}} \subsetneq \mathrm{PSPACE}_{\mathbb{R}} \subseteq \mathrm{MA}\exists_{\mathbb{R}} \cap \mathrm{MA}\forall_{\mathbb{R}}$ *and* $\mathrm{MA}\exists_{\mathbb{R}} \cup \mathrm{MA}\forall_{\mathbb{R}} \subseteq \mathrm{PAT}_{\mathbb{R}}$.

Above, the second containment is trivial because by definition of $\mathrm{PAT}_{\mathbb{R}}$ the alternating quantifiers in a defining formula all range over $\mathbb{R}$. The relation between $\mathrm{MA}\exists_{\mathbb{R}}$ and $\mathrm{MA}\forall_{\mathbb{R}}$ currently is unknown as is the question whether one of the two is strictly contained in $\mathrm{PAT}_{\mathbb{R}}$. In the next section we show our main result: $\mathrm{MA}\exists_{\mathbb{R}}$ is an upper bound for $\mathrm{IP}_{\mathbb{R}}$.

2 This of course only makes sense after $\mathrm{MA}\exists_{\mathbb{R}}$ has been defined precisely.

3 Upper and Lower Bounds for $IP_{\mathbb{R}}$

In this section we prove the main result of this paper, an upper bound for the class $IP_{\mathbb{R}}$. In addition, we deal with a few examples for which an interactive proof can be designed. Unfortunately, our results are far from a characterization of $IP_{\mathbb{R}}$ like Shamir's one in the Turing model. Nevertheless, even the bounds presented here seem not at all obvious and require some efforts. As we shall see, a main obstacle for getting better lower bounds is the fact that now we deal with an uncountable space of information underlying the communications. We shall comment on the results at the end.

3.1 Upper Bound: Recursive Evaluation of Verifier Action

We want to show the inclusion $IP_{\mathbb{R}} \subseteq MA\exists_{\mathbb{R}}$. The proof is based on combining a result in [9] with parts of the proof of Theorem 1 adjusted accordingly. The former gives a recursive procedure for computing the number of random vectors under which a verifier accepts an interactive protocol; in the setting of [9] the procedure runs in additive parallel polynomial time. In our setting, however, due to the fact that in particular the prover can send reals the corresponding procedure is not bounded by $PAR_{\mathbb{R}}$, as has been shown in [9], see also comments below. Instead, dealing with the real information sent introduces real quantifier elimination as part of the task to compute the number of successful random vectors. This naturally leads to considering the class $MA\exists_{\mathbb{R}}$.

To start with let a verifier V and a prover P be given such that a language $L \subseteq \mathbb{R}^\infty$ has an interactive protocol established by (P, V). In order to decide for an input x whether $x \in L$ it is sufficient to count the number of random strings that cause V to accept x and check whether this number is larger than $\frac{1}{2}$ of the strings. We want to show that this task can be accomplished within class $MA\exists_{\mathbb{R}}$. Towards this aim below we recall a recursive procedure in [9] for doing this counting and adapt it to our framework. Note that very similar results were already used in the classical discrete setting.

For technical reasons in the proofs below we first need the notion of a normalized real protocol.

Definition 3. *(a) A real protocol (P, V) is called normalized if the following conditions are satisfied:*

- *for input $x \in \mathbb{R}^n, n \in \mathbb{N}$ there are precisely $p(n)$ rounds of communication for a fixed polynomial p only depending on V;*
- *between receiving a real from P and sending a real back to P the verifier V generates precisely one random bit.*

(b) If in a normalized protocol the verifier generates '0' as random bit in round i and then computes a real w_i as the one to be sent to the prover next, we denote it by $w_i(0)$; similarly for $w_i(1)$ if a '1' was generated.

Notational Convention: For a normalized verifier round i starts with generating random bit r_i, then V computes deterministically $w_i \in \mathbb{R}$ and sends it to P in order to receive a $p_i \in \mathbb{R}$. We shall reflect this order of information flow also below in the arguments of some important functions.

It is easy to see that without loss of generality we can assume each protocol to be normalized.

Definition 4 (cf. [9]). *Let (P, V) be a normalized real protocol. For an input $x \in \mathbb{R}^\infty$ suppose the interaction to last $m = poly(|x|)$ rounds. Let $r = (r_1, \ldots, r_m) \in \{0,1\}^*$ be a sequence of random bits V generates during its computation and let $1 \le j \le m$.*

(a) For $w_1, \ldots, w_j, p_1, \ldots, p_j \in \mathbb{R}$ denote by $Pass(x, r_1, w_1, p_1, r_2, \ldots, r_j, w_j)$ the relation expressing that for j rounds, given x as input and r_i as random bit in round i the w_i's are the correct data sent by V if P answers with p_i for $1 \le i \le j$. Thus, $Pass(x, r_1, w_1, p_1, r_2, \ldots, r_j, w_j) \Leftrightarrow V(x, r_1, w_1, p_1, \ldots, w_{i-1}, p_{i-1}, r_i) = w_i$ for all $1 \le i \le j$.

(b) Define functions Q_j, W_j as follows: $Q_j(x, w_1, p_1, w_2, \ldots, p_{j-1}) :=$ max $|\{r \in \{0,1\}^m \mid Pass(x, r_1, w_1, p_1, \ldots, r_{j-1}, w_{j-1}) \wedge w_m = 1\}|$ and $W_j(x, w_1, p_1, w_2, \ldots, p_{j-1}, w_j) :=$ max $|\{r \in \{0,1\}^m \mid Pass(x, r_1, w_1, p_1, \ldots, p_{j-1}, r_j, w_j) \wedge w_m = 1\}|$. In both cases the max *is taken over all provers that give responses $p_1, \ldots, p_{j-1} \in \mathbb{R}$ as answers to questions $w_1, \ldots, w_{j-1}$ sent by V on input x.*

The only difference between this definition and the corresponding one in [9] is that the w_i and p_i are reals. However, this makes the algorithm behind the next statement more difficult because additional existential quantifiers taking care of the role of the p_i's are needed. Because we deal with normalized protocols the influence of the w_i's now being reals can be controlled without additional quantifiers ranging over $\mathbb{R}$. They basically are determined by the actually generated (discrete) random bit and the previous information of the protocol. This finally is the reason why $\text{MA}\exists_{\mathbb{R}}$ plays an important role.

Lemma 2 (cf. [9], Adapted to Normalized Protocols). *Let $x \in \mathbb{R}^\infty$ and (P, V) be a normalized real protocol accepting a language L. Let m be the number of random bits generated by V during computation on x. Then for all $1 \le j \le m - 1$*

(a) x is accepted iff $Q_1(x) > 2^{m-1}$;
(b) $Q_j(x, w_1, \ldots, p_{j-1}) = W_j(x, w_1, \ldots, p_{j-1}, w_j(0)) + W_j(x, w_1, \ldots, p_{j-1}, w_j(1))$;
(c) $W_j(x, w_1, \ldots, p_{j-1}, w_j) = \max_{p_j \in \mathbb{R}} Q_{j+1}(x, w_1, \ldots, p_{j-1}, w_j, p_j)$;
(d) $W_m(x, w_1, \ldots, p_{m-1}, w_m) = |\{r \in \{0,1\}^m \mid Pass(x, r_1, w_1, p_1, r_2, \ldots, p_{m-1}, r_m, 1)\}|$.

Proof. Except for small changes the proof is similar to the corresponding one in [9]. Item (a) follows from the definition of Q_1 and that of (P, V) accepting an input x. For (b) first note that the verifier at most generates two different real values for w_j given x, r, and the p_i, w_i for $1 \leq i \leq j-1$, namely $w_j(0)$ and $w_j(1)$. In this sense the choice of possible w_j's still is discrete. For each prover P used as candidate for the *max* in the definition of Q_j it is easy to see that the number of accepting r it yields is at most as large as the sum on the right hand side. This holds because the same P is a candidate for both *max* computations related to $W_j(x,\ldots,w_j(0))$ and $W_j(x,\ldots,w_j(1))$. Vice versa, if P_0, P_1 denote the optimal provers in the definitions of $W_j(x,\ldots,w_j(0))$ and $W_j(x,\ldots,w_j(1))$, respectively, an optimal prover for the *max* in Q_j behaves like P_0 if $r_j = 0$ and $w_j(0)$ was sent by the verifier and like P_1 if $r_j = 1$ and $w_j(1)$ was sent. This proves the equality.

For (c) the *max* defining W_j asks for the best continuation of the protocol when $x, r_1, \ldots, r_{j-1}, p_1, \ldots, p_{j-1}, w_1, \ldots, w_j$ are fixed. The next portion of data that can be chosen to achieve the maximum is $p_j \in \mathbb{R}$. The right hand side just asks for the optimal one. Thus both sides are equal. Item (d) holds because the part $(p_1, \ldots, p_{m-1})$ of the argument W_m already fixes the prover. □

The lemma is used in order to compute recursively whether an input x is accepted, i.e., computing $Q_1(x)$ and deciding whether it is larger than 2^{m-1}. In the discrete setting this can be achieved in parallel polynomial time. In our situation, however, due to Theorem 2 in [9] this is not possible;[3] there are problems provably in $\mathrm{IP}_{\mathbb{R}}$ but not in $\mathrm{PAR}_{\mathbb{R}}$. Therefore, the involved max computations increase the problem's difficulty because maximization over an uncountable domain is required. The formal description of the resulting problem introduces $\exists_{\mathbb{R}}$-quantifiers and thus leads to $\mathrm{MA}\exists_{\mathbb{R}}$ as an upper bound for $\mathrm{IP}_{\mathbb{R}}$.

Theorem 2. $\mathrm{IP}_{\mathbb{R}} \subseteq \mathrm{MA}\exists_{\mathbb{R}}$.

Proof. Let $L \in \mathrm{IP}_{\mathbb{R}}$ and (P, V) a corresponding normalized protocol for L. For an input $x \in \mathbb{R}^n$ let $m = poly(n)$ denote the polynomial number of rounds and generated random bits of the interaction. In order to use the above lemma algorithmically it is necessary to compute the maxima occuring in the statements. Though the prover's answers range over real numbers the maxima are integers. Consider $Q_m(x, w_1, p_1, w_2, \ldots, p_{m-1}) = |\{r \in \{0,1\}^m \mid Pass(x, r_1, w_1, p_1, r_2, w_2, \ldots, p_{m-1})\}|$. Obviously, both Q_m and the predicate "$Q_m(x, w_1, \ldots, p_{m-1}) = s$ for given $s \in \{0, 1, \ldots, 2^m\}$" are computable in $\mathrm{PAR}_{\mathbb{R}}$ because for every single r the simulation of (P, V) on x using r needs polynomial time. Now for each $s \in \{0, 1, \ldots, 2^m\}$ the predicate $\exists p_{m-1} \in \mathbb{R} : Q_m(x, w_1, \ldots, p_{m-2}, w_{m-1}, p_{m-1}) = s$ belongs to $\mathrm{MA}\exists_{\mathbb{R}}$: First, Theorem 1 implies that the inner predicate "$Q_m = s$" is in $\mathrm{MA}\exists_{\mathbb{R}}$, then Lemma 1 shows that this class is not left. In order to compute W_{m-1} we can in parallel compute for each $s \in \{0, 1, \ldots, 2^m\}$ whether '$\exists p_{m-1} \in \mathbb{R} : Q_m = s$' holds and finally extract the maximal s for which this is true in polynomial time. Again by Lemma 1 it follows that the predicate $W_{m-1} = s$ can be decided in $\mathrm{MA}\exists_{\mathbb{R}}$. The same holds for the predicate $Q_{m-1}(q, w_1, \ldots, p_{m-1}) = s$ since according to

[3] Though formally the classes in [9] are defined a bit differently it is easy to see that their protocols used to prove the theorem fit into $\mathrm{IP}_{\mathbb{R}}$.

part (b) of Lemma 2 it can be computed using a sum of W_{m-1} when the last component once is $w_{m-1}(0)$ and once $w_{m-1}(1)$.

We continue along the recursion behind Lemma 2. Since its depth m is polynomially bounded in $n = |x|$, by precisely the same arguments as above we see that all predicates $Q_1 = s$ (or, similarly, $Q_1 > s$) for $s \in \{0, 1, \ldots, 2^m\}$ belong to $\text{MA}\exists_{\mathbb{R}}$; the structure $\forall_B \exists_{\mathbb{R}}$ of quantifier prefixes remains the same and for each of the m rounds only a polynomial number of quantifiers is added. Finally, $x \in L$ iff $Q_1 > 2^{m-1}$ finishes the proof. □

3.2 Lower Bounds

In this subsection we report on ongoing research to obtain meaningful lower bounds for $\text{IP}_{\mathbb{R}}$. However, it currently is more a discussion of problems and interesting open questions than a completed project. In particular, so far we have not been able to give a characterization of $\text{IP}_{\mathbb{R}}$ analogue to Shamir's result. Below, we discuss where new difficulties arise and what might be promising ways to go.

Let us first give some interactive proofs for certain restricted subclasses of problems. These results might already shed some light on the difficulties faced when trying to generalize the classical methods to design IP's to the real model.

A major problem here seems to be to obtain suitable arithmetizations of the problems considered in order to apply similar techniques. $\text{PAT}_{\mathbb{R}}, \text{MA}\exists_{\mathbb{R}}, \text{MA}\forall_{\mathbb{R}}$ are classes defined by quantifying a problem $B \in \text{P}_{\mathbb{R}}$ using different sequences of quantifiers of different structures. Another example of such a class is $\text{DPAT}_{\mathbb{R}}$, where all quantifiers are Boolean. It seems natural to expect that at least for sequences of Boolean quantifiers the classical techniques could be adopted. However, this is not obviously true, the reason being the need of a suitable arithmetization of properties $B \in \text{P}_{\mathbb{R}}$. In the following we consider certain subclasses obtained by restricting parts of the general problem definition: either the quantifier structure or the condition in $\text{P}_{\mathbb{R}}$ or both. We shall investigate some cases for which interactive protocols can be designed.

Definition 5. *(a) We denote by* $\text{MA}\forall_{\mathbb{R}}^{=0}$ *the subclass of problems in* $\text{MA}\forall_{\mathbb{R}}$ *where* $B \in \text{P}_{\mathbb{R}}$ *can be chosen to be of the following particular form: There is a multivariate polynomial* F_x *such that* $(x, y, z) \in B$ *if and only if* $F_x(y, z) = 0$. *Moreover, given* (x, y, z) *the value* $F_x(y, z)$ *can be computed in polynomial time in the size of* x.

(b) A problem S *is in class* $\text{DPAT}_{\mathbb{R}}$ *if there is a polynomial* p *and another problem* $B \in \text{P}_{\mathbb{R}}$ *such that* $x \in S$ *if and only if* $\forall_B z_1 \exists_B z_2 \ldots Q_{p(|x|)} z_{p(|x|)} (x, z_1, \ldots, z_{p(|x|)}) \in B$, *where* $Q_{p(|x|)} \in \{\exists_B, \forall_B\}$.

(c) A problem $S \in \text{DPAT}_{\mathbb{R}}$ *belongs to class* $\text{DPAT}_{\mathbb{R}}^{=0}$ *if the condition* $(x, z) \in B$ *in part (b) has the particular form* $F_x(z) = 0$ *for a polynomial* F_x *that can be evaluated in polynomial time in* $|x|$. *Similarly, class* $\text{DPAT}_{\mathbb{R}}^{\neq 0}$ *consists of problems where this condition reads* $F_x(z) \neq 0$.

For the class $\text{MA}\forall_{\mathbb{R}}^{=0}$ of problems defined above we obtain interactive protocols basically by applying the classical Schwartz-Zippel lemma.

Proposition 1. *It holds* $\mathrm{MA}\forall_{\mathbb{R}}^{=0} \subseteq \mathrm{IP}_{\mathbb{R}}$.

Problems in class $\mathrm{DPAT}_{\mathbb{R}}$ are defined using Boolean quantifier prefixes only. Thus, one might expect that the classical discrete technique for designing interactive proofs suffices. However, the problem seems to be finding a suitable arithmetization of the formula. For the subclasses defined above this is possible.

Proposition 2. *It holds* $\mathrm{DPAT}_{\mathbb{R}}^{=0} \subset \mathrm{IP}_{\mathbb{R}}$ *and* $\mathrm{DPAT}_{\mathbb{R}}^{\neq 0} \subset \mathrm{IP}_{\mathbb{R}}$.

Another possible way to extend the class of problems that have a real interactive protocol is the examination of oracle computations and counting problems. In [10] an interactive protocol for verifying the value of a permanent of a 0-1-matrix was given (before Shamir's result was known). Together with Toda's theorem that the polynomial hierarchy PH is included in $P^{\#P}$ and the $\#P$-completeness of the permanent computation this implies an interactive protocol for all problems in the polynomial hierarchy. The protocol for the permanent, as for example described in [1], works as well for real matrices in the BSS model. This implies that real problems that can be decided by a polynomial time BSS algorithm having access to an oracle computing the permanent of real number matrices, i.e., all problems in class $\mathrm{P}_{\mathbb{R}}^{Perm}$, belong to $\mathrm{IP}_{\mathbb{R}}$. However, it is not known whether the permanent plays a similar role for real counting problems as it does in the Turing model. This is an active field of research. Basu and Zell [3] have given a real analogue of Toda's theorem. Instead of the permanent in this approach the computation of so-called Betti numbers of semi-algebraic sets plays a crucial role. The latter express certain topological properties of semi-algebraic sets. They seem to be even more difficult to handle than permanent computations. An intensive study of counting problems has been performed in [5,6] for both the additive and the full real number model. Further topological quantities that turn out to be important are, for example, the topological degree and the Euler characteristic of a set. In both papers several characterizations of real complexity classes via oracle computations as well as completeness results are given. The results in the additive setting actually can be used to prove again the main result of [9].

Theorem 3 ([9]). *In the additive real BSS model the class* $PAR_{\mathbb{R},+}$ *of problems decidable in parallel polynomial time equals the class* $BIP_{\mathbb{R},+}$ *of problems that admit an additive interactive protocol only exchanging boolean messages.*

The prove is technically very similar to the one in [9] in that a crucial inclusion $\mathrm{PAR}_{\mathbb{R},+} \subseteq \mathrm{P}_{\mathbb{R}_+}^{\mathrm{PSPACE}}$ is shown using the existence of small rational points in certain point location tasks. A similar result is central in [9]. This gives the possibility to involve discrete oracles which then can be handled using the classical protocol by Shamir. It is not known whether in the full model discrete oracles play a similarly important role. But the above reasoning makes it interesting to study which real (counting) functions bearing a high complexity can be computed by an interactive protocol in order to use it as an oracle. Another example are so-called resultant functions which are polynomials built from the coefficients of certain polynomial systems and crucial in some of the currently best known algorithms for dealing with the existential theory over the reals like determinants are for the solution of linear systems. Thus we have

Problem 1: Can any of the following problems be solved by an interactive protocol in the full BSS model: given a semi-algebraic set $S \subseteq \mathbb{R}^n$ via a system of polynomial (in-)equalities and a number $k \in \mathbb{N}_0$, verify that the sum of the Betti-numbers of S or the degree or the Euler-characteristic of S equals k. What about verifying the value of resultant polynomials by an interactive protocol?

Even if it is unclear whether positive answers would give the intended characterization of $\mathrm{IP}_{\mathbb{R}}$ it would be a significant step forward. For example, existence of such protocols for the Euler characteristic or the Betti numbers would imply that $\text{co-NP}_{\mathbb{R}} \subseteq \mathrm{IP}_{\mathbb{R}}$ because the latter can be solved using a polynomial time oracle computation that has access to evaluating those function.

Unfortunately, at the moment we do not know how to design an interactive protocol for $\text{co-NP}_{\mathbb{R}}$. It seems unlikely that $\mathrm{MA}\exists_{\mathbb{R}} = \mathrm{MA}\forall_{\mathbb{R}}$. Thus, if $\mathrm{MA}\exists_{\mathbb{R}}$ turns out to equal $\mathrm{IP}_{\mathbb{R}}$ it would not be obvious whether $\mathrm{IP}_{\mathbb{R}}$ is closed under complementation. Classically, this of course holds.

Problem 2: Is $\mathrm{IP}_{\mathbb{R}}$ closed under complementation?

References

1. Arora, S., Barak, B.: Computational Complexity: A Modern Approach. Cambridge University Press, Cambridge (2009)
2. Baartse, M., Meer, K.: The PCP theorem for NP over the reals. Found. Comput. Math. Springer. doi:10.1007/s10208-014-9188-x
3. Basu, S., Zell, T.: Polynomial hierarchy, Betti numbers, and a real analogue of Toda's theorem. Found. Comput. Math. **10**(4), 429–454 (2010)
4. Blum, L., Cucker, F., Shub, M., Smale, S.: Complexity and Real Computation. Springer, New York (1998)
5. Bürgisser, P., Cucker, F.: Counting complexity classes for numeric computations I: semilinear sets. SIAM J. Comput. **33**(1), 227–260 (2003)
6. Bürgisser, P., Cucker, F.: Counting complexity classes for numeric computations. II. algebraic and semialgebraic sets. J. Complex. **22**(2), 147–191 (2006)
7. Cucker, F.: On the complexity of quantifier elimination: the structural approach. Comput. J. **36**(5), 400–408 (1993)
8. Cucker, F., Briquel, I.: A note on parallel and alternating time. J. Complex. **23**, 594–602 (2007)
9. Ivanov, S., de Rougemont, M.: Interactive protocols on the reals. Comput. Complex. **8**, 330–345 (1999)
10. Lund, C., Fortnow, L., Karloff, H., Nisan, N.: Algebraic methods for interactive proof systems. J. ACM **39**(4), 859–868 (1992)
11. Michaux, C.: Une remarque à propos des machines sur $\mathbb{R}$ introduites par Blum. Shub et Smale. C.R. Acad. Sci. Paris, t. 309, Série I, pp. 435–437 (1989)
12. Shamir, A.: IP = PSPACE. J. ACM **39**(4), 869–877 (1992)

Prime Model with No Degree of Autostability Relative to Strong Constructivizations

Nikolay Bazhenov[1,2](✉)

[1] Sobolev Institute of Mathematics, Novosibirsk, Russia
[2] Novosibirsk State University, Novosibirsk, Russia
nickbazh@yandex.ru

Abstract. We build a decidable structure $\mathcal{M}$ such that $\mathcal{M}$ is a prime model of the theory $Th(\mathcal{M})$ and $\mathcal{M}$ has no degree of autostability relative to strong constructivizations.

Keywords: Autostability · Decidable structure · Prime model · Autostability spectrum · Autostability relative to strong constructivizations · Degree of categoricity · Categoricity spectrum · Decidable categoricity

1 Introduction

The study of autostable structures goes back to the works of Fröhlich and Shepherdson [1], and Mal'tsev [2,3]. Since then, the notion of *autostability* has been relativized to the levels of the hyperarithmetical hierarchy, and to arbitrary Turing degrees $\mathbf{d}$, and has been the subject of much study.

Definition 1. Let $\mathbf{d}$ be a Turing degree. A computable structure $\mathcal{A}$ is $\mathbf{d}$-*autostable* if, for every computable structure $\mathcal{B}$ isomorphic to $\mathcal{A}$, there exists a $\mathbf{d}$-computable isomorphism from $\mathcal{A}$ onto $\mathcal{B}$. $\mathbf{0}$-autostable structures are also called *autostable.*

The *autostability spectrum* of the structure $\mathcal{A}$ is the set

$$\mathrm{AutSpec}(\mathcal{A}) = \{\mathbf{d} : \mathcal{A} \text{ is } \mathbf{d}\text{-autostable}\}.$$

A Turing degree $\mathbf{d}_0$ is the *degree of autostability* of $\mathcal{A}$ if $\mathbf{d}_0$ is the least degree in $\mathrm{AutSpec}(\mathcal{A})$.

Autostability spectra and degrees of autostability were introduced by Fokina, Kalimullin, and Miller [4]. Note that much of the literature (see, e.g., [4–8]) uses the terms *categoricity spectrum* and *degree of categoricity* in place of autostability spectrum and degree of autostability, respectively. In this paper, we follow the terminology of [9].

Suppose that n is a natural number and α is a computable ordinal. Fokina, Kalimullin, and Miller [4] proved that every Turing degree $\mathbf{d}$ that is d.c.e. in and above $\mathbf{0}^{(n)}$ is the degree of autostability of a computable structure. This result

A. Beckmann et al. (Eds.): CiE 2015, LNCS 9136, pp. 117–126, 2015.
DOI: 10.1007/978-3-319-20028-6_12

was extended by Csima, Franklin, and Shore [5] to hyperarithmetical degrees. They proved that every degree that is d.c.e. in and above $\mathbf{0}^{(\alpha+1)}$ is a degree of autostability. They also showed that $\mathbf{0}^{(\alpha)}$ is a degree of autostability.

Miller [10] constructed the first example of a computable structure with no degree of autostability. He proved that there exists a computable field F which is not autostable and such that for some $\mathbf{c}_0, \mathbf{c}_1 \in \mathrm{AutSpec}(F)$, $\mathbf{c}_0 \wedge \mathbf{c}_1 = \mathbf{0}$. For more results on autostability spectra, see the survey [11].

Recall that a computable structure $\mathcal{A}$ is *decidable* if its complete diagram $D^c(\mathcal{A})$ is a computable set. The following definition describes the notion of autostability restricted to decidable copies of a structure. This notion is a natural one, as it simply changes the word *computable* to *decidable*.

Definition 2. Suppose that $\mathbf{d}$ is a Turing degree. A decidable structure $\mathcal{A}$ is $\mathbf{d}$-*autostable relative to strong constructivizations* ($\mathbf{d}$-*SC-autostable*) if, for every decidable copy $\mathcal{B}$ of $\mathcal{A}$, there exists a $\mathbf{d}$-computable isomorphism $f\colon \mathcal{A} \to \mathcal{B}$. In case $\mathbf{d} = \mathbf{0}$, we say that $\mathcal{A}$ is *SC-autostable*.

The *autostability spectrum relative to strong constructivizations* (*SC-autostability spectrum*) is the set

$$\mathrm{AutSpec}_{SC}(\mathcal{A}) = \{\mathbf{d} : \mathcal{A} \text{ is } \mathbf{d}\text{-}SC\text{-autostable}\}.$$

A Turing degree $\mathbf{d}_0$ is the *degree of autostability relative to strong constructivizations* (*degree of SC-autostability*) of $\mathcal{A}$ if $\mathbf{d}_0$ is the least degree in the spectrum $\mathrm{AutSpec}_{SC}(\mathcal{A})$.

The study of SC-autostability spectra was initiated by Goncharov [9]. In particular, he proved that every c.e. Turing degree is the degree of SC-autostability of some decidable prime model. In [12] the author announced the following result: for a computable successor ordinal α, every degree $\mathbf{d}$ that is c.e. in and above $\mathbf{0}^{(\alpha)}$ is a degree of SC-autostability.

Suppose that L is a language. If $\mathcal{M}$ is an L-structure, then $Th(\mathcal{M})$ is the first-order theory of $\mathcal{M}$. A structure $\mathcal{M}$ is a *prime model* (of the theory $Th(\mathcal{M})$) if, for every model $\mathcal{N}$ of $Th(\mathcal{M})$, there is an elementary embedding of $\mathcal{M}$ into $\mathcal{N}$. A structure $\mathcal{M}$ is an *almost prime model* if there exists a finite tuple $\bar{c}$ from $\mathcal{M}$ such that $(\mathcal{M}, \bar{c})$ is a prime model.

Our work is concerned with the following problem.

Problem 1. (Goncharov [9]). Suppose that $\mathcal{M}$ is a decidable almost prime model and $\bar{c}$ is a tuple from $\mathcal{M}$ such that $(\mathcal{M}, \bar{c})$ is a prime model of the theory $Th(\mathcal{M}, \bar{c})$. Let $\mathbf{d}$ be the Turing degree of the collection of complete formulas of $Th(\mathcal{M}, \bar{c})$. It is not difficult to see that $\mathbf{d}$ is a c.e. degree and $\mathcal{M}$ is $\mathbf{d}$-SC-autostable. Is it always true that $\mathbf{d}$ is the degree of SC-autostability of $\mathcal{M}$?

We give the negative answer to this question by proving the following result.

Theorem 1. *There exists a decidable structure $\mathcal{M}$ such that $\mathcal{M}$ is a prime model of the theory $Th(\mathcal{M})$ and $\mathcal{M}$ has no degree of SC-autostability.*

2 Preliminaries

Suppose that S is a countable set. A *numbering* of S is a map ν from the set ω of natural numbers *onto* the set S. A numbering ν is a *Friedberg numbering* if ν is 1-1.

γ denotes the standard numbering of the family of all finite subsets of ω. In particular, if $n_0 < n_1 < \ldots < n_k < \omega$, then

$$\gamma(2^{n_0} + 2^{n_1} + \ldots + 2^{n_k}) = \{n_0, n_1, \ldots, n_k\}.$$

For a set $A \subseteq \omega$, we use $|A|$ to denote the cardinality of A. We assume $\{\varphi_e\}_{e\in\omega}$ to be a standard effective enumeration of all unary partial computable functions. We also assume $\langle \cdot, \cdot \rangle$ to be a standard computable pairing function over ω. For a function f, δf denotes the domain of f and ρf denotes the range of f.

An L-structure $\mathcal{M}$ is an *atomic model* if, for any tuple $\bar{a} = a_0, \ldots, a_n$ from $\mathcal{M}$, there exists an L-formula $\phi(x_0, \ldots, x_n)$ such that $\mathcal{M} \models \phi(\bar{a})$, and every L-formula $\psi(x_0, \ldots, x_n)$ satisfies the following condition: if $\mathcal{M} \models \psi(\bar{a})$, then

$$\mathcal{M} \models \forall x_0 \ldots \forall x_n \, (\phi(x_0, \ldots, x_n) \to \psi(x_0, \ldots, x_n)).$$

Such a formula ϕ is called a *complete formula* of the theory $Th(\mathcal{M})$. Recall Vaught's theorem on the relationship of prime and atomic models (see [13]).

Theorem 2. (Vaught). *Suppose that $\mathcal{M}$ is an L-structure. $\mathcal{M}$ is a prime model if and only if $\mathcal{M}$ is a countable atomic model.*

We identify the set $\omega^{<\omega}$ with a tree with the following ordering: $\sigma \preccurlyeq \tau$ iff σ is an initial segment of τ. For any $\sigma, \tau \in \omega^{<\omega}$, we use $\sigma^\frown\tau$ to denote the concatenation of σ and τ. Suppose that T is a subtree of $\omega^{<\omega}$. We use $b(\sigma; T)$ to denote the *branching function* of T which is defined as follows. If $\sigma \in T$, then:

$$b(\sigma; T) = |\{n \in \omega : \sigma^\frown\langle n \rangle \in T\}|.$$

The following is a relativization of the Low Basis Theorem due to Jockusch and Soare (see [14,15]).

Theorem 3. (Jockusch and Soare). *Suppose that $V \subseteq \omega$, and $\mathcal{T}$ is a family of all V-computable finite-branching subtrees T of $\omega^{<\omega}$ with a V-computable branching function $b(\sigma; T)$. Then there exists a Turing degree $\mathbf{d}$ with $\mathbf{d}' \leq \deg_T(V')$ such that every infinite tree $T \in \mathcal{T}$ has a $\mathbf{d}$-computable path. (Such a degree is known as a PA-degree relative to V). Furthermore, there exist two PA-degrees $\mathbf{d}_0$ and $\mathbf{d}_1$ relative to V such that*

$$\forall \mathbf{c} \, ((\mathbf{c} \leq \mathbf{d}_0 \,\&\, \mathbf{c} \leq \mathbf{d}_1) \to \mathbf{c} \leq \deg_T(V)). \quad (1)$$

We refer the reader to [16,17] for further background on computable and decidable structures.

2.1 Colored Algebras

Let L_{BA} be the language $\{\vee^2, \wedge^2, \mathrm{C}^1; 0, 1\}$. We treat Boolean algebras as L_{BA}-structures. If $\mathcal{L}$ is a linear ordering, then $\mathrm{Int}(\mathcal{L})$ denotes the corresponding *interval algebra*. For a Boolean algebra $\mathcal{B}$, $\mathrm{Atom}(\mathcal{B})$ denotes the set of atoms of $\mathcal{B}$. If a is an element of $\mathcal{B}$, then $\widehat{a}_{\mathcal{B}}$ denotes the *relative algebra* with the universe $\{b \in \mathcal{B} : b \leq_{\mathcal{B}} a\}$. For further information on computable Boolean algebras, see [18].

Let k be a non-zero natural number. A *k-partition* of an element a in a Boolean algebra $\mathcal{B}$ is a sequence $b_1, \ldots, b_k$ of pairwise disjoint non-zero elements (i.e., $b_i \wedge b_j = 0$ when $i \neq j$, and $b_i \neq 0$) such that $a = b_1 \vee \ldots \vee b_k$. The formula $(b_1, \ldots, b_k \mid a)$ denotes that $b_1, \ldots, b_k$ is a k-partition of a.

Consider the new computable language $L_0 = L_{BA} \cup \{P_k^1 : k \in \omega\}$, where P_k^1 is a computable predicate.

Definition 3. Let $\mathcal{B}$ be a Boolean algebra. An L_0-structure $\mathcal{B}^c = (\mathcal{B}, P_k)_{k\in\omega}$ is a *colored algebra* if there exists a computable sequence of L_{BA}-formulas $\{\Phi_k(x, \bar{y}_k)\}_{k\in\omega}$ such that for any k, there is a tuple $\bar{b}_k$ from $\mathcal{B}$ with the property

$$\mathcal{B}^c \models \forall x \left(P_k(x) \leftrightarrow \Phi_k(x, \bar{b}_k)\right). \tag{2}$$

Such a sequence $\{\Phi_k\}_{k\in\omega}$ is called a *coloring sequence* of $\mathcal{B}^c$. The Boolean algebra $\mathcal{B}$ is called the *underlying algebra* of $\mathcal{B}^c$.

Colored algebras were introduced in [12]. The informal explanation of the term "colored algebra" is as follows. We treat the predicates P_k as colors and assign these colors to elements of a Boolean algebra $\mathcal{B}$. Note the important difference between our coloring and the graph coloring: we do not require that an element of $\mathcal{B}$ have only one color.

A colored algebra $\mathcal{B}^c$ is *atomic* if its underlying algebra $\mathcal{B}$ is an atomic Boolean algebra. We use $\mathrm{Col}(\mathcal{B}^c)$ to denote the set $\bigcup_{k\in\omega} P_k$ of all colored elements of $\mathcal{B}^c$.

Ershov [19] obtained the following result: a computable atomic Boolean algebra $\mathcal{B}$ is decidable iff the set of atoms $\mathrm{Atom}(\mathcal{B})$ is computable. It is not difficult to show that Ershov's result yields the following corollary.

Proposition 1. *Suppose that $\mathcal{B}^c = (\mathcal{B}, P_k)_{k\in\omega}$ is a computable atomic colored algebra, and $\{\Phi_k(x, \bar{y}_k)\}_{k\in\omega}$ is a coloring sequence of $\mathcal{B}^c$. The structure $\mathcal{B}^c$ is decidable if and only if it satisfies the following conditions:*

(i) the set of atoms $\mathrm{Atom}(\mathcal{B})$ *is computable; and*
(ii) there exists a computable function $g(x)$ such that for any k, the value $g(k)$ is equal to the Gödel number of some tuple $\bar{b}_k$ with the property (2).

3 The Proof of Theorem 1

We will build two decidable atomic colored algebras $\mathcal{A}^c$ and $\mathcal{B}^c$ such that $\mathcal{A}^c$ and $\mathcal{B}^c$ are isomorphic but not computably isomorphic. Lemmas 2, 5, and 8 guarantee that $\mathcal{A}^c$ satisfies Theorem 1. The construction uses the ideas of Miller [10, Theorem 3.4] and Steiner [20, Theorem 2.8].

We fix a computable atomless Boolean algebra $\mathcal{C} = (\omega; \vee, \wedge, \mathrm{C}; 0, 1)$. For clarity, we use $\leq_{\mathcal{C}}$ and $\leq_\omega$ when we need to differentiate between the ordering of the Boolean algebra $\mathcal{C}$ and the standard ordering of ω. We also fix a computable subalgebra $\mathcal{C}^0 \leq \mathcal{C}$ such that $\mathcal{C}^0$ is isomorphic to $\mathrm{Int}(\omega)$ and $\mathcal{C}_0$ has a computable set of atoms $\mathrm{Atom}(\mathcal{C}_0) = \{a_0 <_\omega a_1 <_\omega a_2 <_\omega \ldots\}$. For a set $X \subseteq \omega$, we use $\mathrm{gr}(X)$ to denote the subalgebra of $\mathcal{C}$ generated by X.

Consider a computable language $L_c = \left\{P^1_k : k \in \omega\right\} \cup \left\{Q^1_{k,j} : k, j \in \omega\right\}$. We will construct two L_c-structures $\mathcal{A}^c = \left(\mathcal{A}, P^A_k, Q^A_{k,j}\right)_{k,j\in\omega}$ and $\mathcal{B}^c = \left(\mathcal{B}, P^B_k, Q^B_{k,j}\right)_{k,j\in\omega}$ such that $\mathcal{A}^c \cong \mathcal{B}^c$, $\mathcal{C}^0 \leq \mathcal{A} \leq \mathcal{C}$, and $\mathcal{C}^0 \leq \mathcal{B} \leq \mathcal{C}$.

At stage s we define computable Boolean algebras $\mathcal{A}_s$ and $\mathcal{B}_s$. The universe of $\mathcal{A}_s$ is denoted by A_s, and the universe of $\mathcal{B}_s$ is denoted by B_s. For every $k \in \omega$, we also define the number $f_{k,s}$, elements $c^A_{k,s}, d^A_{k,s}$ from $\mathcal{A}_s$, and elements $c^B_{k,s}, d^B_{k,s}$ from $\mathcal{B}_s$. In addition, we build the predicates P^A_k, P^B_k, $Q^A_{k,j}$, and $Q^B_{k,j}$ in such a way that, for any predicate R, there is a unique stage t which deals with R.

Notation. We say that we *don't change k-parameters at stage $s+1$* if we define $f_{k,s+1} = f_{k,s}$, $c^A_{k,s+1} = c^A_{k,s}$, $d^A_{k,s+1} = d^A_{k,s}$, $c^B_{k,s+1} = c^B_{k,s}$, and $d^B_{k,s+1} = d^B_{k,s}$.

Construction *Stage* 0. Define $\mathcal{A}_0 = \mathcal{B}_0 = \mathcal{C}^0$. For every $k \in \omega$, set $P^A_k = P^B_k = \{a_{2k}, a_{2k+1}\}$, $f_{k,0} = 0$, $c^A_{k,0} = c^B_{k,0} = a_{2k}$, and $d^A_{k,0} = d^B_{k,0} = a_{2k+1}$.

Stage $s+1$. Suppose that $s = \langle k, t\rangle$. Consider the following four cases.

Case 1. Suppose that $f_{k,s} = 0$, and t is the least natural number such that $\varphi_{k,t}(a_{2k})\downarrow = a_{2k}$ and $\varphi_{k,t}(a_{2k+1})\downarrow = a_{2k+1}$. Find the following partitions in the Boolean algebra $\mathcal{C}$: $(c_1, c_2 \mid c^A_{k,s})$, $(d_1, d_2, d_3 \mid d^A_{k,s})$, $(c'_1, c'_2, c'_3 \mid c^B_{k,s})$, and $(d'_1, d'_2 \mid d^B_{k,s})$. Define

$$\mathcal{A}_{s+1} = \mathrm{gr}(A_s \cup \{c_1, c_2, d_1, d_2, d_3\}), \quad \mathcal{B}_{s+1} = \mathrm{gr}(B_s \cup \{c'_1, c'_2, c'_3, d'_1, d'_2\}),$$
$$Q^A_{k,t} = \{c_1, d_1\}, \quad Q^A_{k,t+1} = \{c_2, d_2\}, \quad Q^A_{k,t+2} = \{d_3\},$$
$$Q^B_{k,t} = \{c'_1, d'_1\}, \quad Q^B_{k,t+1} = \{c'_2, d'_2\}, \quad Q^B_{k,t+2} = \{c'_3\},$$
$$Q^A_{k,t+l+3} = Q^B_{k,t+l+3} = \emptyset, \quad l \in \omega.$$

Set $f_{k,s+1} = 1$. For any l, do not change l-parameters (except the parameter $f_{k,s+1}$).

Case 2. Suppose that $f_{k,s} = 0$, and t is the least number such that $\varphi_{k,t}(a_{2k})\downarrow = a_{2k+1}$ and $\varphi_{k,t}(a_{2k+1})\downarrow = a_{2k}$. Find the following partitions in $\mathcal{C}$: $(c_1, c_2 \mid c^A_{k,s})$, $(d_1, d_2, d_3 \mid d^A_{k,s})$, $(c'_1, c'_2 \mid c^B_{k,s})$, and $(d'_1, d'_2, d'_3 \mid d^B_{k,s})$. The definitions of $\mathcal{A}_{s+1}$, $f_{k,s+1}$, $Q^A_{k,t+l}$ (where $l \in \omega$), and $Q^B_{k,t+l}$ (where $l \neq 2$) are the same as in the Case 1. Define

$$\mathcal{B}_{s+1} = \mathrm{gr}(B_s \cup \{c'_1, c'_2, d'_1, d'_2, d'_3\}), \quad Q^B_{k,t+2} = \{d'_3\}.$$

For any l, don't change l-parameters (except $f_{k,s+1}$).

Case 3. Suppose that $f_{k,s} = 0$ and neither of Cases 1 and 2 hold. Find the following partitions in $\mathcal{C}$: $(c_1, c_2 \mid c^A_{k,s})$, $(d_1, d_2 \mid d^A_{k,s})$, $(c'_1, c'_2 \mid c^B_{k,s})$, and $(d'_1, d'_2 \mid d^B_{k,s})$. Set

$$\mathcal{A}_{s+1} = \mathrm{gr}(A_s \cup \{c_1, c_2, d_1, d_2\}), \quad \mathcal{B}_{s+1} = \mathrm{gr}(B_s \cup \{c'_1, c'_2, d'_1, d'_2\}),$$
$$Q^A_{k,t} = \{c_1, d_1\}, \quad c^A_{k,s+1} = c_2, \quad d^A_{k,s+1} = d_2,$$
$$Q^B_{k,t} = \{c'_1, d'_1\}, \quad c^B_{k,s+1} = c'_2, \quad d^B_{k,s+1} = d'_2.$$

For any l, don't change any l-parameters.

Case 4. If $f_{k,s} \neq 0$, then set $\mathcal{A}_{s+1} = \mathcal{A}_s$, $\mathcal{B}_{s+1} = \mathcal{B}_s$, and don't change l-parameters for any l.

We have described the construction. Define Boolean algebras $\mathcal{A} = \mathrm{gr}\left(\bigcup_{s\in\omega} A_s\right)$ and $\mathcal{B} = \mathrm{gr}\left(\bigcup_{s\in\omega} B_s\right)$. It is easy to see that the sets A_s and B_s, $s \in \omega$, are uniformly computable; therefore, we may assume that the structures $\mathcal{A}$ and $\mathcal{B}$ are computable. Consider the structures $\mathcal{A}^c = \left(\mathcal{A}, P^A_k, Q^A_{k,j}\right)_{k,j\in\omega}$ and $\mathcal{B}^c = \left(\mathcal{B}, P^B_k, Q^B_{k,j}\right)_{k,j\in\omega}$. It is not difficult to show that $\mathcal{A}^c$ and $\mathcal{B}^c$ are computable structures.

Verification. It is easy to verify the following properties of the construction.

Lemma 1. (a) *For any $k, j \in \omega$, we have $\left|P^A_k\right| = \left|P^B_k\right| \leq 2$ and $\left|Q^A_{k,j}\right| = \left|Q^B_{k,j}\right| \leq 2$. Moreover, there exist computable functions $f_A(x)$ and $f_B(x)$ such that for any k and j, $P^A_k = \gamma(f_A(\langle k, 0\rangle))$, $Q^A_{k,j} = \gamma(f_A(\langle k, j+1\rangle))$, $P^B_k = \gamma(f_B(\langle k, 0\rangle))$, and $Q^B_{k,j} = \gamma(f_B(\langle k, j+1\rangle))$.*

(b) $\mathrm{Atom}(\mathcal{A}) = \bigcup_{k,j\in\omega} Q^A_{k,j}$ *and* $\mathrm{Atom}(\mathcal{B}) = \bigcup_{k,j\in\omega} Q^B_{k,j}$.

(c) *Suppose that R and S are distinct predicates from the language L_c. Then $R^A \cap S^A = \emptyset$ and $R^B \cap S^B = \emptyset$.*

(d) *Every $k \in \omega$ satisfies one of the following two conditions.*

(d.1) *There exists a number $t \geq 3$ such that each of the algebras $\widehat{(a_{2k})}_\mathcal{A}$, $\widehat{(a_{2k+1})}_\mathcal{A}$, $\widehat{(a_{2k})}_\mathcal{B}$, and $\widehat{(a_{2k+1})}_\mathcal{B}$ is isomorphic either to* $\mathrm{Int}(t)$ *or to* $\mathrm{Int}(t+1)$. *Moreover, $\widehat{(a_{2k})}_\mathcal{A} \not\cong \widehat{(a_{2k+1})}_\mathcal{A}$ and $\widehat{(a_{2k})}_\mathcal{B} \not\cong \widehat{(a_{2k+1})}_\mathcal{B}$.*

(d.2) *Each of the algebras $\widehat{(a_{2k})}_\mathcal{A}$, $\widehat{(a_{2k+1})}_\mathcal{A}$, $\widehat{(a_{2k})}_\mathcal{B}$, and $\widehat{(a_{2k+1})}_\mathcal{B}$ is isomorphic to* $\mathrm{Int}(\omega)$.

(e) $\mathcal{A} = \mathrm{gr}\left(\mathrm{Atom}(\mathcal{A}) \cup \bigcup_{k\in\omega} P^A_k\right)$ *and* $\mathcal{B} = \mathrm{gr}\left(\mathrm{Atom}(\mathcal{B}) \cup \bigcup_{k\in\omega} P^B_k\right)$.

(f) *$\mathcal{A}$ is isomorphic to* $\mathrm{Int}(\omega^2)$.

Lemma 2. *Structures $\mathcal{A}^c$ and $\mathcal{B}^c$ are decidable colored algebras.*

Proof. Consider the function $f_A(x)$ from Lemma 1(a). Define the following sequence of L_{BA}-formulas.

$$\Phi_{k,j}(x, \bar{y}) = \begin{cases} (x = y_1) \vee (x = y_2), & \text{if } |\gamma(f_A(\langle k, j\rangle))| = 2, \\ x = y_1, & \text{if } |\gamma(f_A(\langle k, j\rangle))| = 1, \\ x \neq x, & \text{otherwise.} \end{cases}$$

Lemma 1(a) implies that the sequence $\{\Phi_{k,j}\}_{k,j\in\omega}$ is the coloring sequence for each of the structures $\mathcal{A}^c$ and $\mathcal{B}^c$. Hence, $\mathcal{A}^c$ and $\mathcal{B}^c$ are colored algebras. Lemma 1(a,b) also implies that $\mathcal{A}^c$ and $\mathcal{B}^c$ satisfy the conditions of Proposition 1; therefore, our structures are decidable.

The proof of Lemma 2 actually shows that every computable copy of $\mathcal{A}^c$ satisfies Proposition 1.

Corollary 1. *Every computable copy of $\mathcal{A}^c$ is decidable. In particular, the spectrum* $\mathrm{AutSpec}_{SC}(\mathcal{A}^c)$ *is equal to* $\mathrm{AutSpec}(\mathcal{A}^c)$.

Definition 4. Given an element a from the set $\mathrm{Col}(\mathcal{A}^c)$, we define the L_c-formula $\phi^a(x)$ as follows. First, we find a predicate R from L_c such that $a \in R^A$.

(i) If a is an atom of $\mathcal{A}$, then we define $\phi^a(x) = R(x)$.
(ii) If $a \notin \mathrm{Atom}(\mathcal{A})$, then $R = P_k$ for some $k \in \omega$. Consider the following two cases.
  (ii.a) Suppose that the Boolean algebra $\widehat{a}_{\mathcal{A}}$ is finite and it has exactly t atoms. We define

$$\phi^a(x) = P_k(x) \,\&\, \exists y\,((y \wedge x = y) \,\&\, Q_{k,t-1}(y)) \,\&\, \neg\exists z((z \wedge x = z) \,\&\, Q_{k,t}(z)).$$

  (ii.b) If $\widehat{a}_{\mathcal{A}}$ is an infinite algebra, then we set $\phi^a(x) = P_k(x)$.

Note that Lemma 1(b,c,d) implies that the formulas ϕ^a are well-defined. It is not difficult to prove the following lemma.

Lemma 3. *Suppose that* $\mathcal{A}_1^c = \left(\mathcal{A}_1, P_k^{\mathcal{A}_1}, Q_{k,j}^{\mathcal{A}_1}\right)_{k,j\in\omega}$ *is a colored algebra with the universe contained in ω. Suppose also that F is a bijection from* $\mathrm{Col}(\mathcal{A}^c)$ *onto* $\mathrm{Col}(\mathcal{A}_1^c)$ *with the following properties:*

(a) for any $a, b \in \mathrm{Col}(\mathcal{A}^c)$, $a \leq_{\mathcal{A}} b$ iff $F(a) \leq_{\mathcal{A}_1} F(b)$;
(b) for any $a, b \in \mathrm{Col}(\mathcal{A}^c)$, $\mathcal{A}^c \models \phi^a(b)$ iff $\mathcal{A}_1^c \models \phi^a(F(b))$.

Then there exists a unique isomorphism $F^c\colon \mathcal{A}^c \to \mathcal{A}_1^c$ such that $F^c \supseteq F$. Moreover, F^c can be constructed effectively from F and the atomic diagram of $\mathcal{A}_1^c$.

Lemma 4. *Colored algebras $\mathcal{A}^c$ and $\mathcal{B}^c$ are isomorphic but not computably isomorphic. In particular, $\mathcal{A}^c$ is not SC-autostable.*

Proof. It is easy to construct a $\mathbf{0}'$-computable bijection F from $\mathrm{Col}(\mathcal{A}^c)$ onto $\mathrm{Col}(\mathcal{B}^c)$ satisfying the conditions of Lemma 3. Therefore, $\mathcal{A}^c$ and $\mathcal{B}^c$ are $\mathbf{0}'$-computably isomorphic.

Note that for any $k \in \omega$ and any isomorphism $G\colon \mathcal{A}^c \to \mathcal{B}^c$, G maps a_{2k} to a_{2k} and a_{2k+1} to a_{2k+1}, or vice versa. Therefore, Cases 1 and 2 of the construction guarantee that φ_k is not an isomorphism. For example, Case 1 ensures that if $\varphi_k(a_{2k}) = a_{2k}$ and $\varphi_k(a_{2k+1}) = a_{2k+1}$, then the relative algebras $\widehat{(a_{2k})}_{\mathcal{A}}$ and $\widehat{(a_{2k})}_{\mathcal{B}}$ are not isomorphic.

Lemma 5. *$\mathcal{A}^c$ is a prime model.*

Proof. By Theorem 2, it is sufficient to prove that $\mathcal{A}^c$ is an atomic model. Given a tuple $\bar{a} = a_0, \dots, a_n$ from $\mathcal{A}^c$, we will construct a complete L_c-formula $\Phi(\bar{x})$ such that $\mathcal{A}^c \models \Phi(\bar{a})$. Lemma 1(e) implies that we can choose a tuple $\bar{b} = b_0, \dots, b_m$ from $\mathrm{Col}(\mathcal{A}^c)$ such that $\bar{a} \in \mathrm{gr}(\{b_0, \dots, b_m\})$. For $i \leqslant n$, fix an L_{BA}-term $t_i(\bar{y})$ such that $a_i = t_i(\bar{b})$. We define the formula

$$\Phi(\bar{x}) = \exists y_0 \dots \exists y_m \Psi(\bar{x}, y_0, \dots, y_m),$$

where Ψ is the conjunction of the following formulas:

1. $x_i = t_i(\bar{y})$ for $i \leqslant n$,
2. $\phi^{b_j}(y_j)$ for $j \leqslant m$,
3. $y_j \wedge y_k = y_j$ for all j and k with the property $b_j \leq_C b_k$,
4. $y_j \wedge y_k \neq y_j$ for all j and k with the property $b_j \not\leq_C b_k$.

It is easy to see that $\mathcal{A}^c \models \Phi(\bar{a})$. Suppose that $\mathcal{A}^c \models \Phi(\bar{c})$ for some $\bar{c}$. Using Lemma 3, it is not difficult to show that the structures $(\mathcal{A}^c, \bar{a})$ and $(\mathcal{A}^c, \bar{c})$ are isomorphic. Hence, Φ is a complete formula.

Definition 5. Let $\mathcal{A}_1^c = \left(A_1, P_k^{A_1}, Q_{k,j}^{A_1}\right)_{k,j\in\omega}$ be a copy of $\mathcal{A}^c$ with universe $A_1 \subseteq \omega$. A *special numbering* of $\mathrm{Col}(\mathcal{A}_1^c)$ is a Friedberg numbering of $\mathrm{Col}(\mathcal{A}_1^c)$ with the following properties: ν is a computable function, and for all $x, k, j \in \omega$, if $\nu(x) \in Q_{k,j+1}^{A_1}$, then there exist y_0, y_1, z_0, z_1 such that $y_0 < y_1 < z_0 < z_1 < x$, $P_k^{A_1} = \{\nu(y_0), \nu(y_1)\}$, and $Q_{k,j}^{A_1} = \{\nu(z_0), \nu(z_1)\}$.

Note. If $\mathcal{A}_1^c$ is a computable copy of $\mathcal{A}^c$, then there exists a special numbering ν_1 of $\mathrm{Col}(\mathcal{A}_1^c)$. Moreover, ν_1 can be constructed effectively from the atomic diagram of $\mathcal{A}_1^c$.

We fix a special numbering ν of $\mathrm{Col}(\mathcal{A}^c)$. For a number s, $\nu[s]$ denotes the set $\{\nu(0), \dots, \nu(s)\}$. The following definition is based on [10, Definition 5.1] and [20, Definition 2.16].

Definition 6. Let $\mathcal{A}_1^c$ be a computable copy of $\mathcal{A}^c$ with universe A_1. The universe of the *isomorphism tree* T_{A,A_1} is the set of all functions f with the following properties.

(a) $\delta f = \nu[s]$ for some s, and $\rho f \subseteq A_1$;
(b) Suppose that L_f is a language $\{R \in L_c : \exists a \in \delta f \, (a \in R^A)\}$. Then f is an isomorphic embedding from the L_f-structure $\left(\delta f, L_f^A\right)$ into the L_f-structure $\left(A_1, L_f^{A_1}\right)$.
(c) For every $a, b \in \delta f$, $a \leq_A b$ iff $f(a) \leq_{A_1} f(b)$.

The ordering of the tree T_{A,A_1} is standard, i.e., $f \preccurlyeq g$ iff $f \subseteq g$. We identify the tree T_{A,A_1} with a computable subtree of $\omega^{<\omega}$. We may assume that T_{A,A_1} is built effectively from the atomic diagram of $\mathcal{A}_1^c$.

The following lemma justifies the choice of the term "isomorphism tree."

Lemma 6. *Suppose that $\mathcal{A}_1^c$ is a computable copy of $\mathcal{A}^c$. Let I be the set of all isomorphisms from $\mathcal{A}^c$ onto $\mathcal{A}_1^c$, and P be the set of all paths through the tree T_{A,A_1}. Then there exists a bijection Ψ from P onto I such that for any $\pi \in P$, π is Turing equivalent to $\Psi(\pi)$.*

Proof. Here we omit the details and give only general idea of the proof. Given a path π through T_{A,A_1}, build a bijection $F_\pi \colon \mathrm{Col}(\mathcal{A}^c) \to \mathrm{Col}(\mathcal{A}_1^c)$ such that F_π satisfies the conditions of Lemma 3 and for any $a \in \mathrm{Col}(\mathcal{A}^c)$, there is a finite function $f \prec \pi$ with the property $F_\pi(a) = f(a)$. The function F_π yields an isomorphism $F_\pi^c \colon \mathcal{A}^c \to \mathcal{A}_1^c$. Set $\Psi(\pi) = F_\pi^c$.

It is not difficult to verify the following claim.

Lemma 7. *The tree T_{A,A_1} is a finite-branching tree with a computable branching function $b(\sigma; T_{A,A_1})$. Moreover, for any $\sigma \in T_{A,A_1}$, we have $b(\sigma; T_{A,A_1}) \leq 2$.*

Lemma 8. *(1) Suppose that $\mathbf{d}$ is a PA-degree relative to $\emptyset$. Then $\mathcal{A}^c$ is $\mathbf{d}$-autostable.*
(2) $\mathcal{A}^c$ has no degree of SC-autostability.

Proof. Let $\mathcal{A}_1^c$ be a computable copy of $\mathcal{A}^c$. By Theorem 3 and Lemma 7, the isomorphism tree T_{A,A_1} has a $\mathbf{d}$-computable path π. By Lemma 6, there is a $\mathbf{d}$-computable isomorphism $\Psi(\pi)$ from $\mathcal{A}^c$ onto $\mathcal{A}_1^c$. Therefore, $\mathcal{A}^c$ is $\mathbf{d}$-autostable.

We fix two PA-degrees $\mathbf{d}_0$ and $\mathbf{d}_1$ relative to $\emptyset$ with the property (1) (where $V = \emptyset$). We already proved that $\mathcal{A}^c$ is $\mathbf{d}_0$-SC-autostable and $\mathbf{d}_1$-SC-autostable. Note that (1) implies that if $\mathcal{A}^c$ has a degree of SC-autostability, then $\mathcal{A}^c$ is SC-autostable. Therefore, by Lemma 4, $\mathcal{A}^c$ has no degree of SC-autostability.

This completes the proof of Theorem 1. In conclusion, we formulate some open questions related to Problem 1.

Question 1. Suppose that $\mathcal{M}$ is a decidable almost prime model and $\bar{c}$ is a tuple from $\mathcal{M}$ such that $(\mathcal{M}, \bar{c})$ is a prime model of the theory $Th(\mathcal{M}, \bar{c})$. Let $\mathbf{d}$ be the Turing degree of the collection of complete formulas of $Th(\mathcal{M}, \bar{c})$. Suppose also that $\mathcal{M}$ has the degree of SC-autostability $\mathbf{c}$. Is it possible that $\mathbf{c} < \mathbf{d}$?

Question 2. Is every d.c.e. degree a degree of SC-autostability for some almost prime model?

Note that the positive answer to Question 2 yields the positive answer to Question 1.

Acknowledgements. The author is grateful to Sergey Goncharov and Svetlana Aleksandrova for fruitful discussions on the subject. This work was supported by RFBR (grant 14-01-00376), and by the Grants Council (under RF President) for State Aid of Leading Scientific Schools (grant NSh-860.2014.1).

References

1. Fröhlich, A., Shepherdson, J.C.: Effective procedures in field theory. Philos. Trans. Roy. Soc. London. Ser. A. **248**, 407–432 (1956)
2. Mal'tsev, A.I.: Constructive algebras. I. Russ. Math. Surv. **16**, 77–129 (1961)
3. Mal'tsev, A.I.: On recursive abelian groups. Sov. Math. Dokl. **32**, 1431–1434 (1962)
4. Fokina, E.B., Kalimullin, I., Miller, R.: Degrees of categoricity of computable structures. Arch. Math. Logic. **49**, 51–67 (2010)
5. Csima, B.F., Franklin, J.N.Y., Shore, R.A.: Degrees of categoricity and the hyperarithmetic hierarchy. Notre Dame J. Formal Logic. **54**, 215–231 (2013)
6. Bazhenov, N.A.: Degrees of categoricity for superatomic Boolean algebras. Algebra Logic. **52**, 179–187 (2013)
7. Anderson, B.A., Csima, B.F.: Degrees that are not degrees of categoricity. Notre Dame J. Formal Logic. (to appear)
8. Fokina, E., Frolov, A., Kalimullin, I.: Categoricity spectra for rigid structures. Notre Dame J. Formal Logic. (to appear)
9. Goncharov, S.S.: Degrees of autostability relative to strong constructivizations. Proc. Steklov Inst. Math. **274**, 105–115 (2011)
10. Miller, R.: **d**-computable categoricity for algebraic fields. J. Symb. Log. **74**, 1325–1351 (2009)
11. Fokina, E.B., Harizanov, V., Melnikov, A.: Computable model theory. In: Downey, R. (ed.) Turing's Legacy: Developments from Turing Ideas in Logic. Lecture Notes Logic, vol. 42, pp. 124–194. Cambridge University Press, Cambridge (2014)
12. Bazhenov, N.A.: Autostability spectra for Boolean algebras. Algebra Logic. **53**, 502–505 (2014)
13. Chang, C.C., Keisler, H.J.: Model Theory. North-Holland, Amsterdam (1973)
14. Jockusch, C.G., Soare, R.I.: Π^0_1 classes and degrees of theories. Trans. Amer. Math. Soc. **173**, 33–56 (1972)
15. Cenzer, D.: Π^0_1 classes in computability theory. In: Griffor, E.R. (ed.) Handbook of Computability Theory. Studies Logic Foundations Mathematics, vol. 140, pp. 37–85. Elsevier Science B.V., Amsterdam (1999)
16. Ash, C.J., Knight, J.F.: Computable Structures and the Hyperarithmetical Hierarchy. Elsevier Science B.V, Amsterdam (2000)
17. Ershov, Y.L., Goncharov, S.S.: Constructive Models. Kluwer Academic/Plenum Publishers, New York (2000)
18. Goncharov, S.S.: Countable Boolean Algebras and Decidability. Consultants Bureau, New York (1997)
19. Ershov, Y.L.: Decidability of the elementary theory of distributive lattices with relative complements and the theory of filters. Algebra Logic. **3**, 17–38 (1964)
20. Steiner, R.M.: Effective algebraicity. Arch. Math. Logic. **52**, 91–112 (2013)

Immune Systems in Computer Virology

Guillaume Bonfante[1,2(✉)], Mohamed El-Aqqad[2], Benjamin Greenbaum[3,4], and Mathieu Hoyrup[1]

[1] Loria, B.P. 239, 54506 Vandœuvre-lès-Nancy Cedex, France
Guillaume.Bonfante@loria.fr, Mathieu.Hoyrup@inria.fr
[2] École Nationale Supérieure des Mines de Nancy, Université de Lorraine, Nancy, France
Mohamed.El-aqqad1@etu.univ-lorraine.fr
[3] Icahn School of Medicine at Mount Sinai, New York, NY 10029, USA
Benjamin.Greenbaum@mssm.edu
[4] Institute for Advanced Study, Einstein Drive, Princeton, NJ 08540, USA

The analogy between computer viruses and biological viruses, from which computer viruses get their name [7], has been clear for the past several decades. During that time there has been progress in both understanding the vast diversity of biological viruses, and in abstract approaches to understanding computer viruses. However, there has not been a great deal of effort to see if the formal efforts in theoretical computer science can be of any use to our understanding of biological viruses.

In this work, we use biological viruses as a motivation to extend some well-known results in theoretical computers viruses. In Cohen's [7], a virus is a string which–read by a Turing Machine–reproduces either itself or a variant form of itself. In Adelman's [1] initial formalism, the theory of computer viruses was placed in the theory of recursive functions. One well known result from both theories is that the general problem of viral detection is undecidable, implying that general computer immune systems based on viral detection can always be circumvented and there are no robust ways to modify a detector successfully [5].

In biological systems, there is some notion that a biological virus is less powerful, computationally, than the host it infects. Motivated by that analogy, here we show two cases where viruses, due to diminished computational capacity relative to their hosts, will not always win. But, first, what is a virus? We state after Adelman [1] and Bonfante, Kaczmarek, Marion [3] that a virus $\mathbf{v}$ is a fix point in the sense of Kleene's Second Recursion Theorem:

$$[\![\mathbf{v}]\!](x) = f(\underline{\mathbf{v}}, x)$$

where f is called the propagation function which defines a viruses behavior in regard to its first argument. As justified in [3] or by Case and Moelius in [6], the model is strong enough to capture the virus mutability or even virus "factories". It is shown that the different versions of the Recursion Theorem–weak, strong, extended, double, see Smullyan's [12] for a precise terminology–correspond to different aspects of computer viruses.

G. Bonfante—The first author received the support of ANR-12-INSE-002.

A. Beckmann et al. (Eds.): CiE 2015, LNCS 9136, pp. 127–136, 2015.
DOI: 10.1007/978-3-319-20028-6_13

Fixed points exist as long as the framework is universal [11], that is Turing complete, with a universal function and a specializer. Thus the following defense strategy: to block viruses, one may simply prevent the existence of fixed points. As we will see, the Recursion Theorem holds as long as there is a specializer, projections and composition. It is hard to avoid the two latter criteria. Thus, we focus on systems without specializers. We provide a solution based on cons-free programs as it has been developed in the past by Jones [9]. Whatever the choice of enumeration of programs, there is no specializer for cons-free programs. It is worth noticing that the language is LOGSPACE-complete making it relatively powerful computationally speaking.

We develop an other scenario, perhaps closer to the analogy with biology. Our idea is to strengthen the defense against viruses, not to avoid their existence. Indeed, biological viruses exist, but their hosts have some defense capabilities. In this scenario, we suppose that there is a (finite) set of known viruses. Then, each time a program enters the system, it is submitted to a program (an immune cell) which verifies whether it behaves like one of the viruses and remove it accordingly. Our scenario is close to the strategy of anti-virus software: they (are supposed to) recognize infected programs relative to a malware database which contains the (finite) set of known viruses. The existence of such a detector infringes Rice's Theorem. Indeed, it corresponds to the decidability of program equivalence. Thus, and again, to get such a language, we will loose Turing completeness.

To our knowledge, in computer virology, only "negative" results have been established so far. They state more or less that there is no defenses against viruses. On the theoretical side, we refer the reader to the aforementioned work [1,7] which where followed by Zuo and Zhou [13] and then by Case and Moelius [5]. On a more practical side, there are also many interesting approaches. For instance, Borello and Mé [4] showed how metamorphism can trick anti-virus software. Other escaping techniques involve encryption, self-reproduction and feints, see [8] for a full survey. This contribution is a first attempt to provide "positive" solutions.

1 Introduction

An *alphabet* is a finite set Σ of *letters*. Given an alphabet Σ, let $\mathbb{T}_\Sigma$ be the set of binary trees with leaves in Σ, that is the smallest set containing Σ and $(\mathtt{t}_1 \cdot \mathtt{t}_2)$ whenever $\mathtt{t}_1, \mathtt{t}_2 \in \mathbb{T}_\Sigma$. The two functions π_1 and π_2 are the projections: $\pi_i(\mathtt{t}_1 \cdot \mathtt{t}_2) = \mathtt{t}_i$ and $\pi_i(\mathtt{c}) = \mathtt{c}$ with $i \in \{1, 2\}$ and $\mathtt{c} \in \Sigma$. The size of a tree $\mathtt{t}$ is denoted $|\mathtt{t}|$ and is defined by $|\mathtt{c}| = 1$, $\mathtt{c} \in \Sigma$, and $|(\mathtt{t} \cdot \mathtt{u})| = |\mathtt{t}| + |\mathtt{u}| + 1$.

A word $a_1 \cdot a_2 \cdots a_k$ in Σ^* is encoded in $\mathbb{T}_{\Sigma \cup \{\mathtt{nil}\}}$ as $(a_1 \cdot (a_2 \cdot (\cdots (a_k \cdot \mathtt{nil}) \cdots)))$ where $\mathtt{nil}$ is an atom used as an end-marker. This relates computations over words to the ones over trees.

Definition 1. *Let $\trianglelefteq$ be the sub-tree relation on $\mathbb{T}_\Sigma$, that is the smallest order (reflexive-transitive relation) such that for all $\mathtt{t}, \mathtt{u} \in \mathbb{T}_\Sigma$:*

- $\mathtt{t} \trianglelefteq \mathtt{t}$,
- $\mathtt{t} \trianglelefteq (\mathtt{t} \cdot \mathtt{u})$,
- $\mathtt{t} \trianglelefteq (\mathtt{u} \cdot \mathtt{t})$.

The embedding relation on $\mathbb{T}_\Sigma$ *is defined to be the smallest order such that* $\mathtt{t} \trianglelefteq \mathtt{u} \Rightarrow \mathtt{t} \blacktriangleleft\!\!\!\!\underline{\ \ } \mathtt{u}$, *and closed by context:* $\mathtt{t} \blacktriangleleft\!\!\!\!\underline{\ \ } \mathtt{t}' \wedge \mathtt{u} \blacktriangleleft\!\!\!\!\underline{\ \ } \mathtt{u}' \Rightarrow (\mathtt{t} \cdot \mathtt{u}) \blacktriangleleft\!\!\!\!\underline{\ \ } (\mathtt{u} \cdot \mathtt{u}')$.

Observe that $\mathtt{nil} \trianglelefteq (\mathtt{nil} \cdot \mathtt{nil})$. The difference between the sub-tree relation and the embedding one is exemplified by $(\mathtt{nil} \cdot (\mathtt{nil} \cdot \mathtt{nil})) \blacktriangleleft\!\!\!\!\underline{\ \ } ((\mathtt{nil} \cdot \mathtt{nil}) \cdot (\mathtt{nil} \cdot \mathtt{nil}))$ but $(\mathtt{nil} \cdot (\mathtt{nil} \cdot \mathtt{nil})) \not\trianglelefteq ((\mathtt{nil} \cdot \mathtt{nil}) \cdot (\mathtt{nil} \cdot \mathtt{nil}))$.

Let $\triangleleft$ and $\blacktriangleleft$ denote respectively the strict order relative to $\trianglelefteq$ and $\blacktriangleleft\!\!\!\!\underline{\ \ }$. From the definition, first, it is clear that if $\mathtt{t} \blacktriangleleft \mathtt{u}$, then $\mathtt{u} \not\trianglelefteq \mathtt{t}$. And, second, if $|\mathtt{t}| > |\mathtt{u}|$, then $\mathtt{t} \not\trianglelefteq \mathtt{u}$.

We present (a slight variant of) While, a generic imperative language introduced by Jones [9]. We suppose a given alphabet Σ contains an atom nil. Moreover, we suppose given (a denumerable set of) variables $\mathtt{Var} \ni \mathtt{X}_0, \mathtt{X}_1, \ldots$. In the following, X, Y serve as generic variables. The syntax of While is given by the following grammar:

$$\begin{array}{lll}\mathtt{Expressions} \ni \mathtt{E}, \mathtt{F} & ::= & \mathtt{X} \mid \mathtt{t} \mid \mathtt{cons}\ \mathtt{E}\ \mathtt{F} \mid \mathtt{hd}\ \mathtt{E} \mid \mathtt{tl}\ \mathtt{E} \mid =?\ \mathtt{E}\ \mathtt{F} \\ \mathtt{Commands} \ni \mathtt{C}, \mathtt{D} & ::= & \mathtt{X} := \mathtt{E} \mid \mathtt{C}\ ;\ \mathtt{D} \mid \mathtt{while}\ \mathtt{E}\ \mathtt{do}\ \mathtt{C} \\ \mathtt{Programs} \ni P & ::= & \mathtt{read}\ \mathtt{X}_1, \ldots, \mathtt{X}_n;\ \mathtt{C};\ \mathtt{write}\ \mathtt{Y}\end{array}$$

where $\mathtt{t} \in \mathbb{T}_\Sigma$.

1.1 Semantics of While

A configuration, next called a store, is a function $\sigma : \mathtt{Var} \to \mathbb{T}_\Sigma$. The set of stores is denoted $\mathfrak{S}_\Sigma$, or shorter, $\mathfrak{S}$ when Σ is clear from the context. Given a configuration $\sigma \in \mathfrak{S}$ a variable X and $\mathtt{t} \in \mathbb{T}_\Sigma$, $\sigma[\mathtt{X} \mapsto \mathtt{t}]$ is the store equal to σ on all variables but X for which it is set equal to t.

The semantics of an expression E applied on a configuration σ is denoted $[\![\mathtt{E}]\!]\sigma$ and defined by the equations:

$$\begin{array}{lll}[\![\mathtt{X}]\!]\sigma = \sigma(\mathtt{X}) & [\![\mathtt{hd}\ \mathtt{E}]\!]\sigma = \pi_1([\![\mathtt{E}]\!]\sigma) & [\![\mathtt{cons}\ \mathtt{E}\ \mathtt{F}]\!]\sigma = ([\![\mathtt{E}]\!]\sigma \cdot [\![\mathtt{F}]\!]\sigma) \\ [\![\mathtt{t}]\!]\sigma = \mathtt{t} & [\![\mathtt{tl}\ \mathtt{E}]\!]\sigma = \pi_2([\![\mathtt{E}]\!]\sigma) & [\![=?\ \mathtt{E}\ \mathtt{F}]\!]\sigma = [\![\mathtt{E}]\!]\sigma \simeq [\![\mathtt{F}]\!]\sigma\end{array}$$

where for equality $\simeq$, nil serves as false and $(\mathtt{nil} \cdot \mathtt{nil})$ as true. Each command $\mathtt{C} \in \mathtt{Commands}$ updates the store, that is $[\![\mathtt{C}]\!] : \mathfrak{S} \to \mathfrak{S}$ which is defined recursively as follows:

$$\begin{array}{rll}[\![\mathtt{X} := \mathtt{E}]\!]\sigma = & \sigma[\mathtt{X} \mapsto [\![\mathtt{E}]\!]\sigma] & \\ [\![\mathtt{C};\mathtt{D}]\!]\sigma = & [\![\mathtt{D}]\!]([\![\mathtt{C}]\!]\sigma) & \\ [\![\mathtt{while}\ \mathtt{E}\ \mathtt{do}\ \mathtt{C}]\!]\sigma = & \sigma & \text{if } [\![\mathtt{E}]\!]\sigma = \mathtt{nil} \\ [\![\mathtt{while}\ \mathtt{E}\ \mathtt{do}\ \mathtt{C}]\!]\sigma = & [\![\mathtt{C}\ ;\mathtt{while}\ \mathtt{E}\ \mathtt{do}\ \mathtt{C}]\!]\sigma & \text{otherwise}\end{array}$$

The program $\mathbf{p} \triangleq \mathtt{read}\ \mathtt{X}_1, \ldots, \mathtt{X}_n\ \mathtt{C}\ ;\mathtt{writeY}$ computes the following function. Given $\mathtt{t}_1, \ldots, \mathtt{t}_n$, in the initial configuration $\sigma_0(\mathtt{t}_1, \ldots, \mathtt{t}_n)$, all variables are set to nil, except $\mathtt{X}_1, \ldots, \mathtt{X}_n$ which are respectively set to $\mathtt{t}_1, \ldots, \mathtt{t}_n$. Then $[\![\mathbf{p}]\!](\mathtt{t}_1, \ldots, \mathtt{t}_n)$ is defined to be $([\![\mathtt{C}]\!]\sigma_0(\mathtt{t}_1, \ldots, \mathtt{t}_n))(\mathtt{Y})$.

1.2 While as an Acceptable Language

Let $\{\mathbf{assign}, \mathbf{seq}, \mathbf{while}, \mathtt{Var}, \mathbf{quote}, \mathbf{cons}, \mathbf{hd}, \mathbf{tl}, \mathbf{iseq}, \mathtt{nil}\}$ denote 10 distinct elements of $\mathbb{T}_\Sigma$. The representation $\underline{\mathbf{p}}$ of a program in While is defined recursively:

$$
\begin{aligned}
\underline{0} &= \mathtt{nil} & \underline{\mathtt{X}_i} &= (\mathtt{Var} \cdot \underline{i})\\
\underline{n+1} &= (\mathtt{nil} \cdot \underline{n}) & \underline{\mathtt{t}} &= (\mathbf{quote} \cdot t)\\
\underline{()} &= \mathtt{nil} & \underline{\mathtt{hd\ E}} &= (\mathbf{hd} \cdot \underline{\mathtt{E}})\\
\underline{(x_1, \ldots)} &= (\underline{x_1} \cdot \underline{(\ldots)}) & \underline{\mathtt{tl\ E}} &= (\mathbf{tl} \cdot \underline{\mathtt{E}})\\
\underline{\mathtt{X}_i := \mathtt{E}} &= (\mathbf{assign} \cdot (\underline{\mathtt{X}_i} \cdot \underline{\mathtt{E}})) & \underline{\mathtt{cons\ E\ F}} &= (\mathbf{cons} \cdot (\underline{\mathtt{E}} \cdot \underline{\mathtt{F}}))\\
\underline{\mathtt{C;D}} &= (\mathbf{seq} \cdot (\underline{\mathtt{C}} \cdot \underline{\mathtt{D}})) & \underline{\mathtt{=?\ E\ F}} &= (\mathbf{iseq} \cdot (\underline{\mathtt{E}} \cdot \underline{\mathtt{F}}))\\
\underline{\mathtt{while\ E\ do\ C}} &= (\mathbf{while} \cdot (\underline{\mathtt{E}} \cdot \underline{\mathtt{C}}))
\end{aligned}
$$

and for a program, we define $\underline{\mathtt{read\ X_1, \ldots, X_n;\ C;\ write\ Y}} = ((\underline{\mathtt{X}_1, \ldots, \mathtt{X}_n}) \cdot (\underline{\mathtt{C}} \cdot \underline{\mathtt{Y}}))$.

More generally speaking, the representation of a programming language is an injective function from the set of programs (here While) to its corresponding data set (here $\mathbb{T}_\Sigma$).

As shown by Jones [9], there is a universal program $\mathbf{u} \in$ While, that is a program $\mathbf{u}$ such that for any program $\mathbf{p} \in$ While and any data $\mathtt{t} \in \mathbb{T}_\Sigma$: $[\![\mathbf{u}]\!](\underline{\mathbf{p}}, \mathtt{t}) = [\![\mathbf{p}]\!](\mathtt{t})$. For all $m, n \in \mathbb{N}$, there is a specializer $\mathbf{s\_m\_n}$, that is a program $\mathbf{s\_m\_n}$ such that for all $m+n$-ary program $\mathbf{p}$, for all $\mathtt{t}_1, \ldots, \mathtt{t}_{m+n} \in \mathbb{T}_\Sigma$, $[\![[\![\mathbf{s\_m\_n}]\!](\underline{\mathbf{p}}, \mathtt{t}_1, \ldots, \mathtt{t}_m)]\!](\mathtt{t}_{m+1}, \ldots, \mathtt{t}_{m+n}) = [\![\mathbf{p}]\!](\mathtt{t}_1, \ldots, \mathtt{t}_{m+n})$. Finally, it is Turing-complete. Such a language is said to be acceptable in Jones/Roger's terms. As such, it is isomorphic to any other acceptable language as shown by Rogers:

Theorem 1 (Rogers *[11]***).** *Two acceptable languages are isomorphic.*

That is there is a bijective computable function transforming programs in the first language to programs in the second one with equivalent semantics.

For any acceptable language, Kleene's second recursion theorem is known to hold. We recall:

Theorem 2 (Kleene's Second Recursion Theorem). *For any $k+1$-ary program $\mathbf{p}$, there is a k-ary program $\mathbf{e}$ satisfying for all inputs $\mathtt{t}_1, \ldots, \mathtt{t}_k \in \mathbb{T}_\Sigma$: $[\![\mathbf{e}]\!](\mathtt{t}_1, \ldots, \mathtt{t}_k) = [\![\mathbf{p}]\!](\underline{\mathbf{e}}, \mathtt{t}_1, \ldots, \mathtt{t}_k)$.*

Proof. For later use, we give a proof for $k = 1$. The proof for $k > 1$ follows the same schema. For the specializer $\mathbf{s\_1\_1} \triangleq \mathtt{read\ X_0, X_1; C_{s\_1\_1}; write\ Y}$, for all binary program $\mathbf{p} \in \mathcal{P}$ and $\mathtt{t}, \mathtt{t}' \in \mathbb{T}_\Sigma$, $[\![[\![\mathbf{s\_1\_1}]\!](\underline{\mathbf{p}}, \mathtt{t})]\!](\mathtt{t}') = [\![\mathbf{p}]\!](\mathtt{t}, \mathtt{t}')$. Let $\mathbf{p} = \mathtt{read\ X'_0, X'_1; C_p; write\ Y'}$. By renaming variables, we suppose without loss of generality that it does not share variables with $\mathbf{s\_1\_1}$. Then, let $\mathbf{r_p}$ be the program:

```
read X''_0, X''_1;
X_0 := X''_0; X_1 := X''_0;
C_s_1_1;
X'_0 := Y; X'_1 := X''_1;
C_p;
write Y'
```

with $\mathtt{X}_0'', \mathtt{X}_1''$ some fresh variables. Then, it is clear that for all $\mathbf{q} \in \mathcal{P}$ and all $\mathtt{t} \in \mathbb{T}_\Sigma$, $[\![\mathbf{r_p}]\!](\underline{\mathbf{q}}, \mathtt{t}) = [\![\mathbf{p}]\!]([\![\mathbf{s\_1\_1}]\!](\underline{\mathbf{q}}, \underline{\mathbf{q}}), \mathtt{t})$. Let $\mathbf{e} \triangleq \overline{[\![\mathbf{s\_1\_1}]\!](\underline{\mathbf{r_p}}, \underline{\mathbf{r_p}})}$, we get:

$$\begin{aligned}
[\![\mathbf{e}]\!](t) &= \lceil [\![\mathbf{s_1_1}]\!](\underline{\mathbf{r_p}}, \underline{\mathbf{r_p}}) \rceil(\mathtt{t}) && \text{by def. of } \lceil . \rceil \\
&= [\![\mathbf{r_p}]\!](\underline{\mathbf{r_p}}, \mathtt{t}) && \text{by def. of } \mathbf{s_1_1} \\
&= [\![\mathbf{p}]\!](\underline{[\![\mathbf{s_1_1}]\!](\underline{\mathbf{r_p}}, \underline{\mathbf{r_p}})}, \mathtt{t}) && \text{by remark above} \\
&= [\![\mathbf{p}]\!](\underline{\mathbf{e}}, \mathtt{t}) && \text{since } \underline{\overline{[\![\mathbf{s_1_1}]\!](\underline{\mathbf{r_p}}, \underline{\mathbf{r_p}})}} = [\![\mathbf{s_1_1}]\!](\underline{\mathbf{r_p}}, \underline{\mathbf{r_p}})
\end{aligned}$$

As justified by Bonfante, Kaczmarek and Marion in [3], a virus can be formalized as follows:

Definition 2 (Computer Virus). *Given a computable function B called the propagation function, a virus is a program $\mathbf{v}$ such that $[\![\mathbf{v}]\!](\mathtt{t}) = B(\underline{\mathbf{v}}, \mathtt{t})$ for all $\mathtt{t} \in \mathbb{T}_\Sigma$.*

In other words, it is a fixed point for a propagation function. Thus, as shown in [3], the second recursion theorem of Kleene implies that for any propagation function there is a corresponding virus. In other words, the theorem provides a virus compiler, and there are no general ways to avoid them. In the remaining, we restrict `While` to get around computer viruses. We propose two strategies to that end. First, we delineate a programming language in which the Recursion Theorem does not hold. As shown by the proof of the Recursion Theorem, the existence of a specializer, of composition and projection is sufficient to prove the Theorem. Thus, we find a programming language without specializer.

The second strategy consists in finding a language for which fixed point exists, but viruses can be detected. By detection we mean program equivalence as justified by Adleman in [1].

2 On Cons-Free Programs

$\mathtt{While}_{\backslash\{\mathtt{cons}\}}$ is the language `while` restricted to expressions of the shape:

$$\mathtt{Expressions} \ni \mathtt{E}, \mathtt{F} ::= \mathtt{X} \mid \mathtt{t} \mid \mathtt{hd}\ \mathtt{E} \mid \mathtt{tl}\ \mathtt{E} \mid \mathtt{E} =?\ \mathtt{F}$$

Such programs where initially considered by Jones under a complexity perspective. He proved that they compute exactly LOGSPACE predicates. We show that the Second Recursion Theorem does not hold in $\mathtt{While}_{\backslash\{\mathtt{cons}\}}$.

In this section, when $\mathtt{t} \in \mathbb{T}_\Sigma$ and $S \subseteq \mathbb{T}_\Sigma$, the notation $\mathtt{t} \trianglelefteq S$ means $\exists \mathtt{t}' \in S : \mathtt{t} \trianglelefteq \mathtt{t}'$. When S, S' are two sets, the notation $S \trianglelefteq S'$ states for $\forall \mathtt{t} \in S, \exists \mathtt{t}' \in S' : \mathtt{t} \trianglelefteq \mathtt{t}'$. For a store σ, let $\mathrm{Rg}(\sigma) = \{\sigma(\mathtt{X}) \mid \mathtt{X} \in \mathtt{Var}\}$.

Definition 3. *Let $\mathtt{E}$ be an expression, we denote by $c(\mathtt{E})$ the set of all constants occurring in $\mathtt{E}$; formally, by induction: $c(\mathtt{X}) = \emptyset$, $c(\mathtt{t}) = \{\mathtt{t}\}$, $c(\mathtt{hd}\ \mathtt{E}) = c(\mathtt{tl}\ \mathtt{E}) = c(\mathtt{E})$ and $c(\mathtt{cons}\ \mathtt{E}\ \mathtt{F}) = c(\mathtt{E}\!=?\ \mathtt{F}) = c(\mathtt{E}) \cup c(\mathtt{F})$.*

The definition is extended to commands: $c(\mathtt{X} := \mathtt{E}) = c(\mathtt{E})$, $c(\mathtt{C}\ ;\ \mathtt{D}) = c(\mathtt{C}) \cup c(\mathtt{D})$, $c(\mathtt{while}\ \mathtt{E}\ \mathtt{do}\ \mathtt{C}) = c(\mathtt{E}) \cup c(\mathtt{C})$ and finally to programs by the equation $c(\mathtt{read}\ \mathtt{X}_1, \ldots, \mathtt{X}_n; \mathtt{C};\ \mathtt{write} \mathtt{Y}) = c(\mathtt{C})$.

Proposition 1. *Given a program* $\mathbf{p} \in \mathtt{While}_{\setminus\{\mathtt{cons}\}}$ *of arity* n *and* $\mathtt{t}_1, \cdots, \mathtt{t}_n$ *some elements of* $\mathbb{T}_\Sigma$, $[\![\mathbf{p}]\!](\mathtt{t}_1, \cdots, \mathtt{t}_n) \trianglelefteq c(\mathbf{p}) \cup \{\mathtt{t}_1, \cdots, \mathtt{t}_n\} \cup \{(\mathtt{nil} \cdot \mathtt{nil})\}$ *whenever* $[\![\mathbf{p}]\!](\mathtt{t}_1, \ldots, \mathtt{t}_n)$ *is defined.*

Proof. A very similar result occurs in Jones [9]. It is by induction on the structure of programs.

Proposition 2. *Given a program* $\mathbf{p} \in \mathtt{While}_{\setminus\{\mathtt{cons}\}}$ *and* $t \in \mathbb{T}_\Sigma$ *if* $\mathbf{p}$ *computes the constant function equal to* $\mathtt{t}$*, then either:*

$$\mathtt{t} \trianglelefteq (\mathtt{nil} \cdot \mathtt{nil}) \quad \textit{or} \quad \mathtt{t} \trianglelefteq c(\mathbf{p})$$

Proof. Applying Proposition 1 to the program $\mathbf{p}$: $[\![\mathbf{p}]\!](\mathtt{nil}) \trianglelefteq c(\mathbf{p}) \cup \{\mathtt{nil}, (\mathtt{nil} \cdot \mathtt{nil})\}$. But, again, $\mathtt{nil} \trianglelefteq (\mathtt{nil} \cdot \mathtt{nil})$, thus $[\![\mathbf{p}]\!](\mathtt{nil}) \trianglelefteq c(\mathbf{p}) \cup \{(\mathtt{nil} \cdot \mathtt{nil})\}$.

2.1 $\mathtt{While}_{\setminus\{\mathtt{cons}\}}$ Does Not Contain a Specializer

Theorem 3. *Whatever the choice of a representation, and in particular for the one given in the preceding section, there is no specializer in* $\mathtt{While}_{\setminus\{\mathtt{cons}\}}$*.*

Proof. We assume the existence of a specializer $\mathbf{s\_1\_1}$. Let us define two programs $\mathbf{p}$ and $\mathbf{q}$: $\mathbf{p} \triangleq \mathtt{read}\ \mathtt{X}_1, \mathtt{X}_2;\ \mathtt{Y} = \mathtt{X}_1;\ \mathtt{write}\ \mathtt{Y}$ and $\mathbf{q} \triangleq \mathtt{read}\ \mathtt{X}_1, \mathtt{X}_2;\ \mathtt{if}\ (\mathtt{X}_2 =?\ \mathtt{nil})\ \mathtt{Y} := \mathtt{X}_1$ else $\mathtt{Y} := \mathtt{X}_2;\ \mathtt{write}\ \mathtt{Y}$.

Consider some $\mathtt{t} \in \mathbb{T}_\Sigma$, we apply Proposition 1 to the program $\mathbf{s\_1\_1}$ and the inputs $\underline{\mathbf{p}}, \mathtt{t}$ we obtain:

$$\begin{cases} [\![\mathbf{s\_1\_1}]\!](\underline{\mathbf{p}}, \mathtt{t}) \trianglelefteq C \textit{ or} \\ [\![\mathbf{s\_1\_1}]\!](\underline{\mathbf{p}}, \mathtt{t}) \trianglelefteq \mathtt{t} \end{cases}$$

with $C = ((\mathtt{nil} \cdot \mathtt{nil}) \cdot \underline{\mathbf{p}} \cdot c_1 \cdots c_m)$ and $c(\mathbf{s\_1\_1}) = \{c_1, \cdots, c_m\}$.

Given $\mathtt{t} \neq \mathtt{t}'$, $\lceil [\![\mathbf{s\_1\_1}]\!](\mathbf{p}, \mathtt{t}) \rceil (\mathtt{nil}) = \mathtt{t} \neq \mathtt{t}' = \lceil [\![\mathbf{s\_1\_1}]\!](\mathbf{p}, \mathtt{t}') \rceil (\mathtt{nil})$. Thus, $\mathtt{t} \mapsto [\![\mathbf{s\_1\_1}]\!](\mathbf{p}, \mathtt{t})$ is an injective function. From that, we state that the set $S_C = \{\mathtt{t} \in \mathbb{T}_\Sigma \mid |[\![\mathbf{s\_1\_1}]\!](\mathbf{p}, \mathtt{t})| \leq |C|\}$ is finite. We set $N_1 = \max\{|\mathtt{t}| \mid \mathtt{t} \in S_C\}$. Then, for all $|\mathtt{t}| > N_1$, we can state that $|[\![\mathbf{s\_1\_1}]\!](\mathbf{p}, \mathtt{t})| > |C|$ which in turn means that $[\![\mathbf{s\_1\_1}]\!](\underline{\mathbf{p}}, \mathtt{t}) \trianglelefteq \mathtt{t}$.

Since for any $\mathtt{t} \neq \mathtt{t}'$, $\lceil [\![\mathbf{s\_1\_1}]\!](\mathbf{q}, \mathtt{t}) \rceil (\mathtt{nil}) = \mathtt{t} \neq \mathtt{t}' = \lceil [\![\mathbf{s\_1\_1}]\!](\mathbf{q}, \mathtt{t}') \rceil (\mathtt{nil})$, the function $\mathtt{t} \mapsto [\![\mathbf{s\_1\_1}]\!](\mathbf{q}, \mathtt{t})$ is injective. Thus, we use the same approach: there exists an integer N_2 such that $|\mathtt{t}| > N_2$ implies $[\![\mathbf{s\_1\_1}]\!](\underline{\mathbf{q}}, \mathtt{t}) \trianglelefteq \mathtt{t}$. We set $N = \max(N_1, N_2)$ and we get :

$$|\mathtt{t}| > N \Rightarrow ([\![\mathbf{s\_1\_1}]\!](\underline{\mathbf{p}}, \mathtt{t}) \trianglelefteq \mathtt{t} \ \textit{ and } \ [\![\mathbf{s\_1\_1}]\!](\underline{\mathbf{q}}, \mathtt{t}) \trianglelefteq \mathtt{t})$$

Given $n \in \mathbb{N}$, wet define $n_{\downarrow\trianglelefteq} = \{\mathtt{t} \in \mathbb{T}_\Sigma \mid \mathtt{t} \trianglelefteq \underline{n}\}$ with the encoding of integers defined in the preceding section. Observe that we have the equality:

$$n_{\downarrow\trianglelefteq} = \{\underline{k} \mid 0 \leq k \leq n\}. \tag{1}$$

For all $i, j \in \mathbb{N}$, let k such that $k \neq i$ and $k \neq j$. We have $[\![[\![\mathbf{s\_1\_1}]\!](\underline{\mathbf{p}}, \underline{i})]\!](\underline{k}) = \underline{i} \neq \underline{k} = [\![[\![\mathbf{s\_1\_1}]\!](\underline{\mathbf{q}}, \underline{j})]\!](\underline{k})$ which means that for all $i, j \in \mathbb{N}$: $[\![\mathbf{s\_1\_1}]\!](\underline{\mathbf{p}}, \underline{i}) \neq [\![\mathbf{s\_1\_1}]\!](\underline{\mathbf{q}}, \underline{j})$.

Consider some $M > N$. Recall that both $\mathtt{t} \mapsto [\![\mathbf{s\_1\_1}]\!](\mathbf{p}, \mathtt{t})$ and $\mathtt{t} \mapsto [\![\mathbf{s\_1\_1}]\!](\mathbf{q}, \mathtt{t})$ are injective. Due to the aforementioned remark, the set $S_{MN} = \{[\![\mathbf{s\_1\_1}]\!](\mathbf{p}, \underline{i}) \mid N < i \leq M\} \cup \{[\![\mathbf{s\_1\_1}]\!](\mathbf{q}, \underline{i}) \mid N < i \leq M\}$ contains exactly $2 \times (M - N)$ elements. However, all these elements verify $S_{MN} \trianglelefteq \underline{M} \in M_{\downarrow \trianglelefteq}$, but by Eq. 1, the set $M_{\downarrow \trianglelefteq}$ contains only $M + 1$ elements which leads to $2 \times (M - N) \leq M + 1$. The inequality does not hold for $M = 2 \times (N + 1)$.

The non-existence of a specializer does not mean that there are no fixed points in $\mathtt{While}_{\setminus\{\mathtt{cons}\}}$, for instance the nowhere defined program

$$\mathtt{read}\ \mathtt{X}_1\ ; \mathbf{while}\ (\mathtt{nil} \cdot \mathtt{nil})\ \mathtt{do}\ \mathtt{X}_1 := \mathtt{X}_1\ ;\ \mathtt{write}\ \mathtt{X}_1$$

is a fixed point for the program

$$\mathtt{read}\ \mathtt{X}_1, \mathtt{X}_2\ ; \mathbf{while}\ (\mathtt{nil} \cdot \mathtt{nil})\ \mathtt{do}\ \mathtt{X}_1\ := \mathtt{X}_1\ ; \mathtt{write}\ \mathtt{X}_1.$$

There are other fix-point constructions which are not based on specializers, some are found in Smullyan's [12], but we wish to mention here the approach due to Moss [10] which is based on a very elementary framework, text register machines.

Nevertheless, there are programs for which there is no fixed points. In other words, the Recursion Theorem does not hold. For instance, for the homeomorphic representation of programs presented in $\mathtt{While}_{\setminus\{\mathtt{cons}\}}$:

Proposition 3. *There is no Quine in* $\mathtt{While}_{\setminus\{\mathtt{cons}\}}$.

Proof. Ad absurdum, suppose there is a Quine $\mathbf{q}$ in $\mathtt{While}_{\setminus\{\mathtt{cons}\}}$. It is a fixed point for the program $\mathbf{pi}_1 \triangleq \mathtt{read}\ \mathtt{X}_1, \mathtt{X}_2\ ; \mathtt{X}_1\ := \mathtt{X}_1\ ; \mathtt{write}\ \mathtt{X}_1$. Since $[\![\mathtt{q}]\!](\mathtt{nil})$ $\underline{\mathbf{q}}$, since $|\underline{\mathbf{q}}| > |(\mathtt{nil} \cdot \mathtt{nil})|$ (as it is the case for any programs), we can state with Corollary 1 that $\underline{\mathbf{q}} \trianglelefteq c(\mathbf{q})$. Let $\mathtt{t} \in c(\mathbf{q})$ such that

$$\underline{\mathbf{q}} \trianglelefteq \mathtt{t}. \tag{2}$$

With the homeomorphic encoding we chose, we can state that $(\mathbf{quote} \cdot \mathtt{t}) \trianglelefteq \underline{\mathbf{q}}$. Thus

$$\mathtt{t} \triangleleft (\mathbf{quote} \cdot \mathtt{t}) \trianglelefteq \underline{\mathbf{q}} \tag{3}$$

The two inequalities 2 and 3 are not compatible. The conclusion follows.

Quines are interesting in Adleman's perspective. They correspond to the 'Imitate' scenario. The 'Infection' scenario would not be possible due to Proposition 1. Thus, programs in $\mathtt{While}_{\setminus\{\mathtt{cons}\}}$ cannot be infected in his view. The reader may notice that the proof here depends on the choice of the representation of programs. Indeed, it is not difficult to define an encoding for which there is a Quine. Simply modify $\underline{\ }$ so that the encoding of $\mathbf{nil} \triangleq \mathtt{read}\ \mathtt{X}_1\ ;\ \mathtt{X}_1\ := \mathtt{nil}\ ; \mathtt{write}\ \mathtt{X}_1$ is set to $\mathtt{nil}$. Then, $[\![\mathbf{nil}]\!](\mathtt{t}) = \mathtt{nil}$ which is the required equation. Nevertheless, there are no other Quines. To end the remark, observe that program representation can be on the defense side, not on the virus writer's one.

3 Tiny, a Whippersnapper Programming Language

Tiny is the language While restricted to expressions of the shape:

$$\texttt{Expressions} \ni \texttt{E},\texttt{F} ::= \texttt{X} \mid \texttt{t} \mid \texttt{cons E F} \mid \texttt{hd E} \mid \texttt{tl E}$$

and commands to:

$$\texttt{Commands} \ni \texttt{C},\texttt{D} ::= \texttt{X} := \texttt{E} \mid \texttt{C ; D}.$$

Obviously, Tiny is not Turing complete. Actually, it is a very weak fragment of computable functions: it contains only functions computable in constant time. However, the Recursion Theorem holds, surprisingly, in Tiny. For the representation of programs that we defined, the Recursion Theorem holds:

Theorem 4. *Given a k-ary program* $\mathbf{p} \in \texttt{Tiny}$, *there is a* $k-1$*-ary program* $\mathbf{e} \in \texttt{Tiny}$ *such that for all* $\texttt{t}_1, \ldots, \texttt{t}_k \in \mathbb{T}_\Sigma$: $[\![\mathbf{e}]\!](\texttt{t}_1, \ldots, \texttt{t}_k) = [\![\mathbf{p}]\!](\underline{\mathbf{e}}, \texttt{t}_1, \ldots, \texttt{t}_k)$.

Proof. Again, we give a proof for $k = 1$. The other cases are left to the reader. If we come back to the previous proof of the Recursion Theorem, it is clear that the program $\mathbf{r_p}$ is in Tiny whenever both $\mathbf{p}$ and **s_1_1** are in Tiny. Since $\mathbf{p} \in \texttt{Tiny}$ by hypothesis, $\mathbf{r_p}$ is in Tiny if there is a specializer within Tiny. This is actually the case: define **s_1_1** $\triangleq$

```
read X0, X1;
C := hd (tl X0);               // the representation of the body of X0
X := hd (hd X0);               // the rep. of the first input variable of X0
XL := tl (hd X0);              // the remaining variables of X0
Y := tl (tl X0);               // the rep. of the output variable of X0
E := cons quote X1;            // the rep. of the value t of X1
C0 := cons assign (cons X E);  // the rep of X := t
C := cons seq (cons C0 C);     // the rep. of X := t ; C
P := cons XL(cons C Y);        // the packaging of the new (unary) program
write P
```

This is a specializer. Indeed, let $\mathbf{p} = \texttt{read X}'_0, \texttt{X}'_1; \texttt{C}_\mathbf{p}; \texttt{write Y}'$. For all $\texttt{t} \in \mathbb{T}_\Sigma$, $[\![\mathbf{s\_1\_1}]\!](\underline{\mathbf{p}}, \texttt{t}) = \underline{\texttt{read X}'_1; \texttt{X}'_0 := \texttt{t} ; \texttt{C}_\mathbf{p}; \texttt{write Y}'}$. Thus, for all $\texttt{t}' \in \mathbb{T}_\Sigma$: we have $[\![\overline{[\![\mathbf{s\_m\_n}]\!](\underline{\mathbf{p}}, \texttt{t})}]\!](\texttt{t}') = [\![\mathbf{p}]\!](\texttt{t}, \texttt{t}')$ as required.

So, $\mathbf{r_p}$ is in Tiny. Since the fixed point $\mathbf{e} = \overline{\texttt{S}^1_1(\mathbf{r_p}, \mathbf{r_p})}$ is in Tiny, the proof ends as a corollary of the following Lemma:

Lemma 1. *If* $\mathbf{p} \in \texttt{Tiny}$, *for all* $\texttt{t} \in \mathbb{T}_\Sigma$: $\overline{[\![\mathbf{s\_1\_1}]\!](\underline{\mathbf{p}}, \texttt{t})} \in \texttt{Tiny}$.

Proof. Recall that $[\![\mathbf{s\_1\_1}]\!](\underline{\mathbf{p}}, \texttt{t}) = \underline{\texttt{read X}'_1; \texttt{X}'_0 := \texttt{t} ; \texttt{C}_\mathbf{p}; \texttt{write Y}}$. Since $\mathbf{p}$ is in Tiny, $[\![\mathbf{s\_1\_1}]\!](\underline{\mathbf{p}}, \texttt{t})$ is itself the representation of a program in Tiny.

3.1 Program Equivalence in Tiny

From the above, there are viruses in `Tiny`. However, with the scenario made in the introduction, we can protect a system which is based on `Tiny`. Protection amounts to problem equivalence decision. It is the following. Given two programs $\mathbf{p}, \mathbf{q}$, does $[\![\mathbf{p}]\!] = [\![\mathbf{q}]\!]$? In general—in particular for a Turing-complete Language—, such a decision is not computable (as a direct consequence of Rice's Theorem). But, for `Tiny`, there is a simple decision procedure.

Theorem 5. *Equivalence of programs is computable for programs in* `Tiny`.

Proof. Composing expressions, one may reduce programs in Tiny to just one expression. The semantics of an expression can be explicitly expanded. Then, equivalence is equality of the semantics. We provide in appendix an algorithm that compute the semantics of expressions. However,

Proposition 4. *Equalence of programs in* `Tiny` *is not in* `Tiny`.

Lemma 2. *Any program in* `Tiny` *is monotonic, that is if* $\mathtt{t}_i \trianglelefteq \mathtt{t}'_i$ *for all* $i \leq n$, *then* $[\![\mathbf{p}]\!](\mathtt{t}_1, \ldots, \mathtt{t}_n) \trianglelefteq [\![\mathbf{p}]\!](\mathtt{t}'_1, \ldots, \mathtt{t}'_n)$.

Proof. We have seen above that any program in `Tiny` is equivalent to some program of the shape $\mathtt{read}\ \mathtt{X}_1, \ldots, \mathtt{X}_n; \mathtt{Y} := \mathtt{E}; \mathtt{write}\ \mathtt{Y}$, thus we restrict our attention to these ones. Since A_1 and A_2 are monotonic, the result holds by an immediate induction on E.

Proof (Proposition 4). Let us come back to the proof of the Proposition. Ad absurdum, suppose that there is some program $\mathbf{eq} \in$ `Tiny` such that $[\![\mathbf{eq}]\!](\underline{\mathbf{p}_1}, \underline{\mathbf{p}_2}) \neq \mathtt{nil}$ iff $[\![\mathbf{p}_1]\!] = [\![\mathbf{p}_2]\!]$ for all $\mathbf{p}_1, \mathbf{p}_2 \in$ `Tiny`. Let $\mathbf{p}_1 \triangleq \mathtt{read}\ \mathtt{X}; \mathtt{Y} := \mathtt{nil}; \mathtt{write}\ \mathtt{Y}$. It is in tiny, thus, $[\![\mathbf{eq}]\!](\underline{\mathbf{p}_1}, \underline{\mathbf{p}_1}) \neq \mathtt{nil}$ since $\mathbf{p}_1$ is equivalent to itself. Observe that $\mathbf{p}_2 \triangleq \mathtt{read}\ \mathtt{X}; \mathtt{Y} := \mathtt{cons\ nil\ nil}; \mathtt{write}\ \mathtt{Y}$ which is also in `Tiny` verifies $\underline{\mathbf{p}_1} \trianglelefteq \underline{\mathbf{p}_2}$. By Lemma 2, we can state that $[\![\mathbf{eq}]\!](\underline{\mathbf{p}_1}, \underline{\mathbf{p}_1}) \trianglelefteq [\![\mathbf{eq}]\!](\underline{\mathbf{p}_1}, \underline{\mathbf{p}_2})$. In turn, that means $[\![\mathbf{eq}]\!](\mathbf{p}_1, \mathbf{p}_2) \neq \mathtt{nil}$. But $\mathbf{p}_1$ and $\mathbf{p}_2$ are not equivalent: $[\![\mathbf{p}_1]\!](\mathtt{nil}) \neq [\![\mathbf{p}_2]\!](\mathtt{nil})$, thus a contradiction.

4 Conclusion

Thought it is conceptually deep, the Recursion Theorem can be difficult to utilize for practical applications. In the context of computer viruses, it can often have a negative flavor. To our mind, our work opens a new branch of research which constructively studies fixed point within constrained computation systems. Types, logics and weak arithmetics arise as good candidates for that sake.

We end with a side remark about the efficiency of fixed points. Let us cite Hansen, Nikolajsen, Träff and Jones in [2]: "[...] running a fixed-point program to compute the factorial of n results in n levels of interpretation, each one slowing down execution by a large constant factor". This leads the authors to introduce a self-reflection statement that, supposedly, enables efficient fixed points.

The fixed points presented above never involve any interpretation layer. The construction of the specializer shows that it only introduce a constant time overhead with respect to the initial program. Therefore, the complexity of the fixed point **e** is equal to the one of the program $\mathbf{r_p}$. It involves the code of **s_1_1** which is in `Tiny`, and thus takes constant time and so few assignment which do not increase the size of their inputs. In the end, we see that the program **e** is as efficient as **p** on its input up to a constant factor.

Acknowledgements. The authors would like to thank the Institute for Advanced Study and the organizers of the 2012 Program in Theoretical Physics on Biology and Computation. In particular, we would like to thank Stanislas Leibler for several motivating discussions.

References

1. Adleman, L.M.: An abstract theory of computer viruses. In: Goldwasser, S. (ed.) CRYPTO 1988. LNCS, vol. 403, pp. 354–374. Springer, Heidelberg (1990)
2. Amtoft, T., Thomas, H., Jesper, N., Träff, L., Jones, N.D.: Experiments with implementations of two theoretical constructions. In: Proceedings of the 2007 Workshop on Programming Languages and Analysis for Security, PLAS 2007, San Diego, California, USA, 14 June 2007, pp. 47–52. ACM (2007)
3. Bonfante, G., Kaczmarek, M., Marion, J.-Y.: A classification of viruses through recursion theorems. In: Cooper, S.B., Löwe, B., Sorbi, A. (eds.) CiE 2007. LNCS, vol. 4497, pp. 73–82. Springer, Heidelberg (2007)
4. Borello, J.-M., Mé, L.: Code obfuscation techniques for metamorphic viruses. J. Comput. Virol. **4**(3), 211–220 (2008)
5. Case, J., Moelius, S.E.: Cautious virus detection in the extreme. In: Proceedings of the 2007 Workshop on Programming Languages and Analysis for Security, PLAS 2007, San Diego, California, USA, 14 June 2007, pp. 47–52. ACM (2007)
6. Case, J., Moelius, S.E.: Characterizing programming systems allowing program self-reference. Theory Comput. Syst. **45**(4), 756–772 (2009)
7. Cohen, F.: Computer viruses: theory and experiments. Comput. Secur. **6**(1), 22–35 (1987)
8. Collberg, C., Nagra, J.: Surreptitious Software: Obfuscation, Watermarking, and Tamperproofing for Software Protection, 1st edn. Addison-Wesley Professional, New Jersey (2009)
9. Jones, N.D.: Computability and Complexity, from a Programming Perspective. MIT press, Cambridge (1997)
10. Moss, L.S.: Recursion theorems and self-replication via text register machine programs. Bull. EATCS **89**, 171–182 (2006)
11. Rogers Jr., H.: Theory of Recursive Functions and Effective Computability. McGraw Hill, New York (1967)
12. Smullyan, R.M.: Recursion Theory for Metamathematics. Oxford University Press, Oxford (1993)
13. Zuo, Z., Zhou, M.: Some further theorical results about computer viruses. Comput. J. **47**(6), 627–633 (2004)

$ITRM$-Recognizability from Random Oracles

Merlin Carl(✉)

Fachbereich für Mathematik und Statistik der Universität Konstanz, Konstanz, Germany
merlin.carl@uni-konstanz.de

Abstract. By a theorem of Sacks, if a real x is recursive relative to all elements of a set of positive Lebesgue measure, x is recursive. This statement - and the analogous statement for non-meagerness instead of positive Lebesgue measure - has been shown to carry over to many models of transfinite computations in [7]. Here, we start exploring another analogue concerning recognizability rather than computability. We show that, for Infinite Time Register Machines ($ITRM$s), if a real x is recognizable relative to all elements of a non-meager Borel set Y, then x is recognizable.

1 Introduction

It is well-known (see e.g. [2]) that, if x is a non-recursive real number, the Turing upper-cone of x is a meager set. Intuitively, randomly choosing an oracle is not likely to increase the chance of solving the problem of computing a certain real fixed in advance. In a similar spirit, by a theorem of Sacks (see e.g. [8]), if a real x is recursive relative to all elements of a set Y of positive Lebesgue measure, x is recursive. This statement - along with its analogue where the condition of Y having positive Lebesgue measure is replaced by the condition that Y is Borel and not meager - continues to hold for many machine models of infinitary computations as demonstrated in [7] (for some it is currently still open, while for others it turns out to be independent of ZFC). In particular, it was shown in [7] that, if x is computable by an infinite time register machine ($ITRM$) relative to all elements of a set Y of positive Lebesgue measure or a non-meager Borel set Z, then x is $ITRM$-computable.

Besides computability, there is another way how an infinitary machine can 'determine' a real x: x is recognizable if and only if there is a program that halts on all oracles and outputs 1 when run on the oracle x and otherwise outputs 0. Recognizability is known to be a strictly (and in fact much) weaker property than computability. In [3], a notion of relativized recognizability was considered, resembling computations with oracles. This motivates us to ask whether the 'random oracles are not informative'-intuition is sufficiently stable to still hold in this context, i.e. recognizability relative to all oracles in some 'large' set of reals (i.e. a set of positive Lebesgue measure or a non-meager Borel set) implies recognizability. This paper treats the simplest non-trivial case of this question, namely Infinite Time Register Machines ($ITRM$s) and recognizability from all oracles in a Borel set that is not meager.

A. Beckmann et al. (Eds.): CiE 2015, LNCS 9136, pp. 137–144, 2015.
DOI: 10.1007/978-3-319-20028-6_14

Infinite Time Register Machines (*ITRMs*), introduced in [10] and further studied in [12], work similar to the classical unlimited register machines (*URMs*) described in [6]. In particular, the *ITRM*-programs are the same as the *URM*-programs; moreover, *ITRMs* use finitely many registers each of which can store a single natural number. Also, like a *URM*, an *ITRM* has a special halting state. The difference is that *ITRMs* use transfinite ordinal running time: The state of an *ITRM* at a successor ordinal is obtained as for *URMs*. At limit times, the program line is the inferior limit of the earlier program lines and there is a similar limit rule for the register contents. If the inferior limit of the earlier register contents is infinite, the register is reset to 0.

Notation: If X is a set, $\mathfrak{P}(X)$ denotes its power set. We fix some natural enumeration $(P_i|i \in \omega)$ of the *ITRM*-programs. For P an *ITRM*-program, $x \subseteq \omega$ and $i, j \in \omega$, we write $P^x(i) \downarrow= j$ for the statement that P, when run in the oracle x with i in its first register and 0 in all other registers, halts with j in its first register. $P^x \downarrow$ abbreviates the statement that the computation of P in the oracle x on the input 0 halts. For notions and results on admissible set theory see [1] or [16], for descriptive set theory see [11], concerning forcing [14]. KP denotes Kripke-Platek set theory. $\omega_i^{CK,x}$ denotes the ith x-admissible infinite ordinal, $\omega_\omega^{CK,x} := \sup\{\omega_i^{CK,x} : i \in \omega\}$. δ is the Kronecker symbol, i.e. for $x, y \subseteq \omega$, let $\delta(x, y) = 1$ if and only if $x = y$ and $\delta(x, y) = 0$ otherwise. We say that $A \subseteq [0, 1]$ is non-meager if and only if A is Borel and not meager. When $(A, \in)$ is an extensional $\in$-structure and $f : \omega \to A$ is a bijection, then $c := \{p(i, j) : f(i) \in f(j)\}$ is called a code for $(A, \in)$, where p is Cantor's pairing function.

2 Infinite Time Register Machines

For details on *ITRMs*, we refer to [10,12,13]. Here, we briefly recall some standard notions and facts concerning *ITRMs* that will be used below.

Definition 1. *$x \subseteq \omega$ is ITRM-computable in the oracle $y \subseteq \omega$ if and only if there exists an ITRM-program P such that, for $i \in \omega$, P with oracle y stops for every natural number j in its first register at the start of the computation and returns* 1 *if and only if $j \in x$ and otherwise returns* 0. *A real ITRM-computable in the empty oracle is simply called ITRM-computable.*

It is not hard to see that any *ITRM*-computation either stops or eventually cycles. Moreover, it can be shown (see [12]) that an *ITRM*-computation eventually cycles if and only if some state of the computation, consisting of the active program line index l and the register contents $(r_1, ..., r_n)$ appears at two different times $\alpha < \beta$ such that neither the active program line index nor any of the register contents drops below their corresponding value at time α. This halting criterion can be effectively tested by an *ITRM*, which leads to the following crucial property of *ITRMs*:

Theorem 2. *Let $\mathbb{P}_n$ denote the set of ITRM-programs using at most n registers, and let $(P_{i,n} \mid i \in \omega)$ enumerate $\mathbb{P}_n$ in some natural way. Then the bounded halting problem $H_n^x := \{i \in \omega \mid P_{i,n}^x(0) \downarrow\}$ is computable uniformly in the oracle x by an ITRM-program (using more than n registers, of course).*

Moreover, if $P \in \mathbb{P}_n$, $i \in \omega$, $x \subseteq \omega$ and $P^x(i) \downarrow$, then the computation takes less than $\omega_{n+1}^{CK,x}$ many steps. Consequently, if P is an ITRM-program and $i \in \omega$, $x \subseteq \omega$ are such that $P^x(i) \downarrow$, then $P^x(i)$ stops in less than $\omega_\omega^{CK,x}$ many steps.

Proof. The corresponding results from [12] easily relativize.

Theorem 3. *Let $x, y \subseteq \omega$. Then x is ITRM-computable in the oracle y if and only if $x \in L_{\omega_\omega^{CK,y}}[y]$. Moreover, there is a function $g : \omega \to \omega$ such that any $x \subset L_{\omega_n^{CK,y}}[y]$ is computable in the oracle y by some ITRM-program P using at most $g(n)$ registers.*

Proof. This is a relativization of the main results of [13].

Lemma 1. *There are ITRM-programs $(P_n : n \in \omega)$ and Q such that, for every $x \subseteq \omega$:*

(1) P_n^x computes a real number coding $L_{\omega_{n+1}^{CK,x}+2}[x]$.

(2) Given a natural number n and a natural number m coding a finite set p of natural numbers, $Q^x(m)$ computes a real number $y \supseteq p$ that is Cohen-generic over $L_{\omega_{n+1}^{CK,x}+1}[x]$.

Proof. (1) By standard fine-structural considerations, such a code is contained in $L_{\omega_{n+1}^{CK,x}+3}[x]$ and hence computable by some *ITRM*-program P_x by Theorem 3. Moreover, there is some $k \in \omega$ such that for each x, P_x uses at most k many registers. To compute a code for $L_{\omega_{n+1}^{CK,x}+2}[x]$ uniformly in the oracle x, we search, starting with $i = 0$, through ω in the following way: Given $i \in \omega$, first determine, using Theorem 2, whether $\forall j \in \omega P_{i,k}^x(j) \downarrow \{0,1\}$, i.e. whether $P_{i,k}^x$ computes a real. If so, determine, using the techniques for evaluating proof predicates with *ITRM*s from the proof of the lost melody theorem for *ITRM*s in [10], whether the real computed by P_i^x is a code for a well-founded $\in$-structure of the form $L_\alpha[x]$ such that α is of the form $\beta+2$, $L_\beta[x] \models \text{KP}$ and $L_\beta[x]$ contains exactly n elements of the form $L_\gamma[x]$ such that $L_\gamma[x] \models KP$. If this holds, then a code as desired has been found; otherwise, proceed with $i+1$. As we observed, some program in $\mathbb{P}_k$ computes a code as desired, so this procedure will terminate for some finite value of i.

(2) As $L_{\omega_{n+1}^{CK,x}+1}[x]$ is isomorphic (via the Levy collapsing map) to its own Σ_1-Skolem hull of $\{x\}$, it follows that $L_{\omega_{n+1}^{CK,x}+1}[x]$ is countable in $L_{\omega_{n+1}^{CK,x}+2}[x]$. Hence the proof of the Rasiowa-Sikorski-lemma shows that a real extending p and Cohen-generic over $L_{\omega_{n+1}^{CK,x}+1}[x]$ is contained in $L_{\omega_{n+1}^{CK,x}+2}[x]$. Use the program P_n from (1) to compute a real number c coding $L_{\omega_{n+1}^{CK,x}+2}[x]$. Then search through ω to determine, again using the techniques for evaluating first-order statements with *ITRM*s, some $i \in \omega$ that codes a real

with the desired properties in c. From i and c, the desired real is now easily computable.

We now define relative recognizability and then proceed with stating and proving our theorem.

Definition 4. *Let $x, y \subseteq \omega$. We say that x is ITRM-recognizable from y, written $x \leq_{RECOG} y$, if and only if there is an ITRM-program P such that $P^z \downarrow \in \{0,1\}$ for every $z \subseteq \omega$ and, for all $z \subseteq \omega$, we have $P^{z \oplus y} \downarrow = \delta(x, z)$. For a set $Y \subseteq \mathfrak{P}(\omega)$, we say that x is uniformly recognizable from Y if and only if there is an ITRM-program P such that, for every $y \in Y$ and every $z \subseteq \omega$, we have $P^{z \oplus y} \downarrow = \delta(z, x)$. In this case, we say that x is recognized from Y via P. We say that x is recognizable if and only if $x \leq_{RECOG} 0$. We denote the set of reals recognizable relative to $y \subseteq \omega$ by $RECOG_y$ and abbreviate $RECOG_0$ by $RECOG$.*

Remark: As we are only concerned with *ITRM*s in this paper, we usually drop the prefix '*ITRM*'.

Remark: The condition that P^z stops with output 0 or 1 for every input is introduced merely for the sake of the simplification of further arguments; if P is a program using n registers, we can always use the solvability of the bounded halting problem for *ITRM*s using at most n registers given by Theorem 2 to produce another program P' that, given $z \subseteq \omega$, first tests whether $P^z \downarrow$ with output 0 or 1 and returns the output of P^z if that is the case and otherwise outputs 0. P'^x and P^x will hence produce the same output wherever the output of P is of the required form and P' will satisfy our extra condition.

A typical phenomenon for models of infinitary computations is the existence of reals that are recognizable, but not computable. As computability is easily seen to imply recognizability, it follows that recognizability is a strictly weaker notion than computability. This was first shown in [9] for Infinite Time Turing Machines. Detailed treatments of recognizability for *ITRM*s and for infinitary machines in general can be found in [3–5,10].

We note here that recognizability is computably stable for *ITRM*s, i.e. preserved under *ITRM*-computable equivalence:

Definition 5. *For $x, y \subseteq \omega$, we write $x \equiv_{ITRM} y$ and say that x and y are ITRM-computably equivalent if and only if there are ITRM-programs P and Q such that P^x computes y and Q^y computes x.*

Proposition 6. *Let $x \equiv_{ITRM} y$ be reals. Then $x \in RECOG$ if and only if $y \in RECOG$.*

Proof. Assume that $x \in RECOG$. Let P and Q be *ITRM*-programs such that $P^x \downarrow = y$ and $Q^y \downarrow = x$, and let R be a program for recognizing x, i.e. such that $\forall z \subseteq \omega R^z \downarrow = \delta(x, z)$. To recognize y, we proceed as follows: Assume that z is given in the oracle.

Step 1: Check, using a halting problem solver (see Theorem 2) for Q, whether $Q^z(i)\downarrow$ for all $i \in \omega$. If not, then $z \neq y$, as Q computes x from y and hence $Q^y(i)\downarrow$ for every $i \in \omega$. So in that case, output 0 and stop. Then check whether $Q^z(i)\downarrow\in \{0,1\}$ for all $i \in \omega$ by an exhaustive search. If not, then $z \neq y$, again since $Q^y\downarrow= x$, so in that case, output 0 and stop. Otherwise, proceed with step 2.

Step 2: Let $Q^z\downarrow= a$. Check whether $R^a\downarrow= 1$. If not, then $a \neq x$ as R recognizes x, and hence $z \neq y$ as $Q^y\downarrow= x$. In that case, output 0 and stop. Otherwise, proceed with step 3.

Step 3: At this point, we know that $Q^z\downarrow= a = x$. Check whether $P^a\downarrow= z$ (using a halting problem solver as in step 1). If not, then $z \neq y$ as $P^x\downarrow= y$. In this case, output 0 and stop. Otherwise, $z = y$, so output 1 and stop.

Hence $x \in RECOG$ implies $y \in RECOG$. The reverse direction follows analogously.

Remark: Note that, however, relative recognizability is not transitive (see [4]).

Lemma 2. *Let P be an $ITRM$-program using n registers, let $x \subseteq \omega$ and suppose that g is Cohen-generic over $L_{\omega_{n+1}^{CK,x}+1}[x]$. Then $\omega_i^{CK,x\oplus g} = \omega_i^{CK,x}$ for $i \leq n+1$. Consequently, $P^{x\oplus g}$ halts in less than $\omega_{n+1}^{CK,x}$ many steps or does not halt at all.*

Proof. By Theorem 10.11 of [15], if M is admissible, $\mathbb{P}$ is a forcing in M and G is $\mathbb{P}$-generic over M such that G intersects every subclass of M that is a union of a $\Sigma_1(M)$ and a $\Pi_1(M)$ class, then $M[G]$ is also admissible. Clearly, as $L_{\alpha+1}[x]$ contains all subclasses of $L_\alpha[x]$ definable over $L_\alpha[x]$, we have that, when M is of the form $L_\alpha[x]$ with x-admissible α and $g \subseteq \omega$ Cohen-generic over $L_{\alpha+1}[x]$, then $L_\alpha[x][g]$ is admissible.

As admissible ordinals are indecomposable, it follows from Theorem 9.0 of [15] that the forcing extension $L_{\omega_i^{CK,x}}[x][g]$ agrees with the relativized L-level $L_{\omega_i^{CK,x}}[x \oplus g]$ for all $i \leq n+1$. Consequently, if g is as in the assumption the lemma, then $L_{\omega_i^{CK,x}}[x][g] = L_{\omega_i^{CK,x}}[x \oplus g]$ is admissible for $i \leq n+1$: so $\omega_i^{CK,x}$ is $x \oplus g$-admissible for $i \leq n+1$. Certainly, every $x \oplus g$-admissible ordinal is also x-admissible, so that first $(n+1)$ many x-admissible ordinals agree with the first $(n+1)$ many $x \oplus g$-admissible ordinals. Hence $\omega_{n+1}^{CK,x} = \omega_{n+1}^{CK,x\oplus g}$.

The second claim now follows from Theorem 2.

Theorem 7. *Let $Y \subseteq [0,1]$ be comeager, and let $x \subseteq \omega$ be uniformly recognizable in Y. Then x is recognizable.*

Proof. Let P be an $ITRM$-program that uniformly recognizes x from Y. Suppose that P uses n registers. Now, the set C of reals Cohen-generic over $L_{\omega_{n+1}^{CK,x}+1}[x]$ is comeager and hence has comeager intersection with Y. Assume without loss of generality that $Y = Y \cap C$ and let $z \in Y$. Then, by the forcing theorem for admissible sets (see [15]) there is some forcing condition $p \subseteq z$ such that $p \Vdash P^{x\oplus z}\downarrow= 1$ over $L_{\omega_{n+1}^{CK,x}+1}[x]$. The same hence holds for every real $a \supset p$ which is Cohen-generic over $L_{\omega_{n+1}^{CK,x}+1}[x]$.

We claim that the following procedure recognizes x: Given a real z in the oracle, compute a real $g_z \supseteq p$ Cohen-generic over $L_{\omega_{n+1}^{CK,z}+1}[z]$, using (2) of Lemma 1. Then run $P^{z\oplus g_z}$ and return its output, which must be 0 or 1 by definition of relativized recognizability. This procedure can be carried out by some *ITRM*-program Q. We claim that if the computation of $P^{z\oplus g_z}$ terminates with output 1, then $z = x$, otherwise $z \neq x$.

We saw above that $Q^x \downarrow= 1$. We need to show that $Q^y \downarrow= 0$ if $y \neq x$. To see this, pick $\omega \supseteq y \neq x$ and compute g_y. By definition of relativized recognizability, we have that $P^{y\oplus g_y} \downarrow$ with output 0 or 1. Then, by Lemma 2, P, when run in the oracle $y \oplus g_y$, halts in $< \omega_{n+1}^{CK,y}$ many steps; furthermore, the computation is contained in $L_{\omega_{n+1}^{CK,y}}[y \oplus g_y]$ and is absolute between $L_{\omega_{n+1}^{CK,y}}[y \oplus g_y]$ and the real world. If $P^{y\oplus g_y} \downarrow= 0$, we are done.

So assume otherwise, i.e. we have $P^{y\oplus g_y} \downarrow= 1$. By genericity of g_y, there is some Cohen condition $q \subseteq g_y$ such that $q \Vdash P^{y\oplus g_y} \downarrow= 1$. Hence, we have $P^{y\oplus b} \downarrow= 1$ for every real $b \supseteq q$ which is Cohen-generic over $L_{\omega_{n+1}^{CK,y}+1}[y]$. Now, the set $\hat{C}$ of reals $g \supseteq q$ Cohen-generic over $L_{\omega_{n+1}^{CK,y}+1}[y]$ is non-meager and thus has non-meager (and hence non-empty) intersection with the comeager set Y. Pick $\hat{g} \in Y \cap \hat{C}$. Then, as $\hat{g} \in \hat{C}$, we have $P^{y\oplus \hat{g}} \downarrow= 1$; but on the other hand, $\hat{g} \in Y$ and $y \neq x$, which contradicts the assumption that P recognizes x from Y.

Corollary 1. *Let $Y \subseteq [0,1]$ be non-meager, and let $x \subseteq \omega$ be uniformly recognizable in Y. Then x is recognizable.*

Proof. As Y is non-meager, there is an interval $I = (a,b) \subseteq [0,1]$ such that Y is comeager in I. By shortening I if necessary, we may assume without loss of generality that I is of the form $\{tx | x \in^\omega 2\}$ for $t \in^{<\omega} 2$ (where tx denotes the concatenation of t and x) and (passing to $Y \cap I$ if necessary) that $Y \subseteq I$. Suppose that P recognizes x relative to all elements of Y. We define a program P' that recognizes x from all elements of $Y' := \{x : tx \in I\}$, which is obviously a comeager set. $P'^{a\oplus y}$ works by simply running $P^{a\oplus ty}$. Clearly, P' has the desired properties: For $y \in Y'$, we have $P'^{a\oplus y} \downarrow= 1$ if and only if $P^{a\oplus ty} \downarrow= 1$ which, as $ty \in Y$ by definition of Y', is equivalent with $a = x$. So P' recognizes x from all elements of a comeager set. By Theorem 7, x is recognizable.

We have so far worked with uniform recognizability, i.e. the program recognizing x from $y \in Y$ has to be the same for all elements of Y. If one wants to drop this assumption and allow x to be recognized from y by different programs P for different $y \in Y$, the problem arises that the corresponding subsets $Y_P := \{y \in Y : P \text{ recognizes } x \text{ from } y\}$ might not have the property of Baire and hence not be comeager in some interval so that Theorem 7 is not applicable. At least under some (standard) set-theoretical extra assumption, however, we can strengthen the claim to drop the uniformity condition:

Corollary 2. *Assume that every $\mathbf{\Sigma}_2^1$-set of reals has the Baire property. Let Y be a non-meager set, $x \subseteq \omega$ and assume that, for every $y \in Y$, there is some ITRM-program P such that $P^{z\oplus y} \downarrow= 1$ if and only if $z = x$. Then x is ITRM-recognizable.*

Proof. Y_P is Π^1_2 in x for every *ITRM*-program P: Namely, it is definable by a formula expressing 'For all $z, b \subseteq \omega$: If b codes the computation of P in the oracle z and this computation stops with output 1, then $z = x$'. (Recall that, by the choice of P, the computation of P in the oracle z always terminates and hence is a countable set codable by a real.) As 'b codes the computation of P in the oracle z' is Π^1_1, the negation is Σ^1_1, so the equivalent statement 'For all $z, b \subseteq \omega$: $z = x$ or b does not code the computation of P in the oracle z' is Π^1_2.

Thus, for each *ITRM*-program P, Y_P has the Baire property (as its complement is $\mathbf{\Sigma}^1_2$ and hence Baire by assumption and as complements of Baire sets are again Baire). Now $Y = \bigcup_{i\in\omega} Y_{P_i}$. As Y is not meager, it cannot be the union of countably many meager sets. So there is some $k \in \omega$ such that $\bar{Y} := Y_{P_k}$ is not meager. As $\bar{Y}$ also has the Baire property, there is an interval such that $\bar{Y}$ is comeager relative to that interval. As in the proof of Corollary 1, it follows that x is uniformly *ITRM*-recognizable from all elements of a comeager set of oracles, and hence, by Theorem 7, x is *ITRM*-recognizable.

Remark: The assumption that every $\mathbf{\Sigma}^1_2$-set of reals has the Baire property follows for example from the existence of a measurable cardinal (see e.g. Corollary 14.3 [11]). By Proposition 13.7 of [11], every $\mathbf{\Sigma}^1_2$-set is a union of $\aleph_1$ many Borel sets. By Theorem 2.20 of chapter II of [14], MA_{ω_1} implies that a union of $\aleph_1$ many meager sets is meager (and hence that a union of $\aleph_1$ many sets with the Baire property has the Baire property). As Borel sets have the Baire property, it thus also follows from MA_{ω_1} that all $\mathbf{\Sigma}^1_2$-sets have the Baire property.

Moreover, the statement that all $\mathbf{\Sigma}^1_2$-sets of reals have the Baire property is equivalent to the statement that the set of reals that are Cohen-generic over $L[x]$ is comeager for every $x \subseteq \omega$ (see Theorem 14.2 of [11]).

It well known that L contains $\mathbf{\Sigma}^1_2$-sets of reals that fail to have the Baire property (see e.g. Corollary 13.10 of [11] or observe that in L, the set of reals Cohen-generic over L is empty). On the other hand, MA_{ω_1} holds in a forcing extension of L (see Theorem 10.11 of [11]). The statement that every $\mathbf{\Sigma}^1_2$-set of reals has the Baire property it thus independent of *ZFC*.

We do not know if this non-uniform version of Corollary 1 is provable in *ZFC* alone.

3 Further Work

Since there is no analogue for the stratification of halting times as given by Theorem 2 for Infinite Time Turing Machines (*ITTM*s), which is a crucial ingredient of Theorem 7, the treatment of *ITTM*s will require new ideas. We have a sketch of a considerably more involved proof of the claim corresponding to Corollary 1 for *ITTM*s which we are currently elaborating and plan to cover in future work.

Moreover, it is natural to ask what happens when we replace the condition of non-meagerness by the condition of positive Lebesgue measure. This and other related topics can be dealt with using random forcing over models of *KP*, which will be treated in future work with Philipp Schlicht.

Acknowledgements. We thank Philipp Schlicht for a discussion on Corollary 2 (and the remark following it) as well as various helpful comments on the presentation of the proof of Theorem 7.

References

1. Barwise, J.: Admissible Sets and Structures. Springer, Berlin (1975)
2. Barmpalias, G., Lewis-Pye, A.: The information content of typical reals. In: Sommaruga, G., Strahm, T. (eds.) Turing's Ideas - Their Significance and Impact. Springer, Basel (2014)
3. Carl, M.: The lost melody phenomenon. Festschrift on the occasion of Philip Welch's and Peter Koepke's 60th birthday
4. M. The distribution of $ITRM$-recognizable reals. Annals of Pure and Applied Logic
5. Carl, M.: Optimal results on recognizability by infinite time register machines. J. Symbolic Logic (to appear)
6. Cutland, N.: Computability - An Introduction to Recursive Function Theory. Cambridge University Press, Cambridge (1980)
7. Carl, M., Schlicht, P.: Infinite computations with random oracles. Notre Dame J. Formal Logic (To appear)
8. Downey, R.G., Hirschfeldt, D.: Algorithmic Randomness and Complexity. Theory and Applications of Computability. Springer LLC, New York (2010)
9. Hamkins, J., Lewis, A.: Infinite time turing machines. J. Symbolic Logic **65**(2), 567–604 (2000)
10. Carl, M., Fischbach, T., Koepke, P., Miller, R., Nasfi, M., Weckbecker, G.: The basic theory of infinite time register machines. Arch. Math. Logic **49**(2), 249–273 (2010)
11. Kanamori, A.: The Higher Infinite. Springer, Berlin (2005)
12. Koepke, P., Miller, R.: An enhanced theory of infinite time register machines. In: Beckmann, A., Dimitracopoulos, C., Löwe, B. (eds.) CiE 2008. LNCS, vol. 5028, pp. 306–315. Springer, Heidelberg (2008)
13. Koepke, P.: Ordinal computability. In: Ambos-Spies, K., Löwe, B., Merkle, W. (eds.) CiE 2009. LNCS, vol. 5635, pp. 280–289. Springer, Heidelberg (2009)
14. Kunen, K.: Set Theory. An Introduction to Independence Proofs. Elsevier, Amsterdam (2006)
15. Mathias, A.R.D.: Provident sets and rudimentary set forcing. Preprint. https://www.dpmms.cam.ac.uk/~ardm/fifofields3.pdf
16. Sacks, G.: Higher Recursion Theory. Springer, New York (1990)

P Systems with Parallel Rewriting for Chain Code Picture Languages

Rodica Ceterchi[1(✉)], K.G. Subramanian[2], and Ibrahim Venkat[3]

[1] Faculty of Mathematics and Computer Science,
University of Bucharest, 14 Academiei Street, 010014 Bucharest, Romania
rceterchi@gmail.com
[2] Faculty of Science, Liverpool Hope University, Hope Park, Liverpool L16 9JD, UK
[3] School of Computer Sciences, Universiti Sains Malaysia, 11800 Penang, Malaysia

Abstract. Chain code pictures are composed of unit lines in the plane, drawn according to a sequence of instructions *left, right, up, down* codified by words over $\Sigma = \{l, r, u, d\}$. P systems to generate such languages have been considered in previous work with sequential rewriting in the membranes. We consider here parallel rewriting, with the advantage of reducing the number of membranes. We also consider the problem of generating the finite approximations of space-filling curves, the Hilbert curve and the Peano curve.

1 Introduction

String rewriting P systems and their variants with string objects and context-free rewriting rules have been extensively studied. Rewriting strings in the regions of such rewriting P systems, either sequentially as in Chomsky type of grammars or in parallel as in L systems, have been investigated from the point of view of formal language theory (see, for example, [5,10]). Extension of string rewriting P systems to two-dimensions with array objects (also called picture arrays) and array-rewriting rules applied in a sequential manner as in isometric array grammars, was considered in [1] by introducing array P systems, thus linking P systems with array grammars. Parallel mode of rewriting was incorporated in these array P systems and its power studied in [15]. Improved results in terms of reduction in the number of membranes in the results proved in [15], were obtained in [9].

On the other hand, a picture generating model, called chain code picture grammar, was introduced in [8] as early as 1982 and intensively investigated subsequently in various studies (see, for example, [2,3]). This grammar generates chain code pictures (also called line pictures) that are made of unit lines in the two-dimensional grid and encoded by words over the alphabet $\{l, r, u, d\}$ with the symbols l, r, u, d respectively interpreted as instructions to draw a horizontal or vertical unit line to the left, right, up or down directions from the current position in the chain code picture. Recently, chain code picture grammars were linked with P systems in [16], introducing chain code P system, by considering

A. Beckmann et al. (Eds.): CiE 2015, LNCS 9136, pp. 145–155, 2015.
DOI: 10.1007/978-3-319-20028-6_15

context-free string grammar rules in the regions of the P system with the terminal alphabet $\{l, r, u, d\}$ interpreted as mentioned. The rewriting is done sequentially in the string objects in the chain code P system in [16]. Here we incorporate the parallel mode of rewriting in the application of the rules in the regions of a chain code P system, resulting in Parallel chain code system. We examine the results established in [16], namely comparison with collage grammars [3] restricted to line picture generation. Although not entirely unexpected, the parallel mode of rewriting results in a reduction in the number of membranes in the constructions involved in the results in [16]. We also construct parallel chain code P systems to generate the patterns of approximations of space-filling curves [7,13], the Hilbert curve and the Peano curve.

2 Basic Definitions and Results

We recall the notions of chain code pictures and chain code picture grammars [3,8]. For notions related to Chomsky grammars and languages, we refer to [11] and for P systems, we refer to [1,10,14].

For integers m, n, the left, right, up and down neighbours of the point $z = (m, n)$ in the two-dimensional plane are the points $(m-1, n), (m+1, n), (m, n+1), (m, n-1)$ respectively denoted by $l(z), r(z), u(z), d(z)$. A unit line in the plane, denoted by $(z, n(z))$, connects two neighbouring points z and $n(z)$ for $n \in \{l, r, u, d\}$. A chain code picture or a line picture p is a finite set p_{line} of connected unit lines, which can be made of several disconnected parts in general, but here we consider connected chain code pictures only. Figure 1 shows a star-shaped picture with four arms of equal length (first considered in [4] as a digitized picture) in the form of a chain code picture. Each arm starts at the centre point and its length is the number of unit lines constituting an arm. The leftmost end of the horizontal line is at the origin $(0,0)$. The star-shaped picture in Fig. 1 is made of the unit lines given by the set $S_{line} = \{((0,0),(1,0)),((1,0),(2,0)),((2,0),(3,0)),((3,0),(4,0))\} \cup \{((2,-2),(2,-1)),((2,-1),(2,0)),((2,0),(2,1)),((2,1),(2,2))\}$. A drawing of the picture in Fig. 1, starting from the point $(0,0)$, reaching $(4,0)$ moving right, then moving left to reach $(2,0)$, followed by moving up to reach $(2,2)$ and finally moving down ending at the point $(2,-2)$, can be described by a word $r^4l^2u^2d^4$ over the alphabet $\Sigma = \{l, r, u, d\}$, which is called a *picture description* of the corresponding chain code picture. A picture description of a chain code picture depends on the starting point of the drawing of the picture and so there could be many such picture descriptions.

We are more concerned with the relative positions of the unit lines in a chain code picture rather than absolute positions in the plane. The word w over $\Sigma = \{l, r, u, d\}$ which corresponds to a *drawing* of a chain code picture determines the set of unit lines of the picture. The 'shape' of a chain code picture p is thus given by the set p_{line}. We write $drawing(w) = p_{line}$. The shape of the chain code picture p, which is denoted as (p_{line}, s, e), can then be drawn starting from a given start point s to an end point e of a drawing of p from s to e.

Fig. 1. Star-shaped chain code picture

A context-free chain code grammar ($CFCCG$) [8] is a Chomsky type context-free grammar G with terminal alphabet $\Sigma = \{l, r, u, d\}$. The chain code picture language generated by G is $L(G) = \{drawing(w) \mid w \in L(G)\}$. We denote by $CFCC$, the class of all chain code picture languages generated by $CFCCGs$.

We now recall the context-free rewriting chain code P system of [16] which is a rewriting P system with string objects and internal output.

A context-free rewriting chain code P system of degree $n, n \geq 1$, is a construct $\Pi = (N, \Sigma, \mu, L_1, \cdots, L_n, R_1, \cdots, R_n, i_0)$ where: N is the nonterminal alphabet; $\Sigma = \{l, r, u, d\}$ is the terminal alphabet; $N \cap \Sigma = \emptyset$; μ is a membrane structure with n membranes labelled in a one-to-one way with $1, 2, \ldots, n$; $L_1, \ldots, L_n$ are finite sets of strings over $V = N \cup \Sigma$ associated with the n regions of μ; $R_1, \ldots, R_n$ are finite sets of context-free rewriting rules associated with the n regions of μ; the rules are of the form $A \to \alpha(tar)$, where $A \to \alpha$ is a context-free (CF) rule as in a Chomsky context-free grammar, A being a non-terminal, α, a string of nonterminals and terminals; the rules have attached targets *here, out, in* (in general, *here* is omitted) with the target specifying the region where the result of the rewriting should be placed in the next step: *here* means that the result remains in the same region where the rule was applied, *out* means that the string has to be sent to the region immediately surrounding the region where it has been produced, and *in* means that the string should go to one of the directly inner membranes, if any exists. Each string in a region is processed by at most one rule at a time; if many rules can be used, then one of them, nondeterministically chosen, is used; if no rule can rewrite a string, then it remains unchanged. All strings, from all regions, are rewritten at the same time. A sequence of such steps is called a *computation*. A computation provides a result only if it halts and the system reaches a configuration where no further rule can be applied. Finally, i_o is the label of an elementary membrane of μ, the output membrane. In particular the rules in a region can be regular rules of the forms $A \to wB, A \to w$, where A, B are non-terminals and w is a string of terminals. A *successful computation* in a context-free rewriting chain code P system is a halting computation with the strings over the terminal alphabet Σ collected in the output membrane being the strings accepted. The chain code picture language generated by Π is the set

$$L(\Pi) = \{drawing(w) \mid w \, \text{is a string over} \, \Sigma \, \text{accepted by} \, \Pi\}.$$

The words w over Σ, which are computed by the system Π are interpreted as drawings of chain code pictures, thus giving rise to the picture language $L(\Pi)$. The set of all chain code picture languages computed or generated by context-free rewriting chain code P systems with n membranes is denoted by

$RCCP_n(CF)$. If the rules in the regions are all regular, then the family is denoted by $RCCP_n(REG)$.

3 Parallel Chain Code *P* System

We now consider rewriting the string objects in parallel in the chain code P system of [16].

A context-free parallel chain code P system of degree $n, n \geq 1$, is a construct

$$\Pi = (N, \Sigma, \mu, L_1, \cdots, L_n, R_1, \cdots, R_n, i_0)$$

where all the components of Π are as defined in the chain code P system [16] recalled in the previous section. The only difference is that during a computation, all the nonterminals (if any) in a string object in a region are rewritten at the same time by available and applicable context-free (or regular) rules in the region with the requirement that all the rules used in a string have the same target indication. The words over the terminal alphabet $\Sigma = \{l, r, u, d\}$ collected in the output membrane at the end of a successful halting computation constitute the language generated by the P system

$$L(\Pi) = \{drawing(w) \mid w \text{ is a string over } \Sigma \text{ accepted by } \Pi\}.$$

The set of all chain code picture languages generated by context-free parallel chain code P systems with n membranes is denoted by $PCCP_n(CF)$. If the rules in the regions are all regular, then the family is denoted by $PCCP_n(REG)$.

We shall now illustrate the parallel chain code P system. Let L_s be the set of star shaped chain code pictures with 4 equal arms each of length ≥ 1. A member of this language is shown in Fig. 1. We now define a parallel chain code P system generating the picture description words of the language L_s. Consider the parallel chain code P system with regular rules

$$\Pi_1 = (\{A, B, C, D\}, \{l, r, u, d\}, [_1 [_2]_2]_1, \{ABCD\}, \emptyset, R_1, R_2, 2)$$

with a linear membrane structure having two regions. The rule sets are given by $R_1 = \{A \to r^2A, B \to lB, C \to uC, D \to d^2D, A \to r^2(in), B \to l(in), C \to u(in), D \to d^2(in)\}, R_2 = \emptyset$.

Lemma 1. $L(\Pi_1) = L_s$.

Proof. Initially, the string $ABCD$ is in region 1 and the region 2 has no object. The rules $A \to r^2A, B \to lB, C \to uC, D \to d^2D$ are applicable in parallel as many times as we need and the generated string of the form $r^{2(n-1)}Al^{(n-1)}Bu^{(n-1)}Cd^{2(n-1)}D$ remains in the same region. Once the rules $A \to r^2, B \to l, C \to u, D \to d^2$ with the same target indication *in* are applied in region 1, the resulting string $r^{2n}l^nu^nd^{2n}$ is sent to region 2. The chain code picture language generated by Π_1 is the set $L(\Pi_1) = \{drawing(w) \mid w = r^{2n}l^nu^nd^{2n}, n \geq 1\}$. Note that the symbols $\{l, r, u, d\}$ are interpreted in the manner described earlier so that the words $w = r^{2n}l^nu^nd^{2n}$ yield the star shaped pictures with arms of equal length. □

Remark 1. We note that the chain code P system considered in [16] can not generate the star shaped pictures with four equal arms, using only two membranes and regular rules in the regions, since sequential application of regular rules can correspond to "developing only one arm" at a time in a region and also cannot terminate growth of all the four arms together. On the other hand the following chain code P system Π_2 with sequential application of context free rules and three membranes generates L_s. We define Π_2 as follows:

$\Pi_2 = (\{A, D\}, \{l, r, u, d\}, [_1 [_2 [_3]_3]_2]_1, \{AD\}, \emptyset, R_1, R_2, R_3, 3)$ where $R_1 = \{A \to r^2 Al(in), A \to r^2 A'l(in)\}, R_2 = \{D \to uDd^2(out), D \to uD'd^2(in)\} R_3 = \{A' \to r^2 l, D' \to ud^2\}$ The P system Π_2 generates L_s as follows: Initially, the string AD is in region 1. If the rule $A \to r^2 Al$ is applied then the generated string $r^2 AlD$ is sent to region 2 wherein if the rule $D \to uDd^2$ is applied, then the result $r^2 AluDd^2$ is sent back to region 1 and the process can repeat. Once the rule $A \to r^2 A'l$ is applied in region 1, the resulting string is sent to region 2 and application of the rule $D \to uD'd^2$ sends the result to region 3 where the application of the rules for A' and D' halts the computation and strings of the form $r^{2n} l^n u^n d^{2n}$ are collected in the output membrane. An incorrect application sequence of the rules will make the string get stuck in region 1 or in region 3. The chain code picture language generated by Π_2 is the set $L(\Pi_2) = \{drawing(w) \mid w = r^{2n} l^n u^n d^{2n}, n \geq 1\}$, thus yielding L_s. We believe that the number of membranes in this case cannot be reduced without any additional ingredients to the system.

We now examine the result in [16] on the comparison with context free collage grammars [6], which produce pictures by transforming any sort of basic geometric objects using affine transformations (see [3]). It is shown in [16] that $RCCP_3(REG) - CFCC \neq \emptyset$, and as a consequence $RCCP_3(REG) - CFCL \neq \emptyset$, where $RCCP_3(REG)$ is the class of all chain code picture languages generated by rewriting chain code P systems with 3 membranes and regular rules in the regions applied in the sequential mode, while $CFCL$ denotes the class of languages of context-free collage grammars restricted to chain code picture generation. We now show that the number of membranes in this result is reduced to 2 when parallel mode of rewriting is considered in the regions.

Theorem 1. $PCCP_2(REG) \cap CFCC \neq \emptyset$.

Proof. The set of chain code pictures of "stairs" of equal height, one member of which is shown in Fig. 2, is in $CFCC$.

The language of picture description words of these chain code pictures is $\{(ru)^n r(dr)^n \mid n \geq 1\}$, which is generated by the context-free chain code

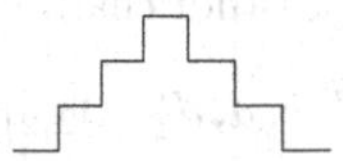

Fig. 2. A chain code picture of two "stairs" of equal height

grammar $(\{S, r, u, d\}, r, u, d, \{S \to ruSdr, S \to ruAdr, A \to r\}, S)$. This language is also in $PCCP_2(REG)$ generated by the parallel chain code P system Π_{stair} of degree 2 with regular rules, given by $\Pi_{stair} = (\{A, B\}, \{r, u, d\}, [_1[_2]_2]_1, \{ArB\}, \emptyset, \emptyset, R_1, R_2, 2)$ where the rule sets are given by $R_1 = \{A \to ruA, B \to drB, A \to ru(in), B \to dr(in)\}$, $R_2 = \emptyset$. Initially, the string ArB is in region 1. The rules $A \to ruA, B \to drB$ are applicable in parallel as many times as needed. Once the rules $A \to ru, B \to dr$ with the same target indication *(in)* are applied in region 1, the resulting string is sent to region 2. The computation halts yielding strings of the form $(ru)^n r(dr)^n$, $n \geq 1$, in the output membrane. □

Since the language of "stairs" is not in $CFCL$ (see Theorem 4 in [3]), we have:

Corollary 1. $PCCP_2(REG) - CFCL \neq \emptyset$.

4 Space-Filling Curves: The Hilbert Curve

Space-filling curves which are continuous but nowhere differentiable as a mapping from the unit interval $[0, 1]$ to the unit square $[0, 1] \times [0, 1]$, are known to have applications to different kinds of problems. Various methods of generating or realizing the sequence of patterns that are iterations or approximations of a space-filling curve have been proposed (See for example [7,13]). A space-filling curve which has been well-studied [12,13] from the point of view of formal language theory, is the Hilbert curve. We now construct a context-free parallel chain code P system with two membranes to generate the sequence of patterns of approximations of the Hilbert curve. The Hilbert curve patterns can be represented by picture description words over the alphabet $\Sigma = \{l, r, u, d\}$ as follows [13]: The first approximation is given by $H_1 = urd$ and for $n > 1$, the subsequent approximations are given by $H_n = g_1(H_{n-1})uH_{n-1}rH_{n-1}dg_2(H_{n-1})$ where g_1, g_2 are homomorphisms on Σ given by $g_1(u) = r, g_1(d) = l, g_1(l) = d, g_1(r) = u, g_2(u) = l, g_2(d) = r, g_2(l) = u, g_2(r) = d$.

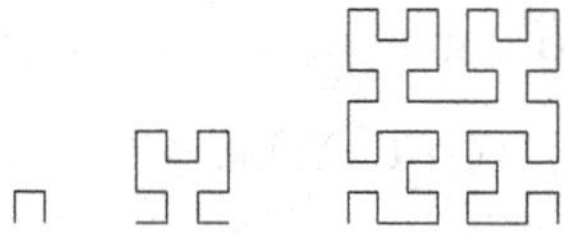

Fig. 3. First three approximations of Hilbert curve

The finite approximations of the Hilbert curve, H_n, $n \geq 1$, are called *Hilbert words* (Fig. 3).

We consider the context-free parallel chain code P system Π_H given by

$$\Pi_H = (\{A, B, C, D\}, \{l, r, u, d\}, [_1[_2]_2]_1, \{A\}, \emptyset, \emptyset, R_1, R_2, 2)$$

where R_1, R_2 are given by $R_1 = \{A \to \beta_1, B \to \beta_2, C \to \beta_3, D \to \beta_4, (here, in)\}$, $R_2 = \{A \to \lambda, B \to \lambda, C \to \lambda, D \to \lambda\}$ Here λ is the empty word and

$\beta_1 = BuArAdC$, $\beta_2 = ArBuBlD$, $\beta_3 = DlCdCrA$, $\beta_4 = CdDlDuB$.

Initially, the region 1 has the string A and the other region is empty. Application in parallel of the rules $A \to \beta_1, B \to \beta_2, C \to \beta_3, D \to \beta_4$ in region 1, expand once the symbols A, B, C, D suitably. This process can be repeated as many times as the target indication *here* is used. But if the target indication *in* is used, then the string enters region 2 where the nonterminals are erased and the resulting string of terminal symbols remains here. These strings are the picture description words of the approximation patterns of the Hilbert curve.

Theorem 2. *The P system Π_H produces in membrane 2 the Hilbert words.*

Proof. We have adapted the proof used in [7] and [12] to show that the Hilbert infinite word is generated by a tag system.

Consider the alphabet $\Theta = \{A, B, C, D\} \cup \Sigma$ and the morphisms $\gamma : \Theta^* \to \Theta^*$ given by $A \to BuArAdC$, $B \to ArBuBlD$, $C \to DlCdCrA$, $D \to CdDlDuB$, $u \to u, d \to d, r \to r, l \to l$ and $f : \Theta^* \to \Sigma^*$ given by $A \to \lambda, B \to \lambda, C \to \lambda, D \to \lambda, u \to u$, $d \to d$, $r \to r$, $l \to l$. Morphism γ is the rewriting in membrane 1, and f the rewriting in membrane 2. We have to prove that $f(\gamma^n(A)) = H_n$. First a technical lemma:

Lemma 2. *The following hold:*

$$f(\gamma^n(A)) = g_1(f(\gamma^n(B))) = g_2(f(\gamma^n(C)))$$
$$f(\gamma^n(B)) = g_1(f(\gamma^n(A))) = g_2(f(\gamma^n(D)))$$
$$f(\gamma^n(C)) = g_1(f(\gamma^n(D))) = g_2(f(\gamma^n(A)))$$
$$f(\gamma^n(D)) = g_1(f(\gamma^n(C))) = g_2(f(\gamma^n(B)))$$

Proof. By induction. For $n = 1$, we have $f(\gamma(A)) = urd$, $f(\gamma(B)) = rul$, $f(\gamma(C)) = ldr$, $f(\gamma(D)) = dlu$, and the rest is straightforward computation. We suppose the equalities hold for n. To prove the first equality:

$$f(\gamma^{n+1}(A)) = f(\gamma^n(BuArAdC))$$
$$= f(\gamma^n(B))uf(\gamma^n(A))rf(\gamma^n(A))df(\gamma^n(C))$$
$$= g_1(f(\gamma^n(A)))ug_1(f(\gamma^n(B)))rg_1(f(\gamma^n(B)))dg_1(f(\gamma^n(D)))$$
$$= g_1(f(\gamma^n(A)))g_1(r)g_1(f(\gamma^n(B)))g_1(u)g_1(f(\gamma^n(B)))g_1(l)g_1(f(\gamma^n(D)))$$
$$= g_1(f(\gamma^n(ArBuBlD))) = g_1(f(\gamma^{n+1}(B))).$$

The other equalities follow by similar computations, and $g_1^2 = g_2^2 = 1_\Sigma$. □

Back to The Proof of Theorem. For $n = 1$ we have $f(\gamma(A)) = urd = H_1$. We suppose that $f(\gamma^n(A)) = H_n$, and we compute for $n + 1$: $f(\gamma^{n+1}(A)) = f(\gamma^n(BuArAdC)) = f(\gamma^n(B))uf(\gamma^n(A))rf(\gamma^n(A))df(\gamma^n(C)) = g_1(f(\gamma^n(A)))$ $uf(\gamma^n(A))rf(\gamma^n(A))dg_2(f(\gamma^n(A))) = g_1(H_n)uH_nrH_ndg_2(H_n) = H_{n+1}$, where we have used the lemma and the induction hypothesis. □

5 The Peano Curve

The first space-filling curve introduced in the literature by G. Peano is the Peano curve. The approximations of the Peano curve are called *Peano words*, and they form an iterative sequence. The first two Peano words are shown in Fig. 4.

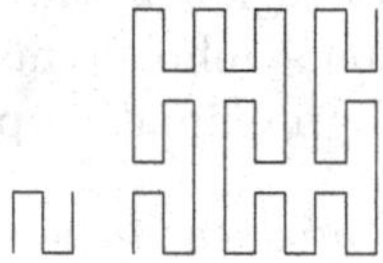

Fig. 4. The first two Peano words

The problem of generating the Peano words has been considered in [16] and a rewriting chain code P system with four membranes and sequential mode of rewriting has been given to generate these patterns. Here we construct a context-free parallel chain code P system with two membranes, completely different from the system of [16], to generate the sequence of Peano words.

The patterns can be represented by picture description words over the alphabet $\Sigma = \{l, r, u, d\}$ as follows [13]: The first approximation is given by $P_1 = uurddruu$ and for $n > 1$, the subsequent approximations are given by

$$P_{n+1} = P_n u h_1(P_n) u P_n r h_2(P_n) d h_3(P_n) d h_2(P_n) r P_n u h_1(P_n) u P_n$$

where h_1, h_2, h_3 are homomorphisms on Σ as given below:

$h_1(u) = u, h_1(d) = d, h_1(l) = r, h_1(r) = l;$
$h_2(u) = d, h_2(d) = u, h_2(l) = l, h_2(r) = r;$
$h_3(u) = d, h_3(d) = u, h_3(l) = r, h_3(r) = l.$

We consider the context-free parallel chain code P system Π_P given by $\Pi_P = (\{U_1, U_2, D_1, D_2, R_1, R_2, L_1, L_2\}, \{l, r, u, d\}, [_1 [_2]_2]_1, \{U_1\}, \emptyset, R_1, R_2, 2)$ where $R_2 = \emptyset$ and R_1 contains the following eight rules with target indication *here*:

$U_1 \to U_1U_2R_1D_1D_2R_2U_1U_2U_1,\ U_2 \to U_2U_1L_1D_2D_1L_2U_2U_1U_2$
$D_1 \to D_1D_2R_2U_1U_2R_1D_1D_2D_1,\ D_2 \to D_2D_1L_2U_2U_1L_1D_2D_1D_2$
$R_1 \to U_1U_2R_1D_1D_2R_2U_1U_2R_1,\ R_2 \to D_1D_2R_2U_1U_2R_1D_1D_2R_2$
$L_1 \to U_2U_1L_1D_2D_1L_2U_2U_1L_1,\ L_2 \to D_2D_1L_2U_2U_1L_1D_2D_1L_2$

and the following eight rules with target indication *in*:

$$U_1 \to u, U_2 \to u, D_1 \to d, D_2 \to d, R_1 \to r, R_2 \to r, L_1 \to l, L_2 \to l.$$

Theorem 3. *The P system Π_P above produces in membrane 2 the words $P_n u$, where P_n is the n-th Peano word.*

Proof. Let us denote by γ the morphism given by the eight rewriting rules with target *here*, and by f the morphism given by the eight rewriting rules with target

in. We will prove by induction that $f(\gamma^n(U_1)) = P_n u$ and $f(\gamma^n(R_1))) = P_n r$.
For $n = 1$ we have $f(\gamma(U_1)) = uurddruuu = P_1 u$, $f(\gamma(R_1)) = uurddruur = P_1 r$.
We also have: $f(\gamma(U_2)) = uulddluuu = h_1(P_1)u = h_1(P_1 u)$,

$\quad f(\gamma(D_1)) = ddruurddd = h_2(P_1)d = h_2(P_1 u)$,
$\quad f(\gamma(D_2)) = ddluulddd = h_3(P_1)d = h_3(P_1 u)$,
$\quad f(\gamma(R_2)) = ddruurddr = h_2(P_1)r = h_2(P_1 r)$,
$\quad f(\gamma(L_1)) = uulddluul = h_1(P_1)l = h_1(P_1 r)$,
$\quad f(\gamma(L_2)) = ddluulddl = h_3(P_1)l = h_3(P_1 r)$.

Lemma 3. *The following relations hold:*

$$f(\gamma^n(U_1)) = h_1(f(\gamma^n(U_2))) = h_2(f(\gamma^n(D_1))) = h_3(f(\gamma^n(D_2)))$$
$$f(\gamma^n(U_2)) = h_1(f(\gamma^n(U_1))) = h_2(f(\gamma^n(D_2))) = h_3(f(\gamma^n(D_1)))$$
$$f(\gamma^n(D_1)) = h_1(f(\gamma^n(D_2))) = h_2(f(\gamma^n(U_1))) = h_3(f(\gamma^n(U_2)))$$
$$f(\gamma^n(D_2)) = h_1(f(\gamma^n(D_1))) = h_2(f(\gamma^n(U_2))) = h_3(f(\gamma^n(U_1)))$$
$$f(\gamma^n(R_1)) = h_1(f(\gamma^n(L_1))) = h_2(f(\gamma^n(R_2))) = h_3(f(\gamma^n(L_2)))$$
$$f(\gamma^n(R_2)) = h_1(f(\gamma^n(L_2))) = h_2(f(\gamma^n(R_1))) = h_3(f(\gamma^n(L_1)))$$
$$f(\gamma^n(L_1)) = h_1(f(\gamma^n(R_1))) = h_2(f(\gamma^n(L_2))) = h_3(f(\gamma^n(R_2)))$$
$$f(\gamma^n(L_2)) = h_1(f(\gamma^n(R_2))) = h_2(f(\gamma^n(L_1))) = h_3(f(\gamma^n(R_1)))$$

Proof. We will prove the relations by induction. For $n = 1$ we have them. Suppose they all hold for n. For $n + 1$ we compute:

$$f(\gamma^{n+1}(U_1)) = f(\gamma^n(U_1 U_2 R_1 D_1 D_2 R_2 U_1 U_2 U_1))$$
$$= f(\gamma^n(U_1))f(\gamma^n(U_2))f(\gamma^n(R_1))f(\gamma^n(D_1))f(\gamma^n(D_2))$$
$$\cdot f(\gamma^n(R_2))f(\gamma^n(U_1))f(\gamma^n(U_2))f(\gamma^n(U_1))$$
$$= h_1(f(\gamma^n(U_2)))h_1(f(\gamma^n(U_1)))h_1(f(\gamma^n(L_1)))h_1(f(\gamma^n(D_2)))$$
$$\cdot h_1(f(\gamma^n(D_1)))h_1(f(\gamma^n(L_2)))h_1(f(\gamma^n(U_2)))h_1(f(\gamma^n(U_1)))h_1(f(\gamma^n(U_2)))$$
$$= h_1(f(\gamma^n(U_2 U_1 L_1 D_2 D_1 L_2 U_2 U_1 U_2))) = h_1(f(\gamma^n(\gamma(U_2))) = h_1(f(\gamma^{n+1}(U_2))).$$

The other 23 relations follow by similar computations, and observing that: $h_1^2 = h_2^2 = h_3^2 = 1_\Sigma$, $h_3 = h_1 \circ h_2 = h_2 \circ h_1$, $h_1 = h_2 \circ h_3 = h_3 \circ h_2$, $h_2 = h_1 \circ h_3 = h_3 \circ h_1$. □

Back to The Proof of Theorem. Suppose the relations hold for n. Then

$$f(\gamma^{n+1}(U_1)) = f(\gamma^n(U_1 U_2 R_1 D_1 D_2 R_2 U_1 U_2 U_1))$$
$$= f(\gamma^n(U_1))f(\gamma^n(U_2))f(\gamma^n(R_1))f(\gamma^n(D_1))f(\gamma^n(D_2))$$
$$\cdot f(\gamma^n(R_2))f(\gamma^n(U_1))f(\gamma^n(U_2))f(\gamma^n(U_1))$$
$$= P_n u h_1(P_n u)P_n r h_2(P_n u)h_3(P_n u)h_2(P_n r)P_n u h_1(P_n u)P_n u$$
$$= P_n u h_1(P_n)u P_n r h_2(P_n)d h_3(P_n)d h_2(P_n)r P_n u h_1(P_n)u P_n u = P_{n+1} u,$$

where we have applied the induction hypothesis and the lemma.

$\quad$ Similarly, $f(\gamma^{n+1}(R_1)) = P_{n+1} r$. □

6 Conclusions

We have introduced P systems with parallel rewriting of strings in order to generate chain code picture languages. Similar P systems, but with sequential rewriting, have been introduced in [16], We have investigated the relation with collage grammars, establishing a result similar to one of [16], but with only two membranes instead of three. Next we have focused on the generation of finite approximations of space-filling curves. We present a P system with parallel rewriting and two membranes which generates the Hilbert words and we prove its correctness. We also present a P system with parallel rewriting and two membranes which generates the Peano words, which is completely different from the one proposed in [16]. The generation of other space-filling curves with P systems can be attacked along the lines introduced in the present paper. Also, the parallel model presented here can be a framework in which to consider other classes of picture languages, for instance stripe picture languages.

Acknowledgments. The third author acknowledges partial support for this research from a RU grant, USM and an FRGS grant from MOHE, Malaysia.

References

1. Ceterchi, R., Mutyam, M., Păun, G., Subramanian, K.G.: Array - rewriting P systems. Nat. Comput. **2**, 229–249 (2003)
2. Dassow, J., Habel, A., Taubenberger, S.: Chain-code pictures and collages generated by hyperedge replacement. In: Cuny, J., Engels, G., Ehrig, H., Rozenberg, G. (eds.) Graph Grammars 1994. LNCS, vol. 1073, pp. 412–427. Springer, Heidelberg (1996)
3. Drewes, F.: Some remarks on the generative power of collage grammars and chain-code grammars. In: Ehrig, H., Engels, G., Kreowski, H.-J., Rozenberg, G. (eds.) TAGT 1998. LNCS, vol. 1764, pp. 1–14. Springer, Heidelberg (2000)
4. Fernau, H., Freund, R., Schmid, M.L., Subramanian, K.G., Wiederhold, P.: Contextual array grammars and array P systems. Ann. Math. Artificial Intell. doi:10.1007/s10472-013-9388-0
5. Ferretti, C., Mauri, G., Paun, G., Zandron, C.: On three variants of rewriting P systems. Theor. Comp. Sci. **301**, 201–215 (2003)
6. Habel, A., Kreowski, H.J.: Collage grammars. In: Ehrig, H., Kreowski, H.-J., Rozenberg, G. (eds.) Graph Grammars 1990. LNCS, vol. 532, pp. 411–429. Springer, Heidelberg (1991)
7. Kitaev, S., Mansour, T., Seebold, P.: The Peano curve and counting occurrences of some patterns. J. Autom. Lang. Combin. **9**(4), 439–455 (2004)
8. Maurer, H.A., Rozenberg, G., Welzl, E.: Using string languages to describe picture languages. Inf. Control **54**, 155–185 (1982)
9. Pan, L., Păun, G.: On parallel array P systems. In: Adamatzky, A. (ed.) Automata, Universality, Computation. ECC, vol. 12, pp. 173–183. Springer, Heidelberg (2015)
10. Păun, G.: Computing with membranes. J. Comp. Syst. Sci. **61**, 108–143 (2000)
11. Salomaa, A.: Formal Languages. Academic Press, London (1973)
12. Seebold, P.: Tag system for the Hilbert curve. Discrete Math. Theor. Comp. Sci. **9**, 213226 (2007)

13. Siromoney, R., Subramanian, K.G.: Space-filling curves and infinite graphs. In: Ehrig, H., Nagl, M., Rozenberg, G. (eds.) Graph Grammars 1982. LNCS, vol. 153, pp. 380–391. Springer, Heidelberg (1983)
14. Subramanian, K.G.: P systems and picture languages. In: Durand-Lose, J., Margenstern, M. (eds.) MCU 2007. LNCS, vol. 4664, pp. 99–109. Springer, Heidelberg (2007)
15. Subramanian, K.G., Isawasan, P., Venkat, I., Pan, L.: Parallel array-rewriting P systems. Rom. J. Inf. Sci. Tech. **17**(1), 103–116 (2014)
16. Subramanian, K.G., Venkat, I., Pan, L.: P Systems generating Chain Code Picture Languages, Proceedings of Asian Conference on Membrane Computing, pp. 115–123 (2012)

Base-Complexity Classifications of QCB_0-Spaces

Matthew de Brecht[1(✉)], Matthias Schröder[2], and Victor Selivanov[3]

[1] National Institute of Information and Communications Technology (NICT), Center for Information and Neural Networks (CiNet), Osaka, Japan
matthew@nict.go.jp

[2] Department of Mathematics, TU Darmstadt, Darmstadt, Germany

[3] A.P. Ershov Institute of Informatics Systems SB RAS, Novosibirsk, Russia

Abstract. We define and study new classifications of qcb_0-spaces based on the idea to measure the complexity of their bases. The new classifications complement those given by the hierarchies of qcb_0-spaces introduced in [7,8] and provide new tools to investigate non-countably based qcb_0-spaces. As a by-product, we show that there is no universal qcb_0-space and establish several apparently new properties of the Kleene-Kreisel continuous functionals of countable types.

Keywords: QCB_0-spaces · Y-based spaces · Hyperspaces · Scott topology · Hyperprojective hierarchy · Kleene-Kreisel continuous functionals

1 Introduction

A basic notion of Computable Analysis [10] is the notion of an *admissible representation* of a topological space X. This is a partial continuous surjection δ from the Baire space $\mathcal{N}$ onto X satisfying a certain universality property. Such a representation of X often induces a reasonable computability theory on X, and the class of admissibly represented spaces is wide enough to include most spaces of interest for Analysis or Numerical Mathematics. This class coincides with the class of the so-called qcb_0-spaces, i.e. T_0-spaces which are quotients of countably based spaces, and it forms a cartesian closed category with the continuous functions as morphisms [5]. Thus, among qcb_0-spaces one meets many important function spaces including the continuous functionals of finite types [3,4] interesting for several branches of logic and computability theory. In addition to being cartesian closed, the category QCB_0 of qcb_0-spaces is also closed under countable limits, countable colimits, and many other important constructions,

M. de Brecht—Supported by JSPS Core-to-Core Program, A. Advanced Research Networks

M. Schröder—Supported by FWF research project "Definability and computability" and by DFG project Zi 1009/4-1.

V. Selivanov—Supported by the DFG Mercator professorship at the University of Würzburg, by the RFBR-FWF project "Definability and computability", by RFBR project 13-01-00015a, and by *7th EU IRSES* project 294962 (COMPUTAL).

A. Beckmann et al. (Eds.): CiE 2015, LNCS 9136, pp. 156–166, 2015.
DOI: 10.1007/978-3-319-20028-6_16

making it a very convenient category of topological spaces. However, along with the benefits of this generality comes the challenge of developing comprehensive theories that provide a deeper understanding of arbitrary qcb$_0$-spaces.

Classical descriptive set theory [2] has proven to be extremely useful for classifying and studying separable metrizable spaces. Every separable metrizable space can be topologically embedded into a Polish space (a complete separable metrizable space), for example by taking the completion of a compatible metric. We can therefore classify a separable metrizable space according to the complexity of defining it as a subspace of some Polish space, where topological complexity can be quantified using natural hierarchies such as the Borel or Luzin (projective) hierarchies. This method of classification is topologically invariant (it does not depend on which Polish space we embed into) because of the remarkable fact that a subspace of a Polish space is Polish if and only if it is of level $\mathbf{\Pi}^0_2$ in the Borel hierarchy. We can even generalize this approach to the entire class of countably based T_0-spaces (abbreviated cb$_0$-spaces) by using quasi-Polish spaces [1], which have the same $\mathbf{\Pi}^0_2$ absoluteness property as Polish spaces. In fact, for classifying cb$_0$-spaces we can restrict ourselves to the algebraic domain $P\omega$ of all subsets of natural numbers (denoted ω), which is quasi-Polish and universal for cb$_0$-spaces.

Unfortunately, this approach to classifying topological spaces does not immediately generalize to the entire category of qcb$_0$-spaces. First of all, as we will see in this paper, there is no universal qcb$_0$-space to serve as a basis for comparing topological complexity. A second critical problem is that the $\mathbf{\Pi}^0_2$ absoluteness property of Polish and quasi-Polish spaces does not apply to subspaces of non-countably based spaces. For example, in [9] it is shown that the space $\mathcal{O}(\mathcal{N})$, the lattice of open subsets of $\mathcal{N}$ with the Scott-topology, contains *singleton* subsets which are $\mathbf{\Pi}^1_1$-complete even though they are trivially Polish with respect to the subspace topology. It is possible to use similar methods to construct qcb$_0$-spaces that have singleton subsets of arbitrarily high complexity in the hyperprojective hierarchy.

Important progress towards classifying qcb$_0$-spaces was made in [7,8], where the Borel, projective, and hyperprojective hierarchies of qcb$_0$-spaces were introduced. The major insight was to classify qcb$_0$-spaces according to the complexity of the equivalence relation on the elements of $\mathcal{N}$ induced by an admissible representation of the space, which elegantly sidesteps the problem of finding a universal space. This approach works well because the universal property of admissible representations causes them to reflect many important topological properties of the underlying space. In fact, it was shown in [7,8] that for cb$_0$-spaces, the newly introduced classification approach using admissible representations is equivalent to the approach described above that uses topological embeddings into $P\omega$.

However, the hierarchies defined in [7,8] do not differentiate between countably based qcb$_0$-spaces and non-countably based spaces. In particular, the problem of placing an upper bound on the relative complexity of even very simple subsets (such as singletons) of non-countably based spaces can not be settled using this approach. Thus, although the Borel, projective, and hyperprojective hierarchies quantify one important aspect of the complexity of qcb$_0$-spaces, there appears to be an additional dimension of complexity that is mostly apparent in the large difference between countably based and non-countably based spaces.

In this paper we attempt to capture this additional dimension of complexity by introducing methods to classify a topological space according to the complexity of defining a basis for its topology. Our hope is that by combining the basis-complexity measures introduced in this paper with the hierarchies defined in [7,8], we can obtain a more complete measure of the topological complexity of qcb_0-spaces.

The basic idea of our approach is a natural generalization of the definition of a countable basis. Given a topological space X, a countable basis for X can be viewed as a mapping ϕ from ω to the set $\mathcal{O}(X)$ of open subsets of X such that the range of ϕ is a basis for the topology of X. As a first approach to generalizing this definition to non-countably based spaces, we can replace the index set ω with an arbitrary topological space Y and consider whether or not a basis for X can be indexed by some mapping $\phi\colon Y \to \mathcal{O}(X)$ which is continuous with respect to the Scott-topology on $\mathcal{O}(X)$. The class of spaces that have such an indexing for a basis will be called *Y-based spaces*, and the complexity of Y according to the hierarchies in [7,8] provides an indication of the complexity of the spaces in this class. This definition is very natural and we will show that it has several useful properties, but unfortunately it can be difficult to use in practice. We therefore also introduce a second related concept that we call *sequentially Y-based* spaces, which requires a more complicated definition but behaves much better when working with sequential spaces. In particular, we will show that universal spaces exist for the class of sequentially Y-based spaces for each qcb_0-space Y. We expect this observation will be useful for future development of a descriptive theory of qcb_0-spaces that avoids the problems mentioned earlier in this introduction.

We will provide a detailed analysis of the relationship between the proposed hierarchies and the previous ones, and provide some applications. The newly introduced basis-complexity classifications can be particularly useful when determining whether one space can be embedded into another space. We will demonstrate this claim by investigating the existence of certain classes of universal qcb_0-spaces, by showing that every qcb_0-space can be embedded into a space with a total admissible representation, and by establishing several apparently new properties of the Kleene-Kreisel continuous functionals of countable types.

In Sect. 2 we discuss the notions of topological and sequential embeddings. In Sects. 3 and 4 we first introduce and study some versions of the notion of a Y-based space, and then define and investigate the two relevant classifications of qcb_0-spaces. In Sect. 5 we study which levels of the the new and old hierarchies have a universal (or sequentially universal) space. Because of the strict space bounds, we omit all proofs and use some notation and notions from [8] without definition.

2 Topological Embeddings Versus Sequential Embeddings

In this section we briefly discuss two notions of embedding for sequential spaces relevant to this paper. The first one is the usual topological embedding which is used in Sect. 3. The second one is a lesser known sequential embedding which is more natural for sequential spaces and results in a more satisfactory theory in Sect. 4 than the theory based on topological embeddings.

We say that a space X *embeds topologically* into Y, if X is homeomorphic to a topological subspace M of Y; the corresponding homeomorphism seen as a function e from X to Y is called a *topological embedding* of X into Y. When dealing with sequential spaces (in particular, qcb$_0$-spaces), it is natural to consider the following modification of the topological embeddings:

Definition 1. Let X, Y be sequential spaces.

(1) The space X is *a sequential subspace* of Y, if $X \subseteq Y$ and, whenever $(x_n)_n$ is a sequence in X and $x_\infty \in X$, convergence of $(x_n)_n$ to x_∞ in X is equivalent to convergence of $(x_n)_n$ to x_∞ in Y.
(2) We say that X *embeds sequentially into* Y, if there is an injection $e\colon X \to Y$ such that convergence of $(x_n)_n$ to x_∞ in X is equivalent to convergence of $(e(x_n))_n$ to $e(x_\infty)$ in Y. In this case we call e a *sequential embedding of X into Y*.

The distinction between topological subspace and sequential subspace is subtle, but very important. It can be shown that if X and Y are sequential spaces, then X embeds sequentially into Y if and only if there is a topological subspace $S \subseteq Y$ such that X is homeomorphic to the sequentialisation of S.

It is easy to check that, for all sequential spaces X, Y, if $e\colon X \to Y$ is a topological embedding then it is also a sequential embedding, but the converse does not hold in general. If $e\colon X \to Y$ is a surjective sequential embedding, then e is a homeomorphism.

3 Y-based Spaces

In this section we introduce and study the notion of a Y-based space, where Y is a topological space. This provides a natural generalization of the notion of a countably based space which can be applied to classifying non-countably based qcb$_0$-spaces.

Let $\mathbb{S}$ be the Sierpinski space and $\mathcal{O}(X)$ be the hyperspace of open subsets of a space X topologised with the ω-Scott topology. If X is a sequential space (in particular a qcb$_0$-space), then $\mathcal{O}(X)$ is homeomorphic to $\mathbb{S}^X$.

Definition 2. Let X, Y be topological spaces. A continuous function $\phi\colon Y \to \mathcal{O}(X)$ is a *Y-indexing* of a basis for X, if the range of ϕ is a basis for the topology on X. The space X is *Y-based* if there is a *Y-indexing* of a basis for X.

These notions are purely topological and apply to arbitrary topological spaces. It is easily shown that if X is Y-based and Y is a continuous image of a space Z then X is Z-based, and that any topological subspace of a Y-based space is Y-based.

The next proposition generalizes the fact that any countably-based T_0-space embeds topologically into $P\omega$, which is homeomorphic to $\mathcal{O}(\omega)$.

Theorem 1. *Let X, Y be sequential T_0-spaces such that X is Y-based. Then X topologically embeds into $\mathcal{O}(Y)$.*

The next basic fact characterizes qcb$_0$-spaces in terms of these notions.

Theorem 2. *The following are equivalent for any sequential T_0-space X:*

(1) *X is Y-based for some zero-dimensional cb$_0$-space Y (i.e., some $Y \subseteq \mathcal{N}$).*
(2) *X is Y-based for some qcb$_0$-space Y.*
(3) *X topologically embeds into $\mathcal{O}(Y)$ for some cb$_0$-space Y.*
(4) *X is a qcb$_0$-space.*

As $\mathcal{O}(Y)$ has a total admissible representation for any cb$_0$-space Y, we obtain:

Corollary 1. *Every qcb$_0$-space topologically embeds into a space with a total admissible representation.*

For any qcb$_0$-space Y, let $Based(Y)$ denote the class of Y-based qcb$_0$-spaces. For a class $\mathcal{S}$ of qcb$_0$-spaces, let $Based(\mathcal{S}) = \bigcup_{Y \in \mathcal{S}} Based(Y)$. Theorem 2 induces some natural classifications of qcb$_0$-spaces. For example, one can relate to any family of pointclasses $\mathbf{\Gamma}$ the classes $Based(\mathbf{\Gamma}(\mathcal{N}))$ and $Based(\mathsf{QCB}_0(\mathbf{\Gamma}))$ and easily check that the classes coincide. Here, the notation $\mathsf{QCB}_0(\mathbf{\Gamma})$ refers to the hierarchies of qcb$_0$-spaces defined in [7,8]. Thus, the classical hierarchies of subsets of the Baire space induce corresponding hierarchies of qcb$_0$-spaces, in particular the "hyperprojective base-hierarchy" $Based(\mathbf{\Sigma}^1_\alpha(\mathcal{N}))$, for which we use the simpler notation $Based(\mathbf{\Sigma}^1_\alpha)$.

We now establish a relationship between $Based(\mathbf{\Sigma}^1_\alpha)$ and the qcb$_0$-space $\mathbb{N}\langle\alpha\rangle$, which is *the space of continuous functionals of type* α over ω. The spaces $\mathbb{N}\langle\alpha\rangle$ are defined by induction on countable ordinals α as follows [8]:

$$\mathbb{N}\langle 0\rangle := \omega,\ \mathbb{N}\langle\beta+1\rangle := \omega^{\mathbb{N}\langle\beta\rangle} \text{ and } \mathbb{N}\langle\lambda\rangle := \textstyle\prod_{\alpha<\lambda}\mathbb{N}\langle\alpha\rangle,$$

where ω denotes the discrete space of natural numbers, $\beta, \lambda < \omega_1$ and λ is a limit ordinal. Obviously, for $k < \omega$ the space $\mathbb{N}\langle k\rangle$ coincides with the space of Kleene-Kreisel continuous functionals of type k extensively studied in the literature, and $\mathbb{N}\langle 1\rangle$ coincides with the Baire space $\mathcal{N}$. Moreover, we consider a standard admissible representation $\delta_\alpha\colon D_\alpha \to \mathbb{N}\langle\alpha\rangle$ for $\mathbb{N}\langle\alpha\rangle$ derived by a natural construction as presented in [8]. We will also deal with the coproduct spaces $\mathbb{N}\langle<\lambda\rangle := \bigoplus_{\alpha<\lambda}\mathbb{N}\langle\alpha\rangle$, where λ is a countable ordinal limit.

Proposition 1. *For any $\alpha < \omega_1$, $Based(D_{\alpha+1}) = Based(\mathbf{\Pi}^1_\alpha) = Based(\mathbf{\Sigma}^1_{\alpha+1}) = Based(\mathbb{N}\langle\alpha+1\rangle)$. For any limit ordinal $\lambda < \omega_1$, $Based(D_\lambda) = Based((\mathbf{\Pi}^1_{<\lambda})_\delta) = Based(\mathbf{\Sigma}^1_\lambda) = Based(\mathbb{N}\langle\lambda\rangle)$.*

By Theorem 1, any space from *Based*(Y) topologically embeds into $\mathcal{O}(Y)$. A principal question is: for which qcb$_0$-spaces Y do we have that the space $\mathcal{O}(Y)$ is Y-based? Clearly, this is equivalent to saying that *Based*(Y) is the class of spaces topologically embeddable into $\mathcal{O}(Y)$. Unfortunately, the assertion does not hold for all Y: one easily checks that the space $\mathcal{O}(\mathbb{Q})$ is not $\mathbb{Q}$-based. Nevertheless, the assertion $\mathcal{O}(Y) \in \mathit{Based}(Y)$ might hold for some natural spaces Y, in particular a positive answer to the following problem would clarify the nature of the hierarchy $\{\mathit{Based}(D_\alpha)\}_{\alpha<\omega_1}$ considerably:

Problem 1. Does the assertion $\mathcal{O}(D_\alpha) \in \mathit{Based}(D_\alpha)$ hold for all $\alpha < \omega_1$?

If the answer is positive, *Based*(D_α) would coincide with the class of spaces topologically embeddable into $\mathcal{O}(D_\alpha)$. For $\alpha = 0$ the assertion holds because $\mathcal{O}(\omega)$ is homeomorphic to $P\omega$, and we will show below that the assertion is also true for $\alpha = 1$. For $\alpha \geq 2$ we still do not know the answer. This is an obstacle to answering the principal question on the non-collapse of the introduced hierarchy $\{\mathit{Based}(D_\alpha)\}_{\alpha<\omega_1}$. By the non-collapse property we mean that the inclusion $\mathit{Based}(D_\alpha) \subseteq \mathit{Based}(D_\beta)$ is proper for each $\alpha < \beta < \omega_1$. The next result (along with the assertion $\mathcal{O}(D_1) \in \mathit{Based}(D_1)$) implies, in particular, that $\mathit{Based}(D_0) \subsetneq \mathit{Based}(D_1)$.

Proposition 2. *For any* $\alpha < \omega_1$, $\mathcal{O}(D_{\alpha+1}) \notin \mathit{Based}(D_\alpha)$. *For any limit ordinal* $\lambda < \omega_1$, $\mathcal{O}(D_\lambda) \notin \mathit{Based}(\bigoplus_{\alpha<\lambda} D_\alpha)$.

The following relation between the hyperprojective hierarchy of qcb$_0$-spaces and the hierarchy $\{\mathit{Based}(D_\alpha)\}_{\alpha<\omega_1}$ is interesting in its own right and also implies a weak non-collapse property:

Proposition 3. *For any* $\alpha < \omega_1$, $\mathsf{QCB}_0(\mathbf{\Pi}^1_\alpha) \subseteq \mathit{Based}(\mathbf{\Pi}^1_{\alpha+1}) = \mathit{Based}(D_{\alpha+2})$.

Conversely, for each ordinal $\alpha < \omega_1$, $\mathsf{QCB}_0(\mathbf{\Pi}^1_\alpha)$ does not even contain all of *Based*(ω), as $D_{\alpha+2} \in \mathit{Based}(\omega) \setminus \mathsf{QCB}_0(\mathbf{\Pi}^1_\alpha)$ by Theorem 2 in [8]. The second item of the next corollary is the weak version of the non-collapse property.

Corollary 2. *Let* $\alpha < \omega_1$. *Then we have* $\mathcal{O}(D_\alpha) \in \mathit{Based}(D_{\alpha+2})$ *and the inclusion* $\mathit{Based}(D_\alpha) \subset \mathit{Based}(D_{\alpha+3})$ *is proper.*

Problem 2. For which $\alpha < \omega_1$ can the inclusion from Proposition 3 be improved to $\mathsf{QCB}_0(\mathbf{\Pi}^1_\alpha) \subseteq \mathit{Based}(D_{\alpha+1})$ or even to $\mathsf{QCB}_0(\mathbf{\Pi}^1_\alpha) \subseteq \mathit{Based}(D_\alpha)$?

Next we further investigate the important class $\mathit{Based}(\mathcal{N}) = \mathit{Based}(D_1)$ of $\mathcal{N}$-based qcb$_0$-spaces, which includes many natural non-countably based spaces. As an example, we state an interesting property of the class of quasi-Polish spaces [1], which includes both Polish spaces and ω-continuous domains.

Proposition 4. *If* X *is quasi-Polish then* $\mathcal{O}(X)$ *is* $\mathcal{N}$*-based.*

For metrizable spaces $X \in \mathsf{CB}_0(\mathbf{\Pi}^1_1)$ we have the following complete characterization of when $\mathcal{O}(X)$ is $\mathcal{N}$-based.

Proposition 5. *Let $X \in \mathsf{CB}_0(\mathbf{\Pi}^1_1)$ be metrizable. Then $\mathcal{O}(X)$ is $\mathcal{N}$-based if and only if X is Polish.*

Corollary 3. *A qcb$_0$-space is $\mathcal{N}$-based if and only if it embeds topologically in $\mathcal{O}(\mathcal{N})$. In particular, Based$(D_0) \subsetneq$ Based(D_1).*

4 Sequentially Y-based Spaces

In this section we consider some modifications of the notion of Y-based spaces from the previous section which are more suitable to the nature of sequential spaces (in particular, qcb$_0$-spaces). This will be sufficient to settle the analogues of the open questions in Sect. 3 for the sequential embeddings in place of topological embeddings.

One could define several modifications of the notion of Y-based space. For instance, for qcb$_0$-spaces X, P we could say that a function $\phi\colon P \to \mathcal{O}(X)$ is a *P-indexed sequential basis* for X, if ϕ is continuous and range(ϕ) is a subbasis for a topology τ on X such that the sequentialisation of τ is the Scott topology in $\mathcal{O}(X)$. Under this definition, some interesting facts may be established, e.g., one can show that for any $\alpha < \omega_1$ the space $\mathbb{N}\langle\alpha+1\rangle$ has an $\mathbb{N}\langle\alpha\rangle$-indexed sequential basis (see Corollary 5). We also consider the following deeper modification:

Definition 3. Let X, P be sequential spaces.

(1) We call a collection $\mathcal{B}$ of open subsets of X *a sequential basis* for X if $\mathcal{B}$ is a subbase of a topology τ on the set X such that the sequentialisation of τ is equal to $\mathcal{O}(X)$.
(2) A function $\phi\colon P \to \mathcal{O}(X)$ is called a P-indexed sequential basis for X if ϕ is continuous and its range $rng(\phi)$ is a sequential basis for X.
(3) For a function $\phi\colon P \to \mathcal{O}(X)$, we define $\mathcal{B}_\phi$ to consist of all intersections of the form $\bigcap_{n\le\infty}\phi(p_n)$, where $(p_n)_n$ converges to p_∞ in P.
(4) A function $\phi\colon P \to \mathcal{O}(X)$ is called a *P-indexed generating system for X* if ϕ is continuous and $\mathcal{B}_\phi$ is a sequential basis for X.
(5) X is called *sequentially P-based* if there is a P-indexed generating system for X.

By Proposition 2.2 in [6] the elements of $\mathcal{B}_\phi$ are open in X, if ϕ is continuous, because $(\phi(p_n))_n$ converges to $\phi(p_\infty)$ in $\mathcal{O}(X)$.

Now we study for which spaces P the existence of a P-indexed generating system implies the existence of a P-indexed sequential basis.

Lemma 1. *Let P be a sequential space such that there is a continuous surjection from P onto $P^{\mathbb{N}_\infty}$. Then any sequential space X is sequentially P-based if, and only if, X has a P-indexed sequential basis.*

The spaces $\mathbb{N}\langle\alpha\rangle$ and $\mathbb{N}\langle{<}\lambda\rangle$ can be shown to fulfill the requirement of Lemma 1. We obtain:

Corollary 4. (1) *For any $\alpha < \omega_1$, a sequential space X is sequentially $\mathbb{N}\langle\alpha\rangle$-based if, and only if, there is an $\mathbb{N}\langle\alpha\rangle$-indexed sequential basis for X.*
(2) *For any limit ordinal $\lambda < \omega_1$, a sequential space X is sequentially $\mathbb{N}\langle<\lambda\rangle$-based if, and only if, there is an $\mathbb{N}\langle<\lambda\rangle$-indexed sequential basis for X.*

Although Definition 3 (4) is very technical, it is justified by several nice properties the main of which is the following theorem:

Theorem 3. *Let X and P be sequential T_0-spaces. Then X is sequentially P-based if, and only if, X embeds sequentially into $\mathcal{O}(P)$.*

Theorem 3 solves in the positive the "sequential analogue" of the question "is $\mathcal{O}(P) \in Based(P)$ for each P?" discussed in the previous section.

For any sequential space P we of course have $Based(P) \subseteq SBased(P)$. An interesting question is "for which P is this inclusion proper?" One example is $\mathbb{Q}$, because $\mathcal{O}(\mathbb{Q})$ is sequentially $\mathbb{Q}$-based by Theorem 3, but not countably based, as $\mathbb{Q}$ is a non-locally-compact metrizable space. This observation can be improved to the following:

Proposition 6. *The space $\mathcal{O}(\mathbb{Q})$ is sequentially $\mathcal{N}$-based, but not $\mathcal{N}$-based.*

We now show how to construct generating systems for countable products and function spaces (formed in the category Seq of sequential spaces).

Proposition 7. *Let X_i, P_i be sequential T_0-spaces such that X_i is sequentially P_i-based. Then the sequential product $\prod_{i\in\omega} X_i$ is sequentially $(\bigoplus_{i\in\omega} P_i)$-based.*

Proposition 8. *Let X, Y, P be sequential T_0-spaces such that Y is sequentially P-based. Then Y^X is sequentially $(P \times X)$-based.*

We obtain the following nice property of the spaces of functionals:

Corollary 5. *For any $\alpha < \omega_1$, the space $\mathbb{N}\langle\alpha + 1\rangle$ is sequentially $\mathbb{N}\langle\alpha\rangle$-based. For any limit ordinal $\lambda < \omega_1$, the space $\mathbb{N}\langle\lambda\rangle$ is sequentially $\mathbb{N}\langle<\lambda\rangle$-based.*

For any qcb$_0$-space Y, let $SBased(Y)$ denote the class of sequentially Y-based qcb$_0$-spaces. For a class $\mathcal{S}$ of qcb$_0$-spaces Y, let $SBased(\mathcal{S}) = \bigcup_{Y\in\mathcal{S}} SBased(Y)$. Obviously, $Based(Y) \subseteq SBased(Y)$ for each qcb$_0$-space Y. Theorem 3 induces some natural classifications of qcb$_0$-spaces. For example, one can relate to any family of pointclasses $\mathbf{\Gamma}$ the classes $SBased(\mathbf{\Gamma}(\mathcal{N}))$, $SBased(\mathsf{QCB}_0(\mathbf{\Gamma}))$ and show that they coincide.

Thus, the classical hierarchies of subsets of the Baire space induce the corresponding hierarchies of qcb$_0$-spaces, in particular the "hyperprojective sequential-based-hierarchy" $SBased(\mathbf{\Sigma}^1_\alpha(\mathcal{N}))$; we simplify the notation to $SBased(\mathbf{\Sigma}^1_\alpha)$ and relate this hierarchy to the admissible representations $\delta_\alpha : D_\alpha \to \mathbb{N}\langle\alpha\rangle$. The next assertion is an analogue of Proposition 1.

Proposition 9. *For any $\alpha < \omega_1$, $SBased(D_{\alpha+1}) = SBased(\mathbf{\Pi}^1_\alpha) = SBased(\mathbf{\Sigma}^1_{\alpha+1}) = SBased(\mathbb{N}\langle\alpha + 1\rangle)$. For any limit ordinal $\lambda < \omega_1$, $SBased(D_\lambda) = SBased(\mathbf{\Sigma}^1_\lambda) = SBased((\mathbf{\Pi}^1_{<\lambda})_\delta) = SBased(\mathbb{N}\langle\lambda\rangle)$.*

Next we solve the principal question on the non-collapse property of the hierarchy $\{SBased(D_\alpha)\}_{\alpha<\omega_1}$. Remember that the corresponding result for the hierarchy $\{Based(D_\alpha)\}_{\alpha<\omega_1}$ remained open.

Proposition 10. *The hierarchy $\{SBased(D_\alpha)\}_{\alpha<\omega_1}$ does not collapse. More precisely, $SBased(D_\alpha) \subsetneqq SBased(D_{\alpha+1})$ for all $\alpha < \omega_1$ and $SBased(\bigoplus_{\alpha<\lambda} D_\alpha) \subsetneqq SBased(D_\lambda)$ for each limit ordinal $\lambda < \omega_1$.*

The next fact shows that the class $SBased(\mathcal{N})$ is rather rich.

Proposition 11. *Let X be a qcb$_0$-space having a total admissible representation $\xi\colon \mathcal{N} \to X$. Then $\mathcal{O}(X)$ embeds sequentially into $\mathcal{O}(\mathcal{N})$.*

Problem 3. We know from Theorem 3 and Proposition 6 that $Based(\mathbb{Q}) \subsetneqq SBased(\mathbb{Q})$ and $Based(\mathcal{N}) \subsetneqq SBased(\mathcal{N})$. We would like to know which sequential spaces X satisfy $Based(X) \subsetneqq SBased(X)$. In particular, we conjecture that $Based(D_\alpha) \subsetneqq SBased(D_\alpha)$ for all non-zero ordinals $\alpha < \omega_1$, and $SBased(\bigoplus_{\alpha<\lambda} D_\alpha) \subsetneqq SBased(\bigoplus_{\alpha<\lambda} D_\alpha)$ for all limit ordinals $\lambda < \omega_1$. Good possible witnesses seem to be $\mathbb{N}\langle\alpha+1\rangle$ and $\mathbb{N}\langle\lambda\rangle$ respectively (see Corollary 5).

By Proposition 10, the spaces $\mathcal{O}(D_\alpha)$ are natural witnesses for the non-collapse property of the hierarchy $\{SBased(D_\alpha)\}_{\alpha<\omega_1}$. Next we observe that the spaces $\mathbb{N}\langle\alpha\rangle$ provide other natural witnesses for this property showing that Corollary 5 is in a sense optimal.

Theorem 4. (1) *For any $\alpha < \omega_1$, $\mathbb{N}\langle\alpha+2\rangle \in SBased(\mathbb{N}\langle\alpha+1\rangle) \setminus SBased(\mathbb{N}\langle\alpha\rangle)$.*
(2) *For any limit ordinal $\lambda < \omega_1$, $\mathbb{N}\langle\lambda+1\rangle \in SBased(\mathbb{N}\langle\lambda\rangle) \setminus SBased(\mathbb{N}\langle<\lambda\rangle)$.*
(3) *For any limit ordinal $\lambda < \omega_1$, $\mathbb{N}\langle\lambda\rangle \in SBased(\bigoplus_{\alpha<\lambda} \mathbb{N}\langle\alpha\rangle) \setminus \bigcup_{\alpha<\lambda} SBased(\mathbb{N}\langle\alpha\rangle)$.*

We also can deduce the following corollary about the continuous functionals.

Corollary 6. *For all $\alpha < \beta < \omega_1$, $\mathbb{N}\langle\beta\rangle$ does not sequentially embed into $\mathbb{N}\langle\alpha\rangle$.*

5 On Universal Spaces

In this section we discuss which classes of qcb$_0$-spaces have and which do not have a universal space. This is of interest because universal spaces are noticeable in several branches of set-theoretic topology.

Definition 4. (1) Let $\mathcal{S}$ be a class of topological spaces. A space X is *universal in $\mathcal{S}$*, if $X \in \mathcal{S}$ and any space from $\mathcal{S}$ embeds topologically in X.
(2) Let $\mathcal{S}$ be a class of sequential spaces. A space X is *sequentially universal in $\mathcal{S}$*, if $X \in \mathcal{S}$ and any space from $\mathcal{S}$ embeds sequentially in X.

The first notion above is well-known in topology. E.g., $P\omega$ is universal in the class of cb$_0$-spaces, while the class of all topological spaces has no universal space. The second notion is a "sequential version" of the first one which is natural when dealing with sequential spaces or qcb_0-spaces. As $P\omega$ is universal in the class of cb_0-spaces and Y-based spaces are designed as a natural generalization of countably based spaces, it is natural to ask for which $Y \subseteq \mathcal{N}$ the class of Y-based spaces has a universal space. At least, we can prove:

Corollary 7. *Let $Y \subseteq \mathcal{N}$ be such that the space $\mathcal{O}(Y)$ is Y-based. Then $\mathcal{O}(Y)$ is universal in the class of Y-based topological spaces. In particular, the space $\mathcal{O}(\mathcal{N})$ is universal in the class of $\mathcal{N}$-based spaces.*

For the sequential version, we derive from Theorem 3:

Corollary 8. *For any $Y \in \mathsf{QCB}_0$, $\mathcal{O}(Y)$ is sequentially universal in $SBased(Y)$.*

It is still open whether or not $Based(D_\alpha)$ contains a universal space when $\alpha > 1$. However, we see that each level of the hierarchy $\{SBased(D_\alpha)\}_{\alpha<\omega_1}$ contains a sequentially universal space $\mathcal{O}(D_\alpha)$ with a total admissible representation. The same applies to the hierarchies of cb_0-spaces in [7] (obviously, $P\omega$ is a universal space in $\mathsf{CB}_0(\mathbf{\Gamma})$ for each family of pointclasses $\mathbf{\Gamma}$ that contains $\mathbf{\Pi}^0_2$).

For the hierarchies of qcb_0-spaces in [7,8] the situation is more complicated. Currently we do not know which of the classes $\mathsf{QCB}_0(\mathbf{\Gamma})$, where $\mathbf{\Gamma}$ is a level of the Borel or hyperprojective hierarchy, have a universal (or a sequentially universal) space. Nevertheless, we can show that the class of all qcb_0-spaces, as well as some natural pointclasses related to the hyperprojective hierarchy of qcb_0-spaces, do not have universal spaces. Recall from [7,8] that $\mathsf{QCB}_0(\mathbf{P}) := \bigcup_{n<\omega} \mathsf{QCB}_0(\mathbf{\Sigma}^1_n)$ and $\mathsf{QCB}_0(\mathbf{HP}) := \bigcup_{\alpha<\omega_1} \mathsf{QCB}_0(\mathbf{\Sigma}^1_\alpha)$ denote the classes of projective and of hyperprojective qcb_0-spaces, respectively.

Theorem 5. (1) *There is no universal (nor a sequentially universal) qcb_0-space.*
(2) *For any limit ordinal $\lambda < \omega_1$, there is no universal (nor a sequentially universal) space in $\mathsf{QCB}_0(\mathbf{\Sigma}^1_{<\lambda})$.*
(3) *There is no universal (nor a sequentially universal) space in $\mathsf{QCB}_0(\mathbf{P})$ (nor in $\mathsf{QCB}_0(\mathbf{HP})$).*

References

1. de Brecht, M.: Quasi-polish spaces. Ann. Pure Applied Logic **164**, 356–381 (2013)
2. Kechris, A.S.: Classical Descriptive Set Theory. Springer, New York (1995)
3. Kleene, S.C.: Countable functionals. In: Heyting, A. (ed.) Constructivity in Mathematics, pp. 87–100. North Holland, Amsterdam (1959)
4. Kreisel, G.: Interpretation of analysis by means of constructive functionals of finite types. In: Heyting, A. (ed.) Constructivity in Mathematics, pp. 101–128. North Holland, Amsterdam (1959)
5. Schröder, M: Admissible representations for continuous computations. Ph.D. thesis Fachbereich Informatik, FernUniversität Hagen (2003)
6. Schröder, M.: A Hofmann-Mislove Theorem for Scott Open Sets (2015). arXiv:1501.06452

7. Schröder, M., Selivanov, V.: Some hierarchies of qcb_0-spaces. Math. Struct. in Comp. Sci. (2014). doi:10.1017/S0960129513000376
8. Schröder, M., Selivanov, V.: Hyperprojective hierarchy of qcb_0-spaces. Computability **4**(1), 1–17 (2015)
9. Selivanov, V.: Total representations. Logical Methods Comput. Sci. **9**(2), 1–30 (2013)
10. Weihrauch, K.: Computable Analysis. Springer, Heidelberg (2000)

New Bounds on Optimal Sorting Networks

Thorsten Ehlers(✉) and Mike Müller

Institut für Informatik, Christian-Albrechts-Universität zu Kiel,
24098 Kiel, Germany
{the,mimu}@informatik.uni-kiel.de

Abstract. We present new parallel sorting networks for 17 to 20 inputs. For 17, 19, and 20 inputs these new networks are faster (i.e., they require fewer computation steps) than the previously known best networks. Therefore, we improve upon the known upper bounds for minimal depth sorting networks on 17, 19, and 20 channels. Furthermore, we show that our sorting network for 17 inputs is optimal in the sense that no sorting network using less layers exists. This solves the main open problem of [D. Bundala & J. Závodný. Optimal sorting networks, Proc. LATA 2014].

1 Introduction

Comparator networks are hardwired circuits consisting of simple gates that sort their inputs. If the output of such a network is sorted for all possible inputs, it is called a *sorting network*. Sorting networks are an old area of interest, and results concerning their size date back at least to the 50's of the last century.

The size of a comparator network in general can be measured by two different quantities: the total number of comparators involved in the network, or the number of layers the network consists of. In both cases, finding optimal sorting networks (i.e., of minimal size) is a challenging task even when restricted to few inputs, which was attacked using different methods.

For instance, Valsalam and Miikkulainen [11] employed evolutionary algorithms to generate sorting networks with few comparators. Minimal depth sorting networks for up to 16 inputs were constructed by Shapiro and Van Voorhis in the 60's and 70's, and by Schwiebert in 2001, who also made use of evolutionary techniques. For a presentation of these networks see Knuth [8, Fig. 51]. However, the optimality of the known networks for 11 to 16 channels was only shown recently by Bundala and Závodný [4], who partitioned the set of first two layers into equivalence classes and reduced the search to extensions of one representative of each class. They then expressed the existence of a sorting network with less layers extending these representatives in propositional logic and used a SAT solver to show that the resulting formulae are unsatisfiable. Codish, Cruz-Filipe, and Schneider-Kamp [5] simplified the generation of the representatives and independently verified Bundala and Závodný's result.

For more than 16 channels, not much is known about the minimal depths of sorting networks. Al-Haj Baddar and Batcher [2] exhibit a network sorting 18 inputs using 11 layers, which also provides the best known upper bound on the

A. Beckmann et al. (Eds.): CiE 2015, LNCS 9136, pp. 167–176, 2015.
DOI: 10.1007/978-3-319-20028-6_17

minimal depth of a sorting network for 17 inputs. The lowest upper bound on the size of minimal depth sorting networks on 19 to 22 channels also stems from a network presented by Al-Haj Baddar and Batcher [1]. For 23 and more inputs, the best upper bounds to date are established by merging the outputs of smaller sorting networks with Batcher's odd-even merge [3], which needs $\lceil \log n \rceil$ layers for this merging step.

The known lower bounds are due to Parberry [10] and Bundala and Závodný. A new insight by Codish, Cruz-Filipe, and Schneider-Kamp [6] into the structure of the last layers of sorting networks lead to a significant further reduction of the search space. Despite all this recently resparked interest in sorting networks, the newly gained insights were insufficient to establish a tight lower bound on the depth of sorting networks for 17 inputs.

We use the SAT approach by Bundala and Závodný to synthesize new sorting networks of small depths, and thus provide better upper bounds for 17, 19, and 20 inputs. Furthermore, our improvements upon their method allow us to raise the lower bound for 17 inputs. Therefore, for the first time after the works of Shapiro, Van Voorhis, and Schwiebert, we present here a new optimal depth sorting network.

An overview of the old and new upper and lower bounds for the minimal depth of sorting networks for up to 20 inputs is presented in Table 1.

Table 1. Bounds on the minimal depth of sorting networks for up to 20 inputs.

Number of inputs	1	2	3	4	5	6	7	8	9	10	11	12	13	14	15	16	17	18	19	20
Old upper bound	0	1	3	3	5	5	6	6	7	7	8	8	9	9	9	9	11	11	12	12
New upper bound	0	1	3	3	5	5	6	6	7	7	8	8	9	9	9	9	**10**	11	**11**	**11**
Old lower bound	0	1	3	3	5	5	6	6	7	7	8	8	9	9	9	9	9	9	9	9
New lower bound	0	1	3	3	5	5	6	6	7	7	8	8	9	9	9	9	**10**	**10**	**10**	**10**

2 Preliminaries

A *comparator* is a gate with two inputs in_1 and in_2 and two outputs $out_{\min}$ and $out_{\max}$, that compares, and if necessary rearranges its inputs such that $out_{\min} = \min\{in_1, in_2\}$ and $out_{\max} = \max\{in_1, in_2\}$. Combining zero or more comparators in a network yields a *comparator network*. Comparator networks are usually visualized in a graphical manner in a so-called *Knuth diagram* as depicted in Fig. 1. Here a comparator connecting two channels is drawn as , where by convention the upper output is $out_{\min}$ and the lower output is $out_{\max}$. A maximal set of comparators with respect to inclusion that can perform parallel comparisons in a comparator network is called a *layer*. The number of layers of a network is called the *depth* of the network. A useful tool to verify that a network is a sorting network is the "0-1-principle" [8], which states that a comparator network is a sorting network, if and only if it sorts all binary inputs.

For more details about sorting networks we refer to Knuth [8, Sect. 5.3.4].

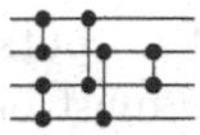

Fig. 1. A comparator network of depth 3 with 5 comparators

3 Improved Techniques

In this section we introduce the new techniques and improvements on existing techniques we used to gain our results. We will stick to the formulation by Bundala and Závodný [4], and introduce new variables if necessary. Furthermore, we will extend a technique introduced in their paper, called *subnetwork optimization*. It is based on the fact that a sorting network must sort all its inputs, but in order to prove non-existence of sorting networks of a certain depth, it is often sufficient to consider only a subset of all possible inputs, which are not all sorted by any network of this restricted depth. Bundala and Závodný chose subsets of the form $T^r = \{0^a x 1^b \mid |x| = r \text{ and } a + b + |x| = n\}$ for $r < n$, which are inputs having a *window* of size r. For an input $0^a x 1^b$ from this set the values on the first a channels at any point in the network will always be 0, and those on the last b channels will always be 1, which significantly reduces the encoding size for these inputs if a and b are sufficiently large.

3.1 Prefix Optimization

It is a well-known fact that permuting the channels of a sorting network, followed by a repair-procedure called untangling yields another feasible sorting network [10]. In fact, Parberry [10] used a first layer with comparators of the form $(2i-1, 2i), 1 \le i \le \lfloor \frac{n}{2} \rceil$, which we will call *Pb-style* whereas Bundala and Závodný used comparators $(i, n+1-i), 1 \le i \le \lfloor \frac{n}{2} \rceil$ in the first layer, which we call *BZ-style*; see Fig. 2. Both versions are equivalent in the sense that if there exists a sorting network C_n^d, then there exist sorting networks of the same depth with either of these prefixes. Nevertheless, for creating networks obeying a certain prefix as well as proving their non-existence, a given prefix may be hard-coded into the SAT formula. Given a prefix P of depth $|P|$, the remaining SAT formula encodes the proposition *"There is a comparator network on n channels of depth $d - |P|$ which sorts all outputs of P"*. Interestingly, the outputs of different prefix styles are not equally handy for the SAT encoding. A Pb-style first layer performs compare-and-swap operations between adjacent channels, thus the presorting performed here is more local than the one done by BZ-style first

Fig. 2. Pb-style first layer (left), and BZ-style (right) for 6 channels

layers. Let $out(P)$ denote the set of outputs of a prefix P on n layers. Then, the number of channels that actually must be considered in the SAT formula is given by

$$\sum_{x \in out(P)} (n - \max\{a \mid x = 0^a x'\} - \max\{b \mid x = x' 1^b\}),$$

i.e., the sum of window-sizes of all outputs of P.

Table 2 shows the impact of these previous deliberations when using a 1-layer-prefix for $2 \leq n \leq 17$ channels.

Table 2. Number of channels to consider in the encoding after the first layer

n	2	3	4	5	6	7	8	9	10	11	12	13	14	15	16	17
Pb-style	0	5	12	44	84	233	408	1016	1704	4013	6564	14948	24060	53585	85296	186992
BZ-style	0	4	10	36	72	196	358	876	1524	3532	5962	13380	22128	48628	79246	171612

Table 3. Impact of prefix style when proving that no sorting network for 16 channels with at most 8 layers exists.

Prefix style	Overall time (s)	Maximum time (s)
Pb	22,241	326
BZ	10,927	150
Opt	5,492	36

Table 3 shows running times when proving that no sorting network on 16 channels with 8 layers exists. In this case, we used 2-layer-prefixes according to [5], and proved unfeasibility for each of the 211 distinct prefixes. In the first case, they were permuted and untangled such that the first layer is in Pb-style, whereas the second case has BZ-style first layers. In the third case, we used an evolutionary algorithm to find a prefix such that the number of channels to consider when using 800 distinct outputs of the respective prefix was minimized. As we have $d - |P|$ Boolean variables for each channel which cannot be hard-coded, this procedure minimizes the number of variables in the resulting SAT formula.

This technique reduced the overall running time by factors of 4.05 and 1.99, and the maximum running times by 9.0 and 4.1, respectively.

3.2 Iterative Encoding

As mentioned above, it is usually not necessary to use all 2^n input vectors to prove lower bounds. In order to take advantage of this fact, we implemented an iterative approach. We start with a formula which describes a feasible comparator network and a (potentially empty) set of initial inputs, and iteratively add inputs until either a feasible sorting network has been found, or no network can be found which is able to sort the given set of inputs, as depicted in Fig. 3. During the iterative process, counter-examples (i.e., inputs that are not sorted by the network created so far) of minimal window-size are chosen.

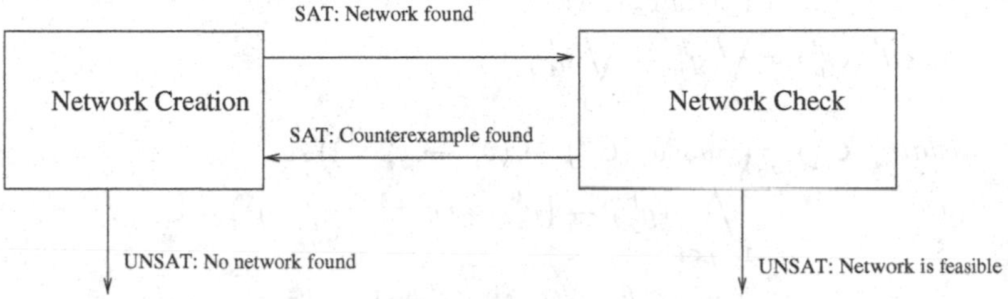

Fig. 3. Iterative generation of new inputs

Using this technique, we tested the impact of different prefix style on both running time, and the number of inputs required. Table 4 shows results for one 2-layer-prefix for 16 channels, used to prove that this cannot be extended to a sorting network with 8 layers. Here, more inputs are required to prove that a BZ-style prefix cannot be extended when compared to a Pb-style prefix. Nevertheless, BZ-style prefixes are superior in terms of running time. Interestingly, the process becomes faster when more inputs than actually required were chosen, this is, using slightly more inputs is beneficial.

Table 4. Impact of prefix style on running time and number of iterations

Initial inputs		0	100	200	300	400	500	600	700	800	900
Pb	Time	157	139	128	86	61	56	45	52	54	59
	Iterations	264	174	72	4	1	1	1	1	1	1
BZ	Time	88	75	64	47	28	14	14	13	13	19
	Iterations	358	259	165	66	10	1	1	1	1	1

Next, we turn to improve the SAT encoding.

3.3 Improved SAT Encoding

We modified the SAT encoding of Bundala and Závodný [4], significantly reducing the number of clauses. A variable $g^k_{i,j}$, with $i < j$, encodes the fact that there is a comparator comparing channels i and j in layer k in the network.

Furthermore, the variable v_i^k stores the value on channel i after layer k. For completeness' sake we list the original encoding completely:

$$once_i^k(C_n^d) = \bigwedge_{1\leq i\neq j\neq l\leq n} \left(\neg g^k_{\min(i,j),\max(i,j)} \vee \neg g^k_{\min(i,l),\max(i,l)}\right)$$

$$valid(C_n^d) = \bigwedge_{1\leq k\leq d, 1\leq 1\leq n} once_i^k(C_n^d)$$

$$used_i^k(C_n^d) = \bigvee_{j<i} g^k_{j,i} \vee \bigvee_{i<j} g^k_{i,j}$$

$$update_i^k(C_n^d) = (\neg used_i^k(C_n^d) \rightarrow (v_i^k \leftrightarrow v_i^{k-1})) \wedge \bigwedge_{1\leq j<i} (g^k_{j,i} \rightarrow (v_i^k \leftrightarrow (v_j^{k-1} \vee v_i^{k-1}))) \wedge \bigwedge_{i<j\leq n} (g^k_{i,j} \rightarrow (v_i^k \leftrightarrow (v_j^{k-1} \wedge v_i^{k-1})))$$

Here, *once* encodes the fact that each channel may be used only once in one layer, and *valid* encodes this constraint for each channel and each layer. The *update*-formula describes the impact of comparators on the values stored on each channel after every layer.

$$sorts(C_n^d, x) = \bigwedge_{1\leq i\leq n} (v_i^0 \leftrightarrow x_i) \wedge \bigwedge_{1\leq k\leq d, 1\leq i\leq n} update_i^k(C_n^d) \wedge \bigwedge_{1\leq i\leq n} (v_i^d \leftrightarrow y_i)$$

The constraint *sorts* encodes if a certain input is sorted by the network C_n^d. For this purpose, the values after layer d (i.e., the outputs of the network) are compared to the vector y, which is a sorted copy of the input x. A sorting network for n channels on d layers exists iff $valid(C_n^d) \wedge \bigwedge_{x\in\{0,1\}^n} sorts(C_n^d, x)$ is satisfiable.

Consider an input $x = 0^a x' 1^b$, and a comparator $g^k_{i,j}$ with $i \leq a$. This is, we have $v_i^k \leftrightarrow 0 \wedge v_j^{k-1} \equiv 0$, and $v_j^k \leftrightarrow 0 \vee v_j^{k-1} \equiv v_j^{k-1}$. As the same holds for $j > n - b$, we have that comparators "leaving" a subnetwork need not be considered for sorting the respective inputs. Furthermore, if $v_i^{k-1} \leftrightarrow 1$ for some k and i, using any comparator $g^k_{j,i}$ will cause $v_i^k \leftrightarrow 1$. Thus, for every channel i we introduce $oneDown^k_{i,j}$- and $oneUp^k_{i,j}$-variables which indicate whether there is a comparator $g^k_{l,j}$ for some $i \leq l < j$ or $g^k_{i,l}$ for some $i < l \leq j$, respectively.

$$oneDown^k_{i,j} \leftrightarrow \bigvee_{i<l\leq j} g^k_{i,l} \qquad noneDown^k_{i,j} \leftrightarrow \neg oneDown^k_{i,j}$$

$$oneUp^k_{i,j} \leftrightarrow \bigvee_{i\leq l<j} g^k_{l,j} \qquad noneUp^k_{i,j} \leftrightarrow \neg oneUp^k_{i,j}$$

To make use of these new predicates, given an input $x = 0^a x' 1^b$, for all $a < i \leq n - b$ we add

$$v_i^{k-1} \wedge noneDown_{i,n-b}^k \rightarrow v_i^k$$
$$\neg v_i^{k-1} \wedge noneUp_{a+1,i}^k \rightarrow \neg v_i^k,$$

to the formula and remove all update-constraints that are covered by these constraints. This offers several advantages: Firstly, it reduces the size of the resulting formula in terms of both the number of clauses, and the overall number of literals. Secondly, this encoding allows for more propagations, thus, conflicts can be found earlier. Thirdly, it offers a more general perspective on the reasons of a conflict. Table 5 shows the impact of both the new encoding and permuting the prefix, which results in an average speed-up of 8.2, and a speed-up of 17.1 for the hardest prefixes.

Table 5. Results for different settings when proving the non-existence of 8-layer sorting-networks for 16 channels

Encoding	Prefix-style	Overall time (s)	Max. time (s)	Variables	Clauses	Literals
Old	Pb	22,241	326	108,802	4,467,201	13,977,393
	BZ	10,927	150	99,850	3,996,902	12,442,522
	Opt	5,492	36	84,028	3,183,363	9,879,588
New	Pb	11,766	196	110,398	2,443,186	8,108,501
	BZ	4,359	54	101,404	2,049,744	6,799,486
	Opt	2,702	19	85,652	1,504,177	4,981,882

4 Obtaining New Lower Bounds

In a first test, we tried to prove that there is no sorting network on 17 channels using at most 9 layers by using a SAT encoding which was almost identical to the one introduced in [4], enriched by constraints on the last layers [6]. Before we broke up this experiment after 48 days, we were able to prove that 381 out of 609 prefixes cannot be extended to sorting networks of depth 9. Showing unsatisfiability of these formulae took $353 \cdot 10^6$ s of CPU time, with a maximum of $3 \cdot 10^6$ s.

In a new attempt, we used a modified encoding as described in the previous section. For every equivalence class of 2-layer-prefixes, we chose a representative which minimizes the number of variables in the SAT encoding when using $2,000$ distinct inputs. This time, we were able to prove that none of the prefixes can be extend to a sorting network using 9 layers. The overall CPU time for all 609 equivalence classes was $27.63 \cdot 10^6$ s with a maximum running time of $97,112$ s. This is a speed up of at least 42.7 concerning the maximum running time, and 20.4 for the average running time. Since the result for all 2-layer-prefixes was unsat, we conclude:

Theorem 1. *Any sorting network for* $n \geq 17$ *channels has at least* 10 *layers.*

5 Finding Faster Networks

Even though SAT encodings for sorting networks as well as SAT solvers themselves have become much better within the last years, generating new, large sorting networks from scratch is still out of their scope. Hence, we extended ideas by Al-Haj Baddar and Batcher [2].

5.1 Using Hand-Crafted Prefixes

A well-known technique for the creation of sorting networks is the generation of partially ordered sets for parts of the input in the first layers. Figure 4 shows comparator networks which create partially ordered sets for 2, 4 and 8 input bits. In the case of $n = 2$, the output will always be sorted. For $n = 4$ bits, the set of possible output vectors is given by

$$\left\{(0\,0\,0\,0)^T, (0\,0\,0\,1)^T, (0\,0\,1\,1)^T, (0\,1\,0\,1)^T, (0\,1\,1\,1)^T, (1\,1\,1\,1)^T\right\},$$

i.e., there are 6 possible outputs. Furthermore, the first output bit will equal zero unless all input bits are set to one, and the last output bit will always be set to one unless all input bits equal zero. Similarly, the network for $n = 8$ inputs allows for 20 different output vectors. Prefixes of sorting networks which consist of such snippets are referred to as *Green filters* [7].

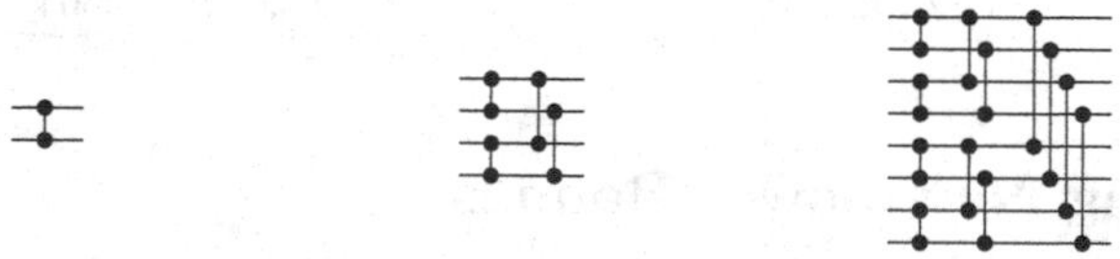

Fig. 4. Generating partially ordered sets for $n \in \{2, 4, 8\}$ inputs.

5.2 Results

We present two sorting networks improving upon the known upper bounds on the minimal depth of sorting networks. The networks presented in Fig. 5 are sorting networks for 17 channels using only 10 layers, which outperform the currently best known network due to Al-Haj Baddar and Batcher [2]. The first three layers of the network on the left are a Green filter on the first 16 channels, the remainder of this network was created by a SAT solver. To create the network on the right, we applied the prefix optimization procedure described earlier to the Green filter prefix. The first version of our solver required 29921 s to find this network, whereas our current solver can find these networks in 282 s, when using the Green filter, and 60 s when using the optimized prefix.

Thus, we can summarize our results in the following theorem:

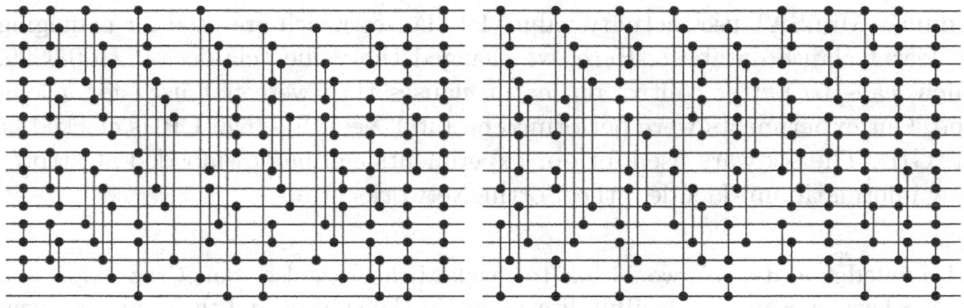

Fig. 5. Sorting networks for 17 channels of depth 10.

Theorem 2. *The optimum depth for a sorting network on* 17 *channels is* 10.

The network displayed in Fig. 6 sorts 20 inputs in 11 parallel steps, which beats the previously fastest network using 12 layers [1]. In the first layer, partially ordered sets of size 2 are created. These are merged to 5 partially ordered sets of size 4 in the second layer. The third layer is used to create partially ordered sets of size 8 for the lowest and highest wires, respectively. These are merged in the fourth layer.

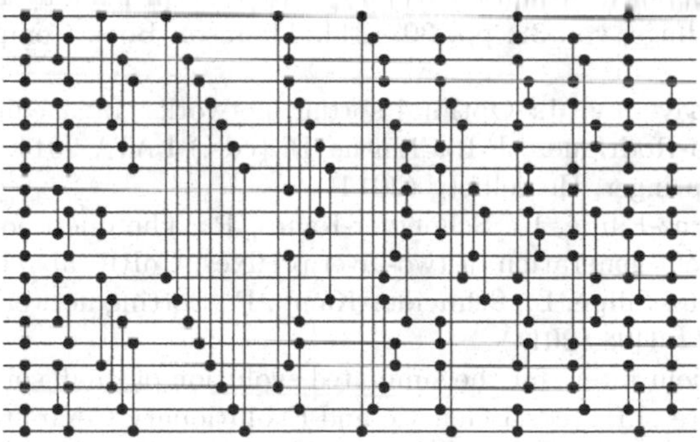

Fig. 6. A sorting network for 20 channels of depth 11.

The wires in the middle of the network are connected in order to totally sort their intermediate output. Using this prefix and the necessary conditions on sorting networks depicted above, we were able to create the remaining layers using our iterative, SAT-based approach. Interestingly, the result was created in 588 iterations, thus 587 different input vectors were sufficient.

6 Tools

Our software is based on the well-known SAT solver MiniSAT 2.20. Before starting a new loop of our network creation process, we used some probing-based preprocessing techniques [9] as they were quite successful on this kind of SAT formulae. MiniSAT uses activity values for clauses which are used for managing the learnt clause database. Here, we changed the value "cla-decay" to 0.9999, which leads to better control on learnt clauses that were not used for a long time. Our experiments were performed on Intel Xeon E5-4640 CPUs clocked at 2.40 GHz. The software used for our experiments can be downloaded at http://www.informatik.uni-kiel.de/~the/SortingNetworks.html.

Acknowledgments. We would like to thank Michael Codish, Luís Cruz-Filipe, and Peter Schneider-Kamp for fruitful discussions on the subject and their valuable comments on an earlier draft of this article. Furthermore, we thank Dirk Nowotka for providing us with the computational resources to run our experiments. We thank Donald E. Knuth and the anonymous reviewers for their helpful comments on this paper.

References

1. Baddar, S.W.A., Batcher, K.E.: A 12-step sorting network for 22 elements. Technical report 2008-05, Department of Computer Science, Kent State University (2008)
2. Baddar, S.W.A., Batcher, K.E.: An 11-step sorting network for 18 elements. Parallel Process. Lett. **19**(1), 97–103 (2009)
3. Batcher, K.E.: Sorting networks and their applications. In: American Federation of Information Processing Societies: AFIPS Conference Proceedings: 1968 Spring Joint Computer Conference, Atlantic City, NJ, USA, 30 April–2 May 1968, AFIPS Conference Proceedings, vol. 32, pp. 307–314. Thomson Book Company, Washington D.C (1968)
4. Bundala, D., Závodný, J.: Optimal sorting networks. In: Dediu, A.-H., Martín-Vide, C., Sierra-Rodríguez, J.-L., Truthe, B. (eds.) LATA 2014. LNCS, vol. 8370, pp. 236–247. Springer, Heidelberg (2014)
5. Codish, M., Cruz-Filipe, L., Schneider-Kamp, P.: The quest for optimal sorting networks: efficient generation of two-layer prefixes. CoRR, abs/1404.0948 (2014)
6. Codish, M., Cruz-Filipe, L., Schneider-Kamp, P.: Sorting networks: the end game. CoRR, abs/1411.6408 (2014)
7. Coles, D.: Efficient filters for the simulated evolution of small sorting networks. In: Soule, T., Moore, J.H. (eds.) Genetic and Evolutionary Computation Conference, GECCO 2012, pp. 593–600. ACM, Philadelphia (2012)
8. Knuth, D.E.: The Art of Computer Programming, Volume 3: Sorting and Searching. Addison-Wesly Professional, Reading (1998)
9. Lynce, I., Silva, J.P.M.: Probing-based preprocessing techniques for propositional satisfiability. In: 15th IEEE International Conference on Tools with Artificial Intelligence (ICTAI 2003), Sacramento, California, USA, 3–5 November 2003, p. 105. IEEE Computer Society (2003)
10. Parberry, I.: A computer-assisted optimal depth lower bound for nine-input sorting networks. Math. Syst. Theor. **24**(2), 101–116 (1991)
11. Valsalam, V.K., Miikkulainen, R.: Using symmetry and evolutionary search to minimize sorting networks. J. Mach. Learn. Res. **14**, 303–331 (2013)

Nonexistence of Minimal Pairs in $L[\mathbf{d}]$

Chengling Fang[1], Jiang Liu[2], Guohua Wu[3](✉), and Mars M. Yamaleev[4]

[1] School of Science, Chongqing Jiaotong University, Chongqing, China
[2] Chongqing Institute of Green Intelligent Technology, Chinese Academy of Sciences, Chongqing, China
[3] School of Physical and Mathematical Sciences, Nanyang Technological University, Singapore, Singapore
guohua@ntu.edu.sg
[4] Institute of Mathematics and Mechanics, Kazan Federal University, Kazan, Russia

Abstract. For a d.c.e. set D with a d.c.e. approximation $\{D_s\}_{s\in\omega}$, the Lachlan set of D is defined as $L(D) = \{s : \exists x \in D_s - D_{s-1} \text{ and } x \notin D\}$. For a d.c.e. degree $\mathbf{d}$, $L[\mathbf{d}]$ is defined as the class of c.e. degrees of those Lachlan sets of d.c.e. sets in $\mathbf{d}$. In this paper, we prove that for any proper d.c.e. degree $\mathbf{d}$, no two elements in $L[\mathbf{d}]$ can form a minimal pair. This result gives another solution to Ishmukhametov's problem, which asks whether for any proper d.c.e. degree $\mathbf{d}$, $L[\mathbf{d}]$ always has a minimal element. A negative answer to this question was first given by Fang, Wu and Yamaleev in 2013.

1 Introduction

For a d.c.e. set D with a d.c.e. approximation $\{D_s\}_{s\in\omega}$, the Lachlan set of D is defined as $L(D) = \{s : \exists x \in D_s - D_{s-1} \text{ and } x \notin D\}$.[1] Note that $L(D)$ is

Fang is partially supported by NSF of China (No. 11401061), SRF for ROCS, SEM and Chongqing Jiaotong University Fund (No. 2012kjc2-018).
Liu is partially supported by NSF of China (No. 61202131), the CAS western light program, Chongqing Natural Science Foundation (No. cstc2014jcsfglyjs0005 and No. cstc2014zktjccxyyB0031).
Wu is partially supported by a grant MOE2011-T2-1-071 (ARC 17/11, M45110030) from Ministry of Education of Singapore and by a grant RG29/14, M4011274 from NTU.
Yamaleev is partially supported by The President grant of Russian Federation (project NSh-941.2014.1), by Russian Foundation for Basic Research (projects 14-01-31200, 15-01-08252), by the subsidy allocated to Kazan Federal University for the project part of the state assignment in the sphere of scientific activities, and by a grant MOE2011-T2-1-071 (ARC 17/11, M45110030) from Ministry of Education of Singapore.

[1] This definition is from Ishmukhametov's articles [2] and [3]. Another definition of the Lachlan set is $L^*(D) = \{\langle x, s\rangle : x \in D_s \text{ and } x \notin D\}$. It is easy to see that $L(D)$ defined by Ishmukhametov and $L^*(D)$ above are Turing equivalent, and hence, make no difference when we consider $L[\mathbf{d}]$, a collection of Turing degrees. In this paper, we will use Ishmukhametov's definition.

A. Beckmann et al. (Eds.): CiE 2015, LNCS 9136, pp. 177–185, 2015.
DOI: 10.1007/978-3-319-20028-6_18

a c.e. set with $L(D) \leq_T D$ and that D is c.e. in $L(D)$. In [2], Ishmukhametov showed that the Turing degree of $L(D)$ doesn't depend on the approximations of D, and that for any X, $L(D) \leq_T X$ if and only if $D \in \Sigma_1^X$. In the same work, Ishmukhametov considered the following class of c.e. predecessors of d.c.e. degree $\mathbf{d}$,

$$R[\mathbf{d}] = \{\deg(W) : W \text{ is c.e. and there exists a } D \in \mathbf{d} \text{ such that } D \in \Sigma_1^W\},$$

and constructed a d.c.e. degree $\mathbf{d}$ such that $R[\mathbf{d}]$ has a minimum element. In [2], Ishmukhametov asked whether $R[\mathbf{d}]$ can always have a minimal element, if $\mathbf{d}$ is proper d.c.e. In [3], Ishmukhametov proved the existence of a d.c.e. degree such that the corresponding Lachlan degrees have no minimum element.

Theorem 1. *(Ishmukhametov [3]) There exists a proper d.c.e. degree* $\mathbf{d}$ *such that for any d.c.e. set* $B \in \mathbf{d}$ *there exists a d.c.e. set* $A \in \mathbf{d}$ *such that* $L(B) \not\leq_T L(A)$.

In [1], Fang, Wu and Yamaleev defined $L[\mathbf{d}]$ as the class of c.e. degrees of those Lachlan sets of d.c.e. sets in $\mathbf{d}$ and also proved that $R[\mathbf{d}] = L[\mathbf{d}]$. After this, Fang, Wu and Yamaleev constructed a proper d.c.e. degree $\mathbf{d}$ such that $R[\mathbf{d}]$ has no minimal element, providing a negative answer to Ishmukhametov's question.

Theorem 2. *(Fang, Wu, and Yamaleev [1]) There exists a proper d.c.e. degree* $\mathbf{d}$ *such that for any d.c.e. set* $B \in \mathbf{d}$ *there exists d.c.e. set* $A \in \mathbf{d}$ *such that* $L(A) <_T L(B)$.

In this paper, we will prove that in $L[\mathbf{d}]$, no two elements can form a minimal pair, if $\mathbf{d}$ is proper d.c.e.

Theorem 3. *If* $\mathbf{d}$ *is a d.c.e. degree then for any d.c.e. sets* $A, B \in \mathbf{d}$, *there exists a d.c.e. set* $D \in \mathbf{d}$ *such that* $L(D) \leq_T L(A), L(B)$. *In particular, if* $\mathbf{d}$ *is proper d.c.e., then no two elements in* $L[\mathbf{d}]$ *can have infimum* $\mathbf{0}$.

Combining Theorems 1 and 3, we can have an alternative proof of Theorem 2, and hence another solution of Ishmukhametov's question via Ishmukhametov's original attempt. Indeed, take $\mathbf{d}$ as constructed in Theorem 1, and assume that $B \in \mathbf{d}$ is a d.c.e. set such that $L(B)$ is a minimal element in $L[\mathbf{d}]$. Then by Theorem 1, there is a d.c.e. set A in $\mathbf{d}$ such that $L(B) \not\leq_T L(A)$. By Theorem 3, there is a d.c.e. set $D \in \mathbf{d}$ such that $L(D) \leq_T L(A), L(B)$ and $L(D)$ incomputable, and hence $L(D) <_T L(B)$. A contradiction.

Our notation is standard and generally follows Soare's textbook [4].

2 Basic Idea of the Construction

We now give some basic idea of the proof of Theorem 3. Let A, B be two d.c.e. sets with $A \equiv_T B$, and let $\{A_s\}_{s\in\omega}$, $\{B_s\}_{s\in\omega}$ be the d.c.e approximations of A and B. We will construct a d.c.e. set $D \equiv_T A$ (hence $D \equiv_T B$) such that $L(D) \leq_T L(A), L(B)$. Assume that $A = \Phi^B$ and $B = \Psi^A$ and we will build partial computable functionals Γ, Δ such that $A = \Gamma^D$ and $D = \Delta^A$, and also partial computable functionals Λ, Θ such that $L(D) = \Lambda^{L(A)}$ and $L(D) = \Theta^{L(B)}$.

Construction of Γ and Δ. For each x, we reserve $2x$ and $2x+1$ to code A in D via a partial computable functional Γ. We define the γ-use $\gamma(x)$ as $\gamma(x) = 2x+1$.

When $\Gamma^D(x)$ is defined and $\Gamma^D(x) \neq A(x)$, then we enumerate $2x$ into D (we can always do so, if it is the first time for $A(x)$ to change, after $\Gamma^D(x)$ is first defined), and redefine $\Gamma^D(x) = A(x)$. If $A(x)$ changes again (the last change), we can enumerate $2x+1$ into D, to undefine it, or remove $2x$ from D to recover $\Gamma^D(x)$ (as 0). *Of course, when we have $\Gamma^D(x) \neq A(x)$ for the first time with $\Gamma^D(x) = 1 \neq 0 = A(x)$ (i.e. $A(x)$ has its first change before $\Gamma^D(x)$ is defined), then the enumeration of $2x$ into D will undefine $\Gamma^D(x)$ (we will then redefine it as 0) and as $A(x)$ will not change later, there is no possibility of removing $2x$ out of D later, to rectify $\Gamma^D(x)$.*

We also define Δ, to make $D(2x) = \Delta^A(2x)$ and $D(2x+1) = \Delta^A(2x+1)$ with use $\delta(2x) = \delta(2x+1) = \psi(\varphi(x))$. Of course, $\delta(2x)$ and $\delta(2x+1)$ should change accordingly, if $A\lceil\psi(\varphi(x))$ changes ($B\lceil\varphi(x)$ may have changes and hence $\varphi(x)$ may be increased), and there will be no such change eventually, as $\Phi^B(x)$ and $\Psi^A(\varphi(x))$ are assumed to be convergent computations.

Removing $2x$ can also happen because of the definition of Δ. It can happen that after the second change of $A(x)$ (as indicated above, $2x$ is enumerated into D, at stage s say, when $A(x)$ has its first change), $A\lceil\psi(\varphi(x))$ changes back to a previous configuration, and hence, according to the definition of Δ^A and to make $D(2x) = \Delta^A(2x)$, we need to keep $D(2x) = 0$, and such a removal of $2x$ from D actually enumerates s, i.e., $s^D(2x)$, into $L(D)$, this enumeration needs to get permissions from $L(A)$ and $L(B)$.

Construction of Λ and Θ. We now describe the construction of Λ and Θ which are built to make $L(D) = \Lambda^{L(A)} = \Theta^{L(B)}$.

Without loss of generality, we assume that x enters A after s_0, the stage when $\Gamma^D(x)$ is defined, and B changes below $\varphi(x)[s_0]$, say at a number y. If y leaves B, then $A_t\lceil\delta(2x)[s_0] \neq A_{s_0}\lceil\delta(2x)[s_0]$ for all $t > s^A(x)$, and $2x$ remains in D forever. So, we assume that y enters B. Note that $s^A(x), s^B(y) > s_0$ and that we put $2x$ into D, with $s^D(2x) > s^A(x), s^B(y)$. We let $s_1 = s^D(2x)$ and define $\Lambda^{L(A)}(s_1)[s_1] = \Theta^{L(B)}(s_1)[s_1]$, with $\lambda(s_1)[s_1] = \theta(s_1)[s_1] = s_1$ (both are bigger than $s^A(x)$ and $s^B(y)$, and it could happen that later, because of the changes of $L(A)$ and $L(B)$, they can become different from each other).

If later x leaves A, say at stage s_2, and $A_{s_2}\lceil\psi(\varphi(x))[s_0] = A_{s_0}\lceil\psi(\varphi(x))[s_0]$, then as $\Delta^A(2x)[s_2] = \Delta^A(2x)[s_0] = 0$, to make $D(2x) = \Delta^A(2x)$, we need to extract $2x$ from D, which means that we need permissions from both $L(A)$ and $L(B)$. A permission from $L(A)$ is just provided, because $s^A(x)$ is enumerated into $L(A)$ at stage s_2. However, if y does not leave B before stage s_2, then it must happen at stage s_2 since $\Psi^A(y)[s_2] = \Psi^A(y)[s_0] = 0$, and hence $s^B(y)$ is enumerated into $L(B)$ at this stage, undefining $\Theta^{L(B)}(s^D(2x))$. Thus, as a consequence, we can extract $2x$ out of D.

The case that y leaves B before stage s_2 is a bit more complicated, as before stage s_2, we may already redefine $\Theta^{L(B)}(s^D(2x))$, and at stage s_2, we have

$\Lambda^{L(A)}(s^D(2x))$ undefined but $\Theta^{L(B)}(s^D(2x))$ defined, and we could not extract $2x$ from D, and in this case, at stage s_2, we put $2x+1$ into D, to code that x leaves A.

Suppose that x leaves A at stage s_2, also assume that B has changes below $\varphi(x)[s_0]$ between stages s_0 and s_2. Then A has changes below $\psi(\varphi(x))[s_0]$. We can assume that some y' enters B and that x' enters A between s_0 and s_2. Recall that at stage s_2, we put $2x+1$ into D to code that x leaves A, and also we redefine

$$\delta(2x)[s_2] = \delta(2x+1)[s_2] = \max_{s_0 \leq t \leq s_2} \{\psi(\varphi(x))[t]\}.$$

As $s^A(x) < s_1$ enters $L(A)$, $\Lambda^{L(A)}(s_1)$ is undefined at stage s_2 and we can refine

$$\lambda(s_1)[s_2] = \lambda(s_2)[s_2] = s_2.$$

Note that when y leaves B (before stage s_2), $\Theta^{L(B)}(s_1)$ is undefined (as $s^B(y)$ enters $L(B)$), and we would have already redefined $\Theta^{L(B)}(s_1)$ with

$$\theta(s_1)[s_2] = \theta(s_2)[s_2] = s_1' > s_1.$$

In particular, $s_2 > s^A(x')$ and that $s_1' > s^B(y')$. This kind of enumerations can be repeated several times, and stop eventually because Φ^B and Ψ^A are assumed to be total. So, without loss of generality, we assume that x', y' are the last numbers in this sequence.

Now suppose that at stage $s_3 > s_2$, $A_{s_3}\lceil\psi(\varphi(x))[s_0] = A_{s_0}\lceil\psi(\varphi(x))[s_0]$, which means that $\Lambda^A(2x)$ should be 0, and to make $D = \Lambda^A$, we need to extract $2x$ and $2x+1$ from D. *Of course, if* $A\lceil\psi(\varphi(x))[s_0]$ *does not recover to* $A_{s_0}\lceil\psi(\varphi(x))[s_0]$, *then we can just leave* $2x$ *and* $2x+1$ *in* D. To extract $2x$ and $2x+1$ from D at stage s_3, we must have permissions from $L(A)$ and $L(B)$, to enumerate s_1, s_2 into $L(D)$. Indeed, as x' leaves A, $s^A(x') \leq s_2$ is enumerated into $L(A)$, which allows to correct $\Lambda^{L(A)}$ at both s_1, s_2. As $B = \Psi^A$, and $A_{s_3}\lceil\psi(\varphi(x))[s_0] = A_{s_0}\lceil\psi(\varphi(x))[s_0]$, $B(y')[s_3] = B(y')[s_0] = 0$, which means that y' also leaves B at stage s_3, and hence $s^B(y') < s_1'$ enters $L(B)$ at stage s_3, an $L(B)$-permission for s_1 and s_2.

3 Construction

First, we define expansionary stages for $A = \Phi^B$ and $B = \Psi^A$ in a standard way. Namely, we define the length agreement as

$$\ell(s) = \max\{y < s : \forall x < y\, [\Phi^B(x)[s] \downarrow = A(x)[s]\ \&\ \forall u < \varphi_s(x)(\Psi^A(u)[s] \downarrow = B(u)[s])]\}.$$

A stage s is expansionary if $s = 0$ or $\ell(s) > \ell(t)$ for all $t < s$.

We consider each stage of the form $4n$ as an expansionary stage, and we call these stage working stages. By this convention, between two stages $4n$ and $4(n+1)$, A and B can have many changes on numbers below $4n$. We will use $\mathbf{4n}$ to denote the n-th working stage. So between two working stages $\mathbf{4n}$ and $4(\mathbf{n}+1)$, there are many stages of enumerations of A and B, and we assume that at each stage, there is at most one x and at most one y, such that $A(x)$ and $B(y)$ change the value. We need this convention, as we need the definition of s^A and s^B well-defined. We may interpret this idea as, after working stage $\mathbf{4n}$, after stage $\mathbf{4n}+3$, at a stage t, we check whether t is an expansionary stage, and if yes, then t is the next working stage, i.e. $4(\mathbf{n}+\mathbf{1})$, and if no, then do nothing and go to stage $t+1$.

We use $\alpha(x)$ to denote the working stage $\mathbf{s}$ at which $\ell(\mathbf{s})$ exceeds x for the first time. We will ensure that at each stage, at most one number is enumerated into D (actually, only at stages $\mathbf{4n}+1, \mathbf{4n}+2, \mathbf{4n}+3$), and thus each stage s, at most one x enters D, and hence different x will have different $s^D(x)$. We clarify this as we are constructing reductions Λ, Θ, reducing $L(D)$ to $L(A)$ and $L(B)$ respectively. We will call $[\mathbf{4n}, \mathbf{4n}+3]$ the n-th working block.

Working Block $[0, 3]$

Let $D_0 = \emptyset$, and initiate all the functionals being constructed.

Working Block $[\mathbf{4s}, \mathbf{4s}+3]$

Stage 4s. Stage $\mathbf{4s}$ is an expansionary stage and at this stage, A and B may have many changes below $\mathbf{4s}$.

1. Let $x_1 < \ell(\mathbf{4s})$ be the least number being enumerated into A at stage $\mathbf{4s}$, if any.
2. Let $x_2 < \ell(\mathbf{4s})$ be the least number being removed from A at stage $\mathbf{4s}$, if any.
3. Check whether there exist some $z < \mathbf{s}$ with $D(2z)[\mathbf{4s}] = 1$ and $\Delta^A(2z)[\mathbf{4s}] = 0$, if any.

Go to next stage.

Stage 4s + 1. Check whether $\Gamma^D(x_1)[\mathbf{4s}]$ is defined or not. If $\Gamma^D(x_1)[\mathbf{4s}]$ is defined, then enumerate $2x_1$ into D. Otherwise do nothing.

(*Thus* $\mathbf{4s}+1$ *is defined as* $s^D(2x_1)$. *The enumeration of* $2x_1$ *undefines* $\Gamma^D(x_1)$, *so we can redefined it as 1 later. It can happen that there are* x' *with* $x_1 < x' < \ell(\mathbf{4s})$ *that enters* A *at stage* $\mathbf{4s}$, *and the enumeration of* $2x_1$ *into* D *at stage* $\mathbf{4s}+1$ *also undefines* $\Gamma^D(x')$. *We may remove* $2x_1$ *from* D *later, at some stage* $\mathbf{4s'}+2$ *say, then at the same stage, we need to enumerate* $2x'$ *into* D, *to make sure that* $\Gamma^D(x')$ *does not recover to* $\Gamma^D(x')[\mathbf{4s}]$. *Because of this, we call* x_1 *the support of those numbers* $x' \geq x_1$, *which are also enumerated into* A *at stage* $\mathbf{4s}$.)

Go to next stage.

Stage 4s + 2. Check whether $2x_2$ has been enumerated into D before stage $\mathbf{4s}$.

If not, then enumerate $2x_2$ into D to undefine $\Gamma^D(x_2)$. ($\mathbf{4s}+2$ *is defined as* $s^D(2x_2)$. *In this case, when we redefine* $\Gamma^D(x_2)$ *after stage* $\mathbf{4s}+2$, *we will define* $\Gamma^D(x_2)$ as 0, *and as* A *is d.c.e.*, $A(x_2)$ *will not have further changes. Thus,* $2x_2$ *will remain in* D *and* $s^D(2x_2)$ *cannot be enumerated into* $L(D)$.)

If $2x_2$ has been enumerated into D before stage $4\mathbf{s}$, then we check whether there is a working stage t with $\alpha(x_2) < t < s^A(x_2) < 4\mathbf{s}$ such that

$$A[4\mathbf{s}]\lceil\delta(2x_2)[t] = A[t]\lceil\delta(2x_2)[t].$$

Here $\alpha(x_2)$ is the first working stage s_0 at which $\ell(s_0) > x_2$.

- If there is such a t, then remove $2x_2$ from D. We also check whether there is some x' with x_2 as a support, and x' is still in $A[4\mathbf{s}]$. If such an x' exists, then we choose x' as the least one, and enumerate $2x'$ into D. For those numbers with x_2 as support before, we now use x' as their support.
  ($4\mathbf{s}+2$ *is defined as* $s^D(2x')$. *As* $2x_2$ *is removed from* D, $s^D(2x_2)$ *is enumerated into* $L(D)$. *Note that the enumeration of* $s^D(2x_2)$ *into* $L(D)$ *is permitted by* $L(A)$ *and* $L(B)$, *by simple argument.*)
- If there is no such a working stage t, then enumerate $2x_2+1$ into D (hence $\Gamma^D(x_2)$ *is undefined, and also note that both* $\Delta^A(2x_2)$ *and* $\Delta^A(2x_2+1)$ *are undefined at this stage, because of our assumption and that* x_2 *leaves* A, *and this allows us to enumerate* $2x_2+1$ *into* D).

Go to next stage.

Stage $4\mathbf{s}+3$. Remove all these $2z$ out of D, together with $2z+1$, if the associated $2z+1$ is also in D. Let x_3 be the least number that has one of these z as its support. Enumerate $2x_3$ into D.

(*These numbers* $2z$ (*also* $2z+1$, *if in* D *already*) *have to be removed from* D, *as we need to keep* $D = \Delta^A$.)

Extending Definitions of $\Gamma, \Delta, \Lambda, \Theta$:

Γ: Find the least x such that $\Gamma^D(x)[4\mathbf{s}+3]$ is not defined, and define $\Gamma^D(x) = A(x)[4\mathbf{s}+3]$ with use $\gamma(x) = 2x+1$.

Δ: Find the least x such that $\Delta^D(2x)[4\mathbf{s}+3]$ is not defined, and define $\Delta^A(2x)[4\mathbf{s}+3] = D(2x)[4\mathbf{s}+3]$, $\Delta^A(2x+1)[4\mathbf{s}+3] = D(2x+1)[4\mathbf{s}+3]$, with use $\delta(2x)[4\mathbf{s}+3] = \delta(2x+1)[4\mathbf{s}+3] = \psi(\varphi(x))[4\mathbf{s}+3]$.

Λ: Find the least k such that $\Lambda^{L(A)}(k)[4\mathbf{s}+3]$ is not defined yet. If k is a stage at which no number is enumerated into D, then define $\Lambda^{L(A)}(k) = 0$, with use $\lambda(k) = \lambda(k-1)[4\mathbf{s}+3]$. If a number n is enumerated into D at stage k, then define $\Lambda^{L(A)}(k) = L(D)(k)[4\mathbf{s}+3]$, with use $\lambda(k) = 4\mathbf{s}+3$, if $\Lambda^{L(A)}(k)$ has not been defined before, or with use $\lambda(k) = s'$, where s' is the last working stage at which some number $x' < \psi(\varphi(n))[s'']$ leaves A, where s'' is the previous use of $\lambda(k)$.

Θ: Find the least k such that $\Theta^{L(B)}(k)[4\mathbf{s}+3]$ is not defined yet. If k is a stage at which no number is enumerated into D, then define $\Theta^{L(B)}(k) = 0$, with use $\theta(k) = \theta(k-1)[4\mathbf{s}+3]$. If a number n is enumerated into D at stage k, then define $\Theta^{L(B)}(k) = L(D)(k)[4\mathbf{s}+3]$, with use $\theta(k) = 4\mathbf{s}+3$, if $\Theta^{L(B)}(k)$ has not been defined before, or with use $\theta(k) = s'$, where s' is the last working stage at which some number $y' < \varphi(n)[s'']$ leaves B, where s'' is the previous use of $\theta(k)$.

This completes the work of block $[4\mathbf{s}, 4\mathbf{s}+3]$. Go to the next block.

4 Verification

We now prove that the constructed set D, and that all the p.c. functionals $\Gamma, \Delta, \Lambda, \Theta$, satisfy all the requirements. Obviously, α is total.

Lemma 1. *Γ^D is well-defined and total, and $\Gamma^D = A$.*

Proof. Fix x. By the construction, $\gamma(x)$ is kept as $2x+1$, and hence $\Gamma^D(x)$ is undefined, only when some $x' \leq x$ enters A or leaves A, which can happen only finitely many times. Also we enumerate $2x$ or $2x+1$ into D, only when (or after) x enter A or leaves A. When $2x$ (maybe together with $2x+1$) is removed from D, x already left A, and removing $2x$ from D recovers $\Gamma^D(x)$ to a previous computation, i.e. a computation with value 0, and hence we have $\Gamma^D(x) = A(x)$. This shows that $\Gamma^D(x)$ is defined and equals to $A(x)$.

This proves that $\Gamma^D = A$.

Lemma 2. *Δ^A is well-defined and total, and $\Delta^A = D$.*

Proof. Fix x. By construction, $\delta(2x)$ and $\delta(2x+1)$ depend on the use $\psi(\varphi(x))$. By our assumption on $A = \Phi^B$ and $B = \Psi^A$, Φ^B and Ψ^A are both total, and hence $\varphi(x)$ and $\psi(u)$ for all $u \leq \varphi(x)$ settle down after a big stage. As we only define $\Delta^A(2x)$ and $\Delta^A(2x+1)$ at a working stage t, when they have no definition at this stage. It can happen that $\Delta^A(2x)[t']$ or $\Delta^A(2x+1)[t']$, with $t' < t$, become valid, after stage t above, at a working stage t'' say, i.e.

$$A_{t''}\lceil\delta(2x)[t'] = A_{t'}\lceil\delta(2x)[t'],$$

then $\Delta^A(2x)[t'']$ and $\Delta^A(2x)[t']$ are actually the same, which means some number $m < \delta(2x)[t']$ leaves A between stage t and stage t'', and hence $\Delta^A(2x)[t]$ become invalid automatically, as A and $A[t]$ can never agree on the initial segment $A_t\lceil\delta(2x)[t']$ after stage t''. This shows that $\Delta^A(2x)$ and $\Delta^A(2x+1)$ are both well-defined.

We now prove that $\Delta^A(2x) = D(2x)$.

If x never enters A, then $2x$ never enters D, which means that whenever $\Delta^A(2x)$ is defined, it is defined as 0. So we assume that x enters A, at a working stage $4\mathbf{s}$ say. Then $\Delta^A(2x)$ is undefined, because of this enumeration. If at this stage, $2x$ is enumerated into D immediately, then we can redefine $\Delta^A(2x)$ as 1. It can also happen that at this stage, a number n (even, or odd) is enumerated into D, and this n is a support of x. Note that such a support can have changes, due to the changes of A, and it could be possible that before x becomes a support of itself, x already leaves A, and if so, whenever we define $\Delta^A(2x)$, we just defined as 0. If x becomes a support of itself, eventually, at a working stage $\mathbf{s}'$ say, then at this stage, $\Delta^A(2x)$ should have no definition, and we can enumerate $2x$ into D. It is actually true, because when x enters A, $\Delta^A(2x)$ is undefined by $x < \delta(2x)$, and then at any stage, when x has a new support, the previous support of x leaves and the new support is still less than x, and these changes, as they are all less than x, actually undefine $\Delta^A(2x)$. Thus, at stage $\mathbf{s}'$, $\Delta^A(2x)$ is not defined, and we are allowed to enumerate $2x$ into D and redefine $\Delta^A(2x)$ as 1.

Now, again because of A-changes, at a working stage $\mathbf{s}'' > \mathbf{s}'$, $\Delta^A(2x)$ could recover to a computation before stage $\mathbf{s}'$, and hence $\Delta^A(2x)[\mathbf{s}''] = 0$. If it happens, then, by construction, $2x$ will be removed out of D at stage $\mathbf{s}''$. We note that at stage $\mathbf{s}''$, $\Delta^A(2x)$ is actually recovered to a computation before $4\mathbf{s}$, because at any stage between $4\mathbf{s}$ and $\mathbf{s}'$, some number less than x, i.e., x's support is in A, and by stage $\mathbf{s}'$, all these numbers are not in A anymore. This means that at stage $\mathbf{s}''$, x is not in A neither, and hence the removal of $2x$ from D is consistent with the definition of Δ^A.

We comment here that after stage $\mathbf{s}''$, $\Delta^A(2x)$ cannot come back to a computation between $4\mathbf{s}$ and $\mathbf{s}''$, because x leaves A, and also some other numbers (at stage $\mathbf{s}''$, all these numbers should already leave A), and as A is d.c.e., these numbers will leave A forever, and hence ensure that $\Delta^A(2x)$ cannot come back to a computation between $4\mathbf{s}$ and $\mathbf{s}''$, which, if true, could require to enumerate $2x$ into D again.

Now we consider $\Delta^A(2x+1) = D(2x+1)$. Without loss of generality, we assume that $2x+1$ is enumerated into D at a working stage $\mathbf{s}'$. Then by construction, we know that by stage $\mathbf{s}'$, x is already a support of itself, and x leaves A by this stage. Because $2x$ is not removed from A, we know $\Delta^A(2x)$ does not recover to any computation before x enters A. That is, before x leaves A, some number z less than $\psi(\varphi(x))$ enters or leaves A. This change allows us to enumerate $2x+1$ into D, which is consistent with the definition of Δ^A.

If z leaves A, then $\Delta^A(2x)$ can never recover to any computation of it before x enters A, and $2x$, $2x+1$ will remain in D forever. That is, after stage $\mathbf{s}'$, whenever we define $\Delta^A(2x)$, $\Delta^A(2x+1)$ again, we just define it as $D(2x), D(2x+1)$, we just let

$$\Delta^A(2x) = D(2x), \quad \Delta^A(2x+1) = D(2x+1),$$

with use $\psi(\varphi(x))$. This ensures $\Delta^A(2x+1) = D(2x+1)$.

If z enters A, and stays in A, then just as discussed above, we can define $\Delta^A(2x) = D(2x)$, $\Delta^A(2x+1) = D(2x+1)$, with use $\psi(\varphi(x))$. This ensures $\Delta^A(2x+1) = D(2x+1)$.

So we assume that z enters A first and later leaves A, and also that at stage $\mathbf{s}''$, $\Delta^A(2x)$ recovers to a previous computation, before x enters A. Then, by construction, $2x, 2x+1$ are removed from D. We have seen why $2x$ can be removed, and for $2x+1$, we know that at stage $\mathbf{s}'$, z enters A, allowing us to enumerate $2x+1$ into D, and when we define $\Delta^A(2x+1)$ again, $\delta(2x+1)$ is bigger than this z. Thus, when z leaves A, $\Delta^A(2x+1)$ is undefined, and especially, at stage $\mathbf{s}''$, $\Delta^A(2x+1)$ is not defined as 1, and hence we can remove $2x+1$ out of D. Again, as discussed for $\Delta^A(2x)$, $\Delta^A(2x+1)$ will not recover to a computation between $\mathbf{s}'$ and $\mathbf{s}''$, and will not require $D(2x+1) = 1$ in the remainder of the construction.

This completes the proof that Δ^D is well-defined and computes A correctly.

Lemma 3. *$\Lambda^{L(A)}$ is well-defined and $L(D) = \Lambda^{L(A)}$.*

Proof. Fix s. Here we consider that s is a general stage, not necessary to be working stage.

If at stage s, no number is enumerated into D, then we always define $\Lambda^{L(A)}(s)$ as 0, with use $\lambda(s)$ the same as $\lambda(s-1)$. Thus, if $\Lambda^{L(A)}(s-1)$ is defined, then so is $\Lambda^{L(A)}(s)$, with value 0. This gives that $\Lambda^{L(A)}(s) = L(D)(s)$, as s can never enter $L(D)$.

So we assume that a number $2x$ is enumerated into D at stage s, i.e. $s^D(2x) = s$. The case that an odd number enters D at stage s can be argued in a similar way, and we leave it to the reader to verify this.

Note that $2x$ is put into D at stage s, x is already in A, and assume that x enters A at stage $s^A(x)$.

We first consider the case when x enters A at stage s. As we also assume that $2x$ also enters D at stage s, x is the support of itself. When we define $\Lambda^{L(A)}(s)$, we define it as 0 with use $\lambda(s) = s$. This use will be kept the same, till s enters $L(A)$ (i.e. x leaves A) at stage s' say. (So, if x remains in A, then such a stage s' does not exist, $\Lambda^{L(A)}(s)$ will have value 0, which equals to $L(D)(s)$.) Then, at stage s', $\Lambda^{L(A)}(s)$ is undefined, and also at (perhaps before) stage s', we do have numbers $x' < \psi(\varphi(x))[s]$ entering A (if some number enters A before stage s and leaves A at stage s', $2x$ will be kept in D forever, by the construction), and when we redefine $\Lambda^{L(A)}(s)$, we define it as 0, with use $\lambda(s) = s'$. Again, the construction ensures that $2x$ will be in D, until x' leaves A, at stage s'', which means that $s^A(x')$ enters $L(A)$ at stage s'', undefining $\Lambda^{L(A)}(s)$. Of course, if $2x$ is not removed from D at stage s'', then another number $x'' < \psi(\varphi(x))[s]$ enters A by stage s'' and thus, when we redefine $\Lambda^{L(A)}(s)$, we define it as 0, with use $\lambda(s) = s''$. As $\psi(\varphi(x))[s]$ is fixed, such a process can be repeated at most finitely many times, and will stop after a certain stage big enough, s^*. Thus, by stage s^*, either s is already in $L(D)$, or s will never be enumerated into $L(D)$. This again gives that $\Lambda^{L(A)}(s)$ is defined and equals to $L(D)(s)$.

Now consider the case that $s^A(x) < s$. Then after stage s, when we check the change below use, the use is $\psi(\varphi(x))[s^A(x)]$, as those computations between stage $s^A(x)$ and stage s can never be recovered after stage s. An almost same argument shows that $\Lambda^{L(A)}(s)$ is defined and equals to $L(D)(s)$.

Lemma 4. *$\Theta^{L(B)}$ is well-defined and $L(D) = \Theta^{L(B)}$.*

Proof. The basic idea of the proof is similar to that in the proof of Lemma 3, but here we consider the changes of B below the use $\varphi(x)$.

This completes the proof of Theorem 3.

References

1. Fang, C.L., Wu, G., Yamaleev, M.M.: On a problem of Ishmuhkametov. Arch. Math. Logic **52**, 733–741 (2013)
2. Ishmukhametov, S.: On the predececcors of d.r.e. degrees. Arch. Math. Logic **38**, 373–386 (1999)
3. Ishmukhametov, S.: On relative enumerability of turing degrees. Arch. Math. Logic **39**, 145–154 (2000)
4. Soare, R.: Recursively Enumerable Sets and Degrees. Springer, Berlin (1987)

Intuitionistic Provability versus Uniform Provability in RCA

Makoto Fujiwara(✉)

Mathematical Institute, Tohoku University, Tohoku, Japan
sb0m29@math.tohoku.ac.jp

Abstract. We provide an exact formalization of uniform provability in RCA and show that for any Π^1_2 sentence of some syntactical form, it is intuitionistically provable if and only if it is uniformly provable in RCA.

Keywords: Reverse mathematics · Constructive mathematics · Computable analysis · Medvedev reducibility

1 Introduction

In the practice of reverse mathematics [15], a lot of mathematical theorems are formalized as Π^1_2 sentences of a form

$$\forall \xi\,(A(\xi) \to \exists \zeta B(\xi, \zeta))$$

(where ξ and ζ may be tuples), and known to be provable in RCA_0. However, in some cases, the construction of the witness ζ from a instance ξ is not uniform. Let's consider the case of well-known intermediate value theorem: if f is a continuous function on the unit interval $0 \leq x \leq 1$ such that $f(0) < 0 < f(1)$, then there exist $x \in (0, 1)$ such that $f(x) = 0$. In fact, it is provable in RCA_0 as follows (see [15, Theorem II.6.6] for details). If there exists $x \in \mathbb{Q}$ such that $0 < x < 1$ and $f(x) = 0$, we are done. Otherwise, one can construct $x \in (0, 1)$ by the method of nested intervals (and the construction is verified) in RCA_0. In this proof, the construction of the intermediate point x from f is depends on whether there is a rational intermediate point (although the construction is trivial if there is). To reveal such a non-uniformity, sequential versions of Π^1_2 statements, which assert to solve infinitely many instances of a particular problem simultaneously, have been investigated. In fact, the sequential version of intermediate value theorem is equivalent to WKL over RCA_0, and hence it is not provable even in RCA (RCA_0+full second-order induction scheme). Consequently, it follows that there is no uniform algorithm to construct an intermediate point x from an arbitrary given continuous function f on the unit interval $0 \leq x \leq 1$ such that $f(0) < 0 < f(1)$. The reason why the above proof in RCA_0 does not work is that one needs to decide in RCA whether there exists a rational intermediate point or not for each problem simultaneously (than just a use of the

M. Fujiwara—The author is supported by a Grant-in-Aid for JSPS fellows.

A. Beckmann et al. (Eds.): CiE 2015, LNCS 9136, pp. 186–195, 2015.
DOI: 10.1007/978-3-319-20028-6_19

law-of-excluded-middle) in the sequentialized case. However, it is not possible in RCA having only Δ^0_1 ($\approx$ computable) set existence scheme. As suggested by this example, uniform provability in RCA (in the presented sense) is closely related to the notions of uniform computability (in the sense of Medvedev reducibility) and constructivity.

Systems of many-sorted arithmetic based on intuitionistic logic serves as base theory formalizing constructive mathematics [18]. Historically constructive mathematics has been developed informally in contrast to formalist foundation of mathematics. Along with the development of reverse mathematics and the discovery of the arithmetical hierarchy of the law-of-excluded-middle principles, however, so-called constructive reverse mathematics [11], which investigates the relationship between mathematical statements and logical principles over an intuitionistic arithmetic, has been carried out in this decade.

In fact, there are several corresponding results between constructive reverse mathematics and classical reverse mathematics of sequential versions. For example, the principle of trichotomy for reals is intuitionistically equivalent to Σ^0_1-PEM whereas its sequential version is equivalent to ACA [5]. On the other hand, the principle of dichotomy for reals is intuitionistically equivalent to Σ^0_1-DML whereas its sequential version is equivalent to WKL [5]. More directly, ACA and WKL are intuitionistically equivalent to Σ^0_1-PEM and Σ^0_1-DML respectively in the presence of choice scheme [11]. Based on these facts, in this paper, we pay strict attention to and analyze the connection between intuitionistic provability and uniform provability in RCA for Π^1_2 statements. In particular, we first provide an exact formalization of uniform provability in RCA and show that for any Π^1_2 sentence of some syntactical form (rich enough), it is intuitionistically provable if and only if it is uniformly provable in RCA. The direction from left to right is shown by formalized realizability with functions (or the Dialectica interpretation). In fact, this direction is just a refinement of [4, Proposition 3.7] (or [9, Theorem 5.6]). Our main contribution is the converse direction, which is shown by means of the negative translation (and the Dialectica interpretation).

From a philosophical point of view, it is remarkable that all of our proofs are constructive, namely, they are just syntactic translations. Thus we constructively (from a meta-perspective) establish the equivalence between constructive provability and classical uniform provability.

Notations. EL is the two-sorted intuitionistic arithmetic (called "intuitionistic analysis") in [16, 1.9.10], which is a conservative extension of Heyting arithmetic HA. EL_0 is its fragment where induction scheme is restricted to quantifier-free formulas (see [4]). Note that EL and EL_0 contain the quantifier-free choice scheme QF-AC^{0,0}. The most popular base system RCA_0 of reverse mathematics, presented in [15], and its extension RCA having full induction scheme use the set-based language (namely, has variables for numbers and sets of numbers) with the membership relation symbol. On the other hand, the systems EL_0 and EL use the function-based language. However, as mentioned in [9] (see also [13]), one can identify EL_0 + LEM with RCA_0 and EL + LEM with RCA respectively in the sense that each is included in a canonical definitional extension of the other. In fact, one can see sets in RCA_0 as their characteristic functions in $\mathsf{EL}_0 + \mathrm{LEM}$ and conversely see functions in $\mathsf{EL}_0 + \mathrm{LEM}$ as their graphs in RCA_0. Therefore, we write RCA_0 and RCA instead of $\mathsf{EL}_0 + \mathrm{LEM}$ and $\mathsf{EL} + \mathrm{LEM}$ under this identification. We basically use Roman small letters e.g. x, y, z for number (type-0) variables and Greek lower case letters e.g. $\alpha, \beta, \xi, \zeta$ for function (type-1) variables. In addition, we sometimes represent the type of a variable by its superscript number. A quantifier-free formula is denoted like A_{qf} with the subscript "qf".

In addition, we use systems of finite type arithmetic (e.g. $\mathsf{E\text{-}HA}^\omega$, $\mathsf{E\text{-}PA}^\omega$) formulated in the language of functionals in all finite types. We employ the notations from [14] for finite type arithmetic. Note that the superscripts on quantified variables indicate their types. We recall that a type-0 functional is a natural number, a type-1 ($= 0(0)$) functional is a function from natural numbers to natural numbers, and a type-1(1) functional is a functional from type 1 functionals to type 1 functionals. $\mathsf{E\text{-}HA}^\omega$ (resp. $\mathsf{E\text{-}PA}^\omega$) is the intuitionistic (resp. classical) arithmetic in all finite types with full extensionality and $\mathsf{WE\text{-}HA}^\omega$ is its weak extensional variant. Note that $\mathsf{E\text{-}HA}^\omega$ is a conservative extension of HA and EL. $\widehat{\mathsf{E\text{-}HA}}^\omega\!\restriction$ (resp. $\widehat{\mathsf{E\text{-}PA}}^\omega\!\restriction$) is Feferman's restriction of $\mathsf{E\text{-}HA}^\omega$ (resp. $\mathsf{E\text{-}PA}^\omega$) to primitive recursion of type 0 and quantifier-free induction. See [14, Chapter 3] for details.

2 Results

Consider a Π^1_2 sentence $\forall \xi (A(\xi) \to \exists \zeta B(\xi, \zeta))$. In terms of computability theory, its provability in RCA corresponds to Muchnik reducibility, namely, the fact that for all ξ satisfying $A(\xi)$, there is an algorithm Φ which computes ζ satisfying $B(\xi, \zeta)$ with the use of ξ as oracle. On the other hand, what one intends to represent by its sequential provability in RCA is that there is an (uniform) algorithm Φ such that for all ξ satisfying $A(\xi)$, Φ computes ζ satisfying $B(\xi, \zeta)$ with the use of ξ as oracle, which corresponds to Medvedev reducibility. We consider the exact formalization of this notion in terms of reverse mathematics, and call that "**uniform provability in** RCA".

Definition 1 (Partially defined application operations, e.g. [16,17]**).** *For* $\alpha, \beta : \mathbb{N} \to \mathbb{N}$,

$$\alpha(\beta) :\, = \begin{cases} \alpha(\bar{\beta}n) - 1 & \textit{where } n \textit{ is the least } n' \textit{ such that } \alpha(\bar{\beta}n') \neq 0, \\ \uparrow & \textit{if there is no such } n', \end{cases}$$

where $\bar{\beta}n'$ *denotes the (code of) finite sequence* $\langle \beta(0), \ldots, \beta(n'-1) \rangle$. *Then*

$$\alpha \,|\, \beta :\, = \lambda n.\, \alpha(\langle n \rangle ^\frown \beta).$$

Definition 2 ([9])**.** RCA^ω *is the finite type system defined as* $\mathsf{E\text{-}PA}^\omega + \mathrm{QF\text{-}AC}^{1,0}$. *Note that* RCA^ω *is a conservative extension of* RCA [9, Theorem 2.8].

As indicated in [4], the provability of sequential versions in RCA seems not to fully represent uniform provability in RCA. In addition, the provability of uniform versions in RCA^ω also seems not to be an "exact" formalization because it suffices for the provability of uniform versions $\exists\Phi\forall\xi(A(\xi) \to B(\xi,\Phi(\xi)))$ in RCA^ω to derive contradiction from the non-existence of Φ rather than the existence of Φ. Using the above notions, we propose the two candidates of the formalization of uniform provability in RCA:

1. There exists a (primitive recursive) closed term t^1 of RCA such that
$$\mathsf{RCA} \vdash \forall\xi\,(A(\xi) \to t\,|\,\xi\downarrow \wedge B(\xi,t\,|\,\xi)).$$
2. There exists a (Gödel primitive recursive) closed term $t^{1(1)}$ of RCA^ω such that
$$\mathsf{RCA}^\omega \vdash \forall\xi\,(A(\xi) \to B(\xi,t(\xi))).$$

In fact, as we show below, these two uniform provability in RCA are equivalent if A is purely universal and B is not so complicated.

On a technical note, thanks to the term existence, the syntactical complexity of the sentence in question is reduced enough to guarantee that the negative translation works (see the proof of Proposition 14). In fact, the proof of Proposition 14 does not work if we interpret uniform provability in RCA by sequential or uniform version.

Definition 3. $\mathrm{N_{KM}}$ *is the class of formulas defined inductively as;*

- A_{qf} *and* $\exists x^\rho A_{qf}$ *are in* $\mathrm{N_{KM}}$, *where* $\rho \in \{0,1\}$.
- *If* A_1, A_2 *are in* $\mathrm{N_{KM}}$, *then* $A_1 \wedge A_2$, $\forall x^\rho A_1$ *are in* $\mathrm{N_{KM}}$, *where* $\rho \in \{0,1\}$.
- *If* A *is in* $\mathrm{N_{KM}}$, *then* $\forall u^\rho \exists v^0 A_{qf} \to A$ *is in* $\mathrm{N_{KM}}$, *where* $\rho \in \{0,1\}$.

Theorem 4. *Let* $\forall\xi\,(A(\xi) \to \exists\zeta B(\xi,\zeta))$ *be a* $\mathcal{L}(\mathsf{EL})$*-sentence where* $A(\xi) \in \mathrm{N_{KM}}$ *and* $B(\xi,\zeta)$ *is equivalent to* $\forall w^\rho \exists s^0 B_{qf}(\xi,\zeta,w,s)$ $(\rho \in \{0,1\})$ *over* $\mathsf{EL} + \mathrm{MP}$, *where* MP *is Markov's principle:*
$$\forall\alpha(\neg\neg\exists x(\alpha(x) = 0) \to \exists x(\alpha(x) = 0)).$$
Then there exists a function term t of RCA *such that*
$$\mathsf{RCA} \vdash \forall\xi\,(A(\xi) \to t\,|\,\xi\downarrow \wedge\, B(\xi,t\,|\,\xi))$$
if and only if
$$\mathsf{EL} + \mathrm{MP} \vdash \forall\xi\,(A(\xi) \to \exists\zeta B(\xi,\zeta)).$$
The corresponding result also holds for RCA_0 *and* EL_0 *instead of* RCA *and* EL.

Proof. By Lemma 15, this follows immediately from Proposition 9 (note that MP is in CN) and Proposition 14. The result for fragments is due to Remark 20. □

Remark 5. The Markov's principle MP is not derivable in EL. However, MP is allowed in Markov-style constructive mathematics (see e.g. [3,18] for details).[1]

---

[1] Troelstra considers $\mathsf{HA} + \mathrm{ECT_0}$(extended Church's thesis)+$\mathrm{MP_{PR}}$(the fragment of MP only for primitive recursive α) to be a formalization of Markov-style constructive mathematics [18, 4.4.12].

Remark 6. A lot of mathematical statements have been investigated in computable analysis [19]. For a theorem S represented as a Π^1_2 sentence $\forall\xi(A(\xi) \rightarrow \exists\zeta B(\xi,\zeta))$, the fact that S is provable in computable analysis (in the sense of [1,2,19]) roughly means that there is a computational program which computes ζ from ξ. This is conceptually the same as intended notion expressed by its sequential provability in RCA. However, there is a crucial difference between provability in computable analysis and uniform provability in RCA. In the former case, the verification that the program works is carried out in the usual mathematical manner. On the other hand, in the latter case, the verification has to be carried out in the restricted mathematical universe having only the Δ^0_1 ($\approx$computable) set existence axiom. In this sense, uniform provability in RCA is more restrictive than provability in computable analysis.[2] On the other hand, the choice of EL as a theory formalizing constructive mathematics is based on considering the meaning of "constructive" as the existence of computational program.[3] Under this interpretation, the provability of S in EL indicates that there is a program which computes ζ from ξ, and in addition, the verification is carried out in a uniformly computable manner. In this sense, provability in EL is seemingly further restrictive than uniform provability in RCA. However, Theorem 4 states that constructive provability is equivalent to uniform provability of in RCA at least for 'practical' Π^1_2 sentences because the syntactical class which our results cover is rich enough to involve most of statements studied in reverse mathematics (under the standard representation). In addition, as we show in Theorem 7 below, even the Markov's principle can be reduced for simpler statements.

Theorem 7. *Let $\forall\xi\,(A(\xi) \rightarrow \exists\zeta B(\xi,\zeta))$ be a $\mathcal{L}(\mathsf{EL})$-sentence where $A(\xi)$ is purely universal (i.e., of the form $\forall u A_{qf}(\xi,u)$) and $B(\xi,\zeta)$ is equivalent to $\forall w^\rho \exists s^0 B_{qf}(\xi,\zeta,w,s)$ ($\rho \in \{0,1\}$) over EL+MP. Then the following are pairwise equivalent.*

1. $\mathsf{EL} \vdash \forall\xi\,(A(\xi) \rightarrow \exists\zeta B(\xi,\zeta))$.
2. *There exists a function term t of RCA such that* $\mathsf{RCA} \vdash \forall\xi(A(\xi) \rightarrow t\,|\,\xi\downarrow \wedge B(\xi, t\,|\,\xi))$.
3. *There exists a term $t^{1(1)}$ of RCA^ω such that* $\mathsf{RCA}^\omega \vdash \forall\xi(A(\xi) \rightarrow B(\xi, t\xi))$ *under the canonical embedding.*

The corresponding result also holds for EL_0, RCA_0 and $\mathsf{RCA}_0^\omega (:= \widehat{\mathsf{E\text{-}PA}}^\omega\restriction + \mathrm{QF\text{-}AC}^{1,0})$ instead of EL, RCA and RCA^ω.

Proof. The equivalence between (1) and (2) follows from Propositions 9 and 18. On the other hand, (1) is equivalent to (3) by Proposition 19. The result for fragments is due to Remark 20. □

Remark 8. 1. The equivalence of (2) and (3) in Theorem 7 reveals that representation of uniform provability by the continuous operator $(\cdot)\,|\,(\cdot)$ and

[2] The exhaustive comparison between these two is in [6].

[3] Troelstra [17] indicates some analogy between Weihrauch's computable analysis and constructive mathematics.

representation by primitive recursive functional in the sense of Gödel are equivalent for a lot of practical statements in reverse mathematics.
2. The related result to the equivalence of (2) and (3) in Theorem 7 for the fragments can be found in [12, Sect. 4], where the relation between the continuous notion in finite type arithmetic and the continuous notion by means of the operation $(\cdot)\,|\,(\cdot)$ has been investigated with respect to higher-order reverse mathematics.
3. The class of type-1(1) functional terms of RCA_0^ω (containing only type-0 recursor R_0) is proper subclass of type-1(1) functional terms of RCA^ω (possibly containing higher type recursors) while the class of function terms of RCA_0 is same as RCA.

Application. For example, by a careful inspection, one can see that Kierstead's effective variant of marriage theorem [10] (see also [8]) is formalized as a Π^1_2 sentence whose premise and conclusion are equivalent to some purely universal formulas over $\mathsf{EL}_0 + \mathrm{MP}$ respectively. In addition, as indicated in [8], it is uniformly provable in RCA_0 (in particular, the verification of the solution-constructing algorithm is carried out in RCA_0). Therefore, by Theorem 7, it follows that Kierstead's effective marriage theorem is provable in EL_0.

Future Work. The author feels that there are at least three possible extensions of this work. One is the attempt to characterize the hierarchy of relativized uniform provability with respect to WKL and ACA by the hierarchy of the law-of-excluded-middle over EL (cf. Sect. 1). Another one is the characterization of uniform provability in stronger systems (like ACA) by (semi-)intuitionistic systems, where the aim is the characterization of computable analysis by constructive mathematics (cf. Remark 6 as well as [17]). The last one is to compare formalized Markov-style constructive mathematics with uniform provability in RCA (cf. Remark 5).

3 Proofs

The following is a refinement of Dorais' result [4, Proposition 3.7] with extracting a witness term, which is based on Kleene's realizability with functions (see [16, Sect. 3.3]). The proof is straightforward by a careful inspection of the proof of [4, Proposition 3.7].[4]

Proposition 9. *Let* CN *be the set of almost negative sentences* φ *such that* $\mathsf{RCA} \vdash \varphi$ *and* GC *be the generalized continuity principle (see [4] for the precise definitions). If*

$$\mathsf{EL} + \mathrm{GC} + \mathrm{CN} \vdash \forall \xi\,(A(\xi) \to \exists \zeta B(\xi,\zeta)),$$

where $A(\xi)$ *is almost negative and* $B(\xi,\zeta)$ *is in* Γ_K *(see [4]), then there exists a function term* t *of* RCA *such that*

$$\mathsf{RCA} \vdash \forall \xi\,(A(\xi) \to t\,|\,\xi\downarrow \wedge\; B(\xi, t\,|\,\xi)).$$

[4] The detailed proof is in author's phD thesis [7].

For the converse direction, we use the well-known negative translation.

Definition 10 (Kuroda's negative translation, see [14]). *A^q is defined as $A^q :\equiv \neg\neg A^*$, where A^* is defined by induction on the logical structure of A:*

- $A^* :\equiv A$, *if A is a prime formula,*
- $(A \Box B)^* :\equiv (A^* \Box B^*)$, *where* $\Box \in \{\wedge, \vee, \rightarrow\}$,
- $(\exists x^\rho A)^* :\equiv \exists x^\rho A^*$,
- $(\forall x^\rho A)^* :\equiv \forall x^\rho \neg\neg A^*$.

Lemma 11. *If* $\mathsf{RCA} \vdash A$, *then* $\mathsf{EL} + \mathrm{MP} \vdash A^q$.

Proof. The proof is straightforward by induction on the length of the derivation as for [14, Proposition 10.3 (ii)]. Note that MP is used only to derive $\left(\text{QF-AC}^{0,0}\right)^q$ intuitionistically from QF-AC^{0,0} (see [14, Proposition 10.6]). □

Definition 12. $\mathrm{N_M}$ *is the class of formulas defined inductively as;*

- A_{qf} *is in* $\mathrm{N_M}$.
- *If* A_1, A_2 *are in* $\mathrm{N_M}$, *then* $A_1 \wedge A_2$, $A_1 \vee A_2$, $\forall x^\rho A_1$, $\exists x^\rho A_1$ *are in* $\mathrm{N_M}$, *where* $\rho \in \{0, 1\}$.
- *If A is in* $\mathrm{N_M}$, *then* $\forall u^\rho \exists v^0 A_{qf} \rightarrow A$ *is in* $\mathrm{N_M}$, *where* $\rho \in \{0, 1\}$.

Lemma 13. *For any formula* $A \in \mathrm{N_M}$, $\mathsf{EL} + \mathrm{MP} \vdash A \rightarrow A^*$.

Proof. The proof is by induction on the structure of $\mathrm{N_M}$. For quantifier-free A_{qf}, this is trivial. Suppose that $A_1, A_2 \in \mathrm{N_M}$ derivable in $\mathsf{EL} + \mathrm{MP}$. Then it is straightforward to see that $(A_1 \wedge A_2)^*$, $(A_1 \vee A_2)^*$, $(\forall x^\rho A_1)^*$ and $(\exists x^\rho A_1)^*$ is derivable in $\mathsf{EL} + \mathrm{MP}$, where $\rho \in \{0, 1\}$. For the case of $\forall u^\rho \exists v^0 A_{qf} \rightarrow A_1$, using MP and the induction hypothesis, one can see that $\forall u^\rho \exists v^0 A_{qf} \rightarrow A_1$ implies $\forall u^\rho \neg\neg \exists v^0 A_{qf} \rightarrow {A_1}^*$, which is identical with $\left(\forall u^\rho \exists v^0 A_{qf} \rightarrow A_1\right)^*$. □

Proposition 14. *Assume that* $A(\xi) \in \mathrm{N_M}$ *and that* $B(\xi, \zeta)$ *is equivalent to* $\forall w^\rho \exists s^0 B_{qf}(\xi, \zeta, w, s)$ $(\rho \in \{0, 1\})$ *over* $\mathsf{EL} + \mathrm{MP}$. *If there exists a function term t of* RCA *such that*

$$\mathsf{RCA} \vdash \forall \xi \, (A(\xi) \rightarrow t \,|\, \xi \downarrow \wedge \, B(\xi, t \,|\, \xi)) \,,$$

then

$$\mathsf{EL} + \mathrm{MP} \vdash \forall \xi \, (A(\xi) \rightarrow \exists \zeta B(\xi, \zeta)) \,.$$

Proof. Suppose $\mathsf{RCA} \vdash \forall \xi \left(A(\xi) \rightarrow t \,|\, \xi \downarrow \wedge \, \forall w^\rho \exists s^0 B_{qf}(\xi, t \,|\, \xi, w, s)\right)$. Expressed more precisely, it asserts that RCA proves (♯):

$$\forall \xi \begin{pmatrix} A(\xi) \rightarrow \forall n \exists m (t(\langle n \rangle ^\frown \overline{\xi} m) > 0) \wedge \\ \forall \gamma^1 \begin{pmatrix} \forall n \exists m \, (t(\langle n \rangle ^\frown \overline{\xi} m) = \gamma(n) + 1 \wedge \forall m' < m (t(\langle n \rangle ^\frown \overline{\xi} m') = 0)) \\ \rightarrow \forall w \exists s B_{qf}(\xi, \gamma, w, s) \end{pmatrix} \end{pmatrix} .$$

By Lemma 11 and standard intuitionistic equivalences, it follows that $\mathsf{EL} + \mathrm{MP}$ proves

$$\forall \xi \left(\begin{array}{l} A^*(\xi) \to \forall n \neg\neg \exists m (t(\langle n \rangle ^\frown \overline{\xi} m) > 0) \wedge \\ \forall \gamma \left(\begin{array}{l} \forall n \neg\neg \exists m \left(t(\langle n \rangle ^\frown \overline{\xi} m) = \gamma(n) + 1 \wedge \forall m' < m (t(\langle n \rangle ^\frown \overline{\xi} m') = 0) \right) \\ \to \forall w \neg\neg \exists s B_{qf}(\xi, \gamma, w, s) \end{array} \right) \end{array} \right).$$

Therefore, using MP and Lemma 13, we have

$$\mathsf{EL} + \mathrm{MP} \vdash (\sharp). \tag{1}$$

In the following, we reason in EL + MP. For ξ satisfying $A(\xi)$, by (1) with the use of $\mathrm{QF\text{-}AC}^{0,0}$, we have g^1 such that $t(\langle n \rangle ^\frown \overline{\xi}(g(n))) > 0$ and $\forall m' < m(t(\langle n \rangle ^\frown \overline{\xi} m') = 0)$ for all n. Then $\zeta := \lambda n.t(\langle n \rangle ^\frown \overline{\xi}(g(n))) \dot{-} 1$ satisfies the condition in ($\sharp$). Thus EL + MP proves $\forall \xi (A(\xi) \to \exists \zeta \forall w \exists s B_{qf}(\xi, \zeta, w, s))$. □

Lemma 15. $\mathrm{N_{KM}} \subset \mathrm{N_K} \cap \mathrm{N_M}$, *where* $\mathrm{N_K}$ *is the class of almost negative formulas (see [4]).*

Proof. Straightforward by induction on the construction of $\mathrm{N_{KM}}$. □

The next proposition is a refinement of Hirst and Mummert's result [9, Theorem 5.6] in the most useful form for our purpose. In the following, AC is the full axiom of choice, $\mathrm{IP}^\omega_\forall$ is the independence of premise for universal formulas and M^ω is the Markov principle in all finite types (see [9] for the precise definitions).

Proposition 16. *For a sentence* $\forall x^\rho (A(x) \to \exists y^\tau B(x, y))$ *of* $\mathcal{L}(\mathsf{WE\text{-}HA}^\omega)$ *where* $A(x)$ *is purely universal and* $B(x, y)$ *is in* Γ_2 *(see [9]), if*

$$\mathsf{WE\text{-}HA}^\omega + \mathrm{AC} + \mathrm{IP}^\omega_\forall + \mathrm{M}^\omega \vdash \forall x^\rho (A(x) \to \exists y^\tau B(x, y)),$$

then there exists a term $t^{\tau(\rho)}$ *of* $\mathsf{WE\text{-}HA}^\omega$ *such that*

$$\mathsf{WE\text{-}HA}^\omega \vdash \forall x^\rho (A(x) \to B(x, t(x))).$$

Proof. By using $\mathrm{IP}^\omega_\forall$, we have that $\forall x \exists y (A(x) \to B(x, y))$ is provable in $\mathsf{WE\text{-}HA}^\omega$ $+ \mathrm{AC} + \mathrm{IP}^\omega_\forall + \mathrm{M}^\omega$. Let $A(x) :\equiv \forall u A_{qf}(x, u)$. Note that $\forall x \exists y (\forall u A_{qf}(x, u) \to B(x, y))$ is in Γ_2 since $B(x, y)$ is in Γ_2. The discussion below is same as in the proof of [9, Theorem 5.6]. Let $(B(x, y))^D \equiv \exists \underline{v} \forall \underline{w} B_D(x, y, \underline{v}, \underline{w})$ (note that $(\forall u A_{qf}(x, u))^D \equiv \forall u A_{qf}(x, u)$). By [14, Theorem 8.6], there exist closed terms $t_Y, \underline{t_V}, t_U$ such that $\mathsf{WE\text{-}HA}^\omega \vdash \forall x, \underline{w} (A_{qf}(x, t_U x \underline{w}) \to B_D(x, t_Y(x), \underline{t_V} x \underline{w}))$. Then, without difficulty, one can see

$$\mathsf{WE\text{-}HA}^\omega \vdash \forall x \exists \underline{v} \forall \underline{w} \exists u (A_{qf}(x, u) \to B_D(x, t_Y(x), \underline{v}, \underline{w})).$$

Since this is equivalent to $\mathsf{WE\text{-}HA}^\omega \vdash \forall x (\forall u A_{qf}(x, u) \to (B(x, t_Y(x)))^D)$ and $B(x, y)$ is in Γ_2, applying [14, Lemma 8.11], we have $\mathsf{WE\text{-}HA}^\omega \vdash \forall x (\forall u A_{qf}(x, u) \to B(x, t_Y(x)))$. □

The following conservation result is an immediate consequence of the previous proposition.

Proposition 17. *For a sentence* $\forall x^\rho(A(x) \to \exists y^\tau B(x,y))$ *of* $\mathcal{L}(\mathsf{WE\text{-}HA}^\omega)$ *where* $A(x)$ *is purely universal and* $B(x,y)$ *is in* Γ_2*, if*

$$\mathsf{WE\text{-}HA}^\omega + \mathrm{AC} + \mathrm{IP}_\forall^\omega + \mathrm{M}^\omega \vdash \forall x^\rho(A(x) \to \exists y^\tau B(x,y)),$$

then

$$\mathsf{WE\text{-}HA}^\omega \vdash \forall x^\rho(A(x) \to \exists y^\tau B(x,y)).$$

Proposition 18. *Assume that* $A(\xi)$ *is purely universal and* $B(\xi,\zeta)$ *is equivalent to* $\forall w^\rho \exists s^0 B_{qf}(\xi,\zeta,w,s)$ $(\rho \in \{0,1\})$ *over* $\mathsf{EL} + \mathrm{MP}$*. If there exists a function term* t *of* RCA *such that*

$$\mathsf{RCA} \vdash \forall \xi\,(A(\xi) \to t\,|\,\xi\downarrow \wedge\, B(\xi, t\,|\,\xi)),$$

then

$$\mathsf{EL} \vdash \forall \xi\,(A(\xi) \to \exists \zeta B(\xi,\zeta)).$$

Proof. Since each purely universal formula is in $\mathrm{N_M}$, by Proposition 14, $\mathsf{EL}+\mathrm{MP}$ proves $\forall\xi(A(\xi) \to \exists\zeta\forall w^\rho\exists s^0 B_{qf}(\xi,\zeta,w,s))$. By identifying $\mathsf{EL} + \mathrm{MP}$ with its canonical embedding into $\mathsf{WE\text{-}HA}^\omega + \mathrm{M}^0$, we have $\mathsf{WE\text{-}HA}^\omega + \mathrm{M}^0 \vdash \forall\xi(A(\xi) \to \exists\zeta\forall w^\rho\exists s^0 B_{qf}(\xi,\zeta,w,s))$. Since $\forall w^\rho\exists s^0 B_{qf}(\xi,\zeta,w,s)$ is in Γ_2, by Proposition 17, we have

$$\mathsf{WE\text{-}HA}^\omega \vdash \forall\xi\left(A(\xi) \to \exists\zeta\forall w^\rho\exists s^0 B_{qf}(\xi,\zeta,w,s)\right).$$

Since $\mathsf{WE\text{-}HA}^\omega$ is conservative over EL for $\mathcal{L}(\mathsf{EL})$ formulas, it follows that EL proves $\forall\xi\left(A(\xi) \to \exists\zeta\forall w^\rho\exists s^0 B_{qf}(\xi,\zeta,w,s)\right)$. □

Proposition 19. *Let* $\forall\xi\,(A(\xi) \to \exists\zeta B(\xi,\zeta))$ *be a* $\mathcal{L}(\mathsf{EL})$*-sentence where* $A(\xi)$ *is purely universal and* $B(\xi,\zeta)$ *is equivalent to* $\forall w^\rho\exists s^0 B_{qf}(\xi,\zeta,w,s)$ $(\rho \in \{0,1\})$ *over* $\mathsf{EL} + \mathrm{MP}$*. Then* $\mathsf{EL} \vdash \forall\xi\,(A(\xi) \to \exists\zeta B(\xi,\zeta))$ *if and only if there exists a term* $t^{1(1)}$ *of* RCA^ω *such that* $\mathsf{RCA}^\omega \vdash \forall\xi\,(A(\xi) \to B(\xi, t(\xi)))$ *under the canonical embedding.*

Proof. Soppose $\mathsf{EL} \vdash \forall\xi(A(\xi) \to \exists\zeta B(\xi,\zeta))$. Then $\mathsf{WE\text{-}HA}^\omega \vdash \forall\xi(A(\xi) \to \exists\zeta B(\xi,\zeta))$ under the canonical embedding. Since $A(\xi)$ is purely universal and $\forall w^\rho\exists s^0 B_{qf}(\xi,\zeta,w,s)$ is in Γ_2, by Proposition 16, there exists a term $t^{1(1)}$ of RCA^ω such that $\mathsf{RCA}^\omega \vdash \forall\xi\,(A(\xi) \to B(\xi,t(\xi)))$.

The converse direction is the same as for Proposition 18 except the use of elimination of extensionality technique. Suppose that RCA^ω proves $\forall\xi(A(\xi) \to B(\xi,t(\xi)))$. Since our sentence contains quantifiers only of type 0 and 1, it follows by [14, Proposition 10.45] that WRCA^ω (i.e. $\mathsf{WE\text{-}PA}^\omega$ + $\mathrm{QF\text{-}AC}^{1,0}$) proves $\forall\xi\,(A(\xi) \to B(\xi,t(\xi)))$. Then using the negative translation [14, Proposition 10.6], the Dialectica interpretation (Proposition 17) and the conservativity just as in the proof of Proposition 18, we have that EL proves $\forall\xi\,(A(\xi) \to \exists\zeta B(\xi,\zeta))$. □

Remark 20. By a careful inspection, one observes that all discussions in the above proofs also work for fragments EL_0, RCA_0, RCA_0^ω instead of EL, RCA, RCA^ω (cf. [4, Remark 3.10], [14, Sect. 8.3] and [14, Sect. 10.5]).

Acknowledgment. The author is grateful to his supervisor Takeshi Yamazaki and also to Ulrich Kohlenbach for helpful discussion.

References

1. Brattka, V., Gherardi, G.: Weihrauch degrees, omniscience principles and weak computability. J. Symb. Log. **76**(1), 143–176 (2011)
2. Brattka, V., Gherardi, G.: Effective choice and boundedness principles in computable analysis. Bull. Symb. Log. **17**(1), 73–117 (2011)
3. Bridges, D.S., Richman, F.: Varieties of constructive mathematics. London Math. Soc. Lecture Notes, vol. 97. Cambridge University Press, Cambridge (1987)
4. Dorais, F.G.: Classical consequences of continuous choice principles from intuitionistic analysis. Notre Dame J. Formal Logic **55**, 25–39 (2014)
5. Dorais, F.G., Hirst, J.L., Shafer, P.: Reverse mathematics, trichotomy, and dichotomy. J. Logic Anal. **4**(13), 1–14 (2012)
6. Dorais, F.G., Dzhafarov, D.D., Hirst, J.L., Mileti, J.R., Shafer, P.: On uniform relationships between combinatorial problems. Trans. AMS. (To appear). http://www.ams.org/journals/tran/0000-000-00/S0002-9947-2015-06465-4/home.html
7. Fujiwara, M.: Intuitionistic and uniform provability in reverse mathematics. Ph.D. thesis, Tohoku University (2015)
8. Fujiwara, M., Higuchi, K., Kihara, T.: On the strength of marriage theorems and uniformity. Math. Logic Q. **60**(3), 136–153 (2014)
9. Hirst, J.L., Mummert, C.: Reverse mathematics and uniformity in proofs without excluded middle. Notre Dame J. Formal Logic **52**(2), 149–162 (2011)
10. Kierstead, H.A.: An effective version of Hall's theorem. Proc. Amer. Math. Soc. **88**, 124–128 (1983)
11. Ishihara, H.: Constructive reverse mathematics: compactness properties. In: Crosilla, L., Schuster, P. (eds.) From Sets and Types to Topology and Analysis: Towards Practicable Foundations for Constructive Mathematics, pp. 245–267. Oxford University Press, Oxford (2005)
12. Kohlenbach, U.: Foundational and mathematical uses of higher types. In: Sieg, W., et al. (eds.) Reflections on the Foundations of Mathematics. Essays in honor of Solomon Feferman. Lecture Notes in Logic, vol. 15, pp. 92–116. A. K. Peters Ltd, Natick, MA (2002)
13. Kohlenbach, U.: Higher order reverse mathematics. In: Simpson, S.G. (ed.) Reverse Mathematics 2001. Association Symbolic Logic. Lecture Notes in Logic, vol. 21, pp. 281–295. A. K. Peters, Wellesley MA (2005)
14. Kohlenbach, U.: Applied Proof Theory: Proof Interpretations and their Use in Mathematics. Springer Monographs in Mathematics. Springer, Berlin (2008)
15. Simpson, S.G.: Subsystems of Second Order Arithmetic. Association for Symbolic Logic, 2nd edn. Cambridge University Press, NY (2009)
16. Troelstra, A.S. (ed.): Metamathematical Investigation of Intuitionistic Arithmetic and Analysis. Lecture notes in Mathematics, vol. 344. Springer, Berlin (1973)
17. Troelstra, A.S.: Comparing the theory of representations and constructive mathematics. In: Börger, E., et al. (eds.) CSL 1991. LNCS, vol. 626, pp. 382–395. Springer, Berlin (1992)
18. Troelstra, A.S., van Dalen, D.: Constructivism in Mathematics: An Introduction (two volumes). North Holland, Amsterdam (1988)
19. Weihrauch, K.: Computable Analysis: An Introduction. Springer, Berlin (2000)

Randomness and Differentiability of Convex Functions

Alex Galicki(✉)

Department of Computer Science, University of Auckland, Auckland, New Zealand
agal629@aucklanduni.ac.nz

Abstract. We study first and second derivatives of computable convex functions on $\mathbb{R}^n$. The main result of the paper is an effective form of Aleksandrov's Theorem: we show that computable randomness implies twice-differentiability of computable convex functions.

1 Introduction

1.1 Overview of the Paper

This paper is concerned with computable analysis [14] and algorithmic randomness (see Nies [9], Downey and Hirschfeldt [5]).

We are mostly interested in computable convex functions on $\mathbb{R}^n$. Convex functions are very well behaved and play an important role in such areas as optimization, control theory and variational analysis. This class has been studied both in classical and in effective contexts.

First derivatives of computable convex functions of one variable have been studied in Dinghzu and Ko [4]. In particular, Dinghzu and Ko noticed that the first derivative is (uniformly) computable on the set of points where it does exist. The second and higher derivatives of computable real functions of one variable were considered for example in Zhong [15].

In Sect. 2 we study computable functions of one variable. Firstly, we characterise sets of differentiability of computable convex functions from $\mathbb{R}$ to $\mathbb{R}$. Next, we introduce a class of *effectively monotone functions* and show that in some precise sense functions from this class are exactly derivatives of computable convex functions. Finally, we prove two new characterisations of computable randomness: in terms of differentiability of effectively monotone functions and in terms of twice-differentiability of computable convex functions.

Computable randomness is one of the more natural algorithmic randomness notions. Originally, it has been defined on the Cantor space in terms of effective betting strategies. It has been generalized to other spaces. For details, see Rute [12].

In Sect. 3 we generalize one of the results from Sect. 2 to $\mathbb{R}^n$. The main result of Sect. 3 (and of this paper) is an effective version of the following classical result, known as Aleksandrov's Theorem [2]:

A. Beckmann et al. (Eds.): CiE 2015, LNCS 9136, pp. 196–205, 2015.
DOI: 10.1007/978-3-319-20028-6_20

Theorem 1 (Aleksandrov, 1939 [2]). *If $f : \mathbb{R}^n \to \mathbb{R}$ is a convex function, then it is twice-differentiable almost everywhere.*

In Theorem 20 we prove that every computable convex function $f : \mathbb{R}^n \to \mathbb{R}$ is twice differentiable at the set of computably random elements of $\mathbb{R}^n$.

1.2 Notation and Conventions

Let $f : \mathbb{R}^n \to \mathbb{R}^m$ be a function. We denote the set of non-differentiability of f, that is the set of points x such that $f'(x)$ does not exist, by N_f.

Working with derivatives often means working with slopes. For functions of one variable, we use the following notation. Let $f : \mathbb{R} \to \mathbb{R}$ be a function. We define

$$S_f(x,h) = \frac{f(x+h) - f(x)}{h}.$$

We denote by I the identity function on $\mathbb{R}^n$.

Computable Real Functions. There were multiple attempts of formalizing the notion of computability of real functions, most of which turned out to be equivalent. In this paper by computability of real functions we mean the Grzegorczyk-Lacombe notion of computability. Recall that a function $f : \mathbb{R}^n \to \mathbb{R}^m$ is computable in the sense of Grzegorczyk-Lacombe, if $f(x)$ is computable uniformly in x, f is continuous and its modulus of continuity satisfies some specific effectivity conditions. For formal details, see Pour-El and Richards [10] and Weihrauch [14].

Computable Randomness. In several of our results we use the notion of computable randomness on $\mathbb{R}^n$. For this purpose we extend the notion of computable randomness on $[0,1]^n$ (see [12]) to $\mathbb{R}^n$.

We say $z = (z_1, \ldots, z_n) \in \mathbb{R}^n$ is computably random if ($z_1 \mod 1, \ldots, z_n \mod 1$) is computably random.

On the real line, computable randomness can be characterised in terms of differentiability of either computable Lipschitz functions or computable monotone functions (see Brattka et al. [3] and Freer et al. [6]).

On $\mathbb{R}^n$, computable randomness implies differentiability of computable Lipshitz functions of several variables and computable monotone functions from $\mathbb{R}^n$ to $\mathbb{R}^n$ (see Galicki and Turetsky [7]).

1.3 Convex, Monotone and Lipschitz Functions

A function $f : \mathbb{R}^n \to \mathbb{R}$ is *convex* if the following condition holds for all $x_0, x_1 \in \mathbb{R}^n$ and all $t \in [0,1]$:

$$f((1-t)x_0 + tx_1) \leq (1-t)f(x_0) + tf(x_1).$$

There are two other classes of functions that are closely related to convex functions: Lipschitz and monotone functions. All three classes play a prominent role in *variational analysis* (see Rockafellar and Wets [11]).

A function $f : \mathbb{R}^n \to \mathbb{R}^m$ is *Lipschitz* if there exists $L \in \mathbb{R}^+$ such that

$$\|f(x) - f(y)\| \leq L \cdot \|x - y\| \text{ for all } x, y \in \mathbb{R}^n.$$

The least such L is called *the Lipschitz constant* for f. We denote it by $\mathbf{Lip}(f)$.

To deal with discontinuous monotone functions, we need to consider set-valued functions from $\mathbb{R}^n$ to $\mathbb{R}^n$, that is functions that map every point in $\mathbb{R}^n$ to a subset of $\mathbb{R}^n$.

We say a set-valued function $u : \mathbb{R}^n \to \mathbb{R}^n$ is *monotone* if

$$\langle y_1 - y_2, x_1 - x_2 \rangle \geq 0 \text{ for all } x_1, x_2 \in \mathbb{R}^n \text{ and all } y_1 \in u(x_1), y_2 \in u(x_2).$$

A set-valued monotone function is said to be *maximal* if its graph is not properly included in the graph of another monotone function.

In this paper we mostly consider single valued functions. Hence, by default, "monotone function" means a single-valued function. Every time we consider set-valued functions, we explicitly say so.

Set-valued monotone functions on $\mathbb{R}^n$ are almost everywhere continuous and are single valued at points of continuity. Moreover, if $g : \mathbb{R}^n \to \mathbb{R}^n$ is a maximal set-valued monotone function and $z \in \mathbb{R}^n$, then $g(z)$ is a closed convex set. Derivatives of convex functions are closely related to monotone set-valued functions. On the real line, a function is monotone if and only if it coincides (almost everywhere) with a derivative of a convex function. In the case of functions on $\mathbb{R}^n$ the situation is somewhat more nuanced: gradients of convex real-valued functions on $\mathbb{R}^n$ form a proper subclass of the class of all monotone functions from $\mathbb{R}^n$ to $\mathbb{R}^n$.

We proceed with recalling a connection between (set-valued) monotone functions and Lipschitz functions discovered by Minty [8] and some of its consequences relevant to our paper.

Minty showed that the so called Cayley transformation

$$\Phi : \mathbb{R}^n \times \mathbb{R}^n \to \mathbb{R}^n \times \mathbb{R}^n \text{ defined by } \Phi(x, y) = \frac{1}{\sqrt{2}}(y + x, y - x)$$

transforms the graph of a set-valued maximal monotone function into a graph of a 1-Lipschitz function. Note that when $n = 1$ this is a clockwise rotation of $\pi/4$. We will rely on the following consequence of the above fact.

Proposition 2 (cf. Proposition 1.2 in [1]**).** *Let $u : \mathbb{R}^n \to \mathbb{R}^n$ be a maximal monotone (set-valued) function. Then $(u + I)$ and $(u + I)^{-1}$ are monotone and $(u + I)^{-1}$ is 1-Lipschitz.*

Furthermore, to use an effective form of Rademacher's Theorem in Sect. 3, we will require the following result.

Proposition 3 (cf. Theorem 12.65 in [11]). *Let $u : \mathbb{R}^n \to \mathbb{R}^n$ be a maximal monotone (set-valued) function. Let $z \in \mathbb{R}^n$ and define $f = (u + I)^{-1}$ and $\hat{z} = u(z) + z$. The following two are equivalent:*

1. *u is differentiable at z, and*
2. *f is differentiable at $\hat{z}$ and $f'(\hat{z})$ is invertible.*

A good exposition of classical results related to this area can be found in Alberti and Ambrosio [1] and in Chap. 12 of [11].

Main Geometric Idea with an Example. Let $u : \mathbb{R}^n \to \mathbb{R}$ be a convex function and let $f = \nabla u$ be its gradient. Denote the maximal extension of f by $\overline{f}$. Since this paper is concerned with those functions only for which a maximal extension exists and it is unique, we will always write about *the maximal extension.* Then $(\overline{f} + I)^{-1}$ is 1-Lipschitz function that, in the sense of Proposition 3, preserves information about non-differentiability points of f. A crucial idea used in this paper is that a question about twice-differentiability of convex functions can be reduced to a question about differentiability of 1-Lipschitz functions. To exploit this, we will prove effective variants of the above described connections between convex, monotone and Lipschitz functions. The following simple example illustrates this idea.

Example 4. Let $u(x) = |x|$. It is a computable convex function with one point of non-differentiability, at $x = 0$. Its derivative $f(x) = u'(x)$ is a step function. f is defined and it is continuous at those points where u is differentiable. Note that f is not a computable function, however $f(x)$ is computable uniformly in x at points where f is continuous. Finally note that $g = (\overline{f} + I)^{-1}$, where $\overline{f}$ is the maximal extension of f, is a computable function, such that if $g'(f(x) + x)$ exists and it is not equal to 0, then f is differentiable at x (Fig. 1).

This example highlights the need to consider set-valued functions: note that $f + I$ is not invertible, while $\overline{f} + I$ is.

We will generalise these observations in Sect. 2.

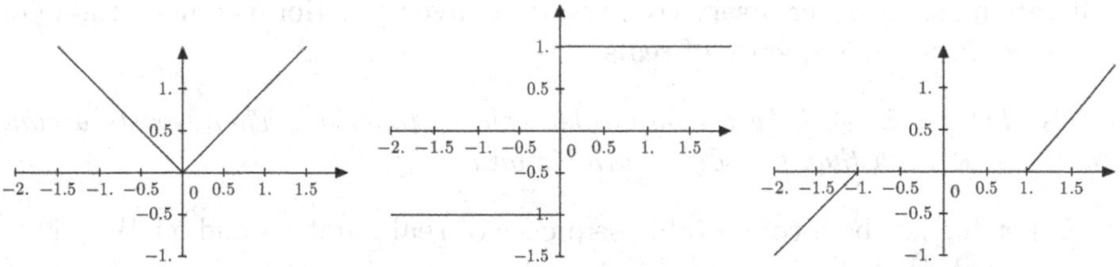

(a) $u(x) = |x|$ (b) $f(x) = u'(x)$ (c) $g = (\overline{f} + I)^{-1}$

Fig. 1. Example 4

2 Differentiability of Convex and Monotone Real Functions of One Variable

2.1 Sets of Non-differentiability of Computable Convex Functions

It is known that the set of points where a convex function $f : \mathbb{R} \to \mathbb{R}$ is not differentiable is countable. In fact, for any countable set of points B on the real line, there exists a convex real function that is not differentiable precisely at elements of B (see [13]).

When f is computable, it is relatively easy to show (and we will do so later) that all points of non-differentiability are computable. However, not all sequences of computable real numbers are sets of non-differentiability of computable convex functions.

Definition 5. *Let $(r_i)_{i \in \mathbb{N}}$ be a sequence of real numbers. We say it is a* cnd-sequence *(convex non-differentiability sequence) if there exists a computable sequence of real numbers $(q_i)_{i \in \mathbb{N}}$ and a c.e. set $W \subseteq \mathbb{N}$ such that*

$$\{r_i : i \in \mathbb{N}\} = \{q_i : i \notin W\}.$$

Now we show that cnd-sequences characterise sets of non-differentiability of computable convex functions on the real line.

Theorem 6. *Let $(r_i)_{i \in \mathbb{N}}$ be a sequence of real numbers. The following two are equivalent:*

1. *$(r_i)_{i \in \mathbb{N}}$ is a cnd-sequence, and*
2. *there exists a computable convex function $f : \mathbb{R} \to \mathbb{R}$ such that $N_f = \{r_i : i \in \mathbb{N}\}$.*

Remark 7. While there is no computable convex function non-differentiable at all computable reals, for any computable sequence of real numbers, there does exist a computable convex function non-differentiable precisely at elements of the sequence. In particular, there is a computable convex function non-differentiable at all rationals. However, every computable convex function is differentiable on a dense computable sequence of reals.

Fact 8. *Let $f : \mathbb{R} \to \mathbb{R}$ be a computable convex function. There exists a computable $r \in \mathbb{R}$ such that f is differentiable on $r + \mathbb{Q}$.*

Proof. Let $(q_i)_{i \in \mathbb{N}}$ be a computable sequence of real numbers and let $W \subseteq \mathbb{N}$ be a c.e. set such that $N_f = \{q_i : i \notin W\}$.

Let $(p_i)_{i \in \mathbb{N}}$ be a computable sequence of real numbers that enumerates all real numbers of the form $q_i + q$ for all $i \in \mathbb{N}$ and all $q \in \mathbb{Q}$.

There exists a computable irrational real number r such that it is not in $(p_i)_{i \in \mathbb{N}}$. Then $\{r+q : q \in \mathbb{Q}\} \cap \{p_i : i \in \mathbb{N}\} = \emptyset$ and hence $\{r+q : q \in \mathbb{Q}\} \cap N_f = \emptyset$.

2.2 Monotone and Convex Functions

From classical analysis we know that there is a strong connection between monotone and convex functions. If $f : \mathbb{R} \to \mathbb{R}$ is convex, its both left and right derivatives are defined everywhere and are monotone and N_f is precisely the set of discontinuity points of both its left or right derivatives. Conversely, if $g : \mathbb{R} \to \mathbb{R}$ is monotone, then $f(x) = \int_0^x g(t)\,\mathrm{d}t$ is convex and N_f is the set of discontinuity points of g. In this subsection we show an analogous result in the effective setting.

We want to characterise derivatives of computable convex functions. Let $f : \mathbb{R} \to \mathbb{R}$ be such a function. We know that f' is monotone and it is defined outside some cnd-sequence. However, it can be discontinuous and thus not computable in the sense of Grzegorczyk (not even relative to any oracle).

The following fact (stated in [4]) shows $f'(x)$ is computable (uniformly in x) where it is defined.

Fact 9. *Let $f : \mathbb{R} \to \mathbb{R}$ be a computable convex function and let $g(x) = f'(x)$. Then g is computable (uniformly in x) on its set of continuity.*

Proof. By convexity of g, we get that $\frac{f(x)-f(x-t)}{t} \leq f'(x) \leq \frac{f(x+t)-f(x)}{t}$ for all $t > 0$ and hence whenever $f'(x) = g(x)$ exists, it can be effectively approximated (uniformly in x) from below and from above.

This fact justifies the following definition.

Definition 10. *We say a monotone function $f : \mathbb{R} \to \mathbb{R}$ is* effectively monotone *if $f(x)$ is computable uniformly in x when f is continuous at x.*

Our notion of effective monotonicity is closely related to several other natural notions of effectiveness when restricted to monotone functions. One of them is the notion of *almost everywhere computable functions* (see Subsects. 7.1 and 7.2 in [12]). Roughly speaking, a function f is a.e. computable if $f(x)$ is computable uniformly in x on some subset of full measure.

The other notion is that of computability on $I_{\mathbb{Q}}$ (see Sect. 7 in [3]). A partial function $f : I \to \mathbb{R}$ is said to be *computable on $I_{\mathbb{Q}}$* if its domain contains $I_{\mathbb{Q}} = [0, 1] \cap \mathbb{Q}$ and $f(q)$ is computable uniformly in $q \in \mathbb{Q}$. Analogously, in this paper we say *f is computable on A* if $f(x)$ is computable uniformly in $x \in A$.

Proposition 11. *Let $f : \mathbb{R} \to \mathbb{R}$ be an a.e. computable monotone function and let $C \subseteq \mathbb{R}$ be its set of continuity. There exists a computable convex function $g : \mathbb{R} \to \mathbb{R}$ such that $g' = f|_C$.*

Proof. We may assume f is a non-decreasing function. Then $u = (\overline{f} + I)^{-1}$ is a 1-Lipschitz function, where $\overline{f}$ is the maximal extension of f. Let us show that it is computable. Since it is Lipschitz, we are only required to show that it is computable on rationals. Let $y \in \mathbb{Q}$, let $s \in \mathbb{N}$ and define $g = \overline{f} + I$. To compute $u(y)$ at stage s, find $x_+, x_- \in \mathbb{Q}$ such that $g(x_-) < y$, $g(x_+) > y$ and $|f(x_-) - f(x_+)| \leq s^{-s}$; declare $u(y)$ at stage s to be x_-.

Since $u(0)$ is computable, without loss of generality we may assume that $u(0) = 0$. Hence we may assume that $f(0) = 0$. Let

$$F(x) = \int_0^x (f(t) + t)\,\mathrm{d}t = x \cdot (f(x) + x) - \int_0^{f(x)+x} u(t)\,\mathrm{d}t.$$

F is convex and a.e. computable. Hence it is computable. It follows that $g(x) = \int_0^x f(t)\,\mathrm{d}t$ is a computable convex function such that $g' = f|_C$.

Now we can prove equivalence between several notions of effectiveness for monotone functions.

Proposition 12. *Let $f : \mathbb{R} \to \mathbb{R}$ be a monotone function and let $C \subseteq R$ be its set of continuity. The following are pairwise equivalent:*

1. *f is effectively monotone,*
2. *f is computable on a dense computable sequence of reals,*
3. *f is computable on $r + \mathbb{Q}$ for some computable r,*
4. *$f|_C$ is a.e. computable, and*
5. *the graph of its maximal extension is a Π^0_1 class.*

Proof. 1. ⇒ 3. follows from Fact 8 and Propositions 11.
3. ⇒ 2. is trivial.
2. ⇒ 1. follows from the fact that $f(x)$ can be effectively approximated from above and from below when f is continuous at x.
1. ⇒ 4. holds since monotone functions are a.e. continuous
4. ⇒ 2. follows from Proposition 11 and Fact 9.
1. ⇒ 5. Let $\overline{f}$ be the maximal extension of f. Then, as we have seen in the proof of Proposition 11, $(\overline{f} + I)^{-1}$ is a computable Lipschitz function, hence its graph is a Π^0_1 class. It follows that the graph of $(\overline{f} + I)$ is a Π^0_1 class and hence the graph of $\overline{f}$ is a Π^0_1 class too.
5. ⇒ 1. Let P be the graph of the maximal extension of f. For every $x \in C$, $\{y|(x, y) \in P\}$ is a Π^0_1 class (uniformly in x) with one element, hence $f(x)$ is computable uniformly in x.

While an effectively monotone function $f : \mathbb{R} \to \mathbb{R}$ is not necessarily computable on $I_\mathbb{Q}$, there exists a computable real r such that $f + r$ is computable on $I_\mathbb{Q}$.

Remark 13. Fact 9 and Proposition 11 show that in a precise sense, effectively monotone functions correspond to derivatives of computable convex functions. Both left and right derivatives of a computable convex functions are effectively monotone functions and every effectively monotone function restricted to its set of continuity is a derivative of some computable convex function.

Finally, note that just like every continuous monotone real function of one variable is computable relative to some oracle, every monotone function is an effectively monotone function relative to some oracle. This suggests that the class of effectively monotone functions is an appropriate class for studying monotone but not necessarily continuous functions in the context of computable analysis.

2.3 New Characterisations of Computable Randomness

The following result shows two new characterisations of computable randomness on $\mathbb{R}$: in terms of differentiability of effectively monotone functions and in terms of twice differentiability of computable convex functions. The result about effectively monotone functions is an extension of a known theorem from [3] to discontinuous monotone functions. The other result can be seen as a bi-directional effective version of Aleksandrov's Theorem. To our knowledge it is the first result that characterises a randomness notion in terms of twice differentiability.

Theorem 14. *Let $z \in \mathbb{R}$. The following are pairwise equivalent:*

(1) z is computably random,
(2) all effectively monotone functions from $\mathbb{R}$ to $\mathbb{R}$ are differentiable at z,
(3) all computable convex functions from $\mathbb{R}$ to $\mathbb{R}$ are twice-differentiable at z.

Remark 15. The (2) $\implies$ (1) and (3) $\implies$ (1) implications follow immediately from known facts from [3]. The other implications follow from Theorems 19 and 20.

It is possible to prove the (3) $\implies$ (1) and (2) $\implies$ (1) implications directly by proving one-dimensional versions of results from the second part of the paper. Alternatively, it seems feasible to combine results from Sect. 7 in (the longer version of) [3] with some of results from this section to prove both implications.

3 Effective Aleksandrov's Theorem

The main result of this section is an effective version of Aleksandrov's theorem. Namely, we show that computable convex functions on $\mathbb{R}^n$ are twice differentiable at computably random elements of $\mathbb{R}^n$. The idea of the proof is straightforward. Suppose $f : \mathbb{R}^n \to \mathbb{R}$ is a computable convex function and $z \in \mathbb{R}^n$ is computably random. It is easy to show that $g = \nabla f$, the gradient of f, is an a.e. computable monotone function. To show that g is differentiable at z, we need to prove a stronger version of Theorem 3.4.1 from [7] which states that computable monotone functions on $\mathbb{R}^n$ are differentiable at computably random points. In what follows we outline the idea behind the original proof and describe points where it has to be modified.

Suppose $g : \mathbb{R}^n \to \mathbb{R}^n$ is a computable monotone function and $z \in \mathbb{R}^n$ is computably random. The proof uses the fact that $(g + I)^{-1}$ is a computable injective 1-Lipschitz function and the fact that computable injective Lipschitz functions satisfy the property that their inverses preserve computable randomness. It follows that $g(z) + z$ is computably random and hence, by an effective version of Rademacher's Theorem proven in [7], $(g + I)^{-1}$ is differentiable at $g(z) + z$. By another classical result, Proposition 3, and by an effective version of Sard's Theorem for Lipschitz functions from [7], g is differentiable at z.

There are two main points where the original proof needs to be strengthened. Let $\overline{g}$ be the maximal extension of g. Firstly, we need to show that $(\overline{g} + I)^{-1}$

is computable when g is an a.e. computable monotone function. Secondly, since g is no longer assumed to be continuous, $(\overline{g} + I)^{-1}$ is no longer guaranteed to be injective. Hence we need to show that the mentioned preservation property holds for all computable 1-Lipschitz function, not just for the injective ones. In the following subsections we deal with those two issues.

3.1 Two Propositions

The following lemma is a stronger version of Lemma 3.2.5 from [7].

Lemma 16. *Let $f : \mathbb{R}^n \to \mathbb{R}^n$ be a computable Lipschitz function and suppose $z \in \mathbb{R}^n$ is not computably random. Then $f(z)$ is not computably random either.*

We also need the following effectively monotone version of Sard's Theorem for Lipschitz functions proven in [7].

Theorem 17 (Theorem 3.3.2, [7]). *Let $f : \mathbb{R}^n \to \mathbb{R}^n$ be a computable Lipschitz function and let $z \in \mathbb{R}^n$. If $f(z)$ is computably random, then $f'(z)$ is invertible.*

3.2 Differentiability of A.E. Computable Monotone Functions

It is known that when $g : \mathbb{R}^n \to \mathbb{R}^n$ is a computable invertible function, its inverse g^{-1} is computable too. This fact, in conjunction with an effective Rademacher's Theorem, was used in [7] to show that computable monotone functions from $\mathbb{R}^n$ to $\mathbb{R}^n$ are differentiable at computably random points.

However, the gradient $g = \nabla f$ of a computable convex function $f : \mathbb{R}^n \to \mathbb{R}$ need not be computable. It is a.e. computable and that is why we need the following result.

Lemma 18. *Let $g : \mathbb{R}^n \to \mathbb{R}^n$ be a monotone a.e. computable function and let $\overline{g}$ be the maximal extension of g. If $f = \overline{g}^{-1}$ is 1-Lipschitz, then f is computable.*

The proof of this lemma relies heavily on geometric properties of monotone and Lipschitz functions and it is not clear whether the existence of a computable inverse can be proven for other (natural) classes of partial computable functions.

We are now ready to formulate and prove our main result concerning monotone a.e. computable functions.

Theorem 19. *Let $z \in [0,1]^n$ be computably random and let $u : \mathbb{R}^n \to \mathbb{R}^n$ be an a.e. computable monotone function. Then u is differentiable at z.*

Proof. Let $\overline{u}$ be the maximal extension of u.

Define $g = (\overline{u}+I)$ and $f = g^{-1}$, then f is a Lipschitz function with $\mathbf{Lip}(f) \leq 1$. Let $y = u(z) + z$ so that $f(y) = z$. By Lemma 18, f is computable. And by Lemma 16, y is computably random. It follows that f is differentiable at y and, by Theorem 17, $f'(y)$ is invertible. Hence, by Proposition 3, u is differentiable at z.

3.3 Main Result

Theorem 20. *Let $f : \mathbb{R}^n \to \mathbb{R}$ be a computable convex function. If $z \in \mathbb{R}^n$ is computably random, then f is twice-differentiable at z.*

Proof. Let $g : \mathbb{R}^n \to \mathbb{R}^n$ be the gradient of f. We know that it is a monotone function. Let us show that g is a.e. computable. To do so, it is sufficient to show that partial derivatives of f are a.e. computable.

Fix $i \leq n$ and let $x \in \mathbb{R}^n$. Consider a convex function $f_x : \mathbb{R} \to \mathbb{R}$ defined by $f_x(h) = f(x + e_i h)$ (here $e_1, \dots, e_n$ denotes the standard basis for $\mathbb{R}^n$). It is differentiable at 0 when $D_i f(x)$ exists and then $f'_x(0) = D_i f(x)$. It is computable (uniformly in x), hence whenever $f'_x(0)$ exists, it is computable (uniformly in x). Since $D_i f(x)$ exists a.e., the function $x \mapsto D_i f(x)$ is a.e. computable.

This shows that g is an a.e. computable monotone function. By Theorem 19, g is differentiable at z.

References

1. Alberti, G., Ambrosio, L.: A geometrical approach to monotone functions in $\mathbb{R}^n$. Math. Z. **230**, 259–316 (1999)
2. Aleksandrov, A.D.: Leningrad Univ. Ann. (Math. Ser.) **6**, 3–35 (1939)
3. Brattka, V., Miller, J., Nies, A.: Randomness and differentiability. Trans. AMS (forthcoming). http://arxiv.org/abs/1104.4465
4. Dingzhu, D., Ko, K.: Computational complexity of integration and differentiation of convex functions. Syst. Sci. Math. Sci. **2**(1), 70–79 (1989)
5. Downey, R., Hirschfeldt, D.: Algorithmic Randomness and Complexity. Springer, Berlin (2010)
6. Freer, C., Kjos-Hanssen, B., Nies, A., Stephan, F.: Algorithmic aspects of lipschitz functions. Computability **3**(1), 45–61 (2014)
7. Galicki, A., Turetsky, D.: Differentiability and randomness in higher dimensions (2014, Submitted)
8. Minty, G.: Monotone nonlinear operators on a Hilbert space. Duke Math J. **29**, 341–346 (1962)
9. Nies, A.: Computability and Randomness. Oxford Logic Guides. Oxford University Press, Oxford (2009)
10. Pour-El, M.B., Richards, I.: Computability in Analysis and Physics. Springer, Berlin (1988)
11. Rockafellar, R.T., Wets, R.J.-B.: Variational Analysis. Grundlehren der Mathematischen Wissenschaften. Springer, Heidelberg (1997)
12. Rute. J.: Computable randomness and betting for computable probability spaces (2012, Submitted)
13. Siksek, S., El Sedy, E.: Points of non-differentiability of convex functions. Appl. Math. Comput. **148**, 725–728 (2004)
14. Weihrauch, K.: Computable Analysis. Springer, Berlin (2000)
15. Zhong, N.: Derivatives of computable functions. Math. Log. Q. **44**, 304–316 (1998)

Weighted Automata on Infinite Words in the Context of Attacker-Defender Games

Vesa Halava[1,2], Tero Harju[1], Reino Niskanen[2(✉)], and Igor Potapov[2]

[1] Department of Mathematics and Statistics, University of Turku, 20014 Turku, Finland
{vesa.halava,harju}@utu.fi

[2] Department of Computer Science, University of Liverpool, Ashton Building, Liverpool L69 3BX, UK
{r.niskanen,potapov}@liverpool.ac.uk

Abstract. We consider several infinite-state Attacker-Defender games with reachability objectives. The results of the paper are twofold. Firstly we prove a new language-theoretic result for weighted automata on infinite words and show its encoding into the framework of Attacker-Defender games. Secondly we use this novel concept to prove undecidability for checking existence of a winning strategy in several low-dimensional mathematical games including vector reachability games, word games and braid games.

1 Introduction

In the last decade there has been a steady, growing interest in the area of infinite-state games and computational complexity of the problem of checking the existence of a winning strategy [1,4,9,11,15,23,31]. Such games provide powerful mathematical framework for a large number of computational problems. In particular they appear in the verification, refinement, and compatibility checking of reactive systems [3], analysis of programs with recursion [11], combinatorial topology and have deep connections with automata theory and logic [23,29,31]. In many cases the most challenging problems appear in low-dimensional models or systems, where it is likely to have a few special cases with decidable problems and open general problem as the system may produce either too complex behaviour for analysis or a lack of "space" to code directly the universal computation for showing undecidability of the problem.

In this paper we present three variants of low-dimensional Attacker-Defender games (i.e. Word Games, Matrix Games and Braid Games) for which it is undecidable to determine whether one of the players has a winning strategy. In addition the proof incorporates new language theoretical result (Theorem 2) about weighted automata on infinite words that can be efficiently used in the context of other reachability games.

R. Niskanen—The author was partially supported by Nokia Foundation Grant.
I. Potapov—The author was partially supported by EPSRC grant "Reachability problems for words, matrices and maps" (EP/M00077X/1).

A. Beckmann et al. (Eds.): CiE 2015, LNCS 9136, pp. 206–215, 2015.
DOI: 10.1007/978-3-319-20028-6_21

The Attacker-Defender game is played in rounds, where in each round the move of Defender (Player 1) is followed by the move of Attacker (Player 2) starting from some initial position. Attacker tries to reach a target position while Defender tries to keep Attacker from reaching the target position. Then we say that Attacker has a winning strategy if it can eventually reach a target position regardless of Defender's moves. We show that in a number of restricted cases of such games it is not possible to decide about existence of the winning strategy for a given set of moves, initial and target positions.

We show that if both players are stateless but the moves correspond to a very restricted linear transformation from $SL(4, \mathbb{Z})$ the problem of existence of winning strategy is undecidable. One can show that using a direct translation from known undecidable reachability games (Robot Games [15]) leads to undecidability for linear transformations in dimension 18. To prove it we first generalize the concept by introducing the *Word Game*, where players are given words over a group alphabet and in alternative way concatenate their words with a goal for Attacker to reach the empty word. The games on words are common for proving results in language theory [21,22,24] over semigroup alphabets, while we formulate a game with much simpler reachability objective for games over a group alphabet.

Later we show that it is possible to stretch the application of the proposed techniques to other models and frameworks even looking at the games on topological objects, which were recently studied in [8,10]. Braids are classical topological objects that attracted a lot of attention due to their connections to topological knots and links as well as their applications to polymer chemistry, molecular biology, cryptography, quantum computations and robotics [12,13,16,17,26]. In this paper we consider games on braids with only 3 or 5 strands, where the braid is modified by composition of braids from a finite set with a target for Attacker to reach a trivial braid. We show that it is undecidable to check the existence of a winning strategy for 3 strands from a given nontrivial braid and for 5 strands starting from a trivial braid, while the reachability with a single player (i.e. with nondeterministic concatenation from a single set) was shown to be decidable for braids with 3 strands in [28].

The whole paper is also based on another important language-theoretic result showing that the universality problem for weighted automata $\mathcal{A}$ having merely five states accepting infinite words is undecidable. The acceptance of an infinite word w means that there exists a finite prefix p of w such that for a word p there is a path in $\mathcal{A}$ that has the zero weight. The problem whether all infinite words are accepted for a given $\mathcal{A}$ is undecidable and corresponds to the fact that there is no solution for infinite PCP.

The considered model of automaton is closely related to the *integer weighted finite automata* as defined in [18] and [2], where finite automata are accepting finite words and having additive integer weights on the transitions. In [18] it was shown that the universality problem is undecidable for integer weighted finite automata on finite words by reduction from Post Correspondence Problem. In the context of a game scenario it is important to have a property of acceptance in relation to infinite words with a finite prefix reaching a target value. Our proof of undecidability in this paper initially follows the idea from [18] for mapping computations on words into weighted (one counter) automata model.

All proofs omitted due to length constrain can be found in [20].

2 Notations and Definitions

An *infinite word* w over a finite alphabet A is an infinite sequence of letters $w = a_0a_1a_2a_3\cdots$ where $a_i \in A$ is a letter for each $i = 0, 1, 2, \ldots$. We denote the set of all infinite words over A by A^ω. The monoid of all finite words over A is denoted by A^*. A word $u \in A^*$ is a *prefix* of $v \in A^*$, denoted by $u \leq v$, if $v = uw$ for some $w \in A^*$. If u and w are both nonempty, then the prefix u is called *proper*, denoted by $u < v$. A *prefix* of an infinite word $w \in A^\omega$ is a finite word $p \in A^*$ such that $w = pw'$ where $w' \in A^\omega$. This is also denoted by $p \leq w$. The length of a finite word w is denoted by $|w|$. For a word w, we denote by $w(i)$ the ith letter of w, i.e., $w = w(1)w(2)\cdots$. Let $w = w(1)\cdots w(n)$, its *reversed word* is denoted by $w^R = w(n)\cdots w(1)$, i.e. the order of the letters is reversed.

Consider a finite integer weighted automaton $\mathcal{A} = (Q, A, \sigma, q_0, F, \mathbb{Z})$ with the set of states Q, the finite alphabet A, the set of transitions $\sigma \subseteq Q \times A \times Q \times \mathbb{Z}$, the initial state q_0, the set of final states $F \subseteq Q$ and the additive group of integers $\mathbb{Z}$ with identity 0, that is $(\mathbb{Z}, +, 0)$. We write the transitions in the form $t = \langle q, a, p, z\rangle \in \sigma$. An edge t we denote $q \xrightarrow{(a,z)} p$. Note that $\mathcal{A}$ is non-deterministic complete automaton in a sense that for each $q \in Q$ and $a \in A$ there is atleast one transition $\langle q, a, p, z\rangle \in \sigma$ for some $p \in Q$ and $z \in \mathbb{Z}$.

Let $\pi = t_{i_0}t_{i_1}\cdots$ be an infinite path of $\mathcal{A}$, where $t_{i_j} = \langle q_{i_j}, a_j, q_{i_{j+1}}, z_j\rangle$ for $j \geq 0$. Define the morphism $\|\cdot\|\colon \sigma^\omega \to A^\omega$ by setting $\|t\| = a$ if $t = \langle q, a, p, z\rangle$. Let $p = t_{i_0}t_{i_1}\cdots t_{i_n}$ for some n be a prefix of π. The *weight of the prefix* p is $\gamma(p) = z_0 + z_1 + \ldots + z_n \in \mathbb{Z}$. The prefix p *reaches* state $q \in Q$ if the last transition of p enters q, i.e., if $t_n = (q_{i_n}, a_n, q_{i_{n+1}}, z_n)$, then $q_{i_{n+1}} = q$. Denote by $R(p)$ the state reached by the finite path p.

An infinite word $w \in A^\omega$ is accepted by $\mathcal{A}$ if there exists an infinite path π such that at least one prefix p of π reaches a state in $R(p) \in F$ and has weight $\gamma(p) = 0$. The language *accepted by* $\mathcal{A}$ is

$$L(\mathcal{A}) = \{w \in A^\omega \mid \exists \pi \in \sigma^\omega\colon \|\pi\| = w \text{ and } \exists \text{ prefix } p \text{ of } \pi\colon \gamma(p) = 0 \text{ and } R(p) \in F\}.$$

We also define *reverse acceptance*, used in undecidability in Attacker-Defender games, in which instead of prefix p we consider p^R and its weight. Now an infinite word w is accepted by the automaton if and only if for corresponding computation π there exists a prefix whose reverse has zero weight. That is,

$$L_R(\mathcal{A}) = \{w \in A^\omega \mid \exists \pi \in \sigma^\omega\colon \|\pi\| = w \text{ and } \exists \text{ prefix } p \text{ of } \pi\colon \gamma(p^R) = 0 \text{ (and } R(p) \in F)\}.$$

A *configuration* of $\mathcal{A}$ is any triple $(q, w, z) \in Q \times A^* \times \mathbb{Z}$. A configuration (q, aw, z_1) is said to *yield* a configuration (p, w, z_1+z_2), denoted by $(q, aw, z_1) \models_{\mathcal{A}} (p, w, z_1 + z_2)$, if there is a transition $t = \langle q, a, p, z_2 \rangle \in \sigma$. Let $\models_{\mathcal{A}}^*$ or simply $\models^*$, if $\mathcal{A}$ is clear from the context, be the reflexive and transitive closure of the relation $\models_{\mathcal{A}}$.

The *Universality Problem* is a problem to decide whether the language accepted by weighted automaton $\mathcal{A}$ is the set of all infinite words. In other words, whether or not $L(\mathcal{A}) = A^\omega$. The problem of *non-universality* is the complement of universality problem, that is, whether or not $L(\mathcal{A}) \neq A^\omega$ or whether there exists $w \in A^\omega$ such that for every path π corresponding to computation of w and every prefix p of π, $\gamma(p) \neq 0$.

An *instance* of the *Post Correspondence Problem* (PCP, for short) consists of two morphisms $g, h : A^* \to B^*$, where A and B are alphabets. A nonempty word $w \in A^*$ is a solution of an instance (g, h) if it satisfies $g(w) = h(w)$. It is undecidable whether or not an instance of the PCP has a solution; see [27]. Also the problem is undecidable for domain alphabets A with $|A| \geq 7$; see [25]. The cardinality of the domain alphabet A is said to be the *size* of the instance.

The *Infinite Post Correspondence Problem*, ω PCP, is a natural extension of the PCP. An infinite word w is a *solution* of the instance (g, h) of the ω PCP if for every finite prefix p of w either $h(p) < g(p)$ or $g(p) < h(p)$. In the ω PCP it is asked whether or not a given instance has an infinite solution or not. Note that in our formulation prefixes have to be proper. It was proven in [19] that the problem is undecidable for domain alphabets A with $|A| \geq 9$ and in [14] it was improved to $|A| \geq 8$. In both proofs more general formulation of ω PCP was used, namely the prefixes did not have to be proper. It is easy to see that adding a new letter α to the alphabets and desynchronizing the morphisms h, g, gives us solution where prefix has to be proper. That is, we add α to the left of each letter in the image under h, to the right of each letter in the image under g and $g(\alpha) = \alpha, h(\alpha) = \varepsilon$. Now the solution has to start with α and images cannot be of equal length because the image under g ends with α but not under h. Note that in fact, both constructions already have this property, see [14,19].

3 Universality for Weighted Automata on A^ω

We prove that the universality problem is undecidable for integer weighted automata on infinite words by reducing the instances of the *infinite Post Correspondence Problem* to the universality problem.

Let (g, h) be a fixed instance of the ω PCP. Then $g, h \colon A^* \to B^*$ where $A = \{a_1, a_2, \ldots, a_{m-1}\}$ and $B = \{b_1, b_2, \ldots, b_{s-1}\}$. We construct an integer weighted automaton $\mathcal{A} = (Q, A, \sigma, q_0, \{q_4\}, \mathbb{Z})$, where $Q = \{q_0, q_1, q_2, q_3, q_4\}$, corresponding to the instance (g, h) such that an infinite word $w \in A^\omega$ is accepted by $\mathcal{A}$ if and only if for some finite prefix p of w, $g(p) \not< h(p)$ and $h(p) \not< g(p)$. Note that our automaton is complete, i.e. there is a transition labeled with (a, z) from each state q_i for every $a \in A$ and some $z \in \mathbb{Z}$.

The idea of encoding ω PCP and proof of undecidability for universality problem is based on computation in weighted automaton that can be partitioned into four parts A,B,C and D. Let us consider the case where the image under h is always longer than the image under g, the other cases are taken into account in the construction of the automaton. In part A, differences of lengths of images under h and g are stored for initial part of the input word. In part B, differences of position k of image of a letter under morphism h and length of image under g are stored together with a natural number j_k representing letter at kth position. In part C, the lengths of images under morphism g catch-up (by subtracting lengths of images under morphism g) to position specified after parts A and B. Finally, in part D, position ℓ in the image of the second morphism is subtracted together with a natural number i_ℓ from a set of natural numbers representing letters not at ℓth position under morphism g.

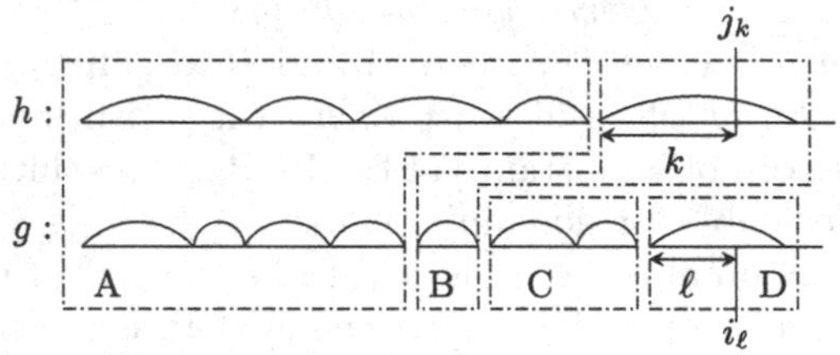

To store two different values (differences of lengths and a code for a stored symbol) at the same time, we can use a single counter since we only store one symbol from a finite alphabet. Let us assume that symbols are encoded as natural numbers in $\{1, \ldots, s-1\}$, where s is larger than the size of image alphabet B. The length n will be defined as $n \cdot s$ and we have enough space to store a single symbol from B by adding its code. If in the part D we refer to the same position the difference of the lengths of images should be 0 and if the letters are the same, the difference of letter codes is non-zero. This is done by allowing to subtract only that number which does not equal the number corresponding to a code of the letter at the ℓth position.

In the above consideration we considered the case where images under h was always longer than images under g. To make the construction work for all cases several computation paths are needed to be implemented in the automaton. The difference in lengths of images is positive when image under h is longer and negative when image under g is longer. For each case there are two possibilities for position of error. Either the difference is small enough that, after reading the next letter, there will be a position in images where letters differ (parts B and D have to be done simultaneously), or the difference is large enough, that image of the second morphism has to catch-up before error can be verified. Also from our formulation of ωPCP, it is possible that images are of equal length which means that the word is not a solution of ωPCP.

Lemma 1. *Let $w \in A^\omega$. Then w is a solution of an instance (g, h) of the* ωPCP *if and only if $w \notin L(\mathcal{A})$.*

Theorem 2. *It is undecidable whether or not $L(\mathcal{A}) = A^\omega$ holds for integer weighted automata $\mathcal{A}$ over its alphabet A.*

Proof. Claim follows from Lemma 1 and the undecidability of infinite PCP, see [30]. □

Corollary 3. *It is undecidable whether or not for weighted automaton $\mathcal{A}$, there exists a word $w \in A^\omega$ such that for its each computation path π and prefix $p \leq \pi$, $\gamma(p) \neq 0$ holds.*

For proving undecidability of finding a winning strategy in Attacker-Defender games we need to utilize slightly different acceptance condition.

Theorem 4. *It is undecidable whether or not $L_R(\mathcal{B}) = A^\omega$ holds for integer weighted automata $\mathcal{B}$ over its alphabet A.*

Proof (Sketch). The proof is based on Theorem 2. It is easy to see that if we reverse all edges and follow the computation path p of original automaton from end to finish, we have a computation path for p^R. The full proof is in [20].

4 Applications to Attacker-Defender Games

Let us consider a two-player Attacker-Defender game which is played in rounds and in each round a move of Defender is followed by a move of Attacker starting from some initial position. Attacker tries to reach a target position while Defender tries to keep Attacker from reaching the target position. Attacker has a winning strategy if it can eventually reach a target position regardless of Defender's moves. The main computational question is to check whether Attacker has a winning strategy for a given set of moves, initial and target positions.

Following the result for the weighted automata on infinite words (Theorem 4) we can now define a simple scenario of undecidable infinite-state game that can be also applied to other game frameworks. Assume that Defender will provide any input letters from a finite alphabet, one by one, to Attacker and Attacker appends dummy symbol # until he chooses to follow computation path of automaton $\mathcal{B}$ of Theorem 4. Attacker has to decide whether provided word (ignoring symbols #) when played according to available transitions is accepted by $\mathcal{B}$.

In the above framework Defender will have a winning strategy if there is a solution for infinite PCP and Attacker will have a winning strategy otherwise.

4.1 Weighted Word Game

Let us define the Attacker-Defender game on words, where the moves of Attacker and Defender correspond to concatenations of words (over free group alphabet) and follows a computation path of weighted automaton. This simplification allows us to use *Word Game* to prove nontrivial results for games with *low-dimensional* linear transformations and topological objects just by using injective homomorphism (i.e. monomorphism) to map words into other mathematical objects.

A *weighted Word Game* consists of two players, *Attacker* and *Defender* having sets of words $\{u_1, \ldots, u_r\} \subseteq \Gamma^*$ and $\{v_1, \ldots, v_s\} \subseteq \Gamma^*$ respectively, where Γ is finite alphabet from a free group, and integers $x_{u_1}, \ldots, x_{u_r}, x_{v_1}, \ldots, x_{v_s}$ corresponding to each word. In each round Defender chooses the word before Attacker, the initial position is the pair $(w, 0)$, where $w \in \Gamma^*$ and 0 is initial value of the counter, and target position of this game is the group identity, i.e. the empty word, with zero weight. The *configuration* of a game at time t is denoted by a word w_t and integer x as a counter. In each round of the game both Defender and Attacker concatenate their words (append from the right) and update the counter value. Clearly $w_t = w \cdot v_{i_1} \cdot u_{i_1} \cdot v_{i_2} \cdot u_{i_2} \cdot \ldots \cdot v_{i_t} \cdot u_{i_t}$ after t rounds of the game, where u_{i_j} and v_{i_j} are words from defined above sets of Attacker and Defender, and the counter value is $\sum_{j=1}^{t}(x_{v_{i_j}} + x_{u_{i_j}})$. The decision problem for the word game is to check whether there exists a winning strategy for Attacker to reach an empty word with zero weight.

Lemma 5. *Let $\Sigma' = \{z_1, z_2, \ldots, z_l\}$ be a group alphabet and $\Sigma_2 = \{c, d, \overline{c}, \overline{d}\}$ be a binary group alphabet. Define the mapping $\alpha : \Sigma' \to \Sigma_2^*$ by: $\alpha(z_i) = c^i d \overline{c}^i, \alpha(\overline{z_i}) = c^i \overline{d} \overline{c}^i$, where $1 \leq i \leq l$. Then α is a monomorphism. Note that α can be extended to domain Σ'^* in the usual way [6, 7].*

Theorem 6. *It is undecidable whether Attacker has a winning strategy in the weighted Word Game with words over a binary free group alphabet.*

The idea of proof is that Defender plays words of $\{v_1, \ldots, v_s\}$ corresponding to $w \in A^\omega$ letter by letter. Attacker either plays # or starts following computation path of the automaton $\mathcal{B}$ and stores weights in a counter. Word $q_0\overline{q_4}$, where q_0 and q_4 are initial and final states of $\mathcal{B}$ respectively, with weight 0 is reached if and only if w is accepted by the automaton. Then we encode the group alphabet using Lemma 5 to have binary group alphabet. The full proof can be found in [20].

4.2 Word Games on Pairs of Group Words

We now modify the game of previous section by encoding counter as a separate word over unary group alphabet $\Gamma' = \{\rho, \overline{\rho}\}$.

Now *Word Game* consists of Attacker and Defender having sets of words $\{(u_1, u_1'), \ldots, (u_r, u_r')\} \subseteq \Gamma^* \times \Gamma'^*$ and $\{(v_1, \varepsilon), \ldots, (v_s, \varepsilon)\} \subseteq \Gamma^* \times \Gamma'^*$ respectively, where Γ is binary group alphabet. Now the configuration of a game after t rounds is a word $w_t = (v_{i_1}, \varepsilon) \cdot (u_{i_1}, u'_{i_1}) \cdot (v_{i_2}, \varepsilon) \cdot (u_{i_2}, u'_{i_2}) \cdot \ldots \cdot (v_{i_t}, \varepsilon) \cdot (u_{i_t}, u'_{i_t})$, where (u_{i_j}, u'_{i_j}) and (v_{i_j}, ε) are words from defined above sets of Attacker and Defender. The initial position is the word (w, ε) and target position of this game is an empty word $(\varepsilon, \varepsilon)$. The decision problem for the word game is to check whether there exists a winning strategy for Attacker to reach an empty word $(\varepsilon, \varepsilon)$.

Theorem 7. *It is undecidable whether Attacker has a winning strategy in the Word Game with one component of words over a binary free group alphabet and the other over unary group alphabet.*

Proof (Sketch). The proof is based on Theorem 6 with encoding of counter x as a word ρ^x over unary group alphabet $\{\rho, \overline{\rho}\}$.

We follow idea of [5] to construct a word game where the initial word is $(\varepsilon, \varepsilon)$. For this we do 4 consequent games over 4 disjoint group alphabets. The games are constructed in such way that $(\varepsilon, \varepsilon)$ is reached if and only if $(\varepsilon, \varepsilon)$ is reached in every game. If none of the games result in $(\varepsilon, \varepsilon)$, then there are at least 4 words, from distinct group alphabets, which are non-canceled. Now if the computation is completed twice (i.e. 8 games have been played in total), the number of non-canceled elements cannot decrease.

4.3 Matrix Games and Braid Games

We extend the domain of the game and a set of rules to the class of linear transformations on integer lattice $\mathbb{Z}^4$ and to the domain of braids considering moves of the game as a concatenation of braids in B_3 (a class of braids with only three strands) and B_5 (a class of braids with only five strands) [28].

A *Matrix Game* consists of two players, *Attacker* and *Defender* having sets of linear transformations $MU_1, \ldots, MU_r \subseteq \mathbb{Z}^{n \times n}$ and $MV_1, \ldots, MV_s \subseteq \mathbb{Z}^{n \times n}$ respectively and an *initial vector* $\mathbf{x}_0 \in \mathbb{Z}^n$ of the game representing a starting position. The *dimension* of the game is clearly the dimension of the integer lattice n. Starting from $\mathbf{x}_0$, players move the current point by applying available linear transformations (by matrix multiplication) from their respective sets in turns. The decision problem of the *Matrix Game* is to check whether there exist a winning strategy for Attacker to return to the starting point (vector in $\mathbb{Z}^n$) of the game.

Theorem 8. *Given two finite sets matrices $MU_1, MU_2, \ldots, MU_r \subseteq \mathbb{Z}^{n \times n}$ and $MV_1, MV_2, \ldots, MV_s \subseteq \mathbb{Z}^{n \times n}$ for* Attacker *and* Defender *players respectively and initial starting vector $\mathbf{x}_0 \in \mathbb{Z}^n$. It is undecidable whether Attacker has a winning strategy in the Matrix Game of dimension four, i.e. when $n = 4$.*

Proof (Sketch). We encode word game on pairs of words into matrices from $SL(4, \mathbb{Z}) = \{M \in \mathbb{Z}^{4 \times 4} \mid det(M) = \pm 1\}$. Identity matrix is reachable if and only if empty word is reachable in Word Game.

Now we translate the Attacker-Defender games into games on topological objects - braids in B_n. We consider very simple games on braids with only 3 or 5 strands (i.e. B_3 or B_5) where the braid is modified by composition with a finite set of braids. We show that it is undecidable to check the existence of a winning strategy in such game, while the reachability with a single player (i.e. with nondeterministic concatenation from a single set) was shown to be decidable for B_3 and undecidable for B_5 in [28].

Definition 9. *The n-strand braid group B_n is the group given by the presentation with $n-1$ generators $\sigma_1, \ldots, \sigma_{n-1}$ and the following relations $\sigma_i \sigma_j = \sigma_j \sigma_i$, for $|i - j| \geq 2$ and $\sigma_i \sigma_{i+1} \sigma_i = \sigma_{i+1} \sigma_i \sigma_{i+1}$ for $1 \leq i \leq n-2$. These relations are called Artin's relations. Words in the alphabet $\{\sigma, \sigma^{-1}\}$ will be referred to as braid words.*

The Braid Game can be defined in the following way. Given a set of words for Attacker $\{a_1, ..., a_r\} \subseteq B_n$ and Defender $\{d_1, ..., d_s\} \subseteq B_n$ will correspond to braids (or braid words in B_n). The game is starting with a given initial braid c and each following configuration of the game is changed by Attacker or Defender by concatenating braids from their corresponding sets. The concatenation (composition) of two braids is defined by putting one after the other making the endpoints of the first one coincide with the starting points of the second one. There is a neutral element for the composition: it is the trivial braid, also called identity braid, i.e. the class of the geometric braid where all the strings are straight.

Finally, the goal of Attacker is to unbraid, i.e. to reach a configuration of the game that is isotopic to the trivial braid (empty word) and Defender tries to keep Attacker from reaching it. Two braids are isotopic if their braid words can be translated one into each other via the relations from the Definition 9 plus the relations $\sigma_i\sigma_i^{-1} = \sigma_i^{-1}\sigma_i = 1$, where 1 is the identity (trivial braid).

Theorem 10. *The Braid Game is undecidable for braids from B_3 starting from non-trivial braid and for braids from B_5 starting from a trivial braid.*

The idea is to encode words of weighted word game into braids of B_3 and weight into central element of B_3. While B_5 contains direct product of two free subgroup and can encode pair of words of word game into braids of B_5. The full proof can be found in [20].

References

1. Abdulla, P.A., Bouajjani, A., d'Orso, J.: Deciding monotonic games. In: Baaz, M., Makowsky, J.A. (eds.) CSL 2003. LNCS, vol. 2803, pp. 1–14. Springer, Heidelberg (2003)
2. Almagor, S., Boker, U., Kupferman, O.: What's decidable about weighted automata? In: Bultan, T., Hsiung, P.-A. (eds.) ATVA 2011. LNCS, vol. 6996, pp. 482–491. Springer, Heidelberg (2011)
3. Alur, R., Henzinger, T.A., Kupferman, O.: Alternating-time temporal logic. J. ACM **49**(5), 672–713 (2002)
4. Arul, A., Reichert, J.: The complexity of robot games on the integer line. In: Proceedings of QAPL 2013, EPTCS, vol. 117, pp. 132–148 (2013)
5. Bell, P.C., Potapov, I.: On the undecidability of the identity correspondence problem and its applications for word and matrix semigroups. Int. J. Found. Comput. Sci. **21**(6), 963–978 (2010)
6. Bell, P.C., Potapov, I.: On the computational complexity of matrix semigroup problems. Fundam. Inform. **116**(1–4), 1–13 (2012)
7. Birget, J.C., Margolis, S.W.: Two-letter group codes that preserve aperiodicity of inverse finite automata. Semigroup Forum. **76**, 159–168 (2008). Springer
8. Bovykin, A., Carlucci, L.: Long games on braids (2006). Preprint. Available online at http://logic.pdmi.ras.ru/~andrey/braids_final3.pdf
9. Brázdil, T., Jančar, P., Kučera, A.: Reachability games on extended vector addition systems with states. In: Abramsky, S., Gavoille, C., Kirchner, C., Meyer auf der Heide, F., Spirakis, P.G. (eds.) ICALP 2010. LNCS, vol. 6199, pp. 478–489. Springer, Heidelberg (2010)

10. Carlucci, L., Dehornoy, P., Weiermann, A.: Unprovability results involving braids. Proc. Lond. Math. Soc. **102**(1), 159–192 (2011)
11. Chatterjee, K., Fijalkow, N.: Infinite-state games with finitary conditions. In: Proceedings of CSL 2013, LIPIcs, vol. 23, pp. 181–196 (2013)
12. Collins, G.P.: Computing with quantum knots. Sci. Am. **294**(4), 56–63 (2006)
13. Dehornoy, P., Dynnikov, I., Rolfsen, D., Wiest, B.: Ordering Braids. Mathematical Surveys and Monographs, vol. 148. American Mathematical Society, Providence (2008)
14. Dong, J., Liu, Q.: Undecidability of infinite post correspondence problem for instances of size 8. RAIRO - Theor. Inf. Appl. **46**(3), 451–457 (2012)
15. Doyen, L., Rabinovich, A.: Robot games. Technical report LSV-13-02, LSV, ENS Cachan (2013)
16. Epstein, D., Paterson, M., Cannon, J., Holt, D., Levy, S., Thurston, W.P.: Word Processing in Groups. AK Peters, Ltd, USA (1992)
17. Garber, D.: Braid group cryptography. In: Braids: Introductory Lectures on Braids, Configurations and Their Applications, vol. 19, pp. 329 (2010)
18. Halava, V., Harju, T.: Undecidability in integer weighted finite automata. Fundam. Inform. **38**(1–2), 189–200 (1999)
19. Halava, V., Harju, T.: Undecidability of infinite post correspondence problem for instances of size 9. ITA **40**(4), 551–557 (2006)
20. Halava, V., Harju, T., Niskanen, R., Potapov, I.: Weighted automata on infinite words in the context of attacker-defender games (2015). CoRR. abs/1411.4796
21. Kunc, M.: Regular solutions of language inequalities and well quasi-orders. Theor. Comput. Sci. **348**(2–3), 277–293 (2005)
22. Kunc, M.: The power of commuting with finite sets of words. Theory Comput. Syst. **40**(4), 521–551 (2007)
23. Kupferman, O., Vardi, M.Y., Wolper, P.: An automata-theoretic approach to branching-time model checking. J. ACM **47**(2), 312–360 (2000)
24. Ly, O., Wu, Z.: On effective construction of the greatest solution of language inequality XA ⊆ BX. Theor. Comput. Sci. **528**, 12–31 (2014)
25. Matiyasevich, Y., Sénizergues, G.: Decision problems for semi-thue systems with a few rules. Theor. Comput. Sci. **330**(1), 145–169 (2005)
26. Panangaden, P., Paquette, É.O.: A categorical presentation of quantum computation with anyons. In: Coecke, B. (ed.) New Structures for Physics. LNP, pp. 983–1025. Springer, Heidelberg (2011)
27. Post, E.L.: A variant of a recursively unsolvable problem. Bull. Am. Math. Soc. **52**(4), 264–268 (1946)
28. Potapov, I.: Composition problems for braids. In: Proceedings of FSTTCS 2003, LIPIcs, vol. 24, pp. 175–187 (2003)
29. Rabin, M.O.: Decidability of second-order theories and automata on infinite trees. Bull. Am. Math. Soc. **74**(5), 1025–1029 (1968)
30. Ruohonen, K.: Reversible machines and post's correspondence problem for biprefix morphisms. J. of Information Processing and Cybernetics **21**(12), 579–595 (1985)
31. Walukiewicz, I.: Pushdown processes: games and model-checking. Inf. Comput. **164**(2), 234–263 (2001)

Turing Jumps Through Provability

Joost J. Joosten(✉)

University of Barcelona, Barcelona, Spain
jjoosten@ub.edu

Abstract. Fixing some computably enumerable theory T, the Friedman-Goldfarb-Harrington (FGH) theorem says that over elementary arithmetic, each Σ_1 formula is equivalent to some formula of the form $\Box_T \varphi$ provided that T is consistent. In this paper we give various generalizations of the FGH theorem. In particular, for $n > 1$ we relate Σ_n formulas to provability statements $[n]_T^{\mathsf{True}}\varphi$ which are a formalization of "provable in T together with all true Σ_{n+1} sentences". As a corollary we conclude that each $[n]_T^{\mathsf{True}}$ is Σ_{n+1}-complete. This observation yields us to consider a recursively defined hierarchy of provability predicates $[n+1]_T^{\Box}$ which look a lot like $[n+1]_T^{\mathsf{True}}$ except that where $[n+1]_T^{\mathsf{True}}$ calls upon the oracle of all true Σ_{n+2} sentences, the $[n+1]_T^{\Box}$ recursively calls upon the oracle of all true sentences of the form $\langle n\rangle_T^{\Box}\phi$. As such we obtain a 'syntax-light' characterization of Σ_{n+1} definability whence of Turing jumps which is readily extended beyond the finite. Moreover, we observe that the corresponding provability predicates $[n+1]_T^{\Box}$ are well behaved in that together they provide a sound interpretation of the polymodal provability logic GLP_ω.

1 Introduction and Preliminaries

In first order arithmetic we have natural syntactical definitions that correspond to finite iterations of the Turing jump. For example, a set of natural numbers is computably enumerable relative to the n-th Turing jump of the empty set if and only if it can be defined by a Σ_{n+1} formula.

In this paper we shall use the fact that various provability predicates are Turing complete in a certain sense so that we can give alternative characterizations for the finite Turing jumps.

We shall work with theories with identity in the language $\{0, 1, \exp, +, \cdot, <\}$ of arithmetic where exp denotes the unary function $x \mapsto 2^x$. We define $\Delta_0 = \Sigma_0 = \Pi_0$ formulas as those where all quantifiers occur bounded, that is, we only allow quantifiers of the form $\forall\, x{<}t$ or $\exists\, x{<}t$ where t is some term not containing x. We inductively define $\Sigma_n, \Pi_n \subset \Pi_{n+1}$ and $\Sigma_n, \Pi_n \subset \Sigma_{n+1}$; if $\phi, \psi \in \Pi_{n+1}$, then $\forall x\ \phi, \phi \wedge \psi, \phi \vee \psi \in \Pi_{n+1}$ and likewise, if $\phi, \psi \in \Sigma_{n+1}$, then $\exists x\ \phi, \phi \wedge \psi, \phi \vee \psi \in \Sigma_{n+1}$.

We shall write $\Sigma_{n+1}!$ for formulas φ of the form $\exists x\ \varphi_0$ with $\varphi_0 \in \Pi_n$. We will work in the absence of strong versions of (bounded) collection B_φ which is defined as $\mathsf{B}_\varphi := \forall z\, \forall \boldsymbol{u}\, \big(\forall\, x{<}z\, \exists y\ \varphi(x, y, \boldsymbol{u}) \to \exists y'\, \forall\, x{<}z\, \exists\, y{<}y'\ \varphi(x, y, \boldsymbol{u})\big)$. Therefore,

A. Beckmann et al. (Eds.): CiE 2015, LNCS 9136, pp. 216–225, 2015.
DOI: 10.1007/978-3-319-20028-6_22

we will consider the formula class $\Sigma_{n+1,1}$ consisting of existentially quantified disjunctions and conjunctions of Σ_{n+1} formulas with bounded quantifiers over them. To be more precise, we first inductively define

$$\Sigma_{n+1,b} := \Sigma_{n+1} \mid (\Sigma_{n+1,b} \circ \Sigma_{n+1,b}) \mid (\mathcal{Q}x{<}y\ \Sigma_{n+1,b})$$

with $\circ \in \{\wedge, \vee\}$ and $\mathcal{Q} \in \{\forall, \exists\}$. Next we define the $\Sigma_{n+1,1}$ formulas to be of the form $\exists x\ \phi$ with $\phi \in \Sigma_{n+1,b}$.

The theory of *elementary arithmetic*, EA, is axiomatized by the defining axioms for $\{0, 1, \exp, +, \cdot, <\}$ together with induction for all Δ_0 formulas. The theory *Peano Arithmetic*, PA, is as EA but now allowing induction axioms for any first order formula. It is well known that PA proves any instance B_φ of collection so that in particular each $\Sigma_{n+1,1}$ sentence is equivalent to some Σ_{n+1} sentence. Clearly we have that $\Sigma_{n+1}! \subset \Sigma_{n+1}$. Using coding techniques, it is clear that each Σ_{n+1} formula is within EA equivalent to a $\Sigma_{n+1}!$ formula.

For us, a computably enumerable (c.e.) theory T is understood to be given by a Δ_0 formula that defines the set of codes of the elementary set of axioms of T. We will employ standard formalizations of meta-mathematical properties like $\mathtt{Proof}_T(x, y)$ for "x is the Gödel number of a proof from the axioms of T of the formula whose Gödel number is y". We shall often refrain from distinguishing a syntactical object φ from its Gödel number $\ulcorner \varphi \urcorner$ or from a syntactical representation of its Gödel number.

We will write $\Box_T \varphi$ for the $\Sigma_1!$ formula $\exists x\ \mathtt{Proof}_T(x, \varphi)$ and $\Diamond_T \varphi$ for $\neg\Box_T \neg\varphi$. By $\Box_T \varphi(\dot{x})$ we will denote a formula which contains the free variable x, that expresses that for each value of x the formula $\varphi(\overline{x})$ is provable in T. Here, $\overline{x}$ denotes a syntactical representation of the number x.

By Σ_1 completeness we refer to the fact that for any true Σ_1 sentence σ we have that $\text{EA} \vdash \sigma$. It is well-known that EA proves a formalized version of this: for any Σ_1 formula $\sigma(x)$ and any c.e. theory T we have $\text{EA} \vdash \sigma(x) \to \Box_T \sigma(\dot{x})$.

2 Witness-Comparisons: Rosser and the FGH Theorem

In this section we shall be dealing with various so-called *witness-comparison arguments* where the order of (least) witnesses to existential sentences is important. The first and most emblematic such argument occurred in the proof of Rosser's theorem which is a strengthening of Gödel's first incompleteness theorem.

Theorem 1 (Rosser's Theorem). *Let T be a consistent c.e. theory extending* EA. *There is some $\rho \in \Sigma_1$ so that $T \nvdash \rho$ and $T \nvdash \neg\rho$.*

For rhetoric reasons we shall below include a standard proof of this celebrated result. Before doing so, we first need some notation.

Definition 1. *For $\phi := \exists x\ \phi_0(x)$ and $\psi := \exists x\ \psi_0(x)$ we define*

$$\phi \leq \psi := \exists x\ \big(\phi_0(x) \wedge \forall y{<}x\ \neg\psi_0(y)\big) \quad \textit{and,}$$
$$\phi < \psi := \exists x\ \big(\phi_0(x) \wedge \forall y{\leq}x\ \neg\psi_0(x)\big).$$

Statements of the form $\phi \leq \psi$ or $\phi < \psi$ with $\phi, \psi \in \Sigma_{n+1}$ are called witness-comparison statements. Let us now collect some easy principles about witness-comparison statements whose elementary proofs we leave as an exercise.

Lemma 1. *For A and B in Σ_{n+1} we have*

1. $\mathrm{EA} \vdash (A < B) \rightarrow (A \leq B)$;
2. $\mathrm{EA} \vdash (A < B) \wedge (B \leq C) \rightarrow (A < C)$;
3. $\mathrm{EA} \vdash (A \leq B) \wedge (B < C) \rightarrow (A < C)$;
4. $\mathrm{EA} \vdash (A \leq B) \wedge (B \leq C) \rightarrow (A \leq C)$;
5. $\mathrm{EA} \vdash (A \leq B) \rightarrow \neg(B < A)$ *and consequently;*
6. $\mathrm{EA} \vdash (A < B) \rightarrow \neg(B \leq A)$;
7. $\mathrm{EA} \vdash [(B \leq B) \vee (A \leq A)] \ \rightarrow \ [(A \leq B) \vee (B < A)]$;
8. $\mathrm{EA} \vdash (A \leq B) \rightarrow A$;
9. $\mathrm{EA} \vdash A \wedge \neg B \rightarrow (A < B)$;
10. $\mathrm{EA} \vdash A \wedge \neg(A \leq B) \rightarrow B$.
11. *Both $C < D$ and $C \leq D$ are of complexity $\Sigma_{n+1,1}$ if $C, D \in \Sigma_{n+1,1}$.*

We can now present a concise proof of Rosser's theorem.

Proof. We consider a fixed point ρ so that $T \vdash \rho \leftrightarrow (\Box_T \neg\rho \leq \Box_T \rho)$.

If $T \vdash \rho$, then for some number n we have $\mathtt{Proof}_T(n, \rho)$. Since T is consistent we also have $\forall\, m{\leq}n \neg \mathtt{Proof}_T(m, \neg\rho)$. Thus, by Σ_1 completeness we have $T \vdash \Box_T \rho < \Box_T \neg\rho$ whence $T \vdash \neg\rho$; a contradiction.

Likewise, if $T \vdash \neg\rho$ we may conclude $\Box_T \neg\rho \leq \Box_T \rho$ so that $T \vdash \rho$.

We now turn our attention to another theorem, a proof of which can succinctly be given using witness comparison arguments: the *FGH theorem.* The initials FGH refer to *Friedman*, *Goldfarb* and *Harrington* who all substantially contributed to the theorem and we refer to [5] for historical details.

The proof we give here is a slight modification of the one presented in [5]. The most important improvement is that we avoid the use of the least-number principle so that the proof becomes amenable for generalizations without a need to increase the strength of the base theory.

Theorem 2 (FGH theorem). *Let T be any computably enumerable theory extending* EA. *For each $\sigma \in \Sigma_1$ we have that there is some $\rho \in \Sigma_1$ so that*

$$\mathrm{EA} \vdash \Diamond_T \top \rightarrow \ \big(\sigma \leftrightarrow \Box_T \rho\big).$$

Proof. As in [5] we consider the fixed point $\rho \in \Sigma_1$ for which $\mathrm{EA} \vdash \rho \leftrightarrow (\sigma \leq \Box_T \rho)$. Without loss of generality we may assume that $\sigma \in \Sigma_1!$ so that both $\sigma \leq \Box_T \rho$ and $\Box_T \rho < \sigma$ are Σ_1. We now reason in EA, assume $\Diamond_T \top$ and set out to prove $\sigma \leftrightarrow \Box_T \rho$.

($\rightarrow$): assume for a contradiction that σ and $\neg\Box_T \rho$. By Lemma 1.9 we conclude $\sigma \leq \Box_T\, \rho$, i.e., ρ. By provable Σ_1 completeness we get $\Box_T \rho$.

($\leftarrow$): assume for a contradiction that $\neg\sigma$ and $\Box_T \rho$. Again, we conclude $\Box_T \rho < \sigma$ so that $\Box_T(\Box_T \rho < \sigma)$ whence $\Box_T \neg\rho$ so that $\Box_T \bot$ contradicting the assumption $\Diamond_T \top$.

A very special feature of ρ from the above proof is that it is of complexity Σ_1 and that $\neg\rho$ is implied a by a related Σ_1 formula. Thus, we clearly provably have $\rho \rightarrow \Box_T\rho$ but in general we do not have $\neg\rho \rightarrow \Box_T\neg\rho$. But, due to the nature of ρ we have that $\Box_T\neg\rho$ follows from the Σ_1 statement that is slightly stronger than $\neg\rho$, namely $\Box_T\rho < \sigma$. The least number principle for Σ_n formulas, $\mathsf{L}\Sigma_n$, says that for any $\psi \in \Sigma_n$ we have $\exists x\,\psi(x) \rightarrow \exists x\,(\psi(x) \wedge \forall\, y{<}x\, \neg\psi(y))$. Of course, using $\mathsf{L}\Sigma_0$ and by Lemma 1.7, $\neg\rho$ and $\Box\rho < \sigma$ are provably equivalent under the assumption that $\Box\rho \vee \sigma$.

We will be interested in generalizing the FGH theorem to Σ_n formulas using ever stronger notions of provability. Visser's proof of the FGH theorem as presented in [5] used an application of the least number principle for Δ_0 formulas in the guise of $A \rightarrow (A \leq A)$. Thus, a direct generalization of Visser's argument to stronger provability notions would call for stronger and stronger arithmetical principles:

Lemma 2. *The schema $A \rightarrow (A \leq A)$ for $A \in \Sigma_{n+1}$! is over* EA *provably equivalent to the least-number principle for Π_n formulas.*

However, since our proof of the FGH theorem did not use the minimal number principle, we shall see in the next section how the above argument generalizes to other provability predicates. We end this section with some remarks on the fixed point used in the proof of the FGH theorem. By no means, this fixed point is the only one that works. The minor change where we consider $\rho \leftrightarrow (\sigma < \Box\rho)$ works with almost the same proof. But we also have two 'dual' versions of our fixed point which gives the desired result.

Lemma 3. *Let T be a c.e. theory and let $\sigma \in \Sigma_1$.*

1. *If* $\mathrm{EA} \vdash \rho \leftrightarrow (\Box_T\neg\rho \leq \sigma)$ *then* $\mathrm{EA} \vdash \sigma \vee \Box_T\bot \leftrightarrow \Box_T\neg\rho$.
2. *If* $\mathrm{EA} \vdash \rho \leftrightarrow \neg(\Box_T\rho < \sigma)$ *then* $\mathrm{EA} \vdash \sigma \vee \Box_T\bot \leftrightarrow \Box_T\rho$.

Proof. (1) We reason in EA assuming $\rho \leftrightarrow (\Box_T\neg\rho \leq \sigma)$. $\rightarrow$: We suppose σ. If ρ, then $\Box_T\neg\rho \leq \sigma$ whence $\Box_T\neg\rho$. If $\neg\rho$, then $\neg(\Box_T\neg\rho \leq \sigma)$. But from our assumption σ we get by the minimal number principle that $\sigma \leq \sigma$ so that $(\Box_T\neg\rho \leq \sigma) \vee (\sigma < \Box_T\neg\rho)$ whence we get $\sigma < \Box_T\neg\rho$. By provable Σ_1-completeness we get $\Box_T(\sigma < \Box_T\neg\rho)$ whence $\Box_T\neg\rho$.

$\leftarrow$: If $\Box_T\neg\rho$ and $\neg\sigma$, then $\Box_T\neg\rho \leq \sigma$ whence ρ and by Σ_1 completeness $\Box_T\rho$ so that $\Box_T\bot$.

(2) Again, we reason in EA now avoiding the minimal number principle. $\rightarrow$: Suppose σ. If $\sigma \leq \Box_T\rho$, then $\Box_T(\sigma \leq \Box_T\rho)$ whence $\Box_T\neg(\Box_T\rho < \sigma)$ so that $\Box\rho$. If $\neg(\sigma \leq \Box_T\rho)$, then since σ we have $\Box_T\rho$.

$\leftarrow$: Suppose $\Box_T\rho$ and $\neg\sigma$. Then, $\Box_T\rho < \sigma$ whence also $\Box_T(\Box_T\rho < \sigma)$. Thus $\Box_T\neg\rho$, which together with the assumption that $\Box_T\rho$ yields $\Box_T\bot$.

3 Generalizations of the FGH Theorem and Applications

By $[n+1]_T^{\mathsf{True}}$ we will denote the formalization of the predicate "provable in T together with all true Σ_{n+2} sentences". For convenience, we set $[0]_T := \Box_T$.

Basically, for $n > 0$, the predicate $[n]_T^{\mathsf{True}}\varphi$ will be a formalization of "there is a sequence $\pi_0, \ldots, \pi_m$ so that each π_i is either an axiom of T, or a true Σ_{n+1} sentence, or a propositional logical tautology, or a consequence of some rule of T using earlier elements in the sequence as antecedents". Thus, it is clear that for recursive theories T we can write $[n]_T^{\mathsf{True}}$ by a $\Sigma_{n+1,1}$-formula. Also, we have provable $\Sigma_{n+1,1}$ completeness for these predicates, that is:

Lemma 4. *Let T be a c.e. theory extending* EA *and let ϕ be a $\Sigma_{n+1,1}$ formula. We have that*

$$\mathrm{EA} \vdash \phi(x) \to [n]_T^{\mathsf{True}}\phi(\dot{x}).$$

Proof. Given $\phi(z, y_1, \ldots, y_k, x) \in \Sigma_{n+1}$, reason in EA, fix $x_1, \ldots, x_k, x$ and, assume

$$\exists z\, \mathcal{Q}_1\, y_1{<}x_1 \ldots \mathcal{Q}_k\, y_k{<}x_k\ \phi(z, y_1, \ldots, y_k, x)$$

where $\mathcal{Q}_1\, y_1{<}x_1 \ldots \mathcal{Q}_k\, y_k{<}x_k$ is some block of bounded quantifiers. Thus, for some a we have $\mathcal{Q}_1\, y_1{<}x_1 \ldots \mathcal{Q}_k\, y_k{<}x_k\ \phi(a, y_1, \ldots, y_k, x)$. Under the box, we can now replace each $\forall\, y_i{<}\overline{x_i}$ by $\bigwedge_{\overline{y_i}<\overline{x_i}}$ and each $\exists\, y_i{<}\overline{x_i}$ by $\bigvee_{\overline{y_i}<\overline{x_i}}$ so that by applying distributivity we see that $\mathcal{Q}_1\, y_1{<}\overline{x_1} \ldots \mathcal{Q}_k\, y_k{<}\overline{x_k}\ \phi(\overline{a}, y_1, \ldots, y_k, \overline{x})$ is equivalent to a disjunctive normal form of bounded substitution instances of $\phi(\overline{a}, y_1, \ldots, y_k, \overline{x})$. Note that this operation in available within EA since it only requires the totality of exponentiation.

Outside the box we know that for some of these big conjunctions of bounded substitution instances of $\phi(a, y_1, \ldots, y_k, x)$ actually all of the conjuncts are true. Since each of those conjuncts is a true Σ_{n+1} sentence, each conjunct is an axiom whence holds under the box. Thus, the whole conjunct is provable under the box whereby we obtain $[n]_T^{\mathsf{True}}\mathcal{Q}_1\, y_1{<}\overline{x_1} \ldots \mathcal{Q}_k\, y_k{<}\overline{x_k}\ \phi(\overline{a}, y_1, \ldots, y_k, \overline{x})$ whence $[n]_T^{\mathsf{True}}\exists z\, \mathcal{Q}_1\, y_1{<}\overline{x_1} \ldots \mathcal{Q}_k\, y_k{<}\overline{x_k}\ \phi(z, y_1, \ldots, y_k, \overline{x})$ as was to be shown. It is clear that this case suffices for the more general form of $\Sigma_{n+1,1}$ formulas.

It is easy to check that the predicate $[n]_T^{\mathsf{True}}$ is well behaved. In particular one can check that all the axioms of the standard provability logic **GL** as defined in the last section hold for it. Over EA, the notion of $[n]_T^{\mathsf{True}}$ can be related to regular provability $\Box_T$ by collecting all oracle axioms in a big conjunction.

By $\texttt{FinSeq}(f)$ we denote a predicate that only holds on numbers that are codes of a finite sequence of Gödel numbers and by $|f|$ we denote the length of such a sequence. Moreover, f_i will denote the ith element of such a sequence f. With $\texttt{True}_{\Sigma_{n+1}}$ we will denote a partial truth predicate for Σ_{n+1} formulas and $\texttt{Tr}_{\Sigma_{n+1}}$ will denote the set of true Σ_{n+1} sentences.

Lemma 5. *For any c.e. theory T, we have that*

$$\mathrm{EA} \vdash [n]_T^{\mathsf{True}}\varphi \leftrightarrow \exists f\, \Big(\texttt{FinSeq}(f) \wedge \forall i{<}|f|\ \texttt{True}_{\Sigma_{n+1}}(f_i) \wedge \Box_T\big((\wedge_{i<|f|} f_i) \to \varphi\big)\Big).$$

Proof. The $\leftarrow$ direction is trivial. For the other direction we observe that we can express that an element of a proof is a true Σ_{n+1} sentence in a Δ_0 fashion so that by Δ_0 induction on the length of a proof we can prove that there is a sequence containing exactly all the true Σ_{n+1} sentences used in the proof.

Note that our definition of $[n]_T^{\mathsf{True}}$ is slightly non-standard since in the literature (e.g. [1]) it is more common to define $[n]_T^{\mathsf{True}}$ using a Π_n oracle rather than a Σ_{n+1} oracle. With a Π_n oracle one gets provable Σ_{n+1} completeness but in the absence of $\mathsf{B}\Sigma_{n+1}$ not necessarily provable $\Sigma_{n+1,1}$ completeness. With our definition of $[n]_T^{\mathsf{True}}$, since $A \leq B \in \Sigma_{n+1,1}$ for $A, B \in \Sigma_{n+1,1}$, and since we provided a proof where the minimal number principle is avoided, the FGH theorem smoothly generalizes to the new setting.

Theorem 3. *Let T be any computably enumerable theory extending* EA *and let $n < \omega$. For each $\sigma \in \Sigma_{n+1,1}$ we have that there is some $\rho_n \in \Sigma_{n+1,1}$ so that*

$$\mathrm{EA} \vdash \langle n \rangle_T^{\mathsf{True}} \top \to \left(\sigma \leftrightarrow [n]_T^{\mathsf{True}} \rho_n\right).$$

Proof. The proof runs entirely analogue to the proof of Theorem 2. Thus, for each number n we consider the fixed point ρ_n so that $\mathrm{EA} \vdash \rho_n \leftrightarrow (\sigma \leq [n]_T^{\mathsf{True}} \rho_n)$. Note that both ρ_n and $[n]_T^{\mathsf{True}} \rho < \sigma$ are $\Sigma_{n+1,1}$ whence by Lemma 4 we can apply provable $\Sigma_{n+1,1}$ completeness to them.

As an easy corollary we get that $[n]_T^{\mathsf{True}}$ formulas are closed not only under conjunction, as is well know, but also under disjunctions. Note that the FGH theorem yields that provably $\langle n \rangle_T^{\mathsf{True}} \top \to (\sigma \leftrightarrow [n]_T^{\mathsf{True}} \rho_n)$. Using the propositional tautology $(\neg A \to C) \to \Big[(A \to (B \leftrightarrow C)) \leftrightarrow ((\neg A \vee B) \leftrightarrow C)\Big]$ and $[n]_T^{\mathsf{True}} \bot \to [n]_T^{\mathsf{True}} \rho_n$ we see that this is equivalent to $\left([n]_T^{\mathsf{True}} \bot \vee \sigma\right) \leftrightarrow [n]_T^{\mathsf{True}} \rho_n$.

Corollary 1. *Let T be a c.e. theory extending* EA *and let $n \in \mathbb{N}$. For each formulas φ, ψ there is some $\sigma \in \Sigma_{n+1}$ so that $T \vdash ([n]_T^{\mathsf{True}} \varphi \vee [n]_T^{\mathsf{True}} \psi) \leftrightarrow [n]_T^{\mathsf{True}} \sigma$.*

Proof. We consider some $\sigma \in \Sigma_{n+1,1}$ so that provably $\sigma \leftrightarrow ([n]_T^{\mathsf{True}} \varphi \vee [n]_T^{\mathsf{True}} \psi)$. By Theorem 3 applied to this σ we provably have that $([n]_T^{\mathsf{True}} \varphi \vee [n]_T^{\mathsf{True}} \psi) \vee [n]_T^{\mathsf{True}} \bot$ is equivalent to $[n]_T^{\mathsf{True}} \varphi \vee [n]_T^{\mathsf{True}} \psi$.

Lemma 6. *Let T be any sound c.e. theory and let $A \subseteq \mathbb{N}$. The following are equivalent*

1. *A is c.e. in $\emptyset^{(n)}$;*
2. *A is many-one reducible to $\emptyset^{(n+1)}$;*
3. *A is definable on the standard model by a Σ_{n+1} formula;*
4. *A is definable on the standard model by a formula of the form $[n]_T^{\mathsf{True}} \rho(\dot{x})$;*
5. *A is definable on the standard model by a formula of the form $[n]_T^{\mathsf{True}} \rho(\dot{x})$ where $\rho(x) \in \Sigma_{n+1,1}$;*

Proof. The equivalence of 1, 2 and 3 is just Post's theorem. The implication $5 \Rightarrow 4$ is trivial, and implication $4 \Rightarrow 3$ holds in virtue of $[n]_T^{\mathsf{True}}$ being $\Sigma_{n+1,1}$ which on $\mathbb{N}$ is equivalent to Σ_{n+1}, so it suffices to prove $3 \Rightarrow 5$.

Thus, let the number n be fixed and, let A be a set of natural numbers so that for some $\sigma(x) \in \Sigma_{n+1}$ we have $m \in A \iff \mathbb{N} \models \sigma(m)$. Using the fixed point lemma we find $\rho_n(x) \in \Sigma_{n+1,1}$ so that $\mathrm{EA} \vdash \forall x \ \left(\rho_n(x) \ \leftrightarrow \ [\sigma(x) \leq [n]_T^{\mathsf{True}} \rho_n(\dot{x})] \ \right)$.

Reasoning in EA, we pick an arbitrary x and repeat the reasoning as in the proof of Theorem 3 to see that $\mathrm{EA} \vdash \langle n\rangle_T^{\mathsf{True}}\top \to \forall x \Big(\sigma(x) \leftrightarrow ([n]_T^{\mathsf{True}}\rho_n(\dot{x})) \Big)$. Since EA is sound and by assumption of T also being sound we have for each n that $\mathbb{N} \models \langle n\rangle_T^{\mathsf{True}}\top$, we may conclude that for any number m, $\mathbb{N} \models \sigma(m) \iff \mathbb{N} \models [n]_T^{\mathsf{True}}\rho_n(\overline{m})$ which was to be proven.

Feferman showed in [2] that for each unsolvable Turing degree $\boldsymbol{d}$ there is a theory U so that the Turing degree of $\{\ulcorner\varphi\urcorner \mid U \vdash \varphi\}$ is $\boldsymbol{d}$. However, the theories that Feferman considered were formulated in the language of identity and in particular did not contain arithmetic. It is not hard to see that for theories that do contain arithmetic we can only attain degrees that arise as Turing jumps.

Lemma 7. *Let $A \subseteq \mathbb{N}$ be definable on $\mathbb{N}$ by $\alpha(x)$ with Turing degree $\mathbf{a}$. Then, the Turing degree of $\{\psi \mid \mathrm{EA} + \{\alpha(\overline{n}) \mid \mathbb{N} \models \alpha(n)\} \vdash \psi\}$ equals $\mathbf{a}'$.*

We thus see that the Turing degrees of theories that are defined by a minimal amount of arithmetic (EA) together with some oracle, are entirely determined by the Turing degree of the corresponding oracle by means of the jump operator. By Friedberg's jump inversion theorem ([4]) we may thus conclude that any Turing degree above $\mathbf{0}'$ can be attained as the decision problem of a theory containing arithmetic. In this light, Lemma 6 should not come as a surprise. However, in the lemma we have provided a natural subsequence of theories so that moreover, all the necessary reasoning for the reductions can be formalized in EA. Lemma 6 is stated in terms of definability and computability but the proof tells us actually a bit more, namely that the FGH theorem is easily formalizable within EA.

Lemma 8. *For any c.e. theory T we have that*

$$\mathrm{EA} \vdash \forall \sigma{\in}\Sigma_{n+1}\, \exists \rho{\in}\Sigma_{n+1} \Big(\langle n\rangle_T^{\mathsf{True}}\top \to \big(\mathsf{True}_{\Sigma_{n+1}}(\sigma) \leftrightarrow [n+1]_T^{\mathsf{True}}\rho\big) \Big).$$

For other notions of provability we get similar generalizations of the FGH theorem. In particular, let $[n]_T^{\mathsf{Omega}}$ denote the formalization of the predicate "provable in T using at most n nestings of the omega rule". Following the recursive scheme $[0]_T^{\mathsf{Omega}}\varphi := \Box_T\varphi$ and,

$$[n+1]_T^{\mathsf{Omega}}\varphi := \exists\psi \Big(\forall x\, [n]_T^{\mathsf{Omega}}\psi(\dot{x}) \wedge \Box_T(\forall x\, \psi(x) \to \varphi) \Big)$$

we see that for c.e. theories T we can write $[n]_T^{\mathsf{Omega}}$ by a Σ_{2n+1}-formula. Also, we have provable Σ_{2n+1} completeness for these predicates, that is:

Proposition 1. *Let T be a computable theory extending* EA *and let ϕ be a Σ_{2n+1} formula. We have that* $\mathrm{EA} \vdash \phi \to [n]_T^{\mathsf{Omega}}\phi$.

Proof. By an external induction on n where each inductive step requires the application of an additional omega-rule.

This proposition is the omega-rule analogue of provable $\Sigma_{n+1,1}$ completeness for the $[n]_T^{\mathsf{True}}$ predicate. As a corollary we get an FGH Theorem for omega-provability.

Corollary 2. *Let T be any sound computably enumerable theory extending* EA *and let $n < \omega$. For each $\sigma \in \Sigma_{2n+1}$ we have that there is some $\rho_n \in \Sigma_{2n+1,1}$ so that* $\mathrm{PA} \vdash \langle n\rangle_T^{\mathsf{Omega}}\top \to \left(\sigma \leftrightarrow [n]_T^{\mathsf{Omega}}\rho_n\right)$.

We have formulated this corollary over PA so that $\Sigma_{2n+1,1}$ sentences are provably equivalent to Σ_{2n+1} sentences using collection. Consequently, we can now also prove a definability result for the $[n]_T^{\mathsf{Omega}}$ predicate.

Lemma 9. *Let T be any c.e. theory, let n be a natural number, and let $A \subseteq \mathbb{N}$. The following are equivalent*

1. *A is c.e. in $\emptyset^{(2n)}$;*
2. *A is definable on the standard model by a Σ_{2n+1} formula;*
3. *A is definable on the standard model by a formula of the form $[n]_T^{\mathsf{Omega}}\rho(\dot{x})$;*
4. *A is definable on the standard model by a formula of the form $[n]_T^{\mathsf{Omega}}\rho(\dot{x})$ where $\rho(x) \in \Sigma_{2n+1,1}$;*

Proof. The proof of this lemma is analogous to the proof of Lemma 6 if one substitutes Σ_{n+1} by Σ_{2n+1}, EA by PA, and $[n]_T^{\mathsf{True}}$ by $[n]_T^{\mathsf{Omega}}$.

By comparing Lemmas 6 and 9 we see that in a sense the hierarchy of formulas of the form $[n]_T^{\mathsf{True}}\varphi$ is more fine-grained than the hierarchy of formulas of the form $[n]_T^{\mathsf{Omega}}\varphi$. The positive feature of the latter hierarchy is that it is defined solely in terms of provability whereas the former needs to call upon partial truth predicates. As such the $[n]_T^{\mathsf{Omega}}$ hierarchy is more amenable to Turing jumps beyond the finite where no clear-cut syntactical characterizations along the lines of Post's correspondence theorem are available (see [3]). The down-side to the $[n]_T^{\mathsf{Omega}}$ hierarchy is that it runs outline with the Turing-jump hierarchy. In a forth-coming paper we propose a transfinite progression of provability notions in second order arithmetic that takes the best of both worlds: it is defined purely in terms of provability as in (1), synchronizes with the Turing-jump hierarchy as in Theorem 4, and can be transfinitely extended along any ordinal Ξ definable in second order logic yielding for a large class of second order theories a sound interpretation of a well-behaved logic called GLP_Ξ as in Theorem 5.

4 Graded Provability via Turing Jumps

We shall now see how the FGH theorem can be used to define graded provability notions $[n]_T^{\Box}$ for $n \in \mathbb{N}$ which are defined using only provability notions yet which are Σ_{n+1} complete. Note that on the natural numbers we have $[n+1]_T^{\mathsf{True}}\varphi \;\Leftrightarrow\; \exists\,\pi\, \Big(\mathsf{True}_{\Pi_{n+1}}(\pi) \wedge \Box_T(\mathsf{True}_{\Pi_{n+1}}(\pi) \to \phi)\Big)$ where $[0]_T^{\mathsf{True}}$ is nothing but $\Box_T$. It is easy to see that this equivalence is provable within PA. The idea now is to replace true Π_{n+1} sentences by consistency statements, that is, by sentences of the form $\langle n\rangle_T\varphi$. This replacement will be done in a recursive fashion. Thus we can consider the following recursive scheme $\overline{[0]}_T^{\Box}\phi := \Box_T\phi$, and $\overline{[n+1]}_T^{\Box}\phi := \Box_T\phi \;\vee\; \exists\,\psi\; \left(\langle\overline{n}\rangle_T^{\Box}\psi \;\wedge\; \Box(\langle\overline{n}\rangle_T^{\Box}\psi \to \phi)\right)$. For this recursive scheme, we can easily prove various desirable properties. However, for the sake of generalizing the definition to the transfinite setting, we choose to consider a more involved recursion.

$$[0]^{\Box}_{T}\phi := \Box_T\phi, \quad \text{and}$$
$$[n+1]^{\Box}_{T}\phi := \Box_T\phi \vee \exists\psi \bigvee_{0\leq m\leq n} \Big(\langle m\rangle^{\Box}_{T}\psi \wedge \Box(\langle m\rangle^{\Box}_{T}\psi \to \phi)\Big). \qquad (1)$$

We shall now prove that this provability notion $[n]^{\Box}_{T}$ is actually *provably* very similar to that of $[n]^{\mathsf{True}}_{T}$ for any natural number n.

Proposition 2. *Let T be a sound c.e. theory extending* EA. *We have for all $n \in \mathbb{N}$ that*

1. $\mathrm{EA} \vdash \ \forall\varphi\ ([n]^{\Box}_{T}\varphi \ \to\ [n]^{\mathsf{True}}_{T}\varphi)$;
2. $\mathrm{PA} \vdash \ \langle n\rangle^{\mathsf{True}}_{T}\top \ \to\ \forall\varphi\ ([n+1]^{\Box}_{T}\varphi \leftrightarrow [n+1]^{\mathsf{True}}_{T}\varphi)$;
3. $\mathrm{PA} \vdash \ [n]^{\mathsf{True}}_{T}\Big(\ \forall\varphi\ ([n]^{\Box}_{T}\varphi \leftrightarrow [n]^{\mathsf{True}}_{T}\varphi)\ \Big)$;
4. $\mathbb{N} \models \ \forall\varphi\ ([n]^{\Box}_{T}\varphi \leftrightarrow [n]^{\mathsf{True}}_{T}\varphi)$.

Proof. For the sake of readability we shall omit the subscripts T in this proof.

Item 1: This direction is easy, since by induction on n we see that each $\langle n\rangle^{\Box}_{T}\psi$ is of complexity Π_{n+1}.

Item 2: We reason in PA, assume $\langle n\rangle^{\mathsf{True}}\top$ and pick φ arbitrary. By Item 1 we only need to prove that $[n+1]^{\mathsf{True}}\varphi \to [n+1]^{\Box}\varphi$.

Thus, we suppose that $[n+1]^{\mathsf{True}}\varphi$ so that for some π we have $\mathsf{True}_{\Pi_{n+1}}(\pi)$ and $\Box\big(\mathsf{True}_{\Pi_{n+1}}(\pi) \to \varphi\big)$. Using the (formalized) FGH theorem we now pick ρ so that $\mathsf{True}_{\Pi_{n+1}}(\pi) \wedge \langle n\rangle^{\mathsf{True}}\top \leftrightarrow \langle n\rangle^{\mathsf{True}}\rho$. Since we work under the assumption of $\langle n\rangle^{\mathsf{True}}\top$ we thus have $\langle n\rangle^{\mathsf{True}}\rho$. Clearly, since we know that $\Box\big(\mathsf{True}_{\Pi_{n+1}}(\pi) \to \varphi\big)$ we also have the weaker $\Box\big(\langle n\rangle^{\mathsf{True}}\rho \to \varphi\big)$. Thus we have $\langle n\rangle^{\mathsf{True}}\rho \wedge \Box\big(\langle n\rangle^{\mathsf{True}}\rho \to \varphi\big)$ so that by applying twice the induction hypothesis we get $\langle n\rangle^{\Box}\rho \wedge \Box\big(\langle n\rangle^{\Box}\rho \to \varphi\big)$ and by definition $[n+1]^{\Box}\varphi$ as was to be shown.

Item 3: For $n = 0$ the statement holds by definition and for $n+1$, the statement follows from the previous item since $\mathrm{EA} \vdash [n+1]^{\mathsf{True}}\langle n\rangle^{\mathsf{True}}\top$.

Item 4: follows directly from the previous from the soundness of EA and T.

We shall see below that $[n]^{\Box}_{T}$ provability is a very decent provability notion.

Theorem 4. *Let T be a c.e. theory. We have for all $A \subseteq \mathbb{N}$ that the following are equivalent*

1. *A is c.e. in $\emptyset^{(n)}$;*
2. *A is many-one reducible to $\emptyset^{(n+1)}$;*
3. *A is definable on the standard model by a formula of the form $[n]^{\Box}_{T}\rho(\dot{x})$.*

Proof. This is a direct consequence of Lemma 6 and a minor generalization of Proposition 2.

As a consequence of this theorem we see that $[n]_T^\Box$ sentences are closed under disjunctions and conjunctions.

Corollary 3. *Let T be a c.e. theory extending* EA *and let $n \in \mathbb{N}$. For each formulas φ, ψ there is some σ so that*

$$T \vdash ([n]_T^\Box \varphi \ \vee [n]_T^\Box \psi) \ \leftrightarrow \ [n]_T^\Box \sigma.$$

Proof. Immediate from Theorem 4 since c.e. sets (with or without oracles) are closed under both conjunctions and disjunctions.

As a matter of fact, it turns out that this corollary can be formalized being one of the corner stones in a proof to the extent that the provability logic concerning the $[n]^\Box$ predicates is nice. In particular, let **GL** be the normal modal logic axiomatized by $\Box(A \to B) \to (\Box A \to \Box B)$ and $\Box(\Box A \to A) \to \Box A$ and all propositional tautologies with rules Modus Ponens and Necessitation: $\frac{A}{\Box A}$. Then GLP_ω is the polymodal logic axiomatized by **GL** for each modality $[n]$ together with the schemas $[n]A \to [n+1]A$ and $\langle n \rangle A \to [n+1]\langle n \rangle A$.

Theorem 5. *Let T be any c.e. theory extending* EA. *The logic* GLP_ω *is sound w.r.t.* PA *if we interpret each $[n]$-modality as $[n]_T^\Box$.*

The proof proceeds by a straightforward induction on n considering the logics GLP_{n+1} that have only modalities up to $[n]$.

Acknowledgements. I would like to thank Lev Beklemishev, Ramon Jansana, Stephen Simpson and Albert Visser for encouragement and fruitful discussions. También quisiera agradecerles a Diego Agulló Castelló y Rosa María Espinosa Jaén, alcaldes de las pedanías de Elche, Maitino y Perleta respectivamente, por facilitarme un sitio donde trabajar durante el verano del 2014. The research was supported by the Generalitat de Catalunya under grant number 2014SGR437 and from the Spanish Ministry of Science and Education under grant numbers MTM2011-26840, and MTM2011-25747.

References

1. Beklemishev, L.D.: Provability algebras and proof-theoretic ordinals, I. Ann. Pure Appl. Logic **128**, 103–124 (2004)
2. Feferman, S.: Degrees of unsolvability associated with classes of formalized theories. J. Symbolic Logic **22**(2), 161–175 (1957)
3. Fernández-Duque, D., Joosten, J.J.: The omega-rule interpretation of transfinite provability logic (submitted). arXiv:1302.5393 [math.LO] (2013)
4. Friedberg, R.M.: A criterion for completeness of degrees of unsolvability. J. Symbolic Logic **22**(2), 159–160 (1957)
5. Visser, A.: Faith and falsity: a study of faithful interpretations and false Σ_1^0-sentences. Ann. Pure Appl. Logic **131**(1–3), 103–131 (2005)

Rice's Theorem in Effectively Enumerable Topological Spaces

Margarita Korovina[1](✉) and Oleg Kudinov[2]

[1] A.P. Ershov Institute of Informatics Systems, SbRAS, Novosibirsk, Russia
rita.korovina@gmail.com
[2] Sobolev Institute of Mathematics, SbRAS, Novosibirsk, Russia
kud@math.nsc.ru

Abstract. In the framework of effectively enumerable topological spaces, we investigate the following question: given an effectively enumerable topological space whether there exists a computable numbering of all its computable elements. We present a natural sufficient condition on the family of basic neighborhoods of computable elements that guarantees the existence of a principal computable numbering. We show that weakly-effective ω–continuous domains and the natural numbers with the discrete topology satisfy this condition. We prove weak and strong analogues of Rice's theorem for computable elements.

1 Introduction

This research is motivated by rapidly increasing interest in continuous data representations with suitable computational properties. In this paper we concentrate on studying properties of computable elements in the framework of effectively enumerable topological spaces. We use numbering theory and index sets as promising techniques merging classical recursion theory and computability in topological spaces. There are several reasons for doing this. One of them is that the theory of index sets provides methods for encoding problems in an effective way by natural numbers, i.e., generate the corresponding index sets which can be used for analysis of their complexity in the settings of arithmetical and analytical hierarchies. Another reason is that the theory of index sets has been already successfully employed in many areas in mathematics and computer science. In recursion theory, index sets have been applied to obtained both new results and new elegant proofs of classical theorems such as the Post's theorem and the density theorem [9,17,19,21]. Many recent advancements in computable model theory are also closely related to index sets [4,5]. In computer science, the Rice-Shapiro theorem provides simple description of effectively enumerable properties of program languages and the complexity of some decision problems

This research was partially supported by Marie Curie Int. Research Staff Scheme Fellowship project PIRSES-GA-2011-294962, DFG/RFBR grant CAVER BE 1267/14-1 and 14-01-91334, RFBR grants 13-01-00015, 14-01-00376.

A. Beckmann et al. (Eds.): CiE 2015, LNCS 9136, pp. 226–235, 2015.
DOI: 10.1007/978-3-319-20028-6_23

in programming have been studied in [23] using index sets. While some important results have been obtained in [7,8,22–24,26] many methods and techniques of numbering theory have not been fully employed in computable analysis. The class of effectively enumerable topological spaces has been proposed in [15]. This is a wide class containing weakly effective ω–continuous domains, computable metric spaces and positive predicate structures [14] that retains certain natural effectivity requirements which allow us to introduce reasonable concepts of (partial) computable functions. The main results of this paper are as follows.

We propose a natural condition on the family of basic neighborhoods of computable elements that guarantees the existence of a principal computable numbering. We show that weakly effective ω–continuous domains and the natural numbers with the discrete topology satisfy this condition. For such spaces we address the problem whether there exists a nontrivial subset of the computable elements with a recursive index set. We give an example which demonstrates the main difficulties in straightforward transfer of Rice's theorem. Nevertheless, employing Branching Lemma [2], for effectively enumerable T_0-spaces satisfying the condition that the family of basic neighborhoods of the computable elements are wn-families, we prove generalisations of Rice's theorem for sets of computable elements. In particular, for such spaces if the subspace X_c of the computable elements is connected then there is no nontrivial subset of X_c with a computable index set.

2 Preliminaries

We refer the reader to [19] and [20] for basic definitions and fundamental concepts of recursion theory. We recall that, in particular, φ_e denotes the partial computable (recursive) function with an index e in the Kleene numbering, φ^s_e denotes the computation of φ_e for s steps such that the function φ^s_e is uniformly primitive recursive. In this paper we also use notations $W_e = \mathrm{dom}(\varphi_e)$, $W^s_e = \mathrm{dom}(\varphi^s_e)$, and $\pi_e = \mathrm{im}(\varphi_e)$. We use D_m to denote the m-th finite subset of ω in the canonical numbering. In the classical computability (recursion) theory and its applications in computable model theory, computable analysis [1,7,8,25] a common situation is the following. We start with the class of computable real numbers $\mathbb{R}_c$ as an example. It is well-known that a real number r is computable if and only if the constraint

$$*(RD,\, LD; r) \rightleftharpoons (RD = \{i | \mathbb{Q}\mathrm{and} q_i > r\} \text{ and } LD = \{i | \text{ and } q_i < r\} \mathrm{arec.e.})$$

holds. Therefore, a criterion for r to belong to $\mathbb{R}_c$ is based on the existence of computably enumerable sets RD, LD satisfying the constraint $*(RD, LD; r)$. This could be generalised to another class K, i.e., a criterion for an element k to belong to the class K is based on the existence of computably enumerable sets $V_1, \dots, V_s$ satisfying a particular constraint $*(\bar{V}; k)$ such that, for any $\bar{V}$ and different k_1, k_2, the constraints $*(\bar{V}; k_1)$ and $*(\bar{V}; k_2)$ don't hold simultaneously. We call such K a *constrained class*. In the case when K is a constrained class we say that $\{k^m\}_{m\in\omega}$ is a *computable sequence* of elements of K if there exist computable sequences $\{V^m_1\}_{m\in\omega}, \dots, \{V^m_s\}_{m\in\omega}$ of computably enumerable sets such that for every $m \in \omega$ the constraint $*(V_1^m, \ldots, V_s^m; k^m)$ holds. In other words, one can say that the property $(\exists \bar{V}) * (\bar{V}; k)$ holds uniformly on the elements of the sequence. For example, $\{r^m\}_{m\in\omega}$ is a computable sequence of computable real numbers if there exist computable sequences $\{LD_m\}_{m\in\omega}$ and $\{RD_m\}_{m\in\omega}$ of computably enumerable sets such that $*(RD_m, LD_m; r_m)$ holds for every $m \in \omega$, i.e., the property $(\exists RD)(\exists LD) * (RD, LD; r)$ holds uniformly on $r \in \mathbb{R}_c$.

For background on numbering theory we refer to [10]. Let A be a set. A *numbering* is a surjection function $\xi : \omega \to A$. Assume additionally that A is a constrained class. A numbering $\xi : \omega \to A$ is called *computable* if $\{\xi(i)\}_{i\in\omega}$ is the computable sequence of all elements of A. A numbering $\alpha : \omega \to A$ is called *principal computable* if it is computable and every computable numbering ξ is computably reducible to α, i.e., there exists a computable function $f : \omega \to \omega$ such that $\xi(i) = \alpha(f(i))$. It is worth noting that these definitions agree with the notions of computable and principal computable numberings of $A \subseteq \mathcal{P}(\omega)$ [10].

Definition 1. *Let K be a constrained class, $\alpha : \omega \to K$ be a principal computable numbering and $L \subseteq K$. The set $Ix(L) = \{n | \alpha(n) \in L\}$ is called an* index set *for the subclass L.*

3 Effectively Enumerable Topological Spaces

Now we recall the notion of effectively enumerable topological space. Let (X, τ, α) be a topological space, where X is a non-empty set, $B \subseteq 2^X$ is a base of the topology τ and $\alpha : \omega \to B$ is a numbering.

Definition 2. *[15] A topological space (X, τ, α) is effectively enumerable if the following conditions hold.*

1. *There exists a computable function $g : \omega \times \omega \times \omega \to \omega$ such that*

$$\alpha(i) \cap \alpha(j) = \bigcup_{n\in\omega} \alpha(g(i, j, n)).$$

2. *The set $\{i | \alpha(i) \neq \emptyset\}$ is computably enumerable.*

In [15] it has been shown that the class of effectively enumerable topological spaces is a natural proper extension of weakly effective ω–continuous domains and computable metric spaces as well, in particular, the real numbers, the continuous real functions belong to this class. Let us note that when a base numbering is changed, computability properties may also change. Therefore we need to formalise reducibility and equivalence between base numberings.

Definition 3. *Let (X, τ, α) and (X, τ, β) be topological spaces. We don't assume that* $\mathrm{im}(\alpha) = \mathrm{im}(\beta)$. *We say that β is* computably reducible *to α, denoted $\beta \leq \alpha$, if there exists a computable sequence $\{A_n\}_{n\in\omega}$ of c.e. sets such that $\beta(n) = \bigcup_{i\in A_n} \alpha(i)$ for all $n \in \omega$. Numberings α and β are equivalent, denoted $\alpha \equiv \beta$, if $\alpha \leq \beta$ and $\beta \leq \alpha$.*

It is worth noting that if $\alpha \equiv \beta$ then (X, τ, α) is an effectively enumerable topological space if and only if (X, τ, β) is an effectively enumerable topological space. As we will see below, any change of base numberings from the same equivalence class preserves the classes of effectively open sets, computable elements and etc. We can relativise the notions from Definition 3 to an appropriate oracle or a Turing degree **a**.

Proposition 1. *For every Turing degree* **a***, there exists an effectively enumerable T_1-space (X, τ, α) such that any base numbering β* **a***-equivalent to α (and even any $\beta \leq^{\mathbf{a}} \alpha$) is not* **a***-decidable, i.e., the set $\{(n,m) | \alpha(n) = \alpha(m)\}$ is not* **a***-computable.*

Recently, various notions of computable topological space have been proposed (see among others [12,22,25]). The following results illustrate differences of our approach and the approaches proposed in [12,25].

Proposition 2. *There exists an effectively enumerable topological space which is not T_0-space.*

Proof. One of the examples is an algebraically closed field with Zariski topology.

Proposition 3. *There exists an effectively enumerable T_1-space which is not a computable topological space in the sense of [25].*

Proof. The claim follows from Proposition 1 under the assumption that $\mathbf{a} = \mathbf{0}'$.

Proposition 4. *There exists an effectively enumerable T_2-space which is not a computable topological space in the sense of [12].*

In this paper we use the following notion of effectively open sets and characterisation of total computable functions between two effectively enumerable topological spaces [15].

Definition 4. *[15] Let (X, τ, α) be an effectively enumerable topological space. A set $A \subseteq X$ is* effectively open *if there exists a computably enumerable set V such that $A = \bigcup_{n \in V} \alpha(n)$.*

Proposition 5. *[15] Let $\mathcal{X} = (X, \tau, \alpha)$ be an effectively enumerable topological space and $\mathcal{Y} = (Y, \lambda, \beta)$ be an effectively enumerable T_0-space. For a total function $F : X \to Y$ F is computable iff F is effectively continuous, i.e., there exists a computable function $h : \omega \times \omega \to \omega$ such that $F^{-1}(\beta(j)) = \bigcup_{i \in \omega} \alpha(h(i,j))$.*

4 Computable elements of Effectively Enumerable Topological Spaces

The notion of computable element is based on the following observation.

Proposition 6. *For an effectively enumerable T_0–space (X, τ, α) and a point $x \in X$ the following assertions are equivalent.*

(1) The function $f : \{0\} \to X$, defined as $f(0) = x$, is computable.
(2) The set $A_x = \{n | x \in \alpha(n)\}$ is computably enumerable.
(3) For any effectively enumerable T_0–space $\mathcal{Y}$ the function $f_Y : Y \to X$, defined as $f_Y(y) = x$, is computable.

Proof. $(1 \to 2)$. First, it is worth noting that, for total functions, computability is equivalent to effective continuity [15]. So, we assume that f is effectively continuous. Therefore there exists a computable function $h : \omega \times \omega \to \omega$ such that $f^{-1}(\alpha(n)) = \bigcup_{i \in \omega} \beta(h(n,i))$, where β is a numbering of the base of the space $\{0\}$. So, $x \in \alpha(n) \leftrightarrow f^{-1}(\alpha(n)) \neq \emptyset \leftrightarrow \exists i\, \beta(h(n,i)) \neq \emptyset$. Therefore A_x is computably enumerable.

$(2 \to 3)$. Let us show that f_Y is effectively continuous. It is easy to see that

$$f_Y^{-1}(\alpha(n)) = \begin{cases} Y, & x \in \alpha(n) \\ \emptyset, & \text{otherwise.} \end{cases}$$

Let β be a numbering of a topology base of the space Y such that $\beta(m) = Y$ and $\beta(k) = \emptyset$ and

$$S_n = \begin{cases} \{k\}, & n \notin A_x \\ \{k, m\}, & \text{otherwise.} \end{cases}$$

It is easy to see that $f_Y^{-1}(\alpha(n)) = \bigcup_{s \in S_n} \beta(s)$ is computable enumerable. So, f_Y is effectively continuous, i.e., computable.

$(3 \to 1)$. It is trivial.

For an effectively enumerable T_0–space, if one of the assertions (1)–(3) holds, x is called a *computable element (point)*. For an arbitrary effectively enumerable topological space, x is called a *computable element* if the assertions (2) holds and, for any $y \in X$, $A_x \neq A_y$. Let $*(V; x) \rightleftharpoons (x \in X$ and $V = \{n | x \in \alpha(n)\}$ is computably enumerable). Let X_c be the set of all computable elements of an effectively enumerable space X. It is clear that X_c is a constrained class. It is easy to see that the definition above agrees with the notions of computable real number, computable element of a computable metric space [3], computable element of a weakly–effective ω-continuous domain [24,26] and computable element of a computable topological space [12]. It is worth noting that there are effectively enumerable topological spaces without computable elements. As example we can consider the real numbers without the computable points. It is an effectively enumerable T_0-space, however there are no computable elements in this space. An element $x \in X$ is called *co-effective* if $X \setminus \{x\}$ is an effectively open set. Let X_{co} be the set of all co-effective elements. We investigate relations between the computable and co-effective elements. It turns out that, in general, even for some complete computable metric spaces $X_c \neq X_{co}$. For instance, let us consider the Baire space ω^ω. There exists non-computable $f \in \omega^\omega$ which is Π^0_1. Therefore, f is non-computable but co-effective.

Let $D = (D; B, \beta, \leq, \perp)$ be a ω–continuous domain where B is a basis, $\beta : \omega \to B$ is a numbering of the basis (see [13]). We recall that (D, β) is a weakly effective if the relation $\beta(i) \ll \beta(j)$ is computably enumerable.

Proposition 7. *Let (X, τ, α) be an effectively enumerable topological space. Define $\alpha^*(m) = \cup_{k \in D_m} \alpha(k)$ and $\mathcal{B}_X = \mathrm{im}(\alpha^*)$. Suppose $(\mathcal{O}_X, \mathcal{B}_X, \alpha^*, \subseteq, \emptyset)$ is a weakly effective ω-continuous domain, where the ordering is the set inclusion and $\perp = \emptyset$. Then, $X_{co} \subseteq X_c$.*

Proof. Let $x \in X$ be co-effective. This means that $X \setminus \{x\} = \bigcup_{n \in \omega} \alpha(\chi(n))$ for a computable function $\chi : \omega \to \omega$. Let us fix some n such that $x \in \alpha(n)$ and $\alpha(n) \ll X$. Then,

$$x \in \alpha(k) \leftrightarrow (\exists s \in \omega)(\exists l_0 \in \omega)\ldots(\exists l_s \in \omega)\left(\alpha(n) \ll \bigcup_{i \leq s} \alpha(\chi(l_i)) \cup \alpha(k)\right).$$

By assumption, $\alpha^*(n) \ll \alpha^*(j)$ is computably enumerable. Therefore, the set A_x is computably enumerable. By definition, x is computable.

Proposition 8. *Let (X, τ, α) be an effectively enumerable T_2-space such that there exists a computable function $g : \omega \times \omega \to \omega$ such that $X \setminus \mathrm{cl}(\alpha(n)) = \bigcup_{n \in \omega} \alpha(g(i, n))$. Then, $X_c \subseteq X_{co}$.*

Proof. Let $x \in X$ be computable. Then, $X \setminus \{x\} = \bigcup_{m \in \omega} \{X \setminus \mathrm{cl}(\alpha(m)) | x \in \alpha(m)\}$ is effectively open. So, x is co-effective.

Corollary 1. *For every Euclidean space X, $X_c = X_{co}$.*

Now we address the natural question whether for an effectively enumerable space there exists a computable numbering of the computable elements. First, we observe that while for the computable real numbers there is no computable numbering [6,18] as well as for the computable points of a complete computable metric space [3], for a weakly–effective ω–continuous domain there is a computable numbering of the computable elements [26]. Below we point out a natural sufficient condition on the family of basic neighborhoods of computable elements that guarantees the existence of a principal computable numbering. We show that weakly-effective ω–continuous domains, ω with the discrete topology satisfy this condition.

Definition 5. [10] *Let S be a family of computable enumerable subsets of ω. S is called a wn-family if there exists a partial computable function $\sigma : \omega \to \omega$ such that (i) if $\sigma(n) \downarrow$ then $W_{\sigma(n)} \in S$ and (ii) if $W_n \in S$ then $n \in \mathrm{dom}(\sigma)$ and $W_n = W_{\sigma(n)}$.*

Theorem 1. *Let (X, τ, α) be an effectively enumerable T_0-space and $S_X = \{A_a | a \in X_c\}$. If S_X is a wn-family then there exists an algorithm to construct a principal computable (canonical) numbering $\bar{\alpha} : \omega \to X_c$.*

Proof. From [10] it follows that the set S_X has a standard principal computable numbering $\gamma : n \mapsto W_{\sigma(h_0(n))}$, where $h_0 : \omega \to \omega$ is a total computable function such that $\mathrm{im}(h_0) = \mathrm{dom}(\sigma)$. Let us define $\bar{\alpha}(n) = a \leftrightarrow A_a = W_{\gamma(n)}$.

Proposition 9. *Let $D = (D; B, \beta, \leq, \bot)$ be a weakly effective ω–continuous domain. Then, the set $S_D = \{\{n|\beta(n) \ll a\}|a \in D_c\}$ is a wn-family.*

Proposition 10. *There exists an effectively enumerable T_0-space $\mathcal{X} = (X, \tau, \alpha)$ such that S_X is a wn-family and $\mathcal{X}$ is not a domain.*

Proof. See Example 1 below.

Below we assume that X is an effectively enumerable T_0-space satisfying the requirements of Theorem 1 and $\bar{\alpha} : \omega \to X_c$ is defined as in Theorem 1. Now we address the problem whether there is a nontrivial subset of the computable elements with a computable index set. The following example is important for understanding.

Example 1. We consider (ω, τ, α), where τ is the discrete topology and α is defined as follows: $\alpha(0) = \emptyset$, $\alpha(k) = \{k - 1\}$. It is clear that $S_\omega = \{\{n\}|n > 0\}$ is a *wn*-family. By Theorem 1 there is a principal computable numbering of the set of computable elements which coincides with ω. There exists $K \subset \omega$, e.g., $K = 2 \cdot \omega$, such that $Ix(K)$ is computable.

Proposition 11. *Let $K \subseteq X_c$. If $Ix(K)$ is computable then K is clopen in X_c.*

The claim follows from the following proposition.

Proposition 12. *Let $K \subseteq X_c$. If $Ix(K)$ is computable then K is open in X_c.*

In order to prove the proposition, for $K \subseteq X_c$, we define $\hat{K} = \{A_a|a \in K\}$, $W = \{n|n \in \mathrm{dom}(\sigma)$ and $W_{\sigma(n)} \in \hat{K}\}$, and $W^1 = \{k|k \in \mathrm{dom}(\sigma)$ and $W_{\sigma(k)} \in S_X \setminus \hat{K}\}$. Below we use the observation that $Ix(K) = \{n|\gamma(n) \in \hat{K}\} = \{n|h_0(n) \in W\}$, $\omega \setminus Ix(K) = \{n|\gamma(n) \in S_X \setminus \hat{K}\} = \{n|h_0(n) \in W^1\}$, $W = h_0(Ix(K))$, and $W^1 = h_0(\omega \setminus Ix(K))$, where h_0 and γ are defined in Theorem 1. It is clear that W and W^1 are computably enumerable. We give a proof using the following lemmas.

Lemma 1 (Branching lemma). *[2] Let V and W be computably enumerable sets such that W contains all computably enumerable indices of V. Let $\lambda p.V^p$ be an enumeration of V and $r : \omega \to \omega$ be a total computable function. Then there are $e \in W$ and $p \in \omega$ such that $W_e = V^p \cup W_{r(p)}$. Furthermore, such e and p are computed uniformly from an computably enumerable index of W and computable indices of the functions $\lambda p.V^p$ and r.*

Lemma 2. *If W is computably enumerable then $\hat{K}$ is monotone, i.e., if $A \subseteq B$, $A \in \hat{K}$ and $B \in S_X$ then $B \in \hat{K}$.*

Proof. Let $A = W_{\sigma(m)}$ and $B = W_{\sigma(n)}$. Define $V = A$ and $r(p) \equiv \sigma(n)$. It is worth noting that if $W_a = W_{\sigma(m)} \in \hat{K}$ then $W_{\sigma(a)} = W_a \in \hat{K}$, so that $a \in W$. Therefore, we can use Lemma 1 to find $e \in W$ and $p \in \omega$ such that $W_e = V^p \cup W_{r(p)} = W^p_{\sigma(m)} \cup W_{\sigma(n)} = W_{\sigma(n)}$. Since $B \in S_X$ and $e \in W$, $W_{\sigma(e)} = W_e \in \hat{K}$. Therefore, $B \in \hat{K}$.

Lemma 3. *Assume $K \subseteq X_c$, $Ix(K)$ is computable and K is not open in X_c. Then, there exist $a \in K$, $m \in \omega$, a computable function $h : \omega \to \omega$ such that $W_{\sigma(m)} = A_a$ and, for all $p \in \omega$, $W^p_{\sigma(m)} \subseteq W_{\sigma(h(p))}$ and $h(p) \in W^1$.*

Proof. (Proposition 12) Assume contrary: $K \subseteq X_c$ and $Ix(K)$ is computable and K is not open in X_c. Then there exist a, m and h satisfying Lemma 3. Let $V = W_{\sigma(m)}$ and $r(p) \equiv \sigma(h(p))$, so, for all $p \in \omega$, $W_{\sigma(p)} \in S_X \setminus \hat{K}$. Using Lemma 1 we find $e \in W$ and $p \in \omega$ such that $W_e = W^p_{\sigma(m)} \cup W_{\sigma(h(p))} = W_{\sigma(h(p))} \in S_X \setminus \hat{K}$. Since, $W_e = W_{\sigma(e)}$, $e \notin W$. This contradicts to $e \in W$.

The following corollary of Proposition 12 is a weak analogue of Rice's theorem.

Proposition 13 (Weak Rice's Theorem). *If X_c is connected then there is no nontrivial $K \subset X_c$ such that $Ix(K)$ is computable.*

It is worth noting that in a weakly effective ω-continuous domains with the bottom, e.g., $\mathcal{P}(\omega)$ with Scott topology, the set of computable elements is connected. In order to prove a strong analogue of Rice's theorem we use the notations and the results from Proposition 12 and the following observations.

Let $S = \{W_{\sigma(n)}\}_{n\in\omega}$ be a *wn*-family with the standard principal computable numbering $\gamma : n \mapsto W_{\sigma(h_0(n))}$, where $h_0 : \omega \to \omega$ is a total computable function such that $\mathrm{im}(h_0) = \mathrm{dom}(\sigma)$. For $\hat{K} \subseteq S$ we define $Ix_S(\hat{K}) = \{m | \gamma(m) \in \hat{K}\}$.

Definition 6. *[19] We say that $\hat{K}$ is effectively discrete in S if there exists a strongly computable family $\{F_n\}_{n\in\omega}$ of finite subsets of ω such that*

$$(\forall A \in S)\left(A \in \hat{K} \leftrightarrow (\exists n \in \omega) A \supseteq F_n\right).$$

In other words, $\hat{K}$ is effectively discrete in S if it is effectively open in S that is a subspace of $\mathcal{P}(\omega)$ with Scott topology.

Proposition 14. *If $Ix_S(\hat{K})$ is computable then $\hat{K}$ and $S \setminus \hat{K}$ are effectively discrete in S.*

Proof. We use the notations $W = \{n | n \in \mathrm{dom}(\sigma) \text{ and } W_{\sigma(n)} \in \hat{K}\}$, and $W^1 = \{k | k \in \mathrm{dom}(\sigma) \text{ and } W_{\sigma(k)} \in S_X \setminus \hat{K}\}$ as in Proposition 12. Let $W = \mathrm{dom}(g)$ for a total computable $g : \omega \to \omega$. First, using uniformisation we construct a partial computable function $h : \omega \times \omega \to \omega$ such that

$$(\forall p \in \omega)(\forall m \in W)\left((\exists k \in W^1) W_{\sigma(k)} \supseteq W^p_{\sigma(m)} \leftrightarrow \left(W_{\sigma(h(m,p))} \supseteq W^p_{\sigma(m)} \wedge h(m,p) \in W^1\right)\right).$$

In particular, $h(m,p) \uparrow$ if $\sigma(m) \uparrow$. Given $m \in W$ and $V = W_{\sigma(m)}$, using completeness of the numbering $\{W_n\}_{n\in\omega}$, we construct a total computable function $r : \omega \to \omega$ such that, for all $p \in \omega$,

$$W_{r(p)} = \begin{cases} W_{\sigma(h(m,p))}, & \text{if } h(m,p) \downarrow \\ \emptyset, & \text{if } h(m,p) \uparrow. \end{cases}$$

Using Lemma 1, from an computably enumerable index of W and computable indices of the functions $\lambda p.V^p$ and r, we uniformly compute e and p such that $W_e = W^p_{\sigma(m)} \cup W_{r(p)}$. It is worth noting that if $h(m,p)\downarrow$ then $W_e = W_{\sigma(h(m,p))} \in S \setminus \hat{K}$, so $W_{\sigma(e)} = W_e$. Since $e \in W$, $W_{\sigma(e)} \in \hat{K}$, a contradiction. It follows that $h(m,p)\uparrow$. Using uniform computability of e and p we construct a partial computable function $p : \omega \to \omega$ such that, for all $m \in \omega$, if $\sigma(m)\downarrow$ then $h(m,p(m))\uparrow$ and $\mathrm{im}(p) = \mathrm{im}(\sigma)$. Then, for $m \in W$, we obtain that $W_{\sigma(k)} \supseteq W^{p(m)}_{\sigma(m)} \to k \in W$ by the choice of p. So, the family $\{F_n\}_{n\in\omega}$, defined as $F_n = W^{p(g(n))}_{\sigma(g(n))}$, testifies that $\hat{K}$ is effectively discrete in S.

Theorem 2 (Strong Rice's Theorem). *Let $K \subseteq X_c$. The set $Ix(K)$ is computable if and only if K and $X_c \setminus K$ are effectively open in X_c.*

Proof ($\leftarrow$). If K is effectively open in X_c then, by definition, $K = \bigcup_{i\in I}(\alpha(i) \cap X_c)$, where I is computably enumerable. Since $\bar{\alpha}$ is computable, we have

$$n \in Ix(K) \leftrightarrow (\exists i \in I)\, i \in A_{\bar{\alpha}(n)}.$$

So, $Ix(K)$ is computable enumerable. Replacing K by $X_c \setminus K$ in the previous reasoning we obtain that $\omega \setminus Ix(K)$ is computable enumerable.
($\to$). We apply Proposition 14 to $\hat{K} = \{A_a | a \in K\}$ and $S = S_X$. We immediately obtain that, for any $m \in W$,

$$\bigcap_{i\in W^{p(m)}_{\sigma(m)}} (\alpha(i) \cap X_c) \subseteq K.$$

Therefore,

$$K = \bigcup_{m\in W}\left(\bigcap_{i\in W^{p(m)}_{\sigma(m)}} (\alpha(i) \cap X_c)\right).$$

This means that K is effectively open in X_c. Replacing K by $X_c \setminus K$ in the previous reasoning we obtain that $X_c \setminus K$ is effectively open in X_c. The proof is complete.

References

1. Arslanov, M.M.: Families of recursively enumerable sets and their degrees of unsolvability. Sov. Math. **29**(4), 13–21 (1985)
2. Berger, U.: Total sets and objects in domain theory. Ann. Pure Appl. Logic. **60**(2), 91–117 (1993)
3. Brattka, V.: Computable versions of baire's category theorem. In: Sgall, J., Pultr, A., Kolman, P. (eds.) MFCS 2001. LNCS, vol. 2136, pp. 224–235. Springer, Heidelberg (2001)

4. Calvert, W., Fokina, E., Goncharov, S.S., Knight, J.F., Kudinov, O.V., Morozov, A.S., Puzarenko, V.: Index sets for classes of high rank structures. J. Symb. Log. **72**(4), 1418–1432 (2007)
5. Calvert, W., Harizanov, V.S., Knight, J.F., Miller, S.: Index sets of computable structures. J. Algebr. Log. **45**(5), 306–325 (2006)
6. Ceitin, G.S.: Mean value theorems in constructive analysis. Trans. Am. Math. Soc. Trans. Ser. **2**(98), 11–40 (1971)
7. Cenzer, D.A., Remmel, J.B.: Index sets for Π_1^0 classes. Ann. Pure Appl. Log. **93**(1–3), 3–61 (1998)
8. Cenzer, D.A., Remmel, J.B.: Index sets in computable analysis. Theor. Comput. Sci. **219**(1—-2), 111–150 (1999)
9. Ershov, Y.L.: Model ℂ of partial continuous functionals. In: Gandy, R.O., Hyland, J.M.E. (eds.) Logic colloquium 76, pp. 455–467. North-Holland, Amsterdam (1977)
10. Ershov, Y.L.: Theory of numberings. In: Griffor, E.R. (ed.) Handbook of Computability Theory, pp. 473–503. Elsevier Science B.V, Amsterdam (1999)
11. Grubba, T., Weihrauch, K.: On computable metrization. Electron. Notes Theor. Comput. Sci. **167**, 345–364 (2007)
12. Grubba, T., Weihrauch, K.: Elementary computable topology. J. UCS. **15**(6), 1381–1422 (2009)
13. Gierz, G., Heinrich Hofmann, K., Keime, K., Lawson, J.D., Mislove, M.W.: Continuous Lattices and Domain. Encyclopedia of Mathemtics and its Applications, vol. 93. Cambridge University Press, Cambridge (2003)
14. Korovina, M.V., Kudinov, O.V.: Positive predicate structures for continuous data. J. Math. Struct. Comput. Sci. (2015, To appear)
15. Korovina, M.V., Kudinov, O.V.: Towards computability over effectively enumerable topological spaces. Electron. Notes Theor. Comput. Sci. **221**, 115–125 (2008)
16. Korovina, M.V., Kudinov, O.V.: Towards computability of higher type continuous data. In: Cooper, S.B., Löwe, B., Torenvliet, L. (eds.) CiE 2005. LNCS, vol. 3526, pp. 235–241. Springer, Heidelberg (2005)
17. Lempp, S.: Hyperarithmetical index sets in recursion theory. Trans. Am. Math. Sot. **303**, 559–583 (1987)
18. Martin-Löf, P.: Notes on Constructive Mathematics. Stockholm, Sweden (1970)
19. Rogers, H.: Theory of Recursive Functions and Effective Computability. McGraw-Hill, New York (1967)
20. Soare, R.I.: Recursively Enumerable Sets and Degrees: A Study of Computable Functions and Computably Generated Sets. Springer, Heidelberg (1987)
21. Shoenfield, J.R.: Degrees of unsolvability. North-Holland Publishing, Amsterdam (1971)
22. Spreen, D.: On effective topological spaces. J. Symb. Log. **63**(1), 185–221 (1998)
23. Spreen, D.: On some decision problems in programming. Inf. Comput. **122**(1), 120–139 (1995)
24. Spreen, D.: On r.e. inseparability of CPO index sets. In: Börger, E., Rödding, D., Hasenjaeger, G. (eds.) Rekursive Kombinatorik 1983. LNCS, vol. 171, pp. 103–117. Springer, Heidelberg (1984)
25. Weihrauch, K.: Computable Analysis. Springer, Heidelberg (2000)
26. Weihrauch, K., Deil, T.: Berechenbarkeit auf cpo-s. Schriften zur Angew. Math. u. Informatik 63. RWTH Aachen, Aachen (1980)

Decidability of Termination Problems for Sequential P Systems with Active Membranes

Michal Kováč(✉)

Faculty of Mathematics, Physics and Informatics, Comenius University, Bratislava, Slovakia
kovac@fmph.uniba.sk

Abstract. We study variants of P systems that are working in the sequential mode. Basically, they are not computationally universal, but there are possible extensions that can increase the computation power. Extensions that implement a notion of zero-checking, are often computationally universal. P systems with an ability to create new membranes are a rare exception as they are known to be computationally universal even in the sequential mode without using a dedicated zero-check operation. In this paper we show a result that seems surprising to us - the existence of an infinite computation for sequential P systems with active membranes is decidable. The standard construction of coverability tree is extended to provide an algorithm for detecting infinite loops. In addition, we show that the existence of a halting computation is undecidable as it can be reduced to reachability of register machines.

1 Introduction

Membrane systems (P systems) [1] were introduced by Păun (see [2]) as distributed parallel computing devices inspired by the structure and functionality of cells. Starting from the observation that there is an obvious parallelism in the cell biochemistry and relying on the assumption that "if we wait enough, then all reactions which may take place will take place", a feature of the P systems is given by the maximal parallel way of using the rules. For various reasons ranging from looking for more realistic models to just more mathematical challenge, the maximal parallelism was questioned, either simply criticized, or replaced with presumably less restrictive assumptions. In some cases, a sequential model may be a more reasonable assumption. In sequential P systems, only one rewriting rule is used in each step of computation. Without priorities, they are equivalent to Petri nets [3], hence not computationally universal. However priorities, inhibitors and other modifications can increase the computation power. It seems that there is a link between universality and ability to zero-check [4].

In this paper we study a variant where universality can be achieved without checking for zero by allowing membranes to be created unlimited number of

Work supported by the grant VEGA 1/1333/12.

A. Beckmann et al. (Eds.): CiE 2015, LNCS 9136, pp. 236–245, 2015.
DOI: 10.1007/978-3-319-20028-6_24

times [3]. Such P systems are called active P systems. Contrary, if we place a limit on the number of times a membrane is created, we get a class of P systems which is only equivalent to Petri nets, hence not computationally universal.

In Sect. 2 we will recall some basic notions from formal languages, multisets and graph theory. Then in Sect. 3 we will introduce membrane structure and formally define membrane configuration and active P system, because standard definitions are not convenient for our formal proofs.

Section 4 contains two main results. The existence of an infinite computation is surprisingly shown to be decidable. On the other hand, the existence of a halting computation is shown to be undecidable.

2 Preliminaries

Here we recall several notions from the classical theory of formal languages.

An **alphabet** is a finite nonempty set of symbols. Usually it is denoted by Σ. A **string** over an alphabet is a finite sequence of symbols from the alphabet. We denote by Σ^* the set of all strings over an alphabet Σ. By $\Sigma^+ = \Sigma^* - \{\varepsilon\}$ we denote the set of all nonempty strings over Σ. A **language** over the alphabet Σ is any subset of Σ^*.

The number of occurrences of a given symbol $a \in \Sigma$ in the string $w \in \Sigma^*$ is denoted by $|w|_a$. $\Psi_\Sigma(w) = (|w|_{a_1}, |w|_{a_2}, \dots, |w|_{a_n})$ is called a Parikh vector associated with the string $w \in \Sigma^*$, where $\Sigma = \{a_1, a_2, \dots, a_n\}$. For a language $L \subseteq \Sigma^*$, $\Psi_\Sigma(L) = \{\Psi_\Sigma(w) | w \in L\}$ is the Parikh image of L. If FL is a family of languages, PsFL denotes the family of Parikh images of languages in FL.

A multiset over a set Σ is a mapping $M : \Sigma \to \mathbb{N}$. We denote by $M(a), a \in \Sigma$ the multiplicity of a in the multiset M. The **support** of a multiset M is the set $supp(M) = \{a \in \Sigma | M(a) \geq 1\}$. It is the set of items with at least one occurrence. A multiset is **empty** when its support is empty. A multiset M with finite support $\{a_1, a_2, \dots, a_n\}$ can be represented by the string $a_1^{M(a_1)} a_2^{M(a_2)} \dots a_n^{M(a_n)}$. We say that multiset M_1 is included in multiset M_2 if $\forall a \in supp(M_1) : M_1(a) \leq M_2(a)$. We denote it by $M_1 \subseteq M_2$. The **difference** of two multisets $M_2 - M_1$ is defined as a multiset where $\forall a \in supp(M_2) : (M_2 - M_1)(a) = \max(M_2(a) - M_1(a), 0)$. The **union** of two multisets $M_1 \cup M_2$ is a multiset where $\forall a \in supp(M_1) \cup supp(M_2) : (M_1 \cup M_2)(a) = M_1(a) + M_2(a)$. The product of multiset M with natural number $n \in \mathbb{N}$ is a multiset where $\forall a \in supp(M) : (n \cdot M)(a) = n \cdot M(a)$.

Next, we recall notions from graph theory.

A **rooted tree** is a tree, in which a particular node is distinguished from the others and called the root node. Let T be a rooted tree. We will denote its root node by r_T. Let d be a node of $T \backslash \{r_T\}$. As T is a tree, there is a unique path from d to r_T. The node adjacent to d on that path is also unique and is called a **parent node** of d and is denoted by $parent_T(d)$. We will denote the set of nodes of T by $V(T)$ and set of its edges by $E(T)$. Let T_1, T_2 be rooted trees. A bijection $f : V(T_1) \to V(T_2)$ is an **isomorphism** iff $\{(f(u), f(v)) | (u, v) \in E(T_1)\} = E(T_2)$ and $f(r_{T_1}) = r_{T_2}$.

3 Active P Systems

The fundamental ingredient of a P system is the **membrane structure** (see [5]). It is a hierarchically arranged set of membranes, all contained in the **skin membrane**. Each membrane determines a compartment, also called region, which is the space delimited from above by it and from below by the membranes placed directly inside, if any exists. Clearly, the correspondence membrane region is one-to-one, that is why we sometimes use interchangeably these terms. The membrane structure can be also viewed as a rooted tree with the skin membrane as the root node.

A P system consists of a membrane structure, where each membrane is labeled with a number from 1 to m. Each membrane contains a multiset of objects. Objects can be transformed into other objects and sent through a membrane according to given rules defined for membrane labels. The rules are known from the beginning for each possible membrane, even for the ones that do not exist yet, or the ones that will never exist.

In this paper we work with P systems with active membranes (Active P systems). The rules can modify the membrane structure by dissolving and creating new membranes. That is why we will define the configuration to include the membrane structure as well.

Let Σ be a set of objects. We denote by $\mathbb{N}^{\Sigma}$ a set of all mappings from Σ to $\mathbb{N}$, so it contains all multisets of objects from Σ. A **membrane configuration** is a tuple (T, l, c), where:

- T is a rooted tree,
- $l \in \mathbb{N}^{V(T)}$ is a mapping that assigns for each node of T a number (label), where $l(r_T) = 1$, so the skin membrane is always labeled with 1,
- $c \in (\mathbb{N}^{\Sigma})^{V(T)}$ is a mapping that assigns for each node of T a multiset of objects from Σ, so it represents the contents of the membrane.

An **active P system** is a tuple $(\Sigma, C_0, R_1, R_2, \ldots, R_m)$, where:

- Σ is a set of objects,
- C_0 is initial membrane configuration,
- $R_1, R_2, \ldots, R_m$ are finite sets of rewriting rules associated with the labels $1, 2, \ldots, m$ and can be of forms:
  - $u \rightarrow w$, where $u \in \Sigma^+$, $w \in (\Sigma \times \{\cdot, \uparrow, \downarrow_j\})^*$ and $1 \leq j \leq m$,
  - a dissolving rule $u \rightarrow w\delta$, where $u \in \Sigma^+$, $w \in (\Sigma \times \{\cdot, \uparrow, \downarrow_j\})^*$ and $1 \leq j \leq m$,
  - a membrane creation $u \rightarrow [_j v]_j$, where $u \in \Sigma^+, v \in \Sigma^*$ and $1 \leq j \leq m$.

Although rewriting rules are defined as strings, u, v and w represent multisets of objects from Σ. For the first two forms, each rewriting rule may specify for each object on the right side, whether it stays in the current region (we will omit the symbol $\cdot$), moves through the membrane to the parent region ($\uparrow$) or to a specific child region ($\downarrow_j$, where j is a label of a membrane). If there are more child membranes with the same label, one is chosen nondeterministically. We denote these transfers with an arrow immediately after the symbol. An example of such rule is the following: $abb \to ab \downarrow_2 c \uparrow c\delta$.

By applying the rule we mean the removal of objects specified on the left side and the addition of the objects on the right side. Symbol $\delta \notin \Sigma$ does not represent an object. It may be present only at the end of the rule, which means that after the application of the rule, the membrane is dissolved and its contents (objects, child membranes) are propagated to the parent membrane.

Active P systems differ from classic (passive) P systems in ability to create new membranes by rules of the third form. Such rule will create new child membrane with a given label j and a given multiset of objects v as its contents.

For an active P system $(\Sigma, C_0, R_1, R_2, \ldots, R_m)$, configuration $C = (T, l, c)$, membrane $d \in V(T)$ the rule $r \in R_{l(d)}$ is **applicable** iff:

- $r = u \to w$ and $u \subseteq c(d)$ and for all $(a, \downarrow_k) \in w$ there exists $d_2 \in V(T)$ such that $l(d_2) = k \wedge parent(d_2) = d$,
- $r = u \to w\delta$ and $u \subseteq c(d)$ and for all $(a, \downarrow_k) \in w$ there exists $d_2 \in V(T)$ such that $l(d_2) = k \wedge parent(d_2) = d$ and $d \neq r_T$,
- $r = u \to [_j v]_j$ and $u \subseteq c(d)$.

In this paper we assume only sequential systems, so in each step of the computation, there is one rule nondeterministically chosen among all applicable rules in all membranes to be applied.

A **computation step** of P system is a relation $\Rightarrow$ on the set of configurations such that $C_1 \Rightarrow C_2$ holds iff there is an applicable rule in a membrane in C_1 such that applying that rule can result in C_2.

An **infinite computation** of a P system is an infinite sequence of configurations $\{C_i\}_{i=0}^{\infty}$, where $\forall i : C_i \Rightarrow C_{i+1}$.

A **finite computation** of a P system is a finite sequence of configurations $\{C_i\}_{i=0}^{n}$, where for all $0 \leq i < n : C_i \Rightarrow C_{i+1}$.

A **halting computation** of a P systems is a finite computation $\{C_i\}_{i=0}^{n}$, where there is no applicable rule in the last configuration C_n.

The P system can work in generating or in accepting mode. For the generating mode there are two possible ways of assigning a result of a computation:

1. By considering the multiplicity of objects present in a designated membrane in a halting configuration. In this case we obtain a vector of natural numbers. We can also represent this vector as a multiset of objects or as Parikh image of a string.
2. By concatenating the objects which leave the system, in the order they are sent out of the skin membrane (if several symbols are expelled at the same time, then any ordering of them is considered). In this case we generate a language.

The result of a single computation is clearly only one multiset or a string, but for one initial configuration there can be multiple possible computations. It follows from the fact that there can be more than one applicable rule in each configuration and they are chosen nondeterministically. For the accepting mode the input multiset is inserted in the skin membrane and it is accepted if and only if a given accepting configuration can be reached [3].

We will now introduce a variant with a global limit upon the membrane structure. We achieve this by restricting the rule application such that if the rule would result in a structure exceeding the limit, the rule will not be applicable.

An **active P system with a limit on the total number of membranes** is a tuple $(\Sigma, L, C_0, R_1, R_2, \ldots, R_m)$, where $(\Sigma, C_0, R_1, R_2, \ldots, R_m)$ is an active P system and $L \in \mathbb{N}$ is a limit on the total number of membranes. Anytime during the computation, a configuration (T, l, c) is not allowed to have more than L membranes, so the following invariant holds: $|V(T)| \leq L$.

This is achieved by adding a constraint for rule of the form $r = u \to [_k v]_k$, which is defined to be applicable iff $u \subseteq c(d)$ and $|V(T)| < L$. If the number of membranes is equal to L, there is no space for newly created membrane, so in that case such rule is not applicable.

4 Termination Problems

In this section we recall the halting problem for Turing machines. The problem is to determine, given a deterministic Turing machine and an input, whether the Turing machine running on that input will halt. It is one of the first known undecidable problems. On the other hand, for non-deterministic machines, there are two possible meanings for halting. We could be interested either in:

- whether there exists an infinite computation (the machine can run forever), or
- whether there exists a finite computation (the machine can halt).

We will prove the (un)decidability of these problems on active P systems with limit on the total number of membranes. These problems are defined for both generating and accepting mode, but in the accepting mode we consider the questions for a given P system along with the input multiset. The results are quite interesting, because:

Theorem 1. *Sequential active P systems with limit on the total number of membranes are computationally universal.*

Proof. The proof of this theorem for sequential active P systems in [3] uses simulation of register machines and during the simulation, every configuration has at most three membranes. Hence the active P system with limit on the total number of membranes exists (e.g. with $L = 3$), so the universality holds. □

This variant is not very realistic from biological point of view. Assuming membranes to know the number of membranes in the whole system is simply not plausible. Although we believe the results also hold for the variant without limit on the total number of membranes, due to technical difficulties it remains an open problem.

4.1 Existence of Infinite Computation

We will propose an algorithm for deciding existence of infinite computation. Basic idea is to consider the minimal coverability graph [6], where nodes are configurations and an edge leads from the configuration C_1 to the configuration C_2, whenever there is a rule applicable in C_1, which results in C_2. The construction in [6] is performed on Petri nets, where the configuration consists just of a vector of natural numbers. The situation is the same for single-membrane sequential P systems [7]. We need to modify the construction for active P systems.

A configuration $C_2 = (T_2, l_2, c_2)$ **covers** configuration $C_1 = (T_1, l_1, c_1)$ iff $\exists$ isomorphism $f : T_1 \rightarrow T_2$ preserving membrane labels and contents: $\forall d \in T_1$ the following properties hold: $l_1(d) = l_2(f(d)) \wedge c_1(d) \subseteq c_2(f(d))$.

We will denote this with $C_1 \leq C_2$.

Lemma 1. *For sequential active P system with limit on the total number of membranes, if $C_2 = (T_2, l_2, c_2)$* **covers** *configuration $C_1 = (T_1, l_1, c_1)$, then there is an isomorphism $f : T_1 \rightarrow T_2$ such that if a rule r is applicable in membrane $d \in T_1$, then r is applicable in $f(d)$.*

Proof. Suppose r is applicable in d. Then the left side u of the rule r is contained within the contents of the membrane $u \subseteq c_1(d)$. Because $C_1 \leq C_2$, then there is an isomorphism $f : T_1 \rightarrow T_2$ such that $c_1(d) \subseteq c_2(f(d))$ and then $u \subseteq c_2(f(d))$.

There are three possible forms of the rule r.

- If $r = u \rightarrow w$, then because r is applicable in d, $\forall (a, \downarrow_k) \in w \exists d_2 \in V(T_1)$: $l_1(d_2) = k \wedge parent_{T_1}(d_2) = d$. Because $C_1 \leq C_2$, then for $f(d_2) \in V(T_2)$ the following holds: $l_2(f(d_2)) = l_1(d_2) = k$ and $parent_{T_2}(f(d_2) = f(d)$. Hence r is applicable in $f(d)$.
- If $r = u \rightarrow w\delta$, then $d \neq r_{T_1}$. Since f is an isomorphism, then also $f(d) \neq r_{T_2}$. Other properties follows from the previous case.
- If $r = u \rightarrow [_k v]_k$, then $|V(T_1)| < L$. Isomorphism preserves number of nodes, hence $|V(T_2)| = |V(T_1)| < L$ and r is applicable in $f(d)$. □

Now, we will define the encoding of a configuration $C = (T, l, c)$ into a tuple of integers.

A membrane $d \in T$ will be encoded as $(n+m)$-tuple $enc(d) \in \mathbb{N}^{(n+m)}$, where first n numbers will be actual counts of objects and next m numbers will encode the membrane label with $m - 1$ zeros and one one:

$$enc(d)_i = \begin{cases} c(d)(a_i) & \text{if } i \leq n \\ 0 & \text{if } n < i \leq m \wedge i - n \neq l(d) \\ 1 & \text{if } n < i \leq m \wedge i - n = l(d). \end{cases}$$

The entire tree will be encoded into concatenated sequences of encoded nodes in the preorder traversal order. This sequence is then padded with zeroes to have length $(n + m)L$ as that is the maximal length of encoded tree.

Example 1. Suppose skin membrane with label 1 and contents $a_1^2 a_2$ with a child membrane with label 2 and contents a_2^2. Then the encoding will be 21100201, where 2110 encodes the skin membrane and 0201 encodes the child membrane.

Since there are only finitely many non-isomorphic trees with at most L nodes [8], there is a constant z such that we can uniquely assign the tree an order number $o(T) \leq z$.

The entire configuration will be encoded in tuple which consists of z parts. All but the part with index $o(T)$ will contain just zeros. The part with index $o(T)$ will contain the encoding of the tree.

We will now show that comparing two encodings corresponds to covering of two configurations. Recall that configurations are encoded into tuples of integers, so the comparison is performed position by position.

Lemma 2. *For configurations* $C_1 = (T_1, l_1, c_1)$ *and* $C_2 = (T_2, l_2, c_2)$, $enc(C_1) \leq enc(C_2) \Rightarrow C_1 \leq C_2$.

Proof. Both $enc(C_1)$ and $enc(C_2)$ contain z parts and exactly one part which contains non-zero values. The non-zero part of $enc(C_1)$ must be non-zero also in $enc(C_2)$, because $enc(C_1) \leq enc(C_2)$. Then $o(T_1) = o(T_2)$, so the trees are isomorphic. Suppose there is an isomorphism $f : T_1 \rightarrow T_2$. For every membrane $d \in T_1$, $l_1(d) = l_2(f(d))$ and $c_1(d) \subseteq c_2(f(d))$. Hence, $C_1 \leq C_2$. □

Lemma 3. *For sequential active P system with limit on the total number of membranes L for every infinite sequence of configurations* $\{C_i\}_{i=0}^{\infty}$ *there is a pair* $i < j$ *such that* $C_i \leq C_j$.

Proof. Suppose an infinite sequence $\{enc(C_i)\}_{i=0}^{\infty}$. We use a variation of Dickson's lemma [9]: Every infinite sequence of tuples from $\mathbb{N}^k$ contains an increasing pair. Applied to our sequence, there are two positions $i < j$ such that $enc(C_i) \leq enc(C_j)$. From Lemma 2, $C_i \leq C_j$. □

Theorem 2. *Existence of infinite computation for active P systems with limit on the total number of membranes is decidable.*

Proof. The algorithm for deciding the problem will traverse the reachability graph. When it encounters a configuration that covers another configuration, from Lemma 1 follows that the same rules can be applied repeatedly, so the algorithm will halt with the answer YES. Otherwise, the algorithm will answer NO. Algorithm will always halt, because if there was an infinite computation, from Lemma 3 there would be two increasing configurations which is already covered in the YES case. □

4.2 Existence of Halting Computation

In this subsection we will focus on the opposite problem: whether there is a computation that is halting. Recall that halting computation has no applicable rule in the last configuration.

The existence of halting configuration seems to be related to the language emptyness problem, which is undecidable. It is the problem to decide, if there is an input that is accepted by the P system. On the other hand, our variant of the halting problem is defined for a given input. It remains an open problem, whether there is a direct reduction between these two problems. If this is the case, the proof could be much simpler.

First, we will reduce this problem to the reachability problem. It is a problem of determining, for a given configuration C, whether there exists a computation from C_0 to C. Then, the reachability of active P systems can be then reduced to the reachabililty of register machines, which is undecidable.

For a given P system Π and a target configuration C we will construct a P system Π' such that there is a halting computation of $\Pi' \Leftrightarrow C$ is reachable for Π. Suppose $\Pi = (\Sigma, C_0, R_1, \dots, R_m)$ and $C = (T, l, c)$. Then we will construct $\Pi' = (\Sigma', C_0', R_1', \dots, R_m')$, where:

- $\Sigma' = \Sigma \cup \{\xi_d | d \in V(T)\}$,
- $C_0' = (T, l, c')$, where $\forall d \in V(T) \setminus r_T : c'(d) = c(d)$ and $c'(r_T) = c(r_T) \cup \{\xi_{r_T}\}$,
- $\forall i \in \{1, \dots, m\} : R_i' = R_i \cup \{\xi_d c(d) \rightarrow \xi_{d'} \downarrow_{l(d')} | d, d' \in V(T), l(d) = i, parent(d') = d\}$.

The ξ_d objects are called verifiers, they are intended to verify if the contents of the membrane corresponds to the contents in the target configuration C. After this verification it descends down into child membranes for the verification of other parts of the membrane structure. Initially, there is an object ξ_{r_T} in the skin membrane. Verification is performed in the rule $\xi_d c(d) \rightarrow \xi_{d'} \downarrow_{l(d')}$, where on the right side there is $\xi_{d'}$ object for every child membrane d' in the target configuration C.

The construction is not complete. The system should not be able to halt unless the verification takes place. That is why we introduce a new object ω to each membrane with a rule $\omega \rightarrow \omega$ and the verifier will erase them with rule $\xi_d \omega c(d) \rightarrow \xi_{d'} \downarrow_{l(d')}$. One application of this rule will erase the ω object and propagate proper ξ object to every child membrane in the target configuration C. We also need to ensure that newly created membranes contain the ω object, so we replace every rule for membrane creation $u \rightarrow [_k v]_k$ with $u \rightarrow [_k v\omega]_k$.

There is still a problem. We actually check, whether a target configuration is contained within the current configuration. If there are additional objects which cannot be erased, so C cannot be reached, we need to ensure Π' will not halt. We will add a rule $a \rightarrow a$ to each membrane for each object $a \in \Sigma$, so Π' can halt only if all objects are erased.

The last issue to solve is the dissolution. It is still possible that in the middle of the verifying, some of already verified membranes got dissolved and all yet unverified membranes will be successfully verified causing Π' to halt, although without that dissolution it would be unable to reach C. We need to ensure all dissolution happen before the verification takes place. We will add a new object σ, which stands as a footprint object. It will be created as a result of the verification rule $\xi_d \omega c(d) \rightarrow \sigma \xi_{d'} \downarrow_{l(d')}$. If a membrane is dissolved after it was verified, then two σs will meet in the same membrane, because the parent membrane also contains σ as it had been verified before. We will add a rule $\sigma\sigma \rightarrow \sigma\sigma$ to prevent Π' from halting.

The final construct is:

- $\Sigma' = \Sigma \cup \{\omega, \sigma\} \cup \{\xi_d | d \in V(T)\}$,
- $C'_0 = (T, l, c')$, where $\forall d \in V(T) \backslash r_T : c'(d) = c(d) \cup \{\omega\}$ and $c'(r_T) = c(r_T) \cup \{\omega, \xi_{r_T}\}$,
- $\forall i \in \{1, \ldots, m\} : R'_i = \{r | r \in R_i, r = u \rightarrow w \vee r = u \rightarrow w\delta\} \cup \{u \rightarrow [_k v\omega]_k | u \rightarrow [_k v]_k \in R_i\} \cup \{a \rightarrow a | a \in \Sigma\} \cup \{\sigma\sigma \rightarrow \sigma\sigma, \omega \rightarrow \omega\} \cup \{\xi_d \omega c(d) \rightarrow \sigma \xi_{d'} \downarrow_{l(d')} | d, d' \in V(T), l(d) = i, parent(d') = d\}$.

We can now state the main theorem.

Theorem 3. *Existence of halting computation for active P systems with limit on the total number of membranes is undecidable.*

We need to prove that C is reachable for Π if and only if there is a halting computation of Π'. We need to prove two implications in order to formally prove correctness of this construction. The proof is quite technical and is present in the appendix.

5 Conclusion

We have studied the termination problems for active sequential P systems. Unlike deterministic systems, the termination problems cannot be simply reduced to the halting problem. We have shown that active P systems with limit on the number of membranes have decidable existence of infinite computation and undecidable existence of halting computation. It is currently unknown whether the same results apply also for a variant without the limit on the number of membranes, so it could be a subject for the future study.

Regarding the open problem stated in [3] about sequential active P systems with hard membranes (without communication between membranes), it could be interesting to find a connection between the universality and decidability of these termination problems.

References

1. Paun, G., Rozenberg, G., Salomaa, A.: The Oxford Handbook of Membrane Computing. Oxford University Press Inc., New York (2010)
2. Păun, G.: Computing with membranes. J. Comput. Syst. Sci. **61**(1), 108–143 (2000)
3. Ibarra, O.H., Woodworth, S., Yen, H.-C., Dang, Z.: On sequential and 1-deterministic P systems. In: Wang, L. (ed.) COCOON 2005. LNCS, vol. 3595, pp. 905–914. Springer, Heidelberg (2005)
4. Alhazov, A.: Properties of membrane systems. In: Gheorghe, M., Păun, G., Rozenberg, G., Salomaa, A., Verlan, S. (eds.) CMC 2011. LNCS, vol. 7184, pp. 1–13. Springer, Heidelberg (2012)

5. Păun, G.: Introduction to membrane computing. In: Ciobanu, G., Păun, G., Pérez-Jiménez, M. (eds.) Applications of Membrane Computing. Natural Computing Series, pp. 1–42. Springer, Heidelberg (2006)
6. Finkel, A.: The minimal coverability graph for petri nets. In: Rozenberg, G. (ed.) Advances in Petri Nets 1993. LNCS, vol. 674, pp. 210–243. Springer, Heidelberg (1993)
7. Dal Zilio, S., Formenti, E.: On the dynamics of PB systems: a petri net view. In: Martín-Vide, C., Mauri, G., Păun, G., Rozenberg, G., Salomaa, A. (eds.) WMC 2003. LNCS, vol. 2933, pp. 153–167. Springer, Heidelberg (2004)
8. Cayley, P.: On the analytical forms called trees. Am. J. Math. **4**(1), 266–268 (1881)
9. Figueira, D., Figueira, S., Schmitz, S., Schnoebelen, P.: Ackermannian and primitive-recursive bounds with dickson's lemma. In: Proceedings of the 2011 IEEE 26th Annual Symposium on Logic in Computer Science, LICS 2011, pp. 269–278. IEEE Computer Society, Washington, DC (2011)

Weihrauch Degrees of Finding Equilibria in Sequential Games

Stéphane Le Roux[1] and Arno Pauly[2](✉)

[1] Département d'informatique, Université libre de Bruxelles, Bruxelles, Belgique
Stephane.Le.Roux@ulb.ac.be
[2] Clare College, University of Cambridge, Cambridge, UK
Arno.Pauly@cl.cam.ac.uk

Abstract. We consider the degrees of non-computability (Weihrauch degrees) of finding winning strategies (or more generally, Nash equilibria) in infinite sequential games with certain winning sets (or more generally, outcome sets). In particular, we show that as the complexity of the winning sets increases in the difference hierarchy, the complexity of constructing winning strategies increases in the effective Borel hierarchy.

1 Overview

We consider questions of (non)computability related to infinite sequential games played by any countable number of players. The best-known example of such games are Gale-Stewart games [10], which are two-player win/lose games. The existence of winning strategies in (special cases of) Gale-Stewart games is often employed to show that truth-values in certain logics are well-determined. The degrees of noncomputability of variations of (Borel) determinacy [17] can be studied using our techniques, and several are fully classified.

This work falls within the research programme to study the computational content of mathematical theorems in the Weihrauch lattice, which was outlined by Brattka and Gherardi in [3]. In particular, it continues the investigation of the Weihrauch degrees of operations mapping games to their equilibria started in [21]. There, finding pure and mixed Nash equilibria in two-player games with finitely many actions in strategic form were classified.

One motivation for this line of inquiry is the general stance that solution concepts in game theory can only be convincing if the players are capable of (at least jointly) computing them, taken e.g. in [22]. Even if we allow for some degree of hypercomputation, or are, e.g., willing to tacitly replace actually attaining a solution concept by some process (slowly) converging to it, we still have to reject solution concepts with too high a Weihrauch degree.

The results for determinacy of specific pointclasses that we provide are a refinement of results obtained in reverse mathematics by Nemoto, MedSalem and Tanaka [20]; the first is also a uniformization of a result by Cenzer and Remmel [7]. For some represented pointclass Γ, let $\mathrm{Det}_\Gamma : \Gamma \rightrightarrows \{0,1\}^{\mathbb{N}}$ be

A full version is available as [16].

A. Beckmann et al. (Eds.): CiE 2015, LNCS 9136, pp. 246–257, 2015.
DOI: 10.1007/978-3-319-20028-6_25

the map taking a Γ-subset A of Cantor space to a (suitably encoded) Nash equilibrium in the sequential two-player game with alternating moves where the first player wins if the induced play is in A, and the second player wins otherwise. Let $\mathcal{A}$ be the closed subsets of Cantor space, and $\mathfrak{D} := \{U \setminus U' \mid U, U' \in \mathcal{A}\}$. Some of our results are:

Theorem. $Det_{\mathcal{A}} \equiv_W C_{\{0,1\}^{\mathbb{N}}}$ *and* $Det_{\mathfrak{D}} \equiv_W C_{\{0,1\}^{\mathbb{N}}} \star \lim$.

We have two remarks. One, by combining the preceding theorem with the main result of [6], we find that $\mathrm{Det}_{\mathfrak{D}}$ is equivalent to the Bolzano-Weierstrass-Theorem. This may be a bit unexpected in particular seeing that $\mathrm{C}_{\{0,1\}^{\mathbb{N}}} \star \lim$ is not (yet) known to contain a plethora of mathematical theorems (unlike, e.g., $\mathrm{C}_{\{0,1\}^{\mathbb{N}}}$). Two, we already need to use a limit operator in order to move up one level of the difference hierarchy – rather than being able to move up one level in the Borel hierarchy as one may have expected naively. Thus, this observation may complement Harvey Friedman's famous result [9] that proving Borel determinacy requires repeated use of the axiom of replacement.

Another group of results is based on inspecting the various results extending Borel determinacy to more general classes of games (and solution concepts) in [13–15]. If we instantiate these generic results with specific determinacy version as above, we can prove for some of them that they are actually optimal w.r.t. Weihrauch reducibility. We shall state two such classifications explicitly.

Consider two-player sequential games with finitely many *outcomes* and antagonistic (inverse of each other) linear preferences over the outcomes. For any upper set of outcomes w.r.t. some player's preference let the corresponding set of plays be open or closed. Let $\mathrm{NE}^{ap}_{\mathcal{O}\cup\mathcal{A}}$ be the operation taking such a game (suitably encoded) and producing a Nash equilibrium. Then:

Theorem. $NE^{ap}_{\mathcal{O}\cup\mathcal{A}} \equiv_W C_{\{0,1\}^{\mathbb{N}}} \times LPO^*$.

Next, we restrict the aforementioned class of games to antagonistic games, that is, games where the preferences of one player are the inverse of the preferences of the other player. Those games will have subgame-perfect equilibria, and we let $\mathrm{SPE}_{\mathcal{O}\cup\mathcal{A}}$ be the operation mapping such games to a subgame-perfect equilibrium.

Theorem. $SPE_{\mathcal{O}\cup\mathcal{A}} \equiv_W \lim$.

2 Fundamentals

We proceed to give brief, informal introductions to represented spaces, Weihrauch reducibility and infinite sequential games. For a formal treatment and further references, we refer to the extended version of the present paper [16].

2.1 Informal Background on Represented Spaces and Weihrauch Reducibility

We use representations to induce computability notions on the spaces of interest to us, in particular on pointclasses (i.e. sets of subsets of Cantor space) and derived from that, infinite sequential games. A represented space is a set X together with a partial surjection $\delta :\subseteq \mathbb{N}^{\mathbb{N}} \to X$. A function between represented spaces is computable, iff there is a matching computable function on Baire space.

An open subset U of Cantor space is represented by some list of finite words $(w_i)_{i\in\mathbb{N}}$ such that $U = \bigcup_{i\in\mathbb{N}} w_i\{0,1\}^{\mathbb{N}}$. For other derived pointclasses their representation follows directly from how they are defined; i.e. closed sets are given via their complement as an open set, a Σ^0_2-set is given by a sequence of closed sets whose union it is, etc. (this idea is explored in more detail in [2]).

Weihrauch reducibility is a relation between multivalued functions on represented spaces, where $f \leq_{\mathrm{W}} g$ means that f can be computed with the help of a single oracle call to g. We make use of some operations on Weihrauch degrees: With $f \times g$ we denote the ability to make a call to f and an independent call to g. Having f^* means being able to make any given finite number of independent calls to f, whereas $\widehat{f}$ allows countably many parallel calls. Given $f \star g$, one can first use g, and then use the answer to choose the query for f. By $f^{(n)}$ we denote the n-th fold iteration of $\star$ on f. Finally, $f^{[n]}$ means that one does not have to provide the input to f explicitly, but merely a sequence converging to a sequence... (n-times) converging to the input to f (so $f^{[0]} = f$, $f^{[1]} = f'$ with f' as in [6]).

There is a zoo of Weihrauch degrees commonly appearing in the classification of theorems. Relevant for us are $\mathrm{C}_{\{0,1\}^{\mathbb{N}}}$, which takes a non-empty closed subset of Cantor space (i.e. an infinite binary tree) and produces a point in it (i.e. an infinite path through the tree), LPO, which decides whether a sequence is constant 0 or not, and lim, which computes the limit of a sequence in Cantor space.

2.2 Informal Background on Infinite Sequential Games

We use the formal definitions of sequential games and related concepts from [15] and [13]. Informally, given a fixed (wlog) set C, we let the players sequentially choose elements in C until an infinite sequence in C^ω is generated. Whose turn it is depends on the finite history of choices. The outcome (from a set O) of the game depends on the generated sequence in C^ω, and each player may compare outcomes via a binary relation over O, called preference. A strategy of a player is an object that fully specifies what the choice of the player would be for each possible finite history that requires this player to play. A combination of one strategy per player is called a strategy profile and it induces one unique infinite sequence in C^ω, and thus one unique outcome. So, preferences may be lifted from outcomes to strategy profiles. A Nash equilibrium is a profile such that no player can unilaterally change strategies and induce a (new) outcome that he or she prefers over the old one. We also consider a refinement of the concept of a Nash equilibrium, namely subgame-perfect Nash equilibria. Intuitively, a strategy profile is subgame-perfect, if it still forms an equilibrium if the game were started at an arbitrary history.

As a important special case we consider win/lose games. These are games with two players a, b and two outcomes w_a, w_b, where a prefers w_a to w_b and b prefers w_b to w_a. We say that a wins the game, if outcome w_a is reached, and call the set of all plays that induce outcome w_a as the winning set for a (likewise for b and w_b).

2.3 Defining the Problems of Interest

Let Γ be a represented pointclass over $\{0,1\}^{\mathbb{N}}$. In a straightforward fashion, we can obtain a representation of the infinite sequential games with countably many agents, countably many outcomes, sets of choices $C = \{0,1\}$ and Γ-measurable valuation function $v : \{0,1\}^{\mathbb{N}} \to O$. The representation encodes the number of agents and outcomes available, for each upper set of outcome the Γ-set of plays resulting in it, the map d as a look-up table, and the relations $\prec_a$ as look-up tables. We always assume that the inverse of any preference relation is well-founded (this guarantees that equilibria exist). Using a canonic isomorphism $\{0,1\}^* \cong \mathbb{N}$, we will pretend that the space of strategy profiles in such a game is $\{0,1\}^{\mathbb{N}}$.

We now consider the following multivalued functions:

1. Det_Γ takes a two-player win/lose game as input, where the first player has a winning set in Γ. Valid outputs are the Nash equilibria, i.e. the pairs of strategies where one strategy is a winning strategy.
2. Win_Γ has the same inputs as Det_Γ, and decides which player (if any) has a winning strategy.
3. FindWS_Γ is the restriction of Det_Γ to games where the first player has a winning strategy.
4. NE_Γ takes as input a game with countably many players, finitely many outcomes, and linear preferences, where each upper set of outcomes (w.r.t. each player preference) comes from a Γ-set. The valid outputs are the Nash equilibria.
5. NE^{ap}_Γ is the restriction of NE_Γ to the two-player games with antagonistic preferences (i.e. $\prec_a = \prec_b^{-1}$).
6. SPE_Γ takes as input a two-player game with finitely many outcomes and antagonistic preferences, where each upper set of outcomes comes from a Γ-set. Valid outputs are the subgame perfect equilibria.

We abbreviate $\overline{\Gamma} := \{U^C \mid U \in \Gamma\}$. Some trivial reducibilities between these problems are: $\mathrm{Win}_\Gamma \equiv_{\mathrm{W}} \mathrm{Win}_{\overline{\Gamma}}$, $\mathrm{Det}_\Gamma \equiv_{\mathrm{W}} \mathrm{Det}_{\overline{\Gamma}}$, $\mathrm{FindWS}_\Gamma \leq_{\mathrm{W}} \mathrm{Det}_\Gamma \leq_{\mathrm{W}} \mathrm{NE}^{ap}_{\Gamma\cup\overline{\Gamma}} \leq_{\mathrm{W}} \mathrm{SPE}_{\Gamma\cup\overline{\Gamma}}$ and $\mathrm{NE}^{ap}_\Gamma \leq_{\mathrm{W}} \mathrm{NE}_\Gamma$.

Throughout the paper we assume that Γ is determined (which implies that all operations are well-defined in the first place), closed under rescaling and finite intersection with clopens, and that $\emptyset, \{0,1\}^{\mathbb{N}} \in \Gamma$. All such closure properties (including those appearing as conditions in the results) are assumed to hold in a uniformly computable way, e.g. given a name for a set in Γ and a clopen, we can compute a name for the intersection of the set with the clopen. With rescaling we refer to the operation $(w, A) \mapsto \{wp \mid p \in A\} : \{0,1\}^* \times \Gamma \to \Gamma$ and its inverse.

2.4 The Difference Hierarchy

The pointclasses we shall study in particular are the levels of the Hausdorff difference hierarchy. Intuitively, these are the sets that can be obtained as boolean combinations of open sets; and their level denotes the least complexity of a suitable term. Roughly following [12, Sect. 22.E], we shall recall the definition of the difference hierarchy. We define a function par from the countable ordinals to $\{0,1\}$ by $\operatorname{par}(\alpha) = 0$, if there is a limit ordinal β and a number $n \in \mathbb{N}$ such that $\alpha = \beta + 2n$; and $\operatorname{par}(\alpha) = 1$ otherwise. For a fixed ordinal α, we let $\mathfrak{D}_\alpha$ be the collection of sets D definable in terms of a family $(U_\lambda)_{\lambda<\alpha}$ of open sets via:

$$x \in D \Leftrightarrow \operatorname{par}\left(\inf\{\beta \mid x \in U_\beta\}\right) \neq \operatorname{par}(\alpha)$$

In the preceding formula, we understand that $\inf \emptyset = \alpha$. In particular, $\mathfrak{D}_0 = \{\emptyset\}$ and $\mathfrak{D}_1 = \mathcal{O}$.

For our constructions, a different characterization is more useful, though: For some pointclass Γ, let $\mathfrak{D}(\Gamma) := \{\bigcup_{i \in I} v_i U_i \mid \forall i,j \in I v_i \in \{0,1\}^* \wedge U_i \in \Gamma \wedge v_i \not\prec v_j\}$.

Lemma 1. $\mathfrak{D}_{\alpha+1} = \mathfrak{D}(\overline{\mathfrak{D}_\alpha})$ *and, more generally,* $\mathfrak{D}_\alpha = \mathfrak{D}\left(\overline{\bigcup_{\lambda<\alpha} \mathfrak{D}_\lambda}\right)$.

Observation 1. *If* A_n *is in* $\mathfrak{D}_\alpha$ *for all* $n \in \mathbb{N}$, *so is* $A := \cup_{n \in \mathbb{N}} 0^n 1 A_n$.

Corollary 1. *If* B_n *is in* $\overline{\mathfrak{D}_\alpha}$ *for all* $n \in \mathbb{N}$, *so is* $B := \{0^{\mathbb{N}}\} \cup \bigcup_{n \in \mathbb{N}} 0^n 1 B_n$.

A fundamental result on the difference hierarchy is the Hausdorff-Kuratowski theorem stating that $\bigcup_{\alpha<\omega_1} \mathfrak{D}_\alpha = \Delta^0_2$ (where ω_1 is the smallest uncountable ordinal), see e.g. [12, Theorem 22.27].

3 The Computational Content of Some Determinacy Principles

We begin by classifying the simplest non-computable games, namely games where the first player wants to reach some closed set. This classification essentially is a uniform version of a result by CENZER and REMMEL [7].

Theorem 2. *FindWS*$_{\mathcal{A}} \equiv_W$ *Det*$_{\mathcal{A}} \equiv_W C_{\{0,1\}^{\mathbb{N}}}$.

Proof. $\mathbf{C}_{\{0,1\}^{\mathbb{N}}} \leq_{\mathbf{W}}$ **FindWS**$_{\mathcal{A}}$. Given a closed subset $A \in \mathcal{A}(\{0,1\}^{\mathbb{N}})$, we can easily obtain the game where only player 1 moves, and player 1 wins iff the induced play falls in A. If A is non-empty, then player 1 has a winning strategy: Play any infinite sequence in A.

$\text{FindWS}_{\mathcal{A}} \leq_{\text{W}} \text{Det}_{\mathcal{A}}$. Trivial.

$\text{Det}_{\mathcal{A}} \leq_{\text{W}} \text{C}_{\{0,1\}^{\mathbb{N}}}$. Given the open winning set of player 2, we can modify the game tree by ending the game once we know for sure that player 2 will win. Now the set of strategy profiles where either player 1 wins and player 2 cannot win, or player 2 wins and player 1 cannot prolong the game, is a closed set effectively obtainable from the game. Moreover, it is non-empty, and any such strategy profile is a Nash equilibrium. □

Proposition 1. *$\mathit{Win}_{\mathcal{A}} \equiv_W \mathit{LPO}$.*

Proof. This follows by combining the constructions from the preceding theorem with the fact that $\text{IsEmpty} : \mathcal{A}(\{0,1\}^{\mathbb{N}}) \to \{0,1\}$ is equivalent to LPO. □

We can use the results for $\mathcal{A}$ as the base case for classifying the strength of determinacy for the difference hierarchy.

Lemma 2 ([1]). *$\mathit{Det}_{\mathfrak{D}(\Gamma)} \leq_W C_{\{0,1\}^{\mathbb{N}}} \star \widehat{(\mathit{Det}_\Gamma \times \mathit{Win}_\Gamma)}$ and $\mathit{Win}_{\mathfrak{D}(\Gamma)} \leq_W \mathit{LPO} \star \widehat{\mathit{Win}_\Gamma}$.*

We will relate deciding the winner and finding a winning strategy for games induced by sets from some level of the difference hierarchy to the *lessor limited principle of omniscience* and the *law of excluded middle* for Σ^0_n-formulae of the corresponding level. These principles were studied in [1,6,11] (among others). Let $(\Sigma^0_n - \text{LLPO}) :\subseteq \{0,1\}^{\mathbb{N}} \times \{0,1\}^{\mathbb{N}} \rightrightarrows \{0,1\}$ be defined via $i \in (\Sigma^0_n - \text{LLPO})(p_0, p_1)$ iff $\forall k_1 \exists k_2 \ldots \natural k_n\ p_i(\langle k_1, \ldots, k_n\rangle) = 1$ (where $\natural = \forall$ if n is odd and $\natural = \exists$ otherwise). Let $(\Sigma^0_n - \text{LEM}) : \{0,1\}^{\mathbb{N}} \to \{0,1\}$ be defined via $(\Sigma^0_n - \text{LEM})(p) = 1$ iff $\forall k_1 \exists k_2 \ldots \natural k_n\ p(\langle k_1, \ldots, k_n\rangle) = 1$ and $(\Sigma^0_n - \text{LEM})(p) = 0$ otherwise. Then:

Proposition 2. *$(\Sigma^0_{n+1} - \mathit{LLPO}) \equiv_W \mathit{LLPO}^{[n]}$ and $(\Sigma^0_{n+1} - \mathit{LEM}) \equiv_W \mathit{LPO}^{[n]}$.*

Lemma 3. *$(\Sigma^0_n - \mathit{LLPO}) \leq_W \mathit{Det}_{\mathfrak{D}_n}$ and $(\Sigma^0_n - \mathit{LEM}) \leq_W \mathit{Win}_{\mathfrak{D}_n}$.*

Partial proof. The game for the first claim works as follows: The second player may pick some $k_1 \in \mathbb{N}$, or refuse to play. If the second player picks a number, then the first player may pick $k_2 \in \mathbb{N}$ or refuse to play. This alternating choice continues until k_{n-1} has been chosen, or a player refuses to pick. A player refusing to pick a number loses. If all numbers are picked, the winner depends on the input p to $\Sigma^0_n - \text{LEM}$ as follows: If n is even and $\exists k_n\ p(\langle k_1, \ldots, k_n\rangle) = 1$, then player 1 wins. If n is odd, and $\exists k_n\ p(\langle k_1, \ldots, k_n\rangle) = 0$, then player 2 wins. Note that this always describes an open component U_{picked} of the winning set of the respective player.

---

[1] This is a generalization of the proof idea for [20, Theorem 3.7] by NEMOTO, MEDSALEM and TANAKA. [20, Theorem 3.7] states that ACA_0 proves determinacy for $\mathfrak{D}(\Sigma^0_1)$.

Furthermore, note that the set of plays U_j where a value for k_j was chosen is always an open set. Now the condition that the second player refused to pick first is $U_1^C \cup (U_2 \cap U_3) \cup (U_4 \cap U_5) \cup \ldots$. This makes for a winning set in $\mathfrak{D}_n$, as required. If player 1 has a winning strategy in the game, the answer to $\left(\Sigma_n^0 - \text{LEM}\right)(p)$ is 1, if player 2 wins, it is 0. □

Combining the results above yields:

Theorem 3. *$Det_{\mathfrak{D}_{n+1}} \equiv_W C_{\{0,1\}^{\mathbb{N}}}^{[n]}$ and $Win_{\mathfrak{D}_{n+1}} \equiv_W LPO^{[n]}$.*

Knowing the Weihrauch degree of a mapping entails some information about the Turing degrees of outputs relative to the Turing degrees of inputs, this was explored in e.g. [4–6,22]. Thus, we can obtain the following corollaries:

Corollary 2. *Any computable game with a winning condition in $\mathfrak{D}_{n+1}$ has a winning strategy s such that s' is computable relative to $\emptyset^{(n+1)}$, and there is a computable game of this type such that any winning strategy computes the n-th Turing jump of a completion of Peano arithmetic.*

Corollary 3. *Let $(G_i)_{i \in \mathbb{N}}$ be an effective enumeration of computable games with winning conditions in $\mathfrak{D}_{n+1}$, and define $w \in \{0,1\}^{\mathbb{N}}$ via $w(i) = 1$ iff the first player has a winning strategy in G_i. Then $w \leq_T \emptyset^{(n+1)}$, and there is an enumeration $(G_i)_{i \in \mathbb{N}}$ such that $w \equiv_T \emptyset^{(n+1)}$.*

Corollary 4. *There is a Σ_{n+1}^0-measurable function mapping games with winning conditions in $\mathfrak{D}_n$ to winning strategies, but no Σ_n^0-measurable such function.*

4 The Complexity of Equilibrium Transfer

In [13–15], various results were provided that transfer Borel determinacy (or, somewhat more general, determinacy for some pointclass), to prove the existence of Nash equilibria (and sometimes even subgame-perfect equilibria) in multi-player multi-outcome infinite sequential games. In this section, we shall inspect those constructions and extract Weihrauch reductions from them.

In [14], the first author gave a very general construction that allows to extend determinacy of win/lose games to the existence of Nash equilibria for two-player games of the same type. For brevity, we only consider the strength of the toy example from [14] here:

Theorem 4 (Equilibrium transfer). *$NE_{\Gamma}^{ap} \leq_W Det_{\Gamma}^* \times Win_{\Gamma}^*$.*

Proof. For any upper set of outcomes (for either players preferences), we construct the win/lose derived game where that player wins, iff he enforces the set, and loses otherwise. There are finitely many such games, so we can use Win_{Γ}^* to decide which are won and which are lost. As shown in [14], there will be a combination of upper sets of outcomes for each player, such that if both players enforce their upper set, this forms a Nash equilibrium. We use Det_{Γ}^* to compute Nash equilibria for all derived games in parallel, and then simply select the suitable strategies. □

Techniques suitable for multiplayer sequential games were then introduced in [13], again by the first author. Again for brevity, we only consider the version with finitely many outcomes:

Theorem 5 (Constructing Nash equilibria). $NE_\Gamma \leq_W \widehat{Win_\Gamma} \times \widehat{Det_\Gamma}$.

A further improvement on the techniques in [13] were provided by the authors in [15]. These techniques in particular suffice to prove the existence of subgame-perfect equilibria in antagonistic games (this implies two players and finitely many outcomes).

Theorem 6. $SPE_\Gamma \leq_W \widehat{Win_\Gamma} \times \widehat{Det_\Gamma}$.

5 Deciding the Winner and Finding Nash Equilibria

The results in Sect. 3 show that for many concrete examples of Γ, the problem Det_Γ is inherently multivalued, i.e. not equivalent to any functions between admissible spaces. On the other hand, the upper bounds provided in Sect. 4 all include Win_Γ, which is of course single-valued. In the current section, we will explore some converse reductions, from deciding the winner to finding Nash equilibria. This generally requires some (rather tame) requirements on the pointclasses involved.

Lemma 4. *Let Γ be obtained by Γ_1 by first closing under finite union, rescaling and union with clopens; and then adding complements. Then:*

$$Win^*_{\Gamma_1} \leq_W NE^{ap}_\Gamma$$

Proof sketch. Given n win/lose games, the first player starts by announcing which of these games she believes she can win. Then the second player can choose one of the listed games to play. If the first player did not claim any winnable games, the game ends and the outcome is 0. If the first player claimed to be able to win k out of n games, then the outcomes of the games subsequently chosen by the second player are scaled up to $k, -k$. Thus, the first player has every reason to list precisely those games she can actually win: If she would not list a game she could win, she trades payoff $k-1$ for payoff k. If she lists a game she cannot win, the second player will subsequently chose and win it, and then the first player is punished by $-k$. □

Lemma 5. *Let Γ be closed under taking unions with Γ_1 and $\overline{\Gamma_1}$. Then:*

$$NE^{ap}_\Gamma \times FindWS_{\Gamma_1} \equiv_W NE^{ap}_\Gamma$$

Lemma 6. *Let Γ be obtained by Γ_1 by first closing under finite union, rescaling and union with clopens; and then adding complements. Then:*

$$FindWS^*_{\Gamma_1} \times Win^*_{\Gamma_1} \leq_W NE^{ap}_\Gamma$$

If we have access to subgame perfect equilibria (and are in a context where they are guaranteed to exist), then we can even decide the winner of countably many games in parallel:

Lemma 7. *Let Γ_1 contain the closed sets and be closed under finite unions and the operation $(A_n)_{n\in\mathbb{N}} \mapsto \left(\{0^{\mathbb{N}}\} \cup \bigcup_{n\in\mathbb{N}} 0^n 1 A_n\right)$. Let Γ be obtained from Γ_1 by closing under complements. Then:*

$$\widehat{Win_{\Gamma_1}} \leq_W SPE_\Gamma$$

Proof sketch. Combine the input games like this:

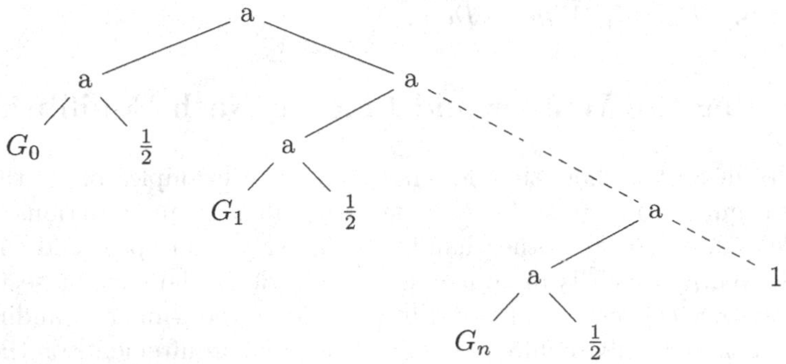

□

6 General Games with Concrete Pointclasses

The general constructions put together with the classifications for specific pointclasses allow us to obtain some concrete Weihrauch degrees. First, we shall see that moving from a win/lose game with closed and open outcomes to a two-player game with several outcomes just complicates the operation of finding Nash equilibria by finitely many uses of LPO in parallel:

Theorem 7. $NE^{ap}_{\mathcal{O}\cup\mathcal{A}} \equiv_W C_{\{0,1\}^{\mathbb{N}}} \times LPO^*$.

Proof. For the reduction $\text{NE}^{ap}_{\mathcal{O}\cup\mathcal{A}} \leq_W \text{C}_{\{0,1\}^{\mathbb{N}}} \times \text{LPO}^*$, instantiate Theorem 4 with the results from Theorem 2 and Proposition 1.

For the other direction, note that $\text{FindWS}_{\mathcal{A}} \equiv_W \text{FindWS}_{\mathcal{O}\cup\mathcal{A}} \equiv_W \text{C}_{\{0,1\}^{\mathbb{N}}}$ as in Theorem 2; and that $\Gamma_1 := \mathcal{A}$ and $\Gamma := \mathcal{O} \cup \mathcal{A}$ satisfy the requirements of Lemma 6, which then provides the desired result. □

The result can actually be strengthened into the following (by noting that the second game constructed in Lemma 3 is always won by the first player):

Theorem 8. $NE^{ap}_{\mathfrak{D}_{n+1}\cup\overline{\mathfrak{D}_{n+1}}} \equiv_W C^{[n]}_{\{0,1\}^{\mathbb{N}}} \times \left(LPO^{[n]}\right)^*$.

If one wishes to have subgame-perfect equilibria instead of mere Nash equilibria, then countably many uses of LPO become necessary, and the problem becomes equivalent to lim. Note that as long as there are at least three distinct outcomes, the number of outcomes has no further impact on the Weihrauch degree (due to the nature of the construction used to prove Lemma 7)– unlike the situation in Theorem 7, where the number of outcomes is related to the number of times that LPO is used.

Theorem 9. $SPE_{\mathfrak{D}_n \cup \overline{\mathfrak{D}_n}} \equiv_W \lim^{(n)}$.

Proof. For $\mathrm{SPE}_{\mathfrak{D}_n \cup \overline{\mathfrak{D}_n}} \leq_W \lim^{(n)}$, instantiate Theorem 6 with the results from Theorem 3, and note that $\widehat{\mathrm{LPO}^{(n)}} \equiv_W \lim^{(n)}$ and $\mathrm{C}^{[n]}_{\{0,1\}^{\mathbb{N}}} \leq_W \lim^{(n+1)}$.

For the other direction, we use Lemma 7 (applicable by Corollary 1) together with Proposition 1. □

Regarding Theorem 5, we do not (yet?) have matching lower bounds for any particular pointclass. The gap is exemplified by the following:

Corollary 5. $C_{\{0,1\}^{\mathbb{N}}} \times LPO^* \leq_W NE_{\mathcal{O} \cup \mathcal{A}} \leq_W \lim$.

7 Conclusions and Outlook

With Theorem 3, we have shown that the computational strength of determinacy provides a tight connection between the difference hierarchy and the Borel hierarchy (in form of Corollary 4). Note that winning sets from the difference hierarchy correspond to Boolean combinations of reachability and safety conditions. Corollary 3 then provides an upper bound and a worst case for corresponding decidability questions for logic. Theorem 3 also shows that the computational powers of the players required to find a winning strategy vastly exceeds the computational power required to determine the outcome, thus casting doubt on the adequateness of winning strategies (or Nash equilibria) as adequate solution concepts for infinite sequential games.

The results in Sect. 4 contrasted with those in Sect. 6 essentially show that the proofs in [13–15] are not too wasteful from a constructive perspective - i.e. the constructions employed are not far less constructive than the theorems proven with them.

There are several immediate avenues for extending the work presented here: The restriction to finite action sets (i.e. finitely branching trees) can mostly be lifted without a significant impact on the proof techniques. Note though that the concrete Weihrauch degrees would change drastically, as in Theorem 2 we would need to replace $\mathrm{C}_{\{0,1\}^{\mathbb{N}}}$ by $\mathrm{C}_{\mathbb{N}^{\mathbb{N}}}$, with the latter residing in a less explored part of the Weihrauch lattice. The results in [15] are more general than covered here, too (with the same proof complexity).

The study of the strength of determinacy for particular pointclasses in reverse mathematics presumably offers further proofs adaptable into the framework of Weihrauch reducibility, e.g. [8,18,19].

Further afield, understanding the Weihrauch degrees of determinacy principles may be a contribution to the development of descriptive set theory in computational/category-theoretical terms as suggested in [23].

Acknowledgements. This work benefited from the Royal Society International Exchange Grant IE111233 and the Marie Curie International Research Staff Exchange Scheme *Computable Analysis*, PIRSES-GA-2011- 294962.

References

1. Akama, Y., Berardi, S., Hayashi, S., Kohlenbach, U.: An arithmetical hierarchy of the law of excluded middle and related principles. In: 19th IEEE Symposium on Logic in Computer Science (LICS 2004), pp. 192–201 (2004)
2. Brattka, V.: Effective Borel measurability and reducibility of functions. Math. Logic Q. **51**(1), 19–44 (2005)
3. Brattka, V., Gherardi, G.: Effective choice and boundedness principles in computable analysis. Bull. Symbolic Logic **1**, 73–117 (2011). arXiv:0905.4685
4. Brattka, V., Gherardi, G.: Weihrauch degrees, omniscience principles and weak computability. J. Symbolic Logic **76**, 143–176 (2011). arXiv:0905.4679
5. Brattka, V., Gherardi, G., Hölzl, R.: Probabilistic computability and choice. arXiv 1312.7305 (2013). http://arxiv.org/abs/1312.7305
6. Brattka, V., Gherardi, G., Marcone, A.: The Bolzano-Weierstrass theorem is the jump of weak König's Lemma. Ann. Pure Appl. Logic **163**(6), 623–625 (2012). arXiv:1101.0792
7. Cenzer, D., Remmel, J.: Recursively presented games and strategies. Math. Soc. Sci. **24**(2–3), 117–139 (1992)
8. Eguchi, N.: Infinite games in the cantor space over admissible set theories. In: Higuchi, K. (ed.) Proceedings of Computability Theory and Foundations of Mathematics (2014)
9. Friedman, H.: Higher set theory and mathematical practice. Ann. Math. Logic **2**(3), 325–357 (1971)
10. Gale, D., Stewart, F.: Infinite games with perfect information. In: Contributions to the Theory of Games, Annals of Mathematical Studies, vol. 28, pp. 245–266. Princeton University Press (1953)
11. Higuchi, K., Kihara, T.: Inside the muchnik degrees I: discontinuity, learnability and constructivism. Ann. Pure Appl. Logic **165**(5), 1058–1114 (2014)
12. Kechris, A.: Classical Descriptive Set Theory. Graduate Texts in Mathematic, vol. 156. Springer, New York (1995)
13. Le Roux, S.: Infinite sequential Nash equilibria. Logical Methods Comput. Sci. **9**(2), 14 (2013)
14. Le Roux, S.: From winning strategy to Nash equilibrium. Math. Logic Q. **60**(4–5), 354–371 (2014). http://dx.doi.org/10.1002/malq.201300034, arXiv 1203.1866
15. Le Roux, S., Pauly, A.: Infinite sequential games with real-valued payoffs. In: CSL-LICS 2014, pp. 62:1–62:10. ACM (2014). http://doi.acm.org/10.1145/2603088.2603120
16. Le Roux, S., Pauly, A.: Weihrauch degrees of finding equilibria in sequential games. arXiv:1407.5587 (2014)
17. Martin, D.A.: Borel determinacy. Ann. Math. **102**(2), 363–371 (1975). http://www.jstor.org/stable/1971035
18. Montalbán, A., Shore, R.A.: The limits of determinacy in second-order arithmetic. Proc. London Math. Soc. **104**(2), 223–252 (2012). http://plms.oxfordjournals.org/content/104/2/223.abstract

19. Nemoto, T.: Determinacy of wadge classes and subsystems of second order arithmetic. Math. Logic Q. **55**(2), 154–176 (2009). http://dx.doi.org/10.1002/malq.200710081
20. Nemoto, T., MedSalem, M.O., Tanaka, K.: Infinite games in the Cantor space and subsystems of second order arithmetic. Math. Logic Q. **53**(3), 226–236 (2007)
21. Pauly, A.: How incomputable is finding Nash equilibria? J. Univ. Comput. Sci. **16**(18), 2686–2710 (2010)
22. Pauly, A.: Computable Metamathematics and its Application to Game Theory. Ph.D. thesis, University of Cambridge (2012)
23. Pauly, A., de Brecht, M.: Towards synthetic descriptive set theory: an instantiation with represented spaces. arXiv 1307.1850

Prefix and Right-Partial Derivative Automata

Eva Maia(✉), Nelma Moreira, and Rogério Reis

CMUP & DCC, Faculdade de Ciências da Universidade do Porto,
Rua do Campo Alegre, 4169-007 Porto, Portugal
{emaia,nam,rvr}@dcc.fc.up.pt

Abstract. Recently, Yamamoto presented a new method for the conversion from regular expressions (REs) to non-deterministic finite automata (NFA) based on the Thompson ε-NFA ($\mathcal{A}_{\mathsf{T}}$). The $\mathcal{A}_{\mathsf{T}}$ automaton has two quotients discussed: the suffix automaton $\mathcal{A}_{\mathsf{suf}}$ and the prefix automaton, $\mathcal{A}_{\mathsf{pre}}$. Eliminating ε-transitions in $\mathcal{A}_{\mathsf{T}}$, the Glushkov automaton ($\mathcal{A}_{\mathsf{pos}}$) is obtained. Thus, it is easy to see that $\mathcal{A}_{\mathsf{suf}}$ and the partial derivative automaton ($\mathcal{A}_{\mathsf{pd}}$) are the same. In this paper, we characterise the $\mathcal{A}_{\mathsf{pre}}$ automaton as a solution of a system of left RE equations and express it as a quotient of $\mathcal{A}_{\mathsf{pos}}$ by a specific left-invariant equivalence relation. We define and characterise the right-partial derivative automaton ($\overleftarrow{\mathcal{A}}_{\mathsf{pd}}$). Finally, we study the average size of all these constructions both experimentally and from an analytic combinatorics point of view.

1 Introduction

Conversion methods from regular expressions to equivalent nondeterministic finite automata have been widely studied. Resulting NFAs can have ε-transitions or not. The standard conversion with ε-transitions is the Thompson automaton ($\mathcal{A}_{\mathsf{T}}$) [15] and the standard conversion without ε-transitions is the Glushkov (or position) automaton ($\mathcal{A}_{\mathsf{pos}}$) [9]. Other conversions such as partial derivative automaton ($\mathcal{A}_{\mathsf{pd}}$) [1,13] or follow automaton ($\mathcal{A}_{\mathsf{f}}$) [10] were proved to be quotients of the $\mathcal{A}_{\mathsf{pos}}$, by specific right-invariant equivalence relations [6,10]. In particular, for REs under special conditions, $\mathcal{A}_{\mathsf{pd}}$ is an optimal conversion method [12]. Moreover, asymptotically and on average, the size of $\mathcal{A}_{\mathsf{pd}}$ is half the size of $\mathcal{A}_{\mathsf{pos}}$ [2]. Reductions on the size of NFAs using left-relations was studied recently by Ko and Han [11].

Yamamoto [16] presented a new conversion method based on the $\mathcal{A}_{\mathsf{T}}$. Given a $\mathcal{A}_{\mathsf{T}}$, two automata are constructed by merging $\mathcal{A}_{\mathsf{T}}$ states: in one, the suffix automaton ($\mathcal{A}_{\mathsf{suf}}$), states with the same right languages and in the other, the prefix automaton ($\mathcal{A}_{\mathsf{pre}}$), states with the same left languages. $\mathcal{A}_{\mathsf{suf}}$ corresponds to $\mathcal{A}_{\mathsf{pd}}$, which is not a surprise because it is known that if ε-transitions are eliminated from $\mathcal{A}_{\mathsf{T}}$, the $\mathcal{A}_{\mathsf{pos}}$ is obtained [8]. $\mathcal{A}_{\mathsf{pre}}$ is a quotient by a left-invariant

This work was partially funded by the European Regional Development Fund through the programme COMPETE and by the Portuguese Government through the FCT under project UID/MAT/00144/2013 and project FCOMP-01-0124-FEDER-020486. Eva Maia was also funded by FCT grant SFRH/BD/78392/2011.

A. Beckmann et al. (Eds.): CiE 2015, LNCS 9136, pp. 258–267, 2015.
DOI: 10.1007/978-3-319-20028-6_26

relation. In this paper, we further study conversions from REs to NFAs based on left-invariant relations. Using the notion of right-partial derivatives introduced by Champarnaud et al. [4], we define the right-partial derivative automaton $\overleftarrow{\mathcal{A}}_{\mathsf{pd}}$, characterise its relation with $\mathcal{A}_{\mathsf{pd}}$ and $\mathcal{A}_{\mathsf{pos}}$, and study its average size. We construct the $\mathcal{A}_{\mathsf{pre}}$ automaton directly from a regular expression without use the $\mathcal{A}_{\mathsf{T}}$ automaton, and we show that it is also a quotient of the $\mathcal{A}_{\mathsf{pos}}$. However, the experimental results suggest that, on average, the reduction on the size of the $\mathcal{A}_{\mathsf{pos}}$ is not large. Considering the framework of analytic combinatorics we study this reduction.

2 Regular Expressions and Automata

Given an alphabet $\Sigma = \{\sigma_1, \sigma_2, \ldots, \sigma_k\}$ of size k, the set RE of *regular expressions* α over Σ is defined by the following grammar:

$$\alpha := \emptyset \mid \varepsilon \mid \sigma_1 \mid \cdots \mid \sigma_k \mid (\alpha + \alpha) \mid (\alpha \cdot \alpha) \mid (\alpha)^\star, \tag{1}$$

where the $\cdot$ is often omitted. If two REs α and β are syntactically equal, we write $\alpha \sim \beta$. The *size* of a RE α, $|\alpha|$, is its number of symbols, disregarding parenthesis, and its *alphabetic size*, $|\alpha|_\Sigma$, is the number of occurrences of letters from Σ. A RE α is *linear* if all its letters occurs only once. The language represented by a RE α is denoted by $\mathcal{L}(\alpha)$. Two REs α and β are *equivalent* if $\mathcal{L}(\alpha) = \mathcal{L}(\beta)$, and we write $\alpha = \beta$. We define the function ε by $\varepsilon(\alpha) = \varepsilon$ if $\varepsilon \in \mathcal{L}(\alpha)$ and $\varepsilon(\alpha) = \emptyset$, otherwise. This function can be naturally extended to sets of REs and languages. We consider REs reduced by the following rules: $\varepsilon\alpha = \alpha = \alpha\varepsilon$, $\emptyset + \alpha = \alpha = \alpha + \emptyset$, and $\emptyset\alpha = \emptyset = \alpha\emptyset$. Given a language $\mathcal{L} \subseteq \Sigma^\star$ and a word $w \in \Sigma^\star$, the *left quotient* of $\mathcal{L}$ w.r.t. w is the language $w^{-1}L = \{x \mid wx \in L\}$, and the *right quotient* of $\mathcal{L}$ w.r.t. w is the language $\mathcal{L}w^{-1} = \{x \mid xw \in \mathcal{L}\}$. The *reversal* of a word $w = \sigma_1\sigma_2 \cdots \sigma_n$ is $w^R = \sigma_n \cdots \sigma_2\sigma_1$. The *reversal* of a language $\mathcal{L}$, denoted by $\mathcal{L}^R$, is the set of words whose reversal is on $\mathcal{L}$. The reversal of $\alpha \in \mathsf{RE}$ is denoted by α^R. The reversal of set of REs is the set of the reversal of its elements. It is not difficult to verify that $\mathcal{L}w^{-1} = ((w^R)^{-1}L^R)^R$.

A *nondeterministic finite automaton* (NFA) is a five-tuple $A = (Q, \Sigma, \delta, I, F)$ where Q is a finite set of states, Σ is a finite alphabet, $I \subseteq Q$ is the set of initial states, $F \subseteq Q$ is the set of final states, and $\delta : Q \times \Sigma \to 2^Q$ is the transition function. The transition function can be extended to words and to sets of states in the natural way. When $I = \{q_0\}$, we use $I = q_0$. Given a state $q \in Q$, the *right language* of q is $\mathcal{L}_q(A) = \{w \in \Sigma^\star \mid \delta(q, w) \cap F \neq \emptyset\}$, and the *left language* is $\overleftarrow{\mathcal{L}}_q(A) = \{w \in \Sigma^\star \mid q \in \delta(I, w)\}$. The language accepted by A is $\mathcal{L}(A) = \bigcup_{q \in I} \mathcal{L}_q(A)$. Two NFAs are *equivalent* if they accept the same language. If two NFAs A and B are isomorphic, we write $A \simeq B$. An NFA is *deterministic* if for all $(q, \sigma) \in Q \times \Sigma$, $|\delta(q, \sigma)| \leq 1$ and $|I| = 1$. The *reversal* of an automaton A is the automaton A^R, where the sets of initial and final states are swapped and all transitions are reversed. Given an equivalence relation $\equiv$ in Q, the *quotient automaton* $A/_{\equiv} = (Q/_{\equiv}, \Sigma, \delta/_{\equiv}, I/_{\equiv}, F/_{\equiv})$ is defined in the usual way. A relation $\equiv$ is *right invariant* w.r.t. A if and only if: $\equiv \subseteq (Q-F)^2 \cup F^2$ and $\forall p, q \in Q, \sigma \in \Sigma$, if $p \equiv q$, then ${}^{\delta(p,\sigma)}/_{\equiv} = {}^{\delta(q,\sigma)}/_{\equiv}$. A relation $\equiv$ is a *left invariant* relation w.r.t. A if and only if it is a right-invariant relation w.r.t. A^R.

The right languages $\mathcal{L}_i$, for $i \in Q = [0,n]$, define a system of right equations, $\mathcal{L}_i = \bigcup_{j=1}^{k} \sigma_j \left(\bigcup_{m \in I_{ij}} \mathcal{L}_m\right) \cup \varepsilon(\mathcal{L}_i)$, where $I_{ij} \subseteq [0,n]$, $m \in I_{ij} \Leftrightarrow m \in \delta(i, \sigma_j)$, and $\mathcal{L}(A) = \bigcup_{i \in I} \mathcal{L}_i$. In the same manner, the left languages of the states of A define a system of left equations $\overleftarrow{\mathcal{L}}_i = \bigcup_{j=1}^{k} \left(\bigcup_{m \in I_{ij}} \overleftarrow{\mathcal{L}}_m\right) \sigma_j \cup \varepsilon(\overleftarrow{\mathcal{L}}_i)$, where $I_{ij} \subseteq [0,n]$, $m \in I_{ij} \Leftrightarrow i \in \delta(m, \sigma_j)$, and $\mathcal{L}(A) = \bigcup_{i \in F} \overleftarrow{\mathcal{L}}_i$.

2.1 Glushkov and Partial Derivative Automata

In the following we review two constructions which define NFAs equivalent to a given regular expression $\alpha \in \mathsf{RE}$. Let $\mathsf{pos}(\alpha) = \{1, 2, \ldots, |\alpha|_\Sigma\}$ be the set of letter positions in α, and let $\mathsf{pos}_0(\alpha) = \mathsf{pos}(\alpha) \cup \{0\}$. We consider the expression $\overline{\alpha}$ obtained by marking each letter with its position in α, i.e. $\mathcal{L}(\overline{\alpha}) \in \overline{\Sigma}^\star$ where $\overline{\Sigma} = \{\sigma_i \mid \sigma \in \Sigma, 1 \leq i \leq |\alpha|_\Sigma\}$. The same notation is used to remove the markings, i.e., $\overline{\overline{\alpha}} = \alpha$. For $\alpha \in \mathsf{RE}$ and $i \in \mathsf{pos}(\alpha)$, let $\mathsf{first}(\alpha) = \{i \mid \exists w \in \overline{\Sigma}^\star, \sigma_i w \in \mathcal{L}(\overline{\alpha})\}$, $\mathsf{last}(\alpha) = \{i \mid \exists w \in \overline{\Sigma}^\star, w\sigma_i \in \mathcal{L}(\overline{\alpha})\}$ and $\mathsf{follow}(\alpha, i) = \{j \mid \exists u, v \in \overline{\Sigma}^\star, u\sigma_i\sigma_j v \in \mathcal{L}(\overline{\alpha})\}$. The *Glushkov automaton* (or position automaton) for α is $\mathcal{A}_{\mathsf{pos}}(\alpha) = (\mathsf{pos}_0(\alpha), \Sigma, \delta_{\mathsf{pos}}, 0, F)$, with $\delta_{\mathsf{pos}} = \{(0, \overline{\sigma_j}, j) \mid j \in \mathsf{first}(\alpha)\} \cup \{(i, \overline{\sigma_j}, j) \mid j \in \mathsf{follow}(\alpha, i)\}$ and $F = \mathsf{last}(\alpha) \cup \{0\}$ if $\varepsilon(\alpha) = \varepsilon$, and $F = \mathsf{last}(\alpha)$, otherwise. We note that the number of states of $\mathcal{A}_{\mathsf{pos}}(\alpha)$ is exactly $|\alpha|_\Sigma + 1$.

The partial derivative automaton of a regular expression was introduced independently by Mirkin [13] and Antimirov [1]. Champarnaud and Ziadi [5] proved that the two formulations are equivalent. For a regular expression $\alpha \in \mathsf{RE}$ and a symbol $\sigma \in \Sigma$, the set of left-partial derivatives of α w.r.t. σ is defined inductively as follows:

$$\begin{aligned} \partial_\sigma(\emptyset) &= \partial_\sigma(\varepsilon) = \emptyset & \partial_\sigma(\alpha + \beta) &= \partial_\sigma(\alpha) \cup \partial_\sigma(\beta) \\ \partial_\sigma(\sigma') &= \begin{cases} \{\varepsilon\} \text{ if } \sigma' = \sigma \\ \emptyset \quad \text{otherwise} \end{cases} & \begin{aligned} \partial_\sigma(\alpha\beta) \\ \partial_\sigma(\alpha^\star) \end{aligned} & \begin{aligned} &= \partial_\sigma(\alpha)\beta \cup \varepsilon(\alpha)\partial_\sigma(\beta) \\ &= \partial_\sigma(\alpha)\alpha^\star \end{aligned} \end{aligned} \tag{2}$$

where for any $S \subseteq \mathsf{RE}$, $S\emptyset = \emptyset S = \emptyset$, $S\varepsilon = \varepsilon S = S$, and $S\beta = \{\alpha\beta | \alpha \in S\}$ if $\beta \neq \emptyset, \varepsilon$ (and analogously for βS). The definition of left-partial derivatives can be extended in a natural way to sets of regular expressions, words, and languages. We have that $w^{-1}\mathcal{L}(\alpha) = \mathcal{L}(\partial_w(\alpha)) = \bigcup_{\tau \in \partial_w(\alpha)} \mathcal{L}(\tau)$, for $w \in \Sigma^\star$. The set of all partial derivatives of α w.r.t. words is denoted by $\mathsf{PD}(\alpha) = \partial_{\Sigma^\star}(\alpha)$. The *partial derivative automaton* of α is $\mathcal{A}_{\mathsf{pd}}(\alpha) = (\mathsf{PD}(\alpha), \Sigma, \delta_{pd}, \alpha, F_{pd})$, where $\delta_{pd} = \{(\tau, \sigma, \tau') \mid \tau \in \mathsf{PD}(\alpha),\ \sigma \in \Sigma,\ \tau' \in \partial_\sigma(\tau)\}$ and $F_{pd} = \{\tau \in \mathsf{PD}(\alpha) \mid \varepsilon(\tau) = \varepsilon\}$.

As noted by Broda et al. [2] and Maia et al. [12], following Mirkin's construction, the partial derivative automaton of α can be inductively constructed. A *(right) support* for α is a set of regular expressions $\{\alpha_1, \ldots, \alpha_n\}$ such that $\alpha_i = \sigma_1\alpha_{i1} + \cdots + \sigma_k\alpha_{ik} + \varepsilon(\alpha_i)$, $i \in [0,n]$, $\alpha_0 \sim \alpha$ and α_{ij} is a linear combination of α_l, $l \in [1,n]$ and $j \in [1,k]$. The set $\pi(\alpha)$ inductively defined below is a right support of α.

$$\begin{aligned} \pi(\emptyset) &= \emptyset & \pi(\alpha+\beta) &= \pi(\alpha) \cup \pi(\beta) \\ \pi(\varepsilon) &= \emptyset & \pi(\alpha\beta) &= \pi(\alpha)\beta \cup \pi(\beta) \\ \pi(\sigma) &= \{\varepsilon\} & \pi(\alpha^\star) &= \pi(\alpha)\alpha^\star. \end{aligned} \tag{3}$$

Champarnaud and Ziadi proved that $\mathsf{PD}(\alpha) = \pi(\alpha) \cup \{\alpha\}$ and the transition function of $\mathcal{A}_{\mathsf{pd}}$ can also be defined inductively from the system of equations above. Let $\varphi(\alpha) = \{(\sigma, \gamma) \mid \gamma \in \partial_\sigma(\alpha), \sigma \in \Sigma\}$ and $\lambda(\alpha) = \{\alpha' \mid \alpha' \in \pi(\alpha),\ \varepsilon(\alpha') = \varepsilon\}$, where both sets can be inductively defined using (2) and (3). We have, $\delta_{pd} = \{\alpha\} \times \varphi(\alpha) \cup F(\alpha)$ where the result of the $\times$ operation is seen as a set of triples and the set F is defined inductively by:

$$\begin{aligned} F(\emptyset) = F(\varepsilon) = F(\sigma) &= \emptyset,\ \sigma \in \Sigma \\ F(\alpha+\beta) &= F(\alpha) \cup F(\beta) \\ F(\alpha\beta) &= F(\alpha)\beta \cup F(\beta) \cup \lambda(\alpha)\beta \times \varphi(\beta) \\ F(\alpha^\star) &= F(\alpha)\alpha^\star \cup (\lambda(\alpha) \times \varphi(\alpha))\alpha^\star. \end{aligned} \tag{4}$$

Note that the concatenation of a transition (α, σ, β) with a RE γ is defined by $(\alpha, \sigma, \beta)\gamma = (\alpha\gamma, \sigma, \beta\gamma)$ (similarly $\gamma(\alpha, \sigma, \beta) = (\gamma\alpha, \sigma, \gamma\beta)$), if $\gamma \notin \{\emptyset, \varepsilon\}$, $(\alpha, \sigma, \beta)\emptyset = \emptyset$ and $(\alpha, \sigma, \beta)\varepsilon = (\alpha, \sigma, \beta)$. Then, $\mathcal{A}_{\mathsf{pd}}(\alpha) = (\pi(\alpha) \cup \{\alpha\}, \Sigma, \{\alpha\} \times \varphi(\alpha) \cup F(\alpha), \alpha, \lambda(\alpha) \cup \varepsilon(\alpha)\{\alpha\})$. In Fig. 1 are represented $\mathcal{A}_{\mathsf{pos}}(\alpha)$ and $\mathcal{A}_{\mathsf{pd}}(\alpha)$, where $\alpha = \beta b$ and $\beta = (a^\star b + a^\star ba + a^\star)^\star$.

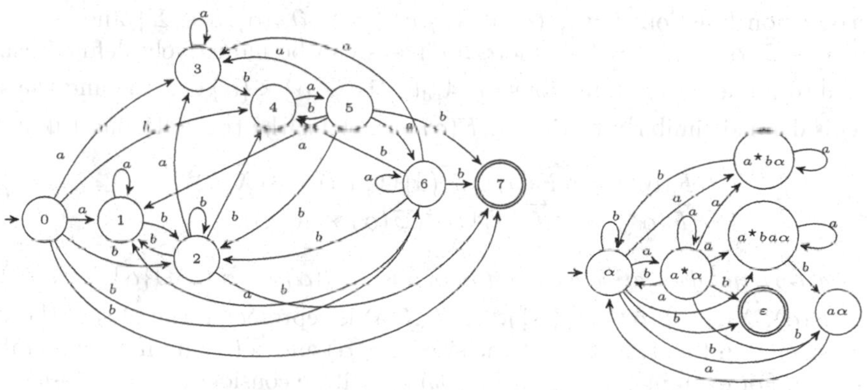

(a) $\mathcal{A}_{\mathsf{pos}}(\alpha)$ (b) $\mathcal{A}_{\mathsf{pd}}(\beta b)$

Fig. 1. Automata for $\alpha = \beta b$ with $\beta = (a^\star b + a^\star ba + a^\star)^\star$.

Champarnaud and Ziadi [6] showed that the partial derivative automaton is a quotient of the Glushkov automaton by the right-invariant equivalence relation $\equiv_c$, such that $i \equiv_c j$ if $\partial_{w\sigma_i}(\overline{\alpha}) = \partial_{w\sigma_j}(\overline{\alpha})$, for $i, j \in \mathsf{pos}_0(\alpha)$ and let $\sigma_0 = \varepsilon$. It is known that $\partial_{w\sigma_i}(\overline{\alpha})$ is either empty or an unique singleton for all $w \in \overline{\Sigma}^\star$.

3 Right-Partial Derivative Automata

The concept of right-partial derivative was introduced by Champarnaud et al. For a regular expression $\alpha \in \mathsf{RE}$ and a symbol $\sigma \in \Sigma$, the set of right-partial derivatives of α w.r.t. σ, $\overleftarrow{\partial}_\sigma(\alpha)$, is defined in the same way as the set of left-partial derivatives except for the following two rules:

$$\overleftarrow{\partial}_\sigma(\alpha\beta) = \alpha\overleftarrow{\partial}_\sigma(\beta) \cup \varepsilon(\beta)\overleftarrow{\partial}_\sigma(\alpha) \qquad \overleftarrow{\partial}_\sigma(\alpha^\star) = \alpha^\star\overleftarrow{\partial}_\sigma(\alpha). \tag{5}$$

This definition can be extended in a natural way to sets of regular expressions, words, and languages. The set of all right-partial derivatives of α w.r.t. words is denoted by $\overleftarrow{\mathsf{PD}}(\alpha) = \overleftarrow{\partial}_{\Sigma^\star}(\alpha)$. The right- and left-partial derivatives of α w.r.t. $w \in \Sigma^\star$ are related by $\overleftarrow{\partial}_w(\alpha) = (\partial_{w^R}(\alpha^R))^R$. Thus, $\mathcal{L}(\overleftarrow{\partial}_w(\alpha)) = \mathcal{L}(\alpha)w^{-1}$. The right-partial derivative automaton of α, $\overleftarrow{\mathcal{A}}_{\mathsf{pd}}(\alpha)$, can be defined inductively as a solution of a left system of expression equations, $\alpha_i = \alpha_{i1}\sigma_1 + \cdots + \alpha_{ik}\sigma_k + \varepsilon(\alpha_i)$, $i \in [0, n]$, $\alpha_0 \sim \alpha$, α_{ij} is a linear combination of α_l, $l \in [1, n]$ and $j \in [1, k]$.

Proposition 1. *The set of regular expressions $\overleftarrow{\pi}(\alpha)$ defined in the same way as the set π, except for the concatenation and Kleene star rules, is a solution of a left system of expression equations,*

$$\overleftarrow{\pi}(\alpha\beta) = \alpha\overleftarrow{\pi}(\beta) \cup \overleftarrow{\pi}(\alpha) \qquad \overleftarrow{\pi}(\alpha^\star) = \alpha^\star\overleftarrow{\pi}(\alpha). \tag{6}$$

Again, the solution of the system of equations also allows to inductively define the transition function. Let $\overleftarrow{\varphi}(\alpha) = \{(\gamma, \sigma) \mid \gamma \in \overleftarrow{\partial}_\sigma(\alpha), \sigma \in \Sigma\}$ and $\overleftarrow{\lambda}(\alpha) = \{\alpha' \mid \alpha' \in \overleftarrow{\pi}(\alpha),\ \varepsilon(\alpha') = \varepsilon\}$, where both sets can be inductively defined using (5) and (6). The set of transitions of $\overleftarrow{\mathcal{A}}_{\mathsf{pd}}(\alpha)$ is $\overleftarrow{\varphi}(\alpha) \times \{\alpha\} \cup \overleftarrow{F}(\alpha)$ and the set $\overleftarrow{F}(\alpha)$ is defined similarly to the set $F(\alpha)$ except for the two following rules:

$$\begin{aligned}\overleftarrow{F}(\alpha\beta) &= \alpha\overleftarrow{F}(\beta) \cup \overleftarrow{F}(\alpha) \cup \varphi(\alpha) \times (\alpha\overleftarrow{\lambda}(\beta))\\ \overleftarrow{F}(\alpha^\star) &= \alpha^\star\overleftarrow{F}(\alpha) \cup \alpha^\star(\overleftarrow{\varphi}(\alpha) \times \overleftarrow{\lambda}(\alpha)).\end{aligned} \tag{7}$$

The *right-partial derivative automaton* of α is $\overleftarrow{\mathcal{A}}_{\mathsf{pd}}(\alpha) = (\overleftarrow{\pi}(\alpha) \cup \{\alpha\}, \Sigma, \overleftarrow{\varphi}(\alpha) \times \{\alpha\} \cup \overleftarrow{F}(\alpha), \overleftarrow{\lambda}(\alpha) \cup \varepsilon(\alpha)\{\alpha\}, \{\alpha\})$. In Fig. 3(a) is represented the $\overleftarrow{\mathcal{A}}_{\mathsf{pd}}$ of the RE βb considered in Fig. 1. Note that the sizes of $\pi(\alpha)$ and $\overleftarrow{\pi}(\alpha)$ are not comparable in general. For example, $|\pi(\beta b)| > |\overleftarrow{\pi}(\beta b)|$, but if we consider $\alpha = b(ba^\star + aba^\star + a^\star)^\star$ then $|\pi(\alpha)| < |\overleftarrow{\pi}(\alpha)|$. The following result relates the functions defined above to the ones used to define the $\mathcal{A}_{\mathsf{pd}}$ is given by the following result.

Proposition 2. *Let α be a regular expression. Then $\overleftarrow{\pi}(\alpha) = (\pi(\alpha^R))^R$, $\overleftarrow{\lambda}(\alpha) = (\lambda(\alpha^R))^R$, $\overleftarrow{\varphi}(\alpha) = (\varphi(\alpha^R))^R$ and $\overleftarrow{F}(\alpha) = (F(\alpha^R))^R$.*

From the previous result and the fact that $\mathcal{A}_{\mathsf{pd}}(\alpha) \simeq {}^{\mathcal{A}_{\mathsf{pos}}(\alpha)}/_{\equiv_c}$ we have

Proposition 3. *For any $\alpha \in \mathsf{RE}$,*

1. $(\mathcal{A}_{\mathsf{pd}}(\alpha^R))^R \simeq \overleftarrow{\mathcal{A}}_{\mathsf{pd}}(\alpha)$.
2. $\mathcal{L}(\overleftarrow{\mathcal{A}}_{\mathsf{pd}}(\alpha)) = \mathcal{L}(\alpha)$.
3. $\overleftarrow{\mathcal{A}}_{\mathsf{pd}}(\alpha) \simeq {}^{(\mathcal{A}_{\mathsf{pos}}(\alpha^R))^R}/_{\equiv_c}$.

4 Prefix Automata

Yamamoto [16] presented a new algorithm for converting a regular expression into an equivalent NFA. First, a labeled version of the usual Thompson NFA $(Q, \Sigma, \delta, I, F)$ is obtained, where each state q is labeled with two regular expressions, one that corresponds to its left language, $LP(q)$, and the other to its right language, $LS(q)$. States which in-transitions are labeled with a letter are called *sym-states*. Then the equivalence relations $\equiv_{pre}$ and $\equiv_{suf}$ are defined on the set of sym-states: for two states $p, q \in Q$, $p \equiv_{pre} q$ if and only if $LP(p) = LP(q)$; and $p \equiv_{suf} q$ if and only if $LS(p) = LS(q)$. The *prefix automaton* $\mathcal{A}_{\mathsf{pre}}$ and the *suffix automaton* $\mathcal{A}_{\mathsf{suf}}$ are the quotient automata by these relations. The final automaton is a combination of these two. The author also shows that $\mathcal{A}_{\mathsf{suf}}$ coincides with $\mathcal{A}_{\mathsf{pd}}$. This relation between $\mathcal{A}_{\mathsf{pd}}$ and $\mathcal{A}_{\mathsf{suf}}$ could lead us to think that $\overleftarrow{\mathcal{A}}_{\mathsf{pd}}$ coincide with $\mathcal{A}_{\mathsf{pre}}$, which is not true. For instance, considering $\alpha = a+b$, the $\overleftarrow{\mathcal{A}}_{\mathsf{pd}}(\alpha)$ has 2 states and the $\mathcal{A}_{\mathsf{pre}}(\alpha)$ has 3 states (see Fig. 2). Note that both automata are obtained from another automaton by merging the states with the same left language: while the $\overleftarrow{\mathcal{A}}_{\mathsf{pd}}(\alpha)$ is obtained from $(\mathcal{A}_{\mathsf{pos}}(\alpha^R))^R$, we will see that the $\mathcal{A}_{\mathsf{pre}}(\alpha)$ is obtained from $\mathcal{A}_{\mathsf{pos}}(\alpha)$.

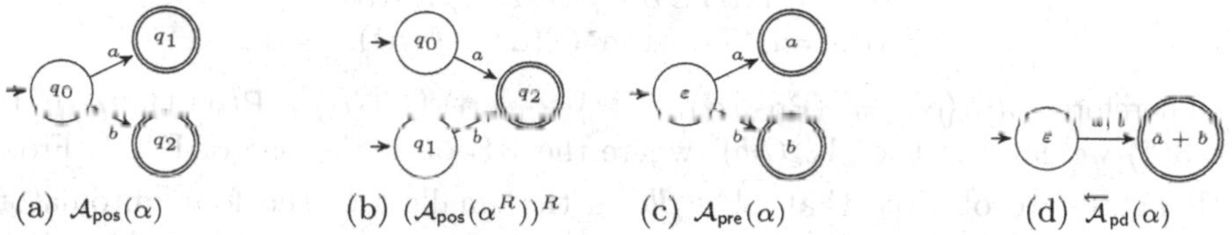

Fig. 2. Automata for $\alpha = a + b$.

The LP labelling scheme proposed by Yamamoto can be obtained as a solution of a system of expression equations for a RE α, as done both for $\mathcal{A}_{\mathsf{pd}}$ and $\overleftarrow{\mathcal{A}}_{\mathsf{pd}}$. Consider a system of left equations $\alpha_i = \alpha_{i1}\sigma_1 + \cdots + \alpha_{ik}\sigma_k$, $i \in [1, n]$, where $\alpha = \sum_{i \in I \subseteq [0,n]} \alpha_i$, $\alpha_{ij} = \sum_{l \in I_{ij} \subseteq [0,n]} \alpha_l$ and $\alpha_0 \sim \varepsilon$.

Proposition 4. *The set* $\mathsf{Pre}(\alpha)$ *inductively defined as follows:*

$$\begin{aligned} &\mathsf{Pre}(\emptyset) = \emptyset && \mathsf{Pre}(\alpha + \beta) = \mathsf{Pre}(\alpha) \cup \mathsf{Pre}(\beta) \\ &\mathsf{Pre}(\varepsilon) = \emptyset && \mathsf{Pre}(\alpha\beta) = \alpha\mathsf{Pre}(\beta) \cup \mathsf{Pre}(\alpha) \\ &\mathsf{Pre}(\sigma) = \{\sigma\} && \mathsf{Pre}(\alpha^\star) = \alpha^\star\mathsf{Pre}(\alpha). \end{aligned} \tag{8}$$

is a solution (left support) of the system of left equations defined above.

The set $\mathsf{Pre}_0(\alpha) = \mathsf{Pre}(\alpha) \cup \{\varepsilon\}$ constitutes the set of states of the prefix automaton $\mathcal{A}_{\mathsf{pre}}(\alpha)$. It also follows from the resolution of the above system of equations, that the set of transitions of $\mathcal{A}_{\mathsf{pre}}(\alpha)$ can be inductively defined. Let $\mathsf{P}(\alpha)$, $\psi(\alpha)$, and $\mathsf{T}(\alpha)$ be defined, respectively, as follows:

$$\begin{aligned} &\mathsf{P}(\emptyset) = \emptyset && \mathsf{P}(\alpha + \beta) = \mathsf{P}(\alpha) \cup \mathsf{P}(\beta) \\ &\mathsf{P}(\varepsilon) = \{\varepsilon\} && \mathsf{P}(\alpha\beta) = \alpha\mathsf{P}(\beta) \cup \varepsilon(\beta)\mathsf{P}(\alpha) \\ &\mathsf{P}(\sigma) = \{\sigma\} && \mathsf{P}(\alpha^\star) = \alpha^\star\mathsf{P}(\alpha). \end{aligned} \tag{9}$$

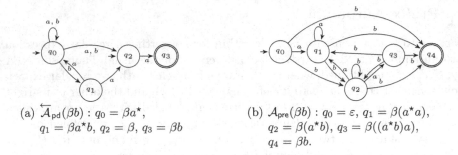

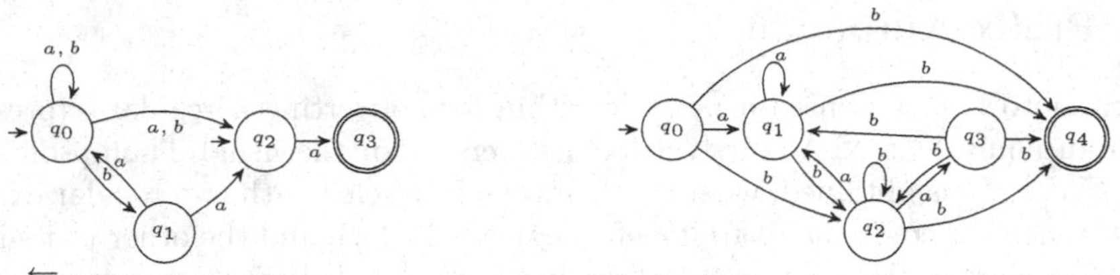

(a) $\overleftarrow{\mathcal{A}}_{\mathsf{pd}}(\beta b) : q_0 = \beta a^\star$, $q_1 = \beta a^\star b$, $q_2 = \beta$, $q_3 = \beta b$

(b) $\mathcal{A}_{\mathsf{pre}}(\beta b) : q_0 = \varepsilon$, $q_1 = \beta(a^\star a)$, $q_2 = \beta(a^\star b)$, $q_3 = \beta((a^\star b)a)$, $q_4 = \beta b$.

Fig. 3. Automata for βb, where $\beta = (a^\star b + a^\star ba + a^\star)^\star$

$$\begin{array}{ll} \psi(\emptyset) = \emptyset & \psi(\alpha + \beta) = \psi(\alpha) \cup \psi(\alpha) \\ \psi(\varepsilon) = \emptyset & \psi(\alpha\beta) = \psi(\alpha) \cup \varepsilon(\alpha)\ \alpha\ \psi(\beta) \\ \psi(\sigma) = \{(\sigma, \sigma)\} & \psi(\alpha^\star) = \alpha^\star \psi(\alpha) \end{array} \tag{10}$$

$$\begin{aligned} \mathsf{T}(\emptyset) &= \mathsf{T}(\varepsilon) = \mathsf{T}(\sigma) = \emptyset,\ \sigma \in \Sigma \\ \mathsf{T}(\alpha + \beta) &= \mathsf{T}(\alpha) \cup \mathsf{T}(\beta) \\ \mathsf{T}(\alpha\beta) &= \mathsf{T}(\alpha) \cup \alpha\mathsf{T}(\beta) \cup \mathsf{P}(\alpha) \times (\alpha\psi(\beta)) \\ \mathsf{T}(\alpha^\star) &= \alpha^\star\mathsf{T}(\alpha) \cup \alpha^\star(\mathsf{P}(\alpha) \times \psi(\alpha)). \end{aligned} \tag{11}$$

Therefore, $\mathcal{A}_{\mathsf{pre}}(\alpha) = (\mathsf{Pre}_0(\alpha), \Sigma, \{\varepsilon\} \times \psi(\alpha) \cup \mathsf{T}(\alpha), \varepsilon, \mathsf{P}(\alpha) \cup \varepsilon(\alpha))$. In Fig. 3(b) we can see the $\mathcal{A}_{\mathsf{pre}}(\beta b)$, where the RE βb is the one of Fig. 1. From both figures we observe that $\overleftarrow{\mathcal{A}}_{\mathsf{pd}}(\beta b)$ is the smallest of the four automaton constructions. We now show that the $\mathcal{A}_{\mathsf{pre}}$ is a quotient of $\mathcal{A}_{\mathsf{pos}}$. If α is a linear regular expression, $\mathcal{A}_{\mathsf{pos}}(\alpha)$ is deterministic and thus all its states have distinct left languages. Therefore, in this case, $\mathcal{A}_{\mathsf{pre}}(\alpha)$ coincides with $\mathcal{A}_{\mathsf{pos}}(\alpha)$ and $|\mathsf{Pre}(\alpha)| = |\alpha|_\Sigma$. For an arbitrary RE α, $\mathcal{A}_{\mathsf{pre}}(\overline{\alpha}) \simeq \mathcal{A}_{\mathsf{pos}}(\overline{\alpha})$. Let $\equiv_l$ be the equivalence relation in $\mathsf{Pre}(\overline{\alpha})$ such that for any regular expression α, $\forall \alpha_1, \alpha_2 \in \mathsf{Pre}(\overline{\alpha})$, $\alpha_1 \equiv_l \alpha_2 \Leftrightarrow \overline{\alpha_1} = \overline{\alpha_2}$. It is not difficult to see that $\equiv_l$ is a left-invariant relation.

Proposition 5. *Let α be a regular expression. Then $\mathcal{A}_{\mathsf{pre}}(\alpha) \simeq {}^{\mathcal{A}_{\mathsf{pos}}(\alpha)}/_{\equiv_l}$.*

By construction, the Glushkov automaton is homogeneous, i.e. the in- transitions of each state are all labelled by the same letter. It follows from Proposition 5 that this property also holds for $\mathcal{A}_{\mathsf{pre}}$.

5 Average-Case Complexity

We conducted some experimental tests in order to compare the sizes of $\mathcal{A}_{\mathsf{pos}}$, $\mathcal{A}_{\mathsf{pd}}$, $\overleftarrow{\mathcal{A}}_{\mathsf{pd}}$ and $\mathcal{A}_{\mathsf{pre}}$ automata. We used the FAdo library[1] that includes implementations of the NFA conversions and also several tools for uniformly random generate regular expressions. In order to obtain regular expressions uniformly

[1] http://fado.dcc.fc.up.pt.

Table 1. Experimental results for uniform random generated regular expressions.

k	$\lvert\alpha\rvert$	$\lvert\mathsf{pos}_0\rvert$	$\lvert\delta_{pos}\rvert$	$\lvert\mathsf{PD}\rvert$	$\lvert\delta_{\pi}\rvert$	$\frac{\lvert\pi\rvert}{\lvert\mathsf{pos}\rvert}$	$\lvert\overleftarrow{\mathsf{PD}}\rvert$	$\lvert\delta_{\overleftarrow{\pi}}\rvert$	$\frac{\lvert\overleftarrow{\pi}\rvert}{\lvert\mathsf{pos}\rvert}$	$\lvert\mathsf{Pre}_0\rvert$	$\lvert\delta_{pre}\rvert$	$\frac{\lvert\mathsf{Pre}\rvert}{\lvert\mathsf{pos}\rvert}$	$1-\eta_k$
2	100	28.9	167.5	15.7	56.0	0.55	15.9	56.4	0.55	20.1	73.7	0.71	0.90
	500	139.9	1486.5	71.6	389.8	0.51	71.5	393.1	0.51	91.9	530.8	0.66	
10	100	42.5	159.4	23.8	73.7	0.56	23.8	72.9	0.56	38.5	130.4	0.91	0.99
	500	207.1	1019.1	113.2	423.8	0.55	112.4	425.6	0.54	186	807.1	0.90	
	1000	412.1	2182.1	223.7	884.1	0.54	223.1	884.5	0.54	369.5	1717.6	0.90	

generated in the size of the syntactic tree, a prefix notation version of the grammar was used. For each alphabet size, k, and $|\alpha|$, samples of 10 000 REs were generated, which is sufficient to ensure a 95 % confidence level within a 1 % error margin. Table 1 presents the average values obtained for $|\alpha| \in \{100, 500, 1000\}$ and $k \in \{2, 10\}$. These experiments suggest that in pratice the $\overleftarrow{\mathcal{A}}_{\mathsf{pd}}$ and the $\mathcal{A}_{\mathsf{pd}}$ have the same size and the $\mathcal{A}_{\mathsf{pre}}$ is not significantly smaller then the $\mathcal{A}_{\mathsf{pos}}$. By Proposition 3, $|\alpha^R|_\Sigma = |\alpha|_\Sigma$ and by the fact that $\varepsilon \in \pi(\alpha)$ if and only if $\varepsilon \in \overleftarrow{\pi}(\alpha)$, the analysis of the average size of $\mathcal{A}_{\mathsf{pd}}(\alpha)$ presented in Broda et al. [3] carries on to $\overleftarrow{\mathcal{A}}_{\mathsf{pd}}(\alpha)$. Thus the average sizes of $\mathcal{A}_{\mathsf{pd}}$ and $\overleftarrow{\mathcal{A}}_{\mathsf{pd}}$ are asymptotically the same. However, $\overleftarrow{\mathcal{A}}_{\mathsf{pd}}(\alpha)$ has only one final state and its number of initial states is the number of final states of $\mathcal{A}_{\mathsf{pd}}(\alpha^R)$. As studied by Nicaud [14], the size of $\mathsf{last}(\alpha)$ tends asymptotically to a constant depending on k and $|\lambda(\alpha)|$ is half that size [2]. Thus, that constant value will be also the number of initial states of $\overleftarrow{\mathcal{A}}_{\mathsf{pd}}$. Following, again, the ideas in Broda et al., we estimate the number of mergings of states that arise when computing $\mathcal{A}_{\mathsf{pre}}$ from $\mathcal{A}_{\mathsf{pos}}$. The $\mathcal{A}_{\mathsf{pre}}$ has at most $|\alpha|_\Sigma + 1$ states and this only occurs when all unions in $\mathsf{Pre}(\alpha)$ are disjoint. However there are cases in which this does not happen. For instance, when $\sigma \in \mathsf{Pre}(\beta) \cap \mathsf{Pre}(\gamma)$, then $|\mathsf{Pre}(\beta+\gamma)| = |\mathsf{Pre}(\beta) \cup \mathsf{Pre}(\gamma)| \leq |\mathsf{Pre}(\beta)| + |\mathsf{Pre}(\gamma)| - 1$ and $|\mathsf{Pre}(\beta^\star\gamma)| = |\beta^\star\mathsf{Pre}(\gamma) \cup \beta^\star\mathsf{Pre}(\beta)| \leq |\mathsf{Pre}(\beta)| + |\mathsf{Pre}(\gamma)| - 1$. In what follows we estimate the number of these non-disjoint unions, which correspond to a lower bound for the number of states merged in the $\mathcal{A}_{\mathsf{pos}}$ automaton. This is done in the framework of analytic combinatorics as expounded by Flajolet and Sedgewick [7]. The methods apply to generating functions $A(z) = \sum_n a_n z^n$ for a combinatorial class $\mathcal{A}$ with a_n objects of size n, denoted by $[z^n]A(z)$, and also bivariate functions $C(u,z) = \sum_\alpha u^{c(\alpha)} z^{|\alpha|}$, where $c(\alpha)$ is some measure of the object $\alpha \in \mathcal{A}$.

The regular expressions α_σ for which $\sigma \in \mathsf{Pre}(\alpha_\sigma)$, $\sigma \in \Sigma$, are generated by following grammar:

$$\alpha_\sigma := \sigma \mid \alpha_\sigma + \alpha \mid \alpha_{\overline{\sigma}} + \alpha_\sigma \mid \alpha_\sigma \cdot \alpha \mid \varepsilon \cdot \alpha_\sigma \tag{12}$$

The regular expressions that are not generated by α_σ are denoted by $\alpha_{\overline{\sigma}}$. The generating function for α_σ, $R_{\sigma,k}(z)$ satisfies

$$\begin{aligned} R_{\sigma,k}(z) = z &+ zR_{\sigma,k}(z)R_k(z) + z(R_k(z) - R_{\sigma,k}(z))R_{\sigma,k}(z) \\ &+ zR_{\sigma,k}(z)R_k(z) + z^2R_{\sigma,k}(z) \end{aligned}$$

From this one gets

$$R_{\sigma,k}(z) = \frac{(z^2 + 3zR_k(z) - 1) + \sqrt{(z^2 + 3zR_k(z) - 1)^2 + 4z^2}}{2z}. \tag{13}$$

where $R_k(z) = \frac{1-z-\sqrt{\Delta_k(z)}}{4z}$ is the generating function for REs given by grammar (1) but omitting the $\emptyset$, $\Delta_k(z) = 1 - 2z - (7 + 8k)z^2$ and following Nicaud,

$$[z^n]R_k(z) \sim \frac{\sqrt{2(1-\rho_k)}}{8\rho_k\sqrt{\pi}}\rho_k^{-n} n^{-3/2}, \text{ where } \rho_k = \frac{1}{1+\sqrt{8k+8}} \tag{14}$$

Using the techniques in Broda et al. and namely Proposition 3 one has

$$[z^n]R_{\sigma,k}(z) \sim \frac{3}{16\sqrt{\pi}}\left(1 - \frac{b(\rho_k)}{\sqrt{a(\rho_k)}}\right)\sqrt{2(1-\rho_k)}\rho_k^{-(n+1)} n^{-\frac{3}{2}}, \tag{15}$$

where $a(z)$ and $b(z)$ are polynomials. Thus, the asymptotic ratio of regular expressions with $\sigma \in \mathsf{Pre}(\alpha)$ is:

$$\frac{[z^n]R_{\sigma,k}(z)}{[z^n]R_k(z)} \sim \frac{3}{2}\left(1 - \frac{b(\rho_k)}{\sqrt{a(\rho_k)}}\right). \tag{16}$$

As $\lim\limits_{k\to\infty} \rho_k = 0$, $\lim\limits_{k\to\infty} a(\rho_k) = 1$, and $\lim\limits_{k\to\infty} b(\rho_k) = 1$, this asymptotic ratio tends to 0 with $k \to \infty$.

Let $i(\alpha)$ be the number of non-disjoint unions appearing during the computation of $\mathsf{Pre}(\alpha)$ originated by the two cases above. Then $i(\alpha)$ verifies

$$\begin{aligned} i(\varepsilon) = i(\sigma) &= 0 & i(\alpha_\sigma^\star\alpha_\sigma) &= i(\alpha_\sigma^\star) + i(\alpha_\sigma) + 1\\ i(\alpha_\sigma + \alpha_\sigma) &= i(\alpha_\sigma) + i(\alpha_\sigma) + 1 & i(\overline{\alpha_\sigma^\star}\alpha_\sigma) &= i(\overline{\alpha_\sigma^\star}) + i(\alpha_\sigma)\\ i(\alpha_\sigma + \alpha_{\overline{\sigma}}) &= i(\alpha_\sigma) + i(\alpha_{\overline{\sigma}}) & i(\alpha\alpha_{\overline{\sigma}}) &= i(\alpha) + i(\alpha_{\overline{\sigma}})\\ i(\alpha_{\overline{\sigma}} + \alpha) &= i(\alpha_{\overline{\sigma}}) + i(\alpha) & i(\alpha^\star) &= i(\alpha). \end{aligned}$$

From these equations we can obtain the cost generating function for the number of mergings:

$$I_{\sigma,k}(z) = \frac{(z + z^2)R_{\sigma,k}(z)^2}{\sqrt{\Delta_k(z)}}. \tag{17}$$

Using again the same Proposition 3 from Broda et al., we conclude that:

$$[z^n]I_{\sigma,k}(z) \sim \frac{1+\rho_k}{64}\frac{\left(a(\rho_k) + b(\rho_k)^2 - 2b(\rho_k)\sqrt{a(\rho_k)}\right)}{\sqrt{\pi}\sqrt{2-2\rho_k}}\rho_k^{-(n+1)} n^{-\frac{1}{2}}. \tag{18}$$

The cost generating function for the number of letters in $\alpha \in \mathsf{RE}$, computed by Nicaud, is $L_k(z) = \frac{kz}{\sqrt{\Delta_k(z)}}$ and $[z^n]L_k(z) \sim \frac{k\rho_k}{\sqrt{\pi(2-2\rho_k)}}\rho_k^{-n} n^{-1/2}$. With these, we get an asymptotic estimate for the average number of mergings given by:

$$\frac{[z^n]I_{\sigma,k}(z)}{[z^n]L_k(z)} \sim \frac{1-\rho_k}{4\rho_k^2}\lambda_k = \eta_k, \tag{19}$$

where $\lambda_k = \frac{(1+\rho_k)}{16(1-\rho_k)}\left(a(\rho_k) + b(\rho_k)^2 - 2b(\rho_k)\sqrt{a(\rho_k)}\right)$. It is not difficult to conclude that $\lim_{k\to\infty} \lambda_k = 0$, therefore $\lim_{k\to\infty} \eta_k = 0$. As it is evident from the last two columns of Table 1, for small values of k, the lower bound η_k does not capture all the mergings that occur in $\mathcal{A}_{\mathsf{pre}}$. Although we must study other contributions for those mergings, it seems that for larger values of k, the average number of states of the $\mathcal{A}_{\mathsf{pre}}$ automaton approaches the number of states of the $\mathcal{A}_{\mathsf{pos}}$ automaton.

References

1. Antimirov, V.M.: Partial derivatives of regular expressions and finite automaton constructions. Theor. Comput. Sci. **155**(2), 291–319 (1996)
2. Broda, S., Machiavelo, A., Moreira, N., Reis, R.: On the average size of Glushkov and partial derivative automata. Int. J. Found. Comput. Sci. **23**(5), 969–984 (2012)
3. Broda, S., Machiavelo, A., Moreira, N., Reis, R.: On the average state complexity of partial derivative automata. Int. J. Found. Comput. Sci. **22**(7), 1593–1606 (2011)
4. Champarnaud, J.M., Dubernard, J.P., Jeanne, H., Mignot, L.: Two-sided derivatives for regular expressions and for hairpin expressions. In: Dediu, A.H., Martín-Vide, C., Truthe, B. (eds.) LATA 2013. LNCS, vol. 7810, pp. 202–213. Springer, Heidelberg (2013)
5. Champarnaud, J.M., Ziadi, D.: From Mirkin's prebases to Antimirov's word partial derivatives. Fundam. Inform. **45**(3), 195–205 (2001)
6. Champarnaud, J.M., Ziadi, D.: Canonical derivatives, partial derivatives and finite automaton constructions. Theor. Comput. Sci. **289**(1), 137–163 (2002)
7. Flajolet, P., Sedgewick, R.: Analytic Combinatorics. CUP, Cambridge (2008)
8. Giammarresi, D., Ponty, J.L., Wood, D.: The Glushkov and Thompson constructions: a synthesis (1998) (unpublished manuscript)
9. Glushkov, V.M.: The abstract theory of automata. Russ. Math. Surv. **16**(5), 1–53 (1961)
10. Ilie, L., Yu, S.: Follow automata. Inf. Comput. **186**(1), 140–162 (2003)
11. Ko, S., Han, Y.: Left is better than right for reducing nondeterminism of NFAs. In: Holzer, M., Kutrib, M. (eds.) CIAA 2014. LNCS, vol. 8587, pp. 238–251. Springer, Heidelberg (2014)
12. Maia, E., Moreira, N., Reis, R.: Partial derivative and position bisimilarity automata. In: Holzer, M., Kutrib, M. (eds.) CIAA 2014. LNCS, vol. 8587, pp. 264–277. Springer, Heidelberg (2014)
13. Mirkin, B.: An algorithm for constructing a base in a language of regular expressions. Eng. Cybern. **5**, 110–116 (1966)
14. Nicaud, C.: On the average size of Glushkov's automata. In: Dediu, A.H., Ionescu, A.M., Martín-Vide, C. (eds.) LATA 2009. LNCS, vol. 5457, pp. 626–637. Springer, Heidelberg (2009)
15. Thompson, K.: Regular expression search algorithm. Com. ACM **11**(6), 410–422 (1968)
16. Yamamoto, H.: A new finite automaton construction for regular expressions. In: Bensch, S., Freund, R., Otto, F. (eds.) NCMA, pp. 249–264. Österreichische Computer Gesellschaft, Kassel (2014). books@ocg.at

A Note on the Computable Categoricity of ℓ^p Spaces

Timothy H. McNicholl(✉)

Iowa State University, Ames, IA 50011, USA
mcnichol@iastate.edu

Abstract. Suppose that p is a computable real and that $p \geq 1$. We show that in both the real and complex case, ℓ^p is computably categorical if and only if $p = 2$. The proof uses Lamperti's characterization of the isometries of Lebesgue spaces of σ-finite measure spaces.

1 Introduction

When p is a positive real number, let ℓ^p denote the space of all sequences of complex numbers $\{a_n\}_{n=0}^{\infty}$ so that

$$\sum_{n=0}^{\infty} |a_n|^p < \infty.$$

ℓ^p is a vector space over $\mathbb{C}$ with the usual scalar multiplication and vector addition. When $p \geq 1$ it is a Banach space under the norm defined by

$$\|\{a_n\}_n\| = \left(\sum_{n=0}^{\infty} |a_n|^p\right)^{1/p}.$$

Loosely speaking, a computable structure is *computably categorical* if all of its computable copies are computably isomorphic. In 1989, Pour-El and Richards showed that ℓ^1 is not computably categorical [10]. It follows from a recent result of A.G. Melnikov that ℓ^2 is computably categorical [8]. At the 2014 Conference on Computability and Complexity in Analysis, A.G. Melnikov asked "For which computable reals $p \geq 1$ is ℓ^p computably categorical?" The following theorem answers this question.

Theorem 1. *Suppose p is a computable real so that $p \geq 1$. Then, ℓ^p is computably categorical if and only if $p = 2$.*

We prove Theorem 1 by proving the following stronger result.

Theorem 2. *Suppose p is a computable real so that $p \geq 1$ and $p \neq 2$. Suppose C is a c.e. set. Then, there is a computable copy of ℓ^p, $\mathcal{B}$, so that C computes a linear isometry of ℓ^p onto $\mathcal{B}$. Furthermore, if an oracle X computes a linear isometry of ℓ^p onto $\mathcal{B}$, then X must also compute C.*

A. Beckmann et al. (Eds.): CiE 2015, LNCS 9136, pp. 268–275, 2015.
DOI: 10.1007/978-3-319-20028-6_27

These results also hold for ℓ^p-spaces over the reals. In a forthcoming paper it will be shown that ℓ^p is Δ_2^0-categorical.

The paper is organized as follows. Section 2 covers background and motivation. Section 3 presents the proof of Theorem 2. Concluding remarks are presented in Sect. 4.

2 Background

2.1 Background from Functional Analysis

Fix p so that $1 \leq p < \infty$. A *generating set* for ℓ^p is a subset of ℓ^p with the property that ℓ^p is the closure of its linear span.

Let e_n be the vector in ℓ^p whose $(n+1)$st component is 1 and whose other components are 0. Let $E = \{e_n : n \in \mathbb{N}\}$. We call E the *standard generating set* for ℓ^p.

Recall that an *isometry* of ℓ^p is a norm-preserving map of ℓ^p into ℓ^p. We will use the following classification of the surjective linear isometries of ℓ^p.

Theorem 3. (Banach/Lamperti). *Suppose p is a real number so that $p \geq 1$ and $p \neq 2$. Let T be a linear map of ℓ^p into ℓ^p. Then, the following are equivalent.*

1. *T is a surjective isometry.*
2. *There is a permutation of $\mathbb{N}$, ϕ, and a sequence of unimodular points, $\{\lambda_n\}_n$, so that $T(e_n) = \lambda_n e_{\phi(n)}$ for all n.*
3. *Each $T(e_n)$ is a unit vector and the supports of $T(e_n)$ and $T(e_m)$ are disjoint whenever $m \neq n$.*

In his seminal text on linear operators, S. Banach stated Theorem 3 for the case of ℓ^p spaces over the reals [2]. He also stated a classification of the linear isometries of $L^p[0, 1]$ in the real case. Banach's proofs of these results were sketchy and did not easily generalize to the complex case. In 1958, J. Lamperti rigorously proved a generalization of Banach's claims to real and complex L^p-spaces of σ-finite measure spaces [7]. Theorem 3 follows from J. Lamperti's work as it appears in Theorem 3.2.5 of [4]. Note that Theorem 3 fails when $p = 2$. For, ℓ^2 is a Hilbert space. So, if $\{f_0, f_1, \ldots\}$ is any orthonormal basis for ℓ^2, then there is a unique surjective linear isometry of ℓ^2, T, so that $T(e_n) = f_n$ for all n.

2.2 Background from Computable Analysis

We assume the reader is familiar with the fundamental notions of computability theory as covered in [3].

Suppose $z_0 \in \mathbb{C}$. We say that z_0 is *computable* if there is an algorithm that given any $k \in \mathbb{N}$ as input computes a rational point q so that $|q - z_0| < 2^{-k}$. This is equivalent to saying that the real and imaginary parts of z_0 have computable decimal expansions.

Our approach to computability on ℓ^p is equivalent to the format in [10] wherein a more expansive treatment may be found.

Fix a computable real p so that $1 \leq p < \infty$. Let $F = \{f_0, f_1, \ldots\}$ be a generating set for ℓ^p. We say that F is an *effective generating set* if there is an algorithm that given any rational points $\alpha_0, \ldots, \alpha_M$ and a nonnegative integer k as input computes a rational number q so that

$$q - 2^{-k} < \left\| \sum_{j=0}^{M} \alpha_j f_j \right\| < q + 2^{-k}.$$

That is, the map

$$\alpha_0, \ldots, \alpha_M \mapsto \left\| \sum_{j=0}^{M} \alpha_j f_j \right\|$$

is computable. Clearly the standard generating set is an effective generating set.

Suppose $F = \{f_0, f_1, \ldots\}$ is an effective generating set for ℓ^p. We say that a vector $g \in \ell^p$ is *computable with respect to* F if there is an algorithm that given any nonnegative integer k as input computes rational points $\alpha_0, \ldots, \alpha_M$ so that

$$\left\| g - \sum_{j=0}^{M} \alpha_j f_j \right\| < 2^{-k}.$$

Suppose $g_n \in \ell^p$ for all n. We say that $\{g_n\}_n$, is *computable with respect to* F if there is an algorithm that given any $k, n \in \mathbb{N}$ as input computes rational points $\alpha_0, \ldots, \alpha_M$ so that

$$\left\| g_n - \sum_{j=0}^{M} \alpha_j f_j \right\| < 2^{-k}.$$

When $f \in \ell^p$ and $r > 0$, let $B(f; r)$ denote the open ball with center f and radius r. When $\alpha_0, \ldots, \alpha_M$ are rational points and r is a positive rational number, we call $B\left(\sum_{j=0}^{M} \alpha_j f_j; r\right)$ a *rational ball.*

Suppose $F = \{f_0, f_1, \ldots\}$ and $G = \{g_0, g_1, \ldots\}$ are effective generating sets for ℓ^p. We say that a map $T : \ell^p \to \ell^p$ is *computable with respect to* (F, G) if there is an algorithm P that meets the following three criteria.

- **Approximation:** Given a rational ball $B(\sum_{j=0}^{M} \alpha_j f_j; r)$ as input, P either does not halt or produces a rational ball $B(\sum_{j=0}^{N} \beta_j g_j; r')$.
- **Correctness:** If $B(\sum_{j=0}^{N} \beta_j g_j; r')$ is the output of P on input $B(\sum_{j=0}^{M} \alpha_j f_j; r)$, then $T(f) \in B(\sum_{j=0}^{N} \beta_j g_j; r')$ whenever $f \in B(\sum_{j=0}^{M} \alpha_j f_j; r)$.
- **Convergence:** If U is a neighborhood of $T(f)$, then f belongs to a rational ball $B_1 = B(\sum_{j=0}^{M} \alpha_j f_j; r)$ so that P halts on B_1 and produces a rational ball that is included in U.

When we speak of an algorithm accepting a rational ball $B(\sum_{j=0}^{M} \alpha_j f_j; r)$ as input, we of course mean that it accepts some representation of the ball such as a code of the sequence $(r, M, \alpha_0, \ldots, \alpha_M)$.

All of these definitions have natural relativizations. For example, if $F = \{f_0, f_1, \ldots\}$ is an effective generating set, then we say that X computes a vector $g \in \ell^p$ with respect to F if there is a Turing reduction that given the oracle X and an input k computes rational points $\alpha_0, \ldots, \alpha_M$ so that $\left\| g - \sum_{j=0}^{M} \alpha_j f_j \right\| < 2^{-k}$.

2.3 Background from Computable Categoricity

For the sake of motivation, we begin by considering the following simple example. Let ζ be an incomputable unimodular point in the plane. For each n, let $f_n = \zeta e_n$. Let $F = \{f_0, f_1, \ldots\}$. Thus, F is an effective generating set. However, the vector ζe_0 is computable with respect to F even though it is not computable with respect to the standard generating set E. In fact, the only vector that is computable with respect to E and F is the zero vector. The moral of the story is that different effective generating sets may yield very different classes of computable vectors and sequences. However, there is a surjective linear isometry of ℓ^p that is computable with respect to (E, F); namely multiplication by ζ. Thus, E and F give the same computability theory on ℓ^p even though they yield very different classes of computable vectors. This leads to the following definition.

Definition 4. *Suppose p is a computable real so that $p \geq 1$. We say that ℓ^p is* computably categorical *if for every effective generating set F there is a surjective linear isometry of ℓ^p that is computable with respect to (E, F).*

The definitions just given for ℓ^p can easily be adapted to any separable Banach space. Suppose $G = \{g_0, g_1 \ldots,\}$ is an effective generating set for a Banach space $\mathcal{B}$. The pair $(\mathcal{B}, G)$ is called a *computable Banach space*. Suppose that $\mathcal{B}$ is linearly isometric to ℓ^p, and let T denote a linear isometric mapping of $\mathcal{B}$ onto ℓ^p. Let $f_n = T(g_n)$, and let $F = \{f_0, f_1, \ldots\}$. Then, F is an effective generating set for ℓ^p, and T is computable with respect to (G, F). Thus, Theorem 2 can be rephrased as follows.

Theorem 5. *Suppose p is a computable real so that $p \geq 1$ and $p \neq 2$. Suppose C is a* c.e. *set. Then, there is an effective generating set for ℓ^p, F, so that with respect to (E, F), C computes a surjective linear isometry of ℓ^p. Furthermore, any oracle that computes a surjective linear isometry of ℓ^p with respect to (E, F) must also compute C.*

A.G. Melnikov and K.M. Ng have investigated computable categoricity questions with regards to the space $C[0, 1]$ of continuous functions on the unit interval with the supremum norm [8,9]. The study of computable categoricity for countable structures goes back at least as far as the work of Goncharov [5]. The text of Ash and Knight has a thorough discussion of the main results of this line of inquiry [1]. The survey by Harizanov covers other directions in the countable computable structures program [6].

3 Proof of Theorems 1 and 2

We begin by noting the following easy consequence of the definitions and Theorem 3.

Proposition 6. *Suppose p is a computable real so that $p \geq 1$ and so that $p \neq 2$. Let F be an effective generating set for ℓ^p. Then, the following are equivalent.*

1. *There is a surjective linear isometry of ℓ^p that is computable with respect to (E, F).*
2. *There is a permutation of $\mathbb{N}$, ϕ, and a sequence of unimodular points $\{\lambda_n\}_n$, so that $\{\lambda_n e_{\phi(n)}\}_n$ is computable with respect to F.*
3. *There is a sequence of unit vectors $\{g_n\}_n$ so that $\{g_n\}_n$ is computable with respect to F, $G = \{g_0, g_1, \ldots\}$ is a generating set for ℓ^p, and so that the supports of g_n and g_m are disjoint whenever $n \neq m$.*

Proof. Parts (2) and (3) just restate each other. It follows from Theorem 3 that (1) implies (2).

Suppose (3) holds. Let T be the unique linear map of the span of E onto the span of G so that $T(e_n) = g_n$ for all n. Since the supports of $g_0, g_1, \ldots$ are pairwise disjoint, and since each g_n is a unit vector, T is isometric. It follows that there is a unique extension of T to a unique linear isometry of ℓ^p; denote this extension by T as well. We claim that T is computable with respect to (E, F). For, suppose a rational ball $B(\sum_{j=0}^{M} \alpha_j e_j; r)$ is given as input. Since $\{g_n\}_n$ is computable with respect to F, it follows that we can compute a non-negative integer N and rational points $\beta_0, \ldots, \beta_N$ so that $\left\| \sum_{j=0}^{M} \alpha_j g_j - \sum_{j=0}^{N} \beta_j f_j \right\| < r$. We then output $B(\sum_{j=0}^{N} \beta_j g_j; 2r)$. It follows that the Approximation, Correctness, and Convergence criteria are satisfied and so T is computable with respect to (E, F). □

We now turn to the proof of Theorem 5 which, as we have noted, implies Theorem 2. Our construction of F is a modification of the construction used by Pour-El and Richards to show that ℓ^1 is not computably categorical [10]. Let C be an incomputable c.e. set. Without loss of generality, we assume $0 \notin C$. Let $\{c_n\}_{n \in \mathbb{N}}$ be an effective one-to-one enumeration of C. Set

$$\gamma = \sum_{k \in C} 2^{-k}.$$

Thus, $0 < \gamma < 1$, and γ is an incomputable real. Set:

$$f_0 = (1 - \gamma)^{1/p} e_0 + \sum_{n=0}^{\infty} 2^{-c_n/p} e_{n+1}$$
$$f_{n+1} = e_{n+1}$$
$$F = \{f_0, f_1, f_2, \ldots\}$$

Since $1 - \gamma > 0$, we can use the standard branch of $\sqrt[p]{\cdot}$.

We divide the rest of the proof into the following lemmas.

Lemma 7. *F is an effective generating set.*

Proof. Since

$$(1-\gamma)^{1/p} e_0 = f_0 - \sum_{n=1}^{\infty} 2^{-c_{n-1}/p} f_n$$

the closed linear span of F includes E. Thus, F is a generating set for ℓ^p. Note that $\|f_0\| = 1$.

Suppose $\alpha_0, \ldots, \alpha_M$ are rational points. When $1 \leq j \leq M$, set

$$\mathcal{E}_j = |\alpha_0 2^{-c_{j-1}/p} + \alpha_j|^p - |\alpha_0|^p 2^{-c_{j-1}}.$$

It follows that

$$\begin{aligned} \|\alpha_0 f_0 + \ldots + \alpha_M f_M\|^p &= |\alpha_0|^p \|f_0\|^p + \mathcal{E}_1 + \ldots + \mathcal{E}_M \\ &= |\alpha_0|^p + \mathcal{E}_1 + \ldots + \mathcal{E}_M. \end{aligned}$$

Since $\mathcal{E}_1, \ldots, \mathcal{E}_M$ can be computed from $\alpha_0, \ldots, \alpha_M$, $\|\alpha_0 f_0 + \ldots + \alpha_M f_M\|$ can be computed from $\alpha_0, \ldots, \alpha_M$. Thus, F is an effective generating set. □

Lemma 8. *Every oracle that with respect to F computes a scalar multiple of e_0 whose norm is 1 must also compute C.*

Proof. Suppose that with respect to F, X computes a vector of the form λe_0 where $|\lambda| = 1$. It suffices to show that X computes $(1-\gamma)^{-1/p}$.

Fix a rational number q_0 so that $(1-\gamma)^{-1/p} \leq q_0$. Let $k \in \mathbb{N}$ be given as input. Compute k' so that $2^{-k'} \leq q_0 2^{-k}$. Since X computes λe_0 with respect to F, we can use oracle X to compute rational points $\alpha_0, \ldots, \alpha_M$ so that

$$\left\| \lambda e_0 - \sum_{j=0}^{M} \alpha_j f_j \right\| < 2^{-k'}. \tag{1}$$

We claim that $|(1-\gamma)^{-1/p} - |\alpha_0|| < 2^{-k}$. For, it follows from (1) that $|\lambda - \alpha_0(1-\gamma)^{1/p}| < 2^{-k'}$. Thus, $|1 - |\alpha_0|(1-\gamma)^{1/p}| < 2^{-k'}$. Hence,

$$|(1-\gamma)^{-1/p} - |\alpha_0|| < 2^{-k'}(1-\gamma)^{-1/p} \leq 2^{-k'} q_0 \leq 2^{-k}.$$

Since X computes α_0 from k, X computes $(1-\gamma)^{-1/p}$. □

Lemma 9. *If X computes a surjective linear isometry of ℓ^p with respect to (E, F), then X must also compute C.*

Proof. By Lemma 8 and the relativization of Proposition 6. □

Lemma 10. *With respect to F, C computes e_0.*

Proof. Fix an integer M so that $(1-\gamma)^{-1/p} < M$.

Let $k \in \mathbb{N}$. Using oracle C, we can compute an integer N_1 so that $N_1 \geq 3$ and

$$\left\| \sum_{n=N_1}^{\infty} 2^{-c_{n-1}/p} e_n \right\| \leq \frac{2^{-(kp+1)/p}}{2^{-(kp+1)/p} + M}.$$

We can use oracle C to compute a rational number q_1 so that $|q_1 - (1-\gamma)^{-1/p}| \leq 2^{-(kp+1)/p}$. Set

$$g = q_1 \left[f_0 - \sum_{n=1}^{N_1 - 1} 2^{-c_{n-1}/p} f_n \right].$$

It suffices to show that $\|e_0 - g\| < 2^{-k}$. Note that since $1-\gamma < 1$, $|q_1(1-\gamma)^{1/p} - 1| \leq 2^{-(kp+1)/p}$. Note also that $|q_1| < M + 2^{-(kp+1)/p}$. Thus,

$$\begin{aligned} \|e_0 - g\|^p &= \left\| e_0 - q_1(1-\gamma)^{1/p} e_0 - q_1 \sum_{n=N_1}^{\infty} 2^{-c_{n-1}/p} e_n \right\|^p \\ &\leq |q_1(1-\gamma)^{1/p} - 1|^p + |q_1|^p \left\| \sum_{n=N_1}^{\infty} 2^{-c_{n-1}/p} e_n \right\|^p \\ &< 2^{-(kp+1)} + 2^{-(kp+1)} = 2^{-kp} \end{aligned}$$

Thus, $\|e_0 - g\| < 2^{-k}$. This completes the proof of the lemma. □

Lemma 11. *With respect to* (E, F), C *computes a surjective linear isometry of* ℓ^p.

Proof. By Lemma 10 and the relativization of Proposition 6. □

4 Concluding Remarks

We note that all of the steps in the above proofs work just as well over the real field.

Lamperti's result on the isometries of L^p spaces hold when $0 < p < 1$. For these values of p, ℓ^p is a metric space under the metric

$$d(\{a_n\}_n, \{b_n\}_n) = \sum_{n=0}^{\infty} |a_n - b_n|^p.$$

The steps in the above proofs can be adapted to these values of p as well.

In a forthcoming paper it will be shown that ℓ^p is Δ_2^0-categorical.

Acknowledgement. The author thanks the anonymous referees who made helpful comments. The author's participation in CiE 2015 was funded by a Simons Foundation Collaboration Grant for Mathematicians.

References

1. Ash, C.J., Knight, J.: Computable Structures and the Hyperarithmetical Hierarchy. Studies in Logic and the Foundations of Mathematics, vol. 144. North-Holland Publishing Co., Amsterdam (2000)
2. Banach, S.: Theory of Linear Operations. North-Holland Mathematical Library, vol. 38. North-Holland Publishing Co., Amsterdam (1987). Translated from the French by F. Jellett, With comments by A. Pełczyński and Cz. Bessaga
3. Cooper, S.B.: Computability Theory. Chapman & Hall/CRC, Boca Raton (2004)
4. Fleming, R.J., Jamison, J.E.: Isometries on Banach Spaces: Function Spaces. Chapman & Hall/CRC Monographs and Surveys in Pure and Applied Mathematics, vol. 129. Chapman & Hall/CRC, Boca Raton (2003)
5. Goncharov, S.: Autostability and computable families of constructivizations. Algebr. Log. **17**, 392–408 (1978). English translation
6. Harizanov, V.S.: Pure computable model theory. Handbook of Recursive Mathematics. Volume 1, Studies in Logic and the Foundations of Mathematics, vol. 138, pp. 3–114. North-Holland, Amsterdam (1998)
7. Lamperti, J.: On the isometries of certain function-spaces. Pac. J. Math. **8**, 459–466 (1958)
8. Melnikov, A.G.: Computably isometric spaces. J. Symb. Log. **78**(4), 1055–1085 (2013)
9. Melnikov, A.G., Ng, K.M.: Computable structures and operations on the space of continuous functions. Available at https://dl.dropboxusercontent.com/u/4752353/Homepage/C[0,1]_final.pdf
10. Pour-El, M.B., Richards, J.I.: Computability in Analysis and Physics. Springer, Berlin (1989)

On the Computational Content of Termination Proofs

Georg Moser and Thomas Powell(✉)

Institute of Computer Science, University of Innsbruck, Innsbruck, Austria
{georg.moser,thomas.powell}@uibk.ac.at

Abstract. Given that a program has been shown to terminate using a particular proof, it is natural to ask what we can infer about its complexity. In this paper we outline a new approach to tackling this question in the context of term rewrite systems and recursive path orders. From an inductive proof that recursive path orders are well-founded, we extract an explicit realiser which bounds the derivational complexity of rewrite systems compatible with these orders. We demonstrate that by analysing our realiser we are able to derive, in a completely uniform manner, a number of results on the relationship between the strength of path orders and the bounds they induce on complexity.

1 Introduction

Proof theory emphasises *proofs* over *theorems*, as put most succinctly by Kreisel's famous question "What more do we know if we have proved a theorem by restricted means than if we merely know that it is true?". One application of this quest in the context of program analysis is the link between termination and complexity. Is it possible to derive computational content from a given termination argument, so that we can automatically deduce bounds on the complexity of our programs?

We study this question in the abstract framework of term rewrite systems and recursive path orders, which we take to encompass multiset path orders, lexicographic path orders, and recursive path orders with status. Our main contribution is to analyse the proof that recursive path orders are well-founded and extract an explicit term in Gödel's system T which bounds the derivational complexity of rewrite systems reducing under these orders. Our framework is uniform in the sense that our term applies to any variant of recursive path order studied, by just adapting its parameters. We then demonstrate that a simple analysis of our term allows us to uniformly derive the well-known primitive recursive bounds on the derivational complexity of multiset path orders [1] (see Theorem 1) and the multiple recursive bounds on lexicographic path orders [2] (see Theorem 2).

The emphasis of this work is less on the technical results achieved, but in the method used to achieve them. Our re-derivation of the standard bounds for multiset and lexicographic path orders contrasts greatly to the somewhat ad-hoc

This work is supported by FWF (Austrian Science Fund) project P-25781.

A. Beckmann et al. (Eds.): CiE 2015, LNCS 9136, pp. 276–285, 2015.
DOI: 10.1007/978-3-319-20028-6_28

originally carried out by Hofbauer and Weiermann, and are much more closely related to the study of Buchholz [3], which forms the starting point of our work. However, whereas in [3] complexity bounds are obtained via a suitable formalisation of termination proofs in fragments of arithmetic and rely on Parson's fundamental work [4], our focus is on extracting an explicit subrecursive bound. Therefore, not only is our proof completely elementary and self-contained, but our concrete realising term is amenable to a much finer analysis of complexity for restricted path orders.

In addition to the aforementioned results, we obtain a novel derivational complexity analysis of recursive path orders with status [5], where we confirm that the induced complexity is multiple recursive, which follows as a corollary to our general boundedness result Theorem 2. This general bound is not surprising in the context of the well-known multiple-recursive bound on the lexicographic path orders and follows with relative ease from earlier work [6]. However, our smooth framework allows us to get rid of the technicalities involved in earlier work.

Throughout the history of term rewriting, a general link has been sought between the strength of a termination argument and the complexity of rewrite systems it admits. An early attempt at such a correspondence is the so-called *Cichon's principle*, which states that the derivational complexity function of a TRS $\mathcal{R}$ for which termination is provable using a termination order of order type α is eventually dominated by a function from the slow-growing hierarchy G_α along α, cf. [7]. Unfortunately, while this principle holds for the standard recursive path orders, it is not true in general, even for its relaxed version as proposed by Touzet [8] - see [9] for a proof. It is now accepted that the link between termination orders and complexity is dependent in a much more subtle way on the structure of the termination proof. Therefore, we believe that applying proof-theoretic techniques to analyse the computational meaning of path orders could provide some important insight into the relationship between termination and complexity. This approach has already been successfully pioneered in e.g. [10,11], and we hope that the work outlined here constitutes a first step towards a similarly successful program in the context of rewriting.

2 Recursive Path Orders

We assume familiarity with term rewriting [5,12], and recall only some basic notation. Let $\mathcal{V}$ denote a countable infinite set of variables, $\mathcal{F}$ a finite set of function symbols, and $\mathcal{T}(\mathcal{F},\mathcal{V})$ ($\mathcal{T}$ for short) the set of terms constructed from these. A *term rewrite system* (*TRS*) $\mathcal{R}$ over $\mathcal{T}(\mathcal{F},\mathcal{V})$ is a *finite* set of rewrite rules $l \to r$. For a given term t, $|t|$ denotes its size (the total number of variables and function symbols in t), $\mathsf{dp}(t)$ its depth (the maximal number of nesting function symbols) and $\mathsf{Var}(t)$ the set of variables in t. The rewrite relation is denoted as $\to_{\mathcal{R}}$ and we use the standard notations for its transitive and reflexive closure. The *derivation height* of a term s with respect to a well-founded, finitely branching relation $\to$ is defined as: $\mathsf{dh}(s,\to) = \max\{n \mid \exists t\ s \to^n t\}$. The *derivational complexity function* $\mathsf{dc}_{\mathcal{R}}$ is defined as follows: $\mathsf{dc}_{\mathcal{R}}(n) := \max\{\mathsf{dh}(t,\to_{\mathcal{R}}) \mid |t| \leqslant n\}$.

Well-founded path orders are a powerful method for proving the termination of rewrite systems, and recursive path orders are one of the best known of these.

For an arbitrary relation $>$ defined on some set X, we let $>_{\mathsf{mul}}$ and $>_{\mathsf{lex}}$ denote respectively the multiset and lexicographic extensions of $>$ to finite tuples X^n. We write $<_{\mathsf{mul}}$ for the reverse of $>_{\mathsf{mul}}$, and analogously for all other annotated inequality symbols used below.

Definition 1 (Recursive path order). *Let $\succ$ be a well-founded precedence (i.e. a proper order) on a finite signature $\mathcal{F}$. The* recursive path order (RPO) *$\succ_{\mathsf{rpo}}$ with respect to some status function $\tau\colon \mathcal{F} \to \{\mathsf{mul}, \mathsf{lex}\}$ is defined recursively as follows: we say that $t = f(t_1, \ldots, t_n) \succ_{\mathsf{rpo}} s$ if one of the following holds:*

(a) $t_i \succeq_{\mathsf{rpo}} s$ for some $i = 1, \ldots, m$;
(b) $s = g(s_1, \ldots, s_m)$ with $f \succ g$ and $t \succ_{\mathsf{rpo}} s_i$ for all $i = 1, \ldots, n$;
(c) $s = f(s_1, \ldots, s_n)$, $t \succ_{\mathsf{rpo}} s_i$ for all $i = 1, \ldots, n$ and $(t_1, \ldots, t_n) \succ_{\mathsf{rpo},\tau(f)} (s_1, \ldots, s_n)$.

Here $\succeq_{\mathsf{rpo}}$ denotes the reflexive closure of $\succ_{\mathsf{rpo}}$.

Recall that the *multiset path order* $\succ_{\mathsf{mpo}}$ is the instance of RPO for which $\tau(f) = \mathsf{mul}$ for all f, and analogously for the *lexicographic path-order*. We say that a TRS $\mathcal{R}$ is compatible with $\succ_{\mathsf{rpo}}$ for some suitable choice of $\succ$ and τ if $\mathcal{R} \subseteq \succ_{\mathsf{rpo}}$. It is easy (but tedious) to show that $\succ_{\mathsf{rpo}}$ is closed under both substitutions and contexts, and therefore from compatibility we obtain that $\to_{\mathcal{R}} \subseteq \succ_{\mathsf{rpo}}$. In the latter case, we say that $\mathcal{R}$ is *reducing* with respect to $\succ_{\mathsf{rpo}}$. It is well-known that $\succ_{\mathsf{rpo}}$ is well-founded, so any TRS compatible with $\succ_{\mathsf{rpo}}$ is terminating.

There are two well-known basic strategies to show that a RPO is well-founded. One can either appeal to some variant of the minimal-bad-sequence argument to show that there cannot exist an infinite descending chain of terms $t_0 \succ_{\mathsf{rpo}} t_1 \succ_{\mathsf{rpo}} \ldots$, either in the form of Kruskal's theorem or directly applied to path orders as in [13], or alternatively one can take what is essentially the contrapositive of this statement and proceed via a series of nested inductions on terms, as in e.g. [3,14]. This second approach is most amenable for the purposes of program extraction, so we sketch the inductive proof below.

Theorem 1. *$\succ_{\mathsf{rpo}}$ is well-founded.*

Proof. Let $t \in \mathrm{WF}$ denote that t is a well-founded term, i.e. there are no infinite descending sequences starting from t. We prove that

$$\forall f \in \mathcal{F} \underbrace{\forall t_1, \ldots, t_n \in \mathrm{WF} \,.\, f(t_1, \ldots, t_n) \in \mathrm{WF}}_{A(f)} . \tag{1}$$

Then, since we trivially have $x \in \mathrm{WF}$ for all variables x, we obtain $\forall t(t \in \mathrm{WF})$ from (1) by well-founded induction over the structure of terms. Therefore it remains to prove $\forall f A(f)$. Let us fix, for now, $f \in \mathcal{F}$ and $t_1, \ldots, t_n \in \mathrm{WF}$ and make the following assumptions:

(A) $\forall g \prec f A(g)$
(B) $\forall (s_1, \ldots, s_n) \prec_{\mathsf{rpo},\tau(f)} (t_1, \ldots, t_n)(s_1, \ldots, s_n \in \mathrm{WF} \to f(s_1, \ldots, s_n) \in \mathrm{WF})$,

where we note that $\prec_{\mathsf{rpo},\tau(f)}$ is only ever applied to tuples of well-founded terms. We prove that $f(t_1,\ldots,t_n) \in \mathrm{WF}$ is well-founded by showing that

$$\forall s(\underbrace{s \prec_{\mathsf{rpo}} f(t_1,\ldots,t_n) \to s \in \mathrm{WF}}_{B(s)}),$$

using induction over $\blacktriangleleft$, where $\blacktriangleleft$ denotes the immediate subterm relation. Let us fix s and assume that $\forall s' \blacktriangleleft s\, B(s')$. Then if $t = f(t_1,\ldots,t_n) \succ_{\mathsf{rpo}} s$ there are three possibilities:

(a) $t_i \succeq_{\mathsf{rpo}} s$ for some $i = 1,\ldots,n$, in which case $s \in \mathrm{WF}$ by assumption that $t_i \in \mathrm{WF}$;
(b) $s = g(s_1,\ldots,s_m)$ with $f \succ g$ and $t \succ_{\mathsf{rpo}} s_i$ for all i. Then by our induction hypothesis we must have $s_1,\ldots,s_m \in \mathrm{WF}$, and therefore by assumption (A) we have $g(s_1,\ldots,s_m) \in \mathrm{WF}$ too;
(c) $s = f(s_1,\ldots,s_n)$ with $t \succ_{\mathsf{rpo}} s_i$ for all i and $(t_1,\ldots,t_n) \succ_{\mathsf{rpo},\tau(f)} (s_1,\ldots,s_n)$. Then again $s_1,\ldots,s_n \in \mathrm{WF}$, and this time by assumption (B) we have $f(s_1,\ldots,s_n) \in \mathrm{WF}$.

This establishes $B(s)$, and thus by $\blacktriangleleft$-induction we obtain $\forall s B(s)$ and hence well-foundedness of $f(t_1,\ldots,t_n)$. We now carry out two further inductions to eliminate the assumptions (A) and (B) in turn. First, from (A) $\to$ (B)$\to$ $(t_1,\ldots,t_n \in \mathrm{WF} \to f(t_1,\ldots,t_n) \in \mathrm{WF})$ and well-founded induction over $(\mathrm{WF}, \succ_{\mathsf{rpo},\tau(f)})$ we obtain (A)$\to$ $A(f)$, and this yields $\forall f A(f)$ by induction on $(\mathcal{F}, \succ)$, and we're done. □

2.1 A Finitary Formulation of Theorem 1

In general, we know that an arbitrary rewrite system $\mathcal{R}$ compatible with some $\succ_{\mathsf{rpo}}$ must terminate by well-foundedness of $\succ_{\mathsf{rpo}}$. However, for a fixed rewrite system, the full strength of Theorem 1 is never used, since unlike $\succ_{\mathsf{rpo}}$ the rewrite relation $\to_{\mathcal{R}}$ is only finitely branching. Rather, $\mathcal{R}$ will always lie in some *finitary approximation* of $\succ_{\mathsf{rpo}}$, where the size of this approximation will depend in some suitable sense on the 'size' of $\mathcal{R}$. Thus, in order to successfully analyse the complexity of the termination proof, we are not interested in analysing the well-foundedness of $\succ_{\mathsf{rpo}}$ itself, but only these finitary approximations to it.

A precise characterisation of the approximation of $\succ_{\mathsf{rpo}}$ needed to prove well-foundedness of a given TRS is established by Buchholz in [3], and we reformulate his idea below in a slightly simplified way (the simplification being possible because here we do not consider varyadic function symbols).

In what follows, we use the abbreviation $\mathsf{RPO}(>, t, s)$ for the statement that one of the conditions (a)–(c) in Definition 1 holds for s, t and $>$. Thus one defines $\succ_{\mathsf{rpo}}$ by $f(t_1,\ldots,t_n) \succ_{\mathsf{rpo}} s$ iff $\mathsf{RPO}(\succ_{\mathsf{rpo}}, f(t_1,\ldots,t_n), s)$.

Definition 2. *The approximation $\succ_k$ of $\succ_{\mathsf{rpo}}$ is recursively defined as follows: we have $t = f(t_1,\ldots,t_n) \succ_k s$ iff*

$$\mathsf{RPO}(\succ_k, f(t_1,\ldots,t_n), s) \wedge \mathsf{dp}(s) \leqslant k + \mathsf{dp}(t) \wedge \mathsf{Var}(s) \subseteq \mathsf{Var}(t),$$

where $\mathsf{dp}(t)$ denotes the depth of t.

We call $\succ_k$ finitary because by definition for each t there are only finitely many s for which $t \succ_k s$. The proof of the next theorem can essentially be read-off from Buchholz's proof in [3]. However, our later proof extraction is based on the simplified proof given here.

Theorem 2 (Buchholz [3]**).** *Any $\mathcal{R}$ compatible with an $\succ_{\mathsf{rpo}}$ is contained in $\succ_k$ for some k depending on $\mathcal{R}$.*

Proof. It is first shown that

(i) $t \succ_{\mathsf{rpo}} s$ implies $t\sigma \succ_{\mathsf{dp}(s)} s\sigma$ for any substitution σ,
(ii) $t_j \succ_k s$ implies $f(t_1,\dots,t_n) \succ_k f(t_1,\dots,t_{j-1},s,t_{j+1},\dots,t_n)$ for any f.

Property (i) is most easily established as in [3]; as usual $t \succ_{\mathsf{rpo}} s$ implies that $\mathsf{Var}(s\sigma) \subseteq \mathsf{Var}(t\sigma)$. Furthermore, induction on $\succ_{\mathsf{rpo}}$ yields that $t \succ_{\mathsf{rpo}} s$ implies $\mathsf{dp}(s\sigma) \leqslant \mathsf{dp}(s) + \mathsf{dp}(t\sigma)$. Yet another induction over $\succ_{\mathsf{rpo}}$ derives (i), and property (ii) is similarly straightforward. Now, for a given $\mathcal{R}$ let $k := \max\{\mathsf{dp}(r) : l \to r \in \mathcal{R}\}$. It is then clear that if $\mathcal{R}$ is compatible with $\succ_{\mathsf{rpo}}$ then $t \to_{\mathcal{R}} s$ implies $t \succ_k s$, since by (i) we have $l\sigma \succ_k r\sigma$ for all rules $l \to r$, and therefore $C[l\sigma] \succ_k C[r\sigma]$ by induction on (ii). □

3 Bounding the Derivational Complexity of $\mathcal{R}$

We now construct a term which forms a recursive analogue to Theorem 1, but takes into account the fact that we only need to consider finitary approximations. Let $\mathcal{R}$, a RPO $\succ_{\mathsf{rpo}}$ compatible with $\mathcal{R}$ and a suitable approximation $\succ_k$ of $\succ_{\mathsf{rpo}}$ be fixed for the remainder of the paper.

3.1 Term and Tree Encodings, Gödel's System T

We assume that the terms $\mathcal{T}$ of our rewrite system can be primitive recursively encoded into $\mathbb{N}$, and write $s <_{\mathcal{T}} t$ to denote that the code of s is less than the code of t. Let $\mathcal{T}^\star$ denote the set of all finite trees of terms. For $T \in \mathcal{T}^\star$ we write $\mathsf{rt}(T)$ to denote the root of T, and write $S \sqsubset T$ if S is an immediate subtree of T. Again, we assume that $\mathcal{T}^\star$ has been primitive recursively encoded into $\mathbb{N}$.

In what follows we work in the standard language of system T in all finite types ρ: terms are built from the usual arithmetic constants, λ-abstraction and application, and Gödel's primitive recursor $\mathsf{R}^h_\rho(n) = hn(\lambda m < n \,.\, \mathsf{R}^h_\rho(m))$ whose output can have arbitrary type ρ. It is clear that recursion over the decidable relations $\blacktriangleleft$ and $\sqsubset$ can be defined in terms of the recursor of base type $\mathbb{N}$, since without loss of generality we can assume that if $s \blacktriangleleft t$ then $s <_{\mathcal{T}} t$, and similarly for $\sqsubset$. Therefore terms build up from these forms of recursion are primitive recursive in the usual sense. On the other hand, given a well-founded lifting $\sqsubset_*$ of $\sqsubset$ to tuples $(\mathcal{T}^\star)^n$ (where in the sequel $\sqsubset_*$ will be either the multiset or lexicographic lifting) we let $\mathsf{TR}_{\sqsubset_*}$ denote the transfinite recursor over $\sqsubset_*$ of base output type. We leave open for now how this can be formally defined within system T as this will depend on the lifting.

3.2 Computing Derivation Trees

Let t be a term and let $\Phi_k(t)$ denote the finite tree T with root t, whose branches $(t, t_1, \ldots, t_n)$ are precisely $\succ_k$-derivations $t \succ_k t_1 \succ_k \ldots \succ_k t_n$ from t; terms will be denoted by lower-case letters and derivation trees by upper-case letters. Finiteness of $\Phi_k(t)$ follows since $\succ_k$ is well-founded and finitely branching. We now show how $\Phi_k(t)$ can be computed. Let $t = f(t_1, \ldots, t_n)$ for some $f \in \mathcal{F}$ and terms $t_1, \ldots, t_n$, and suppose that for each $g \prec f$ we have a function $F_g \colon (\mathcal{T}^\star)^m \to \mathcal{T}^\star$ where $m = \mathsf{ar}(g)$ that satisfies

$$F_g(\Phi_k(s_1), \ldots, \Phi_k(s_m)) = \Phi_k(g(s_1, \ldots, s_m)) \text{ for all } s_1, \ldots, s_m. \qquad \text{(A)}$$

Suppose, in addition, that we have a function $G_{t_1,\ldots,t_n} \colon (\mathcal{T}^\star)^n \to \mathcal{T}^\star$ satisfying

$$G_{t_1,\ldots,t_n}(\Phi_k(s_1), \ldots, \Phi_k(s_n)) = \Phi_k(f(s_1, \ldots, s_n)) \text{ for all } \boldsymbol{s} \prec_{k,\tau(f)} \boldsymbol{t}. \qquad \text{(B)}$$

Here $\prec_{k,\tau(f)}$ abbreviates the $\tau(f)$-extension of the approximation $\prec_k$ and $\boldsymbol{t} = t_1, \ldots, t_n$.

Lemma 1. *Given $T_1, \ldots, T_n \in \mathcal{T}^\star$ define the function $H^{F_{g\prec f}, G_{\boldsymbol{t}}, T_1, \ldots, T_n} \colon \mathcal{T} \to \mathcal{T}^\star$ (where $F_{g\prec f}, G_{\boldsymbol{t}}, T_1, \ldots, T_n$ are treated as parameters) by subterm recursion as follows (suppressing parameters):*

$$H(s) := \begin{cases} T_i[s] & \begin{array}{l}\text{for the least } i \text{ such that } s \text{ is equal to} \\ \text{either } \mathsf{rt}(T_i) \text{ or some child of } \mathsf{rt}(T_i) \text{ in } T_i, \\ \text{if such an } i \text{ exists}\end{array} \\ F_g(H(s_1), \ldots, H(s_m)) & \text{if } s = g(s_1, \ldots, s_m) \text{ for } g \prec f \\ G_{\boldsymbol{t}}(H(s_1), \ldots, H(s_n)) & \text{if } s = f(s_1, \ldots, s_n) \\ [] & \text{otherwise,} \end{cases}$$

where [] denotes the empty tree and $T[s]$ the subtree of T with root s. Then

$$H^{F_{g\prec f}, G_{\boldsymbol{t}}, \Phi_k(t_1), \ldots, \Phi_k(t_n)}(s) = \Phi_k(s),$$

for all $s \prec_k t$, assuming (A) and (B).

Proof. By induction on $\blacktriangleleft$. If $s \prec_k t$ there are three possibilities. First, if $s \preceq_k t_i$ for some i then either $s = t_i$ or s is a child of t_i in $\Phi_k(t_i)$, and so $H(s) = \Phi_k(t_i)[s] = \Phi_k(s)$. Otherwise, suppose $s = g(s_1, \ldots, s_m)$ for $g \prec f$ and $s_i \prec_k t$ for all i, by the induction hypothesis we obtain $H(s_i) = \Phi_k(s_i)$ and therefore

$$H(s) = F_g(\Phi_k(s_1), \ldots, \Phi_k(s_m)) = \Phi_k(g(s_1, \ldots, s_m)),$$

by assumption (A). Similarly, if $s = f(s_1, \ldots, s_n)$, where for all i, $s_i \prec_k t$ and $(s_1, \ldots, s_n) \prec_{k,\tau(f)} (t_1, \ldots, t_n)$, then the induction hypothesis together with (B) yields

$$H(s) = G_{\boldsymbol{t}}(\Phi_k(s_1), \ldots, \Phi_k(s_n)) = \Phi_k(f(s_1, \ldots, s_n)).$$

From this the lemma follows. □

Let t be a term and let $(S_i)_{i\in I}$ be a finite collection of trees. Then the tree $t * \prod_{i\in I} S_i$ is the finite tree with root t and immediate subtrees S_i.

Lemma 2. *Define the function* $K^{F_{g\prec f},G_t}\colon (\mathcal{T}^\star)^n \to \mathcal{T}^\star$ *as follows:*

$$K^{F_{g\prec f},G_t}(T_1,\dots,T_n) := t' * \prod_{s\prec_k t'} H^{F_{g\prec f},G_t,T_1,\dots,T_n}(s),$$

where $t' = f(\mathsf{rt}(T_1),\dots,\mathsf{rt}(T_n))$. *Then assuming* (A) *and* (B) *we have*

$$K^{F_{g\prec f},G_t}(\Phi_k(t_1),\dots,\Phi_k(t_n)) = \Phi_k(t).$$

Proof. First, we remark that $K^{F_{g\prec f},G_t}$ is primitive recursive in $F_{g\prec f}$ and G_t, since $s \prec_k t$ is a primitive recursive predicate, and $\prod_{s\prec_k t}$ a finite search bounded by some primitive recursive term, by definition of $\prec_k$. Now, due to Lemma 1 we have

$$\begin{aligned} K^{F_{g\prec f},G_t}(\Phi_k(t_1),\dots,\Phi_k(t_n)) &= t * \prod_{s\prec_k t} H^{F_{g\prec f},G_t,\Phi_k(t_1),\dots,\Phi_k(t_n)}(s) \\ &= t * \prod_{s\prec_k t} \Phi_k(s), \end{aligned}$$

and this is just the tree whose branches are $\succ_k$-derivations $t \succ_k s \succ_k \dots \succ_k s_n$, which is exactly $\Phi_k(t)$. ☐

Lemma 3. *Define function* $F_f^{F_{g\prec f}}\colon (\mathcal{T}^\star)^n \to \mathcal{T}^\star$ *using the transfinite recursor over* $\sqsubset_{\tau(f)}$ *as follows*

$$F_f^{F_{g\prec f}}(T_1,\dots,T_n) := K^{F_{g\prec f},F_f\restriction(S_1,\dots,S_n)\sqsubset_{\tau(f)}(T_1,\dots,T_n)}(T_1,\dots,T_n).$$

Then assuming (A), *for all* $t_1,\dots,t_n$ *we have*

$$F_f^{F_{g\prec f}}(\Phi_k(t_1),\dots,\Phi_k(t_n)) = \Phi_k(f(t_1,\dots,t_n)).$$

Proof. By induction on $\prec_{k,\tau(f)}$; suppose for all $(s_1,\dots,s_n) \prec_{k,\tau(f)} (t_1,\dots,t_n)$ the lemma is true. Then (B) holds for $G_{t_1,\dots,t_n} = F_f\restriction_{(S_1,\dots,S_n)\sqsubset_{\tau(f)}(\Phi(t_1),\dots,\Phi(t_n))}$ and by Lemma 2 we have

$$F_f(\Phi_k(t_1),\dots,\Phi_k(t_n)) = \Phi_k(f(t_1,\dots,t_n)).$$

This completes the induction step, so the result holds for all arguments $t_1,\dots,t_n$. ☐

Lemma 4. *For each* $f \in \mathcal{F}$ *there exists a function* $F_f\colon (\mathcal{T}^\star)^n \to \mathcal{T}^\star$ *for* $n = \mathsf{ar}(f)$ *satisfying*

$$F_f(\Phi_k(t_1),\dots,\Phi_k(t_n)) = \Phi_k(f(t_1,\dots,t_n)).$$

Proof. By Lemma 3 we construct F_f in terms of F_g for $g \prec f$ assuming (A), and since $\prec$ is well-founded this construction is well-defined and correct for all f. ☐

Theorem 3. *There exists a function* $F\colon \mathcal{T} \to \mathcal{T}^\star$ *primitive recursive in* $\mathsf{TR}_{\sqsubset_{\tau(f)}}$ *for* $f \in \mathcal{F}$ *such that*

$$F(t) = \Phi_k(t),$$

for all terms t.

Proof. Define F using subterm recursion as

$$F(t) := \begin{cases} [x] & \text{if}\, t \;=\; x \\ F_f(F(t_1), \ldots, F(t_n)) & \text{if}\, t \;=\; f(t_1, \ldots, t_n). \end{cases}$$

Then theorem follows by Lemma 4 and induction over the structure of t. □

3.3 Derivational Complexity

Let $|\cdot|\colon \mathcal{T}^\star \to \mathbb{N}$ denote the recursive function which returns the length of the longest branch of trees in $\mathcal{T}^\star$.

Theorem 4. *Suppose that the TRS* $\mathcal{R}$ *is compatible with RPO for some suitable status function* τ. *Then its derivational complexity is bounded by a function primitive recursive in* $\mathsf{TR}_{\sqsubset_{\tau(f)}}$ *for* $f \in \mathcal{F}$.

Proof. By Theorem 2, if $\mathcal{R}$ is compatible with RPO then it is compatible with $\succ_k$ for some k, and so by Theorem 3 $\to_{\mathcal{R}}$ derivations from t are contained in the tree $F(t)$, where F is primitive recursive in the $\mathsf{TR}_{\sqsubset_{\tau(f)}}$. In particular, $\mathsf{dh}(t, \to_{\mathcal{R}}) \leqslant |F(t)|$, and therefore

$$\mathsf{dc}_{\mathcal{R}}(n) \leqslant \max_{|t| \leqslant n} |F(t)|$$

which is primitive recursive in F since we can bound the search $|t| \leq n$ because we only need to search over a finite number of variables. □

We can now re-derive, in a completely uniform way, some of the well-known complexity results concerning recursive path orders. To do this, first let $\mathsf{TR}_{\mathsf{mul}(n)}$ denote multiset recursion of lowest type over tuples $(x_1, \ldots, x_n)\colon \mathbb{N}^n$ of size n, and $\mathsf{TR}_{\mathsf{lex}(n)}$ lexicographic recursion. Then we have the following.

Lemma 5. *(a)* $\mathsf{TR}_{\mathsf{mul}(n)}$ *is definable from the Gödel recursor of lowest type; (b)* $\mathsf{TR}_{\mathsf{lex}(n)}$ *is definable from the Gödel type* 1 *recursor.*

Proof. Part (a) is straightforward, as one can easily find a (primitive recursive) encoding of $\mathbb{N}^n$ into $\mathbb{N}$ that preserves the multiset order.

For (b) we use induction on n. It's clear that $\mathsf{TR}_{\mathsf{lex}(1)}$ is just primitive recursion in the usual sense. Now, assuming that $\mathsf{TR}_{\mathsf{lex}(n-1)}$ has been defined, use this to construct the functional $h^H\colon \mathbb{N} \to (\mathbb{N} \to \mathbb{N}^{n-1} \to \mathbb{N}) \to \mathbb{N}^{n-1} \to \mathbb{N}$, parametrised by $H\colon \mathbb{N} \to \mathbb{N}^{n-1} \to (\mathbb{N} \to \mathbb{N}^{n-1} \to \mathbb{N}) \to \mathbb{N}$, and defined by

$$h^H xF\boldsymbol{x} := Hx\boldsymbol{x}\left(\lambda y, \boldsymbol{y} \,.\, \begin{cases} h^H xF\boldsymbol{y} & if\, y = x \wedge \boldsymbol{y} <_{\mathsf{lex}(n-1)} \boldsymbol{x} \\ Fy\boldsymbol{y} & if\, y < x \end{cases}\right).$$

Now by unwinding definitions we see that the term $\mathsf{R}^h_1 \colon \mathbb{N} \to \mathbb{N}^{n-1} \to \mathbb{N}$ satisfies

$$\mathsf{R}^h(x_1)(x_2,\ldots,x_n) = H(x_1)(x_2,\ldots,x_n)(\lambda \boldsymbol{y} <_{\mathsf{lex}(n)} \boldsymbol{x} \,.\, \mathsf{R}^h(y_1)(y_2,\ldots,y_n)),$$

where now $\boldsymbol{x} = (x_1,\ldots,x_n)$. But this is just recursion over $\mathsf{lex}(n)$. □

Corollary 1 (Hofbauer [1]**).** *If $\mathcal{R}$ is compatible with MPO then $\mathcal{R}$ has primitive recursive derivational complexity.*

Proof. This follows from Lemma 5(a) and the observation that $\mathsf{TR}_{\sqsubset_{\mathsf{mul}}}$ is definable from $\mathsf{TR}_{\mathsf{mul}}$ since $(\mathcal{T}^\star, \sqsubset)$ can be recursively encoded in $(\mathbb{N}, <)$. □

Corollary 2 (Weiermann [2]**).** *If $\mathcal{R}$ is compatible with RPO then $\mathcal{R}$ has multiply recursive derivational complexity.*

Proof. This follows analogously to Corollary 1, this time using Lemma 5(b). The fact that type one functions definable from the Gödel level 1 recursor are multiply recursive is folklore, see e.g. [15]. □

4 Conclusion

The most important feature of our work is not the rederivation of known complexity bounds, but in the manner in which we were able to do this. By constructing a concrete realising term F as a computational analogue to Theorem 1 which computes finitary $\succ_k$-derivation trees, we provided a bridge which relates the proof-theoretic complexity of well-founded recursive path orders to the derivational complexity of rewrite systems compatible with these orders.

A crucial point that we want to explore in future work is that our realising term is uniformly dependent on the parameters of the recursive path order used to prove termination, along with the size of the rewrite system, and any restriction in these parameters will cause a corresponding restriction in the complexity of F. Therefore a further, more detailed analysis of the structure of the realiser should enable us to obtain more refined complexity bounds.

For example, it follows from Weiermann's original derivational complexity analysis of the lexicographic path order that the induced multiple recursive bound allows parametrisation in the maximal arity of the function symbols, cf. [2], see also [16, Chap. 8]. Similar results follow from Hofbauer's analysis of the multiset path order, cf. [1]. We expect that these and similar finer characterisations of the derivational complexity induced by specific parameters of the recursive path orders can be obtained with relative ease in our context. More generally, we hope to extend these results and in particular derive new criteria on path orders which guarantee feasible complexity of rewrite systems.

As another example, one could study restricted variants of the lexicographic lifting on tuples which do not require type 1 recursion to define the corresponding recursor, giving us strengthenings of the multiset path order which allow us to prove interesting closure properties for the primitive recursive functions, an idea initiated by Cichon and Weiermann in [17].

References

1. Hofbauer, D.: Termination proofs by multiset path orderings imply primitive recursive derivation lengths. TCS **105**, 129–140 (1992)
2. Weiermann, A.: Termination proofs for term rewriting systems with lexicographic path ordering imply multiply recursive derivation lengths. TCS **139**, 355–362 (1995)
3. Buchholz, W.: Proof-theoretic analysis of termination proofs. APAL **75**, 57–65 (1995)
4. Parsons, C.: On a number theoretic choice schema and its relation to induction. In: Proceedings of the Intuitionism and Proof Theory, pp. 459–473 (1970)
5. Baader, F., Nipkow, T.: Term Rewriting and All That. Cambridge University Press, Cambridge (1998)
6. Moser, G., Weiermann, A.: Relating derivation lengths with the slow-growing hierarchy directly. In: Nieuwenhuis, R. (ed.) RTA 2003. LNCS, vol. 2706, pp. 296–310. Springer, Heidelberg (2003)
7. Cichon, E.A.: Termination orderings and complexity characterisations. In: Aczel, P., Simmons, H., Wainer, S.S. (eds.) Proof Theory, pp. 171–193. Cambridge University Press, Cambridge (1992)
8. Touzet, H.: Encoding the hydra battle as a rewrite system. In: Brim, L., Gruska, J., Zlatuška, J. (eds.) MFCS 1998. LNCS, vol. 1450, p. 267. Springer, Heidelberg (1998)
9. Moser, G.: KBOs, ordinals, subrecursive hierarchies and all that. JLC (2015, to appear)
10. Figueira, D., Figueira, S., Schmitz, S., Schnoebelen, P.: Ackermannian and primitive-recursive bounds with dickson's lemma. In: Proceedings of the 26th LICS, pp. 269–278. IEEE (2011)
11. Berardi, S., Oliva, P., Steila, S.: Proving termination with transition invariants of height ω. In: Proceedings of the 15th ICTCS, pp. 237–240 (2014)
12. Terese, : Term Rewriting Systems. Cambridge Tracks in Theoretical Computer Science. Cambridge University Press, Cambridge (2003)
13. Ferreira, M.C.F., Zantema, H.: Well-foundedness of term orderings. In: Lindenstrauss, N., Dershowitz, N. (eds.) CTRS 1994. LNCS, vol. 968, pp. 106–123. Springer, Heidelberg (1995)
14. Goubault-Larrecq, J.: Well-founded recursive relations. In: Fribourg, L. (ed.) CSL 2001 and EACSL 2001. LNCS, vol. 2142, p. 484. Springer, Heidelberg (2001)
15. Weiermann, A.: How is it that infinitary methods can be applied to finitary mathematics? Gödel's T: a case study. JSL **63**, 1348–1370 (1998)
16. Arai, T.: Some results on cut-elimination, provable well-orderings, induction, and reflection. APAL **95**, 93–184 (1998)
17. Cichon, E.A., Weiermann, A.: Term rewriting theory for the primitive recursive functions. APAL **83**, 199–223 (1997)

Local Compactness for Computable Polish Metric Spaces is Π^1_1-complete

André Nies[1](✉) and Slawomir Solecki[2]

[1] Department of Computer Science, University of Auckland, Auckland, New Zealand
andre@cs.auckland.ac.nz

[2] Department of Mathematics, University of Illinois at Urbana-Champaign, Champaign, IL, USA

Abstract. We show that the property of being locally compact for computable Polish metric spaces is Π^1_1 complete. We verify that local compactness for Polish metric spaces can be expressed by a sentence in $L_{\omega_1,\omega}$.

1 Introduction

Computable model theory is a well-established field of research that studies effectiveness aspects of countable structures and of model-theoretic concepts. Structures occurring in mathematics are often of size the continuum; in particular, separable complete metric spaces (also called Polish) play a central role in analysis, measure theory, and other areas. In order to carry out studies similar to computable model theory in the metric setting, computable Polish metric spaces have been introduced. Recall that a pseudo-metric satisfies symmetry and the triangle inequality, but allows pairs of distinct points to have distance 0.

Definition 1. (i) We represent a Polish metric space as follows. A point $V = \langle v_{i.k} \rangle_{i,k \in \mathbb{N}} \in \mathbb{R}^{\mathbb{N}\times\mathbb{N}}$ is a *distance matrix* if V is a pseudo-metric on $\mathbb{N}$. Let M_V denote the completion of the corresponding pseudo-metric space. In M_V we have a distinguished dense sequence of points $\langle p_i \rangle$ and present the space by giving their distances.
(ii) Let $\langle \phi_e \rangle$ be an effective listing of the rational-valued partial computable functions. A *computable presentation* of a Polish metric space is a distance matrix as in (i) where $|v_{i,k} - \phi_e(\langle i, k, t \rangle)| \leq 2^{-t}$. We call e an index for the space, and write V_e for the distance matrix given by ϕ_e in case ϕ_e is total.

Melnikov and Nies [2] studied the complexity of isomorphism for compact computable metric spaces. They showed that being compact is a Π^0_3 property of (an index for) a computable metric space, and that the complexity of isomorphism is Π^0_2 within that Π^0_3 class.

They also studied the complexity of a more general class. Recall that a topological space is *locally compact* if every point has a compact neighbourhood. For computable Polish metric spaces, this property is Π^1_1: one has to express that for every point x, there is a positive rational r such that the closed ball B of radius

A. Beckmann et al. (Eds.): CiE 2015, LNCS 9136, pp. 286–290, 2015.
DOI: 10.1007/978-3-319-20028-6_29

r around x is compact (which is Π^0_3 in a Cauchy name for x by relativizing the bound for compactness mentioned above).

Melnikov and Nies [2] asserted in Proposition 9 that being locally compact is Π^1_1 complete. Unfortunately, the proof sketch for the Π^1_1-hardness given there was incorrect (the reduction introduced is not Borel). In this note we give a proof of that result.

One can ask for the descriptive complexity of other important classes of (computable) Polish metric space. As an example we mention connectedness, where the only bound known is the trivial one, namely Π^1_2. To be compact and connected (i.e., a continuum) is Π^0_3 by [2, Proposition 11].

2 Main Result

We need a few well-known facts from topology. Firstly, let X be a Hausdorff space, $Y \subseteq X$ be a subspace, and $K \subseteq Y$. Then K is compact in Y iff K is compact in X. Next, suppose also that Y is dense in X. If V is open in X and $V \cap Y \subseteq K$ where $K \subseteq Y$ is compact, then $V \subseteq K$. (Otherwise, $V - K \neq \emptyset$ is open in X, so that $Y \cap (V - K) \neq \emptyset$, a contradiction.)

Lemma 2. *Let X be a Hausdorff space. Suppose Y is dense in X. If Y is locally compact as a subspace of X, then Y is open in X.*

Proof. Let $z \in Y$. There is compact $K \subseteq Y$ such that

$$\exists V \text{ open in } X\, [z \in V \cap Y \subseteq K].$$

Then $V \subseteq K \subseteq Y$.

Lemma 3. *Let X be a compact space, Y dense in X. Then*

$$Y \text{ is locally compact} \Leftrightarrow Y \text{ is open in } X.$$

Proof. $\Rightarrow$: This follows from previous lemma.

$\Leftarrow$: Let $v \in Y$. As a compact space, X is regular. So there are open sets U, V such that $v \in U$ and $X - Y \subseteq V$. Then $X - V \subseteq Y$ is a compact neighbourhood of v. This shows the lemma.

It is easy for a closed subset of $I_{\mathbb{Q}} = \mathbb{Q} \cap [0, 1]$ to be non-compact: for instance, the set of members of any sequence converging to an irrational is closed but not compact. On the other hand, for any successor ordinal α, the range of an embedding of a countable well-order of type α into $\mathbb{Q}$ is a compact set of Cantor-Bendixson (CB)-rank α. By an index for a computable subset R of $I_{\mathbb{Q}}$ we mean a number e such that ϕ_e, interpreted as a function $I_{\mathbb{Q}} \to \mathbb{N}$, is the characteristic function of S. We write $R = R_e$. The following is a straightforward effectivization of the classic result of Hurewicz from descriptive set theory (e.g. [1, Exercise 27.4]) that the compact subsets of $I_{\mathbb{Q}}$ form a Π^1_1-complete set. For detail on the effective version, see the proof of the main result in [3]. Let $O \subseteq \omega$ denote a Π^1_1 complete set.

Fact 4. *The set of indices for compact computable subsets of $I_{\mathbb{Q}}$ is Π^1_1-complete. Moreover, there is a computable function g such that $R_{g(e)}$ is closed in $I_{\mathbb{Q}}$ for each e, and $e \in O \leftrightarrow R_{g(e)}$ is compact.*

We will effectively assign to a closed subset of $I_{\mathbb{Q}}$ a Polish metric space in order to show:

Theorem 5. *(i) $\{V\colon M_V$ is locally compact $\}$ is properly Π^1_1.*
(ii) $\{i\colon M_{V_i}$ is locally compact $\}$ is a Π^1_1-complete set.

Proof. Write $\mathcal{N} = [0,1] - \mathbb{Q}$ (as this space is homeomorphic to Baire space ${}^\omega\omega$). Since $\mathbb{Q}$ is F_σ, by Alexandrov's result (see [1, 3.11]) we have a compatible complete metric on $\mathcal{N}$ given by

$$d(x,y) = |x-y| + \sum_{k=0}^{\infty} \min\{2^{-k-1}, |\frac{1}{|x-q_k|} - \frac{1}{|y-q_k|}|\},$$

where $\langle q_k \rangle_{k\in\mathbb{N}}$ list $I_{\mathbb{Q}}$ without repetitions in some effective way.

For topological space X and sets $S \subseteq Y \subseteq X$, denote by $C_Y(S) = \bar{S} \cap Y$ the closure of S in Y with the subspace topology.

(i) We describe the coding procedure turning a closed subset of $I_{\mathbb{Q}}$ into a representation of a Polish metric space as above so that compactness of the subset corresponds to local compactness of the space. Let R be a closed subset of $I_{\mathbb{Q}}$. For each $v \in R$ the set $\Theta(R)$ contains a certain sequence of irrationals converging to v, as follows:

$$\Theta(R) = [0,1] \cap \{q_k - 2^{-m}\sqrt{2}\colon\ q_k \in R \wedge m \geq k\}.$$

To obtain a representation of $\Theta(R)$ as a Polish metric space, let $\Delta(R)$ be a sequence that lists $[0,1] \cap \{q_k - 2^{-m}\sqrt{2}\colon\ m \geq k\}$ without repetitions. Let $v_{i,j} = d(\Delta(R)_i, \Delta(R)_j)$ where d is the distance on $\mathcal{N}$ defined above. Recall Definition 1 and note that $(v_{i,j})_{i,j\in\mathbb{N}} \in \mathbb{R}^{\mathbb{N}\times\mathbb{N}}$ is a metric on $\mathbb{N}$ and $M_V \cong C_{\mathcal{N}}(\Theta(R))$.

Claim. $C_{[0,1]}(\Theta(R)) = R \,\dot{\cup}\, C_{\mathcal{N}}(\Theta(R))$.

The inclusion "$\supseteq$" is clear. For the inclusion "$\subseteq$", suppose that $x \in C_{[0,1]}(\Theta(R))$. There are sequences of numbers $\langle k_t \rangle_{t\in\mathbb{N}}$ and $\langle m_t \rangle_{t\in\mathbb{N}}$ such that $q_{k_t} - 2^{-m_t}\sqrt{2} \in \Theta(R)$ and $x = \lim_{t\to\infty}[q_{k_t} - 2^{-m_t}\sqrt{2}]$. Clearly $q_{k_t} \in R$ for each t.

If the sequence $\langle k_t \rangle$ is bounded, by passing to a subsequence we may assume that it is constant. Then $m_t \to \infty$, so $x \in R$.

Otherwise, by passing to a subsequence we may assume that $k_t \to \infty$. Then $m_t \to \infty$ because we chose $m \geq k$ in the definition of $\Theta(R)$, and so $x = \lim_t q_{k_t}$. If $x \in \mathbb{Q}$ then $x \in R$ because R is closed. Otherwise, $x \in C_{\mathcal{N}}(\Theta(R))$. This proves the claim.

Claim. Let $R \subseteq I_{\mathbb{Q}}$ be closed. Then

$$R \text{ is compact} \Leftrightarrow C_{\mathcal{N}}(\Theta(R)) \text{ is locally compact.}$$

First suppose that R is compact. By the first claim,

$$C_{\mathcal{N}}(\Theta(R)) = C_{[0,1]}(\Theta(R)) - R.$$

Since R is closed in $[0,1]$, the set $C_{\mathcal{N}}(\Theta(R))$ is open in its closure which is a compact set, so it is locally compact.

Now suppose that $C_{\mathcal{N}}(\Theta(R))$ is locally compact. This set is dense in $C_{[0,1]}(\Theta(R))$, so it is open in $C_{[0,1]}(\Theta(R))$ by Lemma 3. By the first claim again, this means that R is closed in $C_{[0,1]}(\Theta(R))$, and hence compact. This establishes (i).

For (ii), we use the function g from Fact 4. Uniformly in e, we can obtain a computable sequence $\Delta(R_{g(e)})$ as above. For any effective listing $\langle x_e \rangle$ of a countable subset of $\mathcal{N}$, the function $e, i \rightarrow d(x_e, x_i)$ is computable. Hence we can determine an index $i = p(e)$ such that $M_{V_i} \cong C_{\mathcal{N}}(\Theta(R_{g(e)}))$. So $e \in O \leftrightarrow M_{V_{p(e)}}$ is locally compact.

3 Expressive Power of $L_{\omega_1,\omega}$

Recall that for a signature S, the language $L_{\omega_1,\omega}(S)$ is the extension of first-order language that allows countable conjunctions and disjunctions over a set of formulas with a shared finite reservoir of free variables. In the setting of countable structures, the class of models for a sentence in $L_{\omega_1,\omega}(S)$ is Borel. In this section we use our main result to show that this fails when countability is replaced by separability in the context of complete metric spaces.

A metric space (M,d) can be turned into a structure in the classical sense by introducing a binary relation $R_q xy$ for each positive rational q, with the intended meaning that $d(x,y) < q$. Let S denote the signature consisting of these relation symbols.

Proposition 6. *Local compactness among Polish metric spaces can be described by an $L_{\omega_1,\omega}(S)$ sentence α.*

Proof. The sentence α expresses

$$\forall x \bigvee\nolimits_{t \in \mathbb{Q}^+} [\overline{B_t(x)} \text{ is compact}].$$

Recall that for a complete metric space M, compactness is the same as total boundedness: for each rational $r > 0$, M is the union of k many balls of radius r for some k. We can use this to express that $M = \overline{B_t(x)}$ is compact by a conjunction over rationals $r > 0$, of a disjunction over the number of balls k, of first-order sentences of $\exists\forall$ type asserting that there are k balls covering M. This shows the proposition.

Thus, the set of Polish metric models of α is not Borel.

Question 7. Determine the descriptive complexity of being connected among [locally compact] Polish metric spaces.

Question 8. Can connectedness be expressed by an $L_{\omega_1,\omega}(S)$ sentence?

As suggested by T. Tsankov, it would also be interesting to study the meaning of the Π^1_1-rank for the class of locally compact spaces.

Acknowledgment. This work was carried out at the Hausdorff Institute for Mathematics in October 2013, and at the Research Centre Whiritoa in December 2014.

References

1. Kechris, A.S.: Classical descriptive set theory, vol. 156. Springer, New York (1995)
2. Melnikov, A.G., Nies, A.: The Classification Problem for Compact Computable Metric Spaces. In: Bonizzoni, P., Brattka, V., Löwe, B. (eds.) CiE 2013. LNCS, vol. 7921, pp. 320–328. Springer, Heidelberg (2013)
3. Naulin, R., Aylwin, C.: On the complexity of the family of compact subsets of $\mathbb{Q}$. Notas de Mat. **5**(2), 283 (2009)

Iterative Forcing and Hyperimmunity in Reverse Mathematics

Ludovic Patey(✉)

Laboratoire PPS, Université Paris Diderot, Paris, France
ludovic.patey@computability.fr

Abstract. The separation between two theorems in reverse mathematics is usually done by constructing a Turing ideal satisfying a theorem P and avoiding the solutions to a fixed instance of a theorem Q. Lerman, Solomon and Towsner introduced a forcing technique for iterating a computable non-reducibility in order to separate theorems over omega-models. In this paper, we present a modularized version of their framework in terms of preservation of hyperimmunity and show that it is powerful enough to obtain the same separations results as Wang did with his notion of preservation of definitions.

1 Introduction

Reverse mathematics is a mathematical program which aims to capture the provability content of ordinary (i.e. non set-theoretic) theorems. It uses the framework of subsystems of second-order arithmetic, with a base theory RCA_0 which is composed of the basic axioms of Peano arithmetic together with the Δ^0_1 comprehension scheme and the Σ^0_1 induction scheme. Thanks to the equivalence between Δ^0_1-definable sets and computable sets, RCA_0 can be thought as capturing "computational mathematics". See [8] for a good introduction.

Many theorems are Π^1_2 statements $(\forall X)(\exists Y)\Phi(X,Y)$ and come with a natural class of *instances* X. The sets Y such that $\Phi(X,Y)$ holds are *solutions* to X. For example, König's lemma (KL) states that every infinite, finitely branching tree has an infinite path. An instance of KL is an infinite, finitely branching tree T. A solution to T is an infinite path through T. Given two Π^1_2 statements P and Q, proving an implication Q $\rightarrow$ P over RCA_0 consists in taking a P-instance X and constructing a solution to X through a computational process involving several applications of the Q statement. Empirically, many proofs of implications are in fact *computable reductions* [9].

Definition 1. (Computable reducibility). *Fix two Π^1_2 statements* P *and* Q*. We say that* P *is* computably reducible *to* Q *(written* P $\leq_c$ Q*) if every* P-*instance I computes a* Q-*instance J such that for every solution X to J, $X \oplus I$ computes a solution to I.*

If the computable reduction between from P to Q can be formalized over RCA_0, then $\mathsf{RCA}_0 \vdash$ Q $\rightarrow$ P. However, P may not be computably reducible to Q

A. Beckmann et al. (Eds.): CiE 2015, LNCS 9136, pp. 291–301, 2015.
DOI: 10.1007/978-3-319-20028-6_30

while $\mathsf{RCA}_0 \vdash \mathsf{Q} \rightarrow \mathsf{P}$. Indeed, one may need more than one application of Q to solve the instance of P. This is for example the case of Ramsey's theorem for pairs with n colors (RT^2_n) which implies RT^2_{n+1} over RCA_0, but $\mathsf{RT}^2_{n+1} \not\leq_c \mathsf{RT}^2_n$ for $n \geq 1$ (see [21]).

In order to prove the non-implication between P and Q, one needs to iterate the computable non-reducibility in order to build a model of Q which is not a model of P. This is the purpose of the framework developed by Lerman, Solomon and Towsner in [14]. They successfully used their framework for separating the Erdős-Moser theorem (EM) from the stable ascending descending sequence principle (SADS) and separating the ascending descending sequence (ADS) from the stable chain antichain principle (SCAC). Their approach has been reused by Flood & Towsner [5] and the author [19] on diagonal non-computability statements.

However, their framework suffers some drawbacks. In particular the forcing notions involved are heavy and the deep combinatorics witnessing the non-implications are hidden by the complexity of the proof. Moreover, the P-instance chosen in the ground forcing depends on the forcing notion used in the iteration forcing and therefore the overall construction is not modular. On the other hand, Wang [23] recently introduced the notion of preservation of definitions and made independent proofs of preservations for various statements included EM. Then he deduced that the conjunction of those statements does not imply SADS, therefore strengthening the result of Lerman, Solomon & Towsner in a modular way. Variants of this notion have been reused by the author [21] for separating the free set theorem (FS) from RT^2_2.

In this paper, we present a modularized version of the framework of Lerman, Solomon & Towsner and use it to reprove the separation results obtained by Wang [23]. We thereby show that this framework is a viable alternative to the notion introduced by Wang for separating statements in reverse mathematics. In particular, we reprove the following theorem, in which COH is the cohesiveness principle, WKL_0 is weak König's lemma, RRT^2_2 the rainbow Ramsey theorem for pairs, $\Pi^0_1\mathsf{G}$ the Π^0_1-genericity principle and STS^2 the stable thin set theorem for pairs.

Theorem 2. (Wang [23]**)** *Let Φ be the conjunction of* COH, WKL_0, RRT^2_2, $\Pi^0_1\mathsf{G}$, *and* EM. *Over* RCA_0, *Φ does not imply any of* SADS *and* STS^2.

In Sect. 2, we introduce the framework of Lerman, Solomon & Towsner in its original form and detail its drawbacks. Then, in Sect. 3, we develop a modularized version of their framework. In Sect. 4, we establish basic preservation results, before reproving in Sect. 5 Wang's theorem. Last, we reprove in Sect. 6 the separation obtained by the author in [21].

1.1 Notation

String, sequence. Fix an integer $k \in \omega$. A *string* (over k) is an ordered tuple of integers $a_0, \ldots, a_{n-1}$ (such that $a_i < k$ for every $i < n$). The empty string is written ε. A *sequence* (over k) is an infinite listing of integers $a_0, a_1, \dots$ (such that $a_i < k$ for every $i \in \omega$). Given $s \in \omega$, k^s is the set of strings of length s over k and $k^{<\omega}$ is the set of finite strings over k. Given a string $\sigma \in k^{<\omega}$, we denote by $|\sigma|$ its length. Given two strings $\sigma, \tau \in k^{<\omega}$, σ is a *prefix* of τ (written $\sigma \preceq \tau$) if there exists a string $\rho \in k^{<\omega}$ such that $\sigma\rho = \tau$. A *binary string* (resp. real) is a *string* (resp. sequence) over 2. We may equate a real with a set of integers by considering that the real is its characteristic function.

Tree, path. A tree $T \subseteq \omega^{<\omega}$ is a set downward-closed under the prefix relation. The tree T is *finitely branching* if every node $\sigma \in T$ has finitely many immediate successors. A *binary* tree is a tree $T \subseteq 2^{<\omega}$. A set $P \subseteq \omega$ is a *path* though T if for every $\sigma \prec P$, $\sigma \in T$. A string $\sigma \in k^{<\omega}$ is a *stem* of a tree T if every $\tau \in T$ is comparable with σ. Given a tree T and a string $\sigma \in T$, we denote by $T^{[\sigma]}$ the subtree $\{\tau \in T : \tau \preceq \sigma \vee \tau \succeq \sigma\}$.

Sets. Given two sets X and Y, $X \subseteq^* Y$ means that X is almost included into Y, $X =^* Y$ means $X \subseteq^* Y \wedge Y \subseteq^* X$ and $X \subseteq_{\texttt{fin}} Y$ means that X is a finite subset of Y. Given some $x \in \omega$, $A > x$ denotes the formula $(\forall y \in A)[y > x]$.

2 The Iteration Framework

An *ω-structure* is a structure $\mathscr{M} = (\omega, S, +, \cdot, <)$ where ω is the set of standard integers, $+$, $\cdot$ and $<$ are the standard operations over integers and S is a set of reals such that $\mathscr{M}$ satisfies the axioms of RCA_0. Friedman [7] characterized the second-order parts S of ω-structures as those forming a *Turing ideal*, that is, a set of reals closed under Turing join and downward-closed under the Turing reduction.

Fix two Π^1_2 statements P and Q. The construction of an ω-model of P which is not a model of Q consists in creating a Turing ideal $\mathscr{I}$ together with a fixed Q-instance $I \in \mathscr{I}$, such that every P-instance $J \in \mathscr{I}$ has a solution in $\mathscr{I}$, whereas I contains no solution in $\mathscr{I}$. In the first place, let us just focus on the one-step case, that is, a proof that $\mathsf{Q} \not\leq_c \mathsf{P}$. To do so, one has to choose carefully some Q-instance I such that every I-computable P-instance has a solution X which does not I-compute a solution to I. The construction of a solution X to some I-computable P-instance J will have to satisfy the following scheme of requirements for each index e:

$$\mathscr{R}_e : \Phi_e^{X \oplus I} \text{ infinite } \rightarrow \Phi_e^{X \oplus I} \text{ is not a solution to } I$$

Such requirements may not be satisfiable for an arbitrary Q-instance I. The choice of the instance and the satisfaction of the requirement is strongly dependent on the combinatorics of the statement Q and the forcing notion used for constructing a solution to J. A recurrent approach in the framework of Lerman, Solomon & Towsner consists in constructing a Q-instance I which satisfies some fairness property. The forcing notion $\mathbb{P}^I$ used in the construction of a solution to J is usually designed so that

(i) There exists an I-computable set encoding (at least) every condition in $\mathbb{P}^I$

(ii) Given some forcing condition in $\mathbb{P}^I$, one can uniformly find in a c.e. search a finite set of candidate extensions such that one of them is in $\mathbb{P}^I$ (e.g. the notion of split pair in [14], of finite cover for a tree forcing, ...).

The fairness property states the following:

"For every condition in $\mathbb{P}^I$, if for every $x \in \omega$, there exists a *finite* Q-instance $A > x$ and a finite set of candidate extensions $d_0, \dots, d_m$ such that $\Phi_e^{d_i \oplus I}$ is not a solution to A for each $i \leq m$, then one of the A's is a subinstance of I."

This property is designed so that we can satisfy it by taking each condition $c \in \mathbb{P}^I$ one at a time, find some finite Q-instance A on which I is not yet defined, and define I over A. One can think of the instance I as a fair adversary who, if we have infinitely often the occasion to beat him, will be actually beaten at some time.

Suppose now we want to extend this computable non-reducibility into a separation over ω-structures. One may naturally try to make the instance I satisfy the fairness property at every level of the iteration forcing. At the first iteration with an I-computable P-instance J, the property is unchanged. At the second iteration, the P-instance J_1 is $X_0 \oplus I$-computable, but the set X_0 is not yet constructed. Hopefully, the fairness property requires a finite piece of oracle X_0. Therefore we can modify the fairness property which becomes

"For every condition $c_0 \in \mathbb{P}^I$ and every condition $c_1 \in \mathbb{P}^{c_0 \oplus I}$, if for every $x \in \omega$, there exists a Q-instance $A > x$, a finite set of candidate extensions $d_0, \dots, d_m \in \mathbb{P}^I$ and $d_{0,i}, \dots, d_{n_i,i} \in \mathbb{P}^{d_i \oplus I}$ for each $i \leq m$ such that $\Phi_e^{d_{j,i} \oplus d_i \oplus I}$ is not a solution to A for each $i \leq m$ and $j \leq n_i$, then one of the A's is a subinstance of I."

Since this property becomes overly complicated in the general case, Lerman, Solomon and Towsner abstracted the notion of requirement and made it a $\Sigma_1^{0,I}$ black box which takes as parameters a condition and a finite Q-instance. Instead of making the instance I in charge of satisfying the fairness property at every level of the iteration forcing, the instance I satisfies the property only at the first level. Then, by encoding a requirement at the next level into a requirement at the current level, the iteration forcing ensures the propagation of this fairness property from the first level to every level. The property in its abstracted form is then

"For every condition in $\mathbb{P}^I$ and every $\Sigma_1^{0,I}$ predicate $\mathscr{K}^I$, if for every $x \in \omega$, there is a *finite* Q-instance $A > x$ and a finite set of candidate extensions $d_0, \dots, d_m$ such that $\mathscr{K}^I(A, d_i)$ is satisfied for each $i \leq m$, then one of the A's is a subinstance of I."

In particular, by letting $\mathscr{K}^I(A, c)$ be the predicate "$\Phi_e^{d_i \oplus I}$ is not a solution to A", the requirements $\mathscr{R}_e$ will be satisfied.

The problem of such an approach is that the construction of the Q-instance strongly depends on the forcing notion used in the iteration forcing. A slight modification of the latter requires to modify the ground forcing. Moreover, if someone wants to prove that the conjunction of two statements does not imply a third one, we need to construct an instance I which will satisfy the fairness property for the two statements, and in each iteration forcing, we will need to ensure that both properties are propagated to the next iteration. The size of the overall construction explodes when trying to make a separation of the conjunction of several statements at the same time.

3 Preservation of Hyperimmunity

In this section, we propose a general simplification of the framework of Lerman, Solomon & Towsner [14] and illustrate it in the case of the separation of EM from SADS. The corresponding fairness property happens to coincide with the notion of hyperimmunity. The underlying idea ruling this simplification is the following: since each condition in the iteration forcing can be given an index and since the finite set of candidate extensions of a condition c, can be found in a c.e. search, given a $\Sigma_1^{0,I}$ predicate $\mathscr{K}^I$, the following formula is again $\Sigma_1^{0,I}$:

$\varphi(U) =$ "there exists a finite set of candidate extensions $d_0, \dots, d_m$ of c such that $\mathscr{K}^I(U, d_i)$ is satisfied for each $i \leq m$"

We can therefore abstract the iteration forcing and ask the instance I to satisfy the following property:

"For every $\Sigma_1^{0,I}$ predicate $\varphi(U)$, if for every $x \in \omega$, there exists a finite Q-instance $A > x$ such that $\varphi(A)$ is satisfied, then one of the A's is a subinstance of I."

Let us illustrate how this simplification works by reproving the separation of the Erdős-Moser theorem from the ascending descending sequence principle.

Definition 3. (Ascending descending sequence). *A linear order is* stable *if it is of order type $\omega + \omega^*$.* ADS *is the statement "Every linear order admits an infinite ascending or descending sequence".* SADS *is the restriction of* ADS *to stable linear orders.*

The ascending descending sequence principle has been studied within the framework of reverse mathematics by Hirschfeldt & Shore [10]. Lerman, Solomon & Towsner [14] constructed an infinite stable linear order I with ω and ω^* parts respectively B_0 and B_1, such that for every condition c and every $\Sigma_1^{0,I}$ predicate $\mathscr{K}^I$, if for every $x \in \omega$, there exists a finite set $A > x$ and a finite set of candidate extensions $d_0, \dots, d_m$ of c such that $\mathscr{K}^I(A, d_i)$ is satisfied for each $i \leq m$, then one of the A's will be included in B_0 and another one will be included in B_1. In particular, taking $\mathscr{K}^I(A, c) = \Phi_e^{c \oplus I} \cap A \neq \emptyset$, no infinite solution to the constructed tournament I-computes a solution to I. After abstraction, we obtain the following property:

"For every $\Sigma_1^{0,I}$ predicate $\varphi(U)$, if for every $x \in \omega$, there exists a finite set $A > x$ such that $\varphi(A)$ is satisfied, one of the A's is included in B_0 and one of the A's is included in B_1."

Following the terminology of [14], we say that a formula $\varphi(U)$ is *essential* if for every $x \in \omega$, there exists some finite set $A > x$ such that $\varphi(A)$ holds. This fairness property coincides with the notion of hyperimmunity for $\overline{B_0}$ and $\overline{B_1}$.

Definition 4. (Preservation of hyperimmunity).

1. *Let $D_0, D_1, \ldots$ be a computable list of all finite sets and let f be computable.* A c.e. array $\{D_{f(i)}\}_{i \geq 0}$ *is a c.e. set of mutually disjoint finite sets $D_{f(i)}$. A set B is* hyperimmune *if for every c.e. array $\{D_{f(i)}\}_{i \geq 0}$, $D_{f(i)} \cap B = \emptyset$ for some i.*
2. *A Π^1_2 statement* P admits preservation of hyperimmunity *if for each set Z, each countable collection of Z-hyperimmune sets $A_0, A_1, \ldots$, and each* P*-instance $X \leq_T Z$ there exists a solution Y to X such that the A's are $Y \oplus Z$-hyperimmune.*

The following lemma establishes the link between the fairness property for SADS and the notion of hyperimmunity.

Lemma 5. *Fix a set Z. A set B is Z-hyperimmune if and only if for every essential $\Sigma^{0,Z}_1$ predicate $\varphi(U)$, $\varphi(A)$ holds for some finite set $A \subseteq \overline{B}$.*

Hirschfeldt, Shore & Slaman constructed in [11, Theorem 4.1] a stable computable linear order such that both the ω and the ω^* part are hyperimmune. As every ascending (resp. descending) sequence is an infinite subset of the ω (resp. ω^*) part of the linear order, we deduce the following theorem.

Theorem 6. SADS *does not admit preservation of hyperimmunity.*

A slight modification of the forcing in [14] gives preservation of hyperimmunity of the Erdős-Moser theorem. We will however reprove it in a later section with a simpler forcing notion. As expected, the notion of preservation of hyperimmunity can be used to separate statements in reverse mathematics.

Lemma 7. *Fix two Π^1_2 statements* P *and* Q*. If* P *admits preservation of hyperimmunity and* Q *does not, then* P *does not imply* Q *over* RCA_0*.*

Before starting an analysis of preservations of hyperimmunity for basic statements, we state another negative preservation result which enables to reprove that the Erdős-Moser theorem does not imply the stable thin set theorem for pairs [15].

Definition 8. (Thin set theorem). *Let $k \in \omega$ and $f : [\omega]^k \to \omega$. A set A is f-thin if $f([A]^n) \neq \omega$, that is, if the set A "avoids" at least one color.* TS^k *is the statement "every function $f : [\omega]^k \to \omega$ has an infinite f-thin set".* STS^2 *is the restriction of* TS^2 *to stable functions.*

Introduced by Friedman in [6], the basic reverse mathematics of the thin set theorem has been settled by Cholak, Hirst & Jockusch in [2]. Its study has been continued by Wang [24], Rice [22] and the author [17,21]. The author constructed in [20] an infinite computable stable function $f : [\omega]^2 \to \omega$ such that the sets $B_i = \{n \in \omega : \lim_s f(n, s) \neq i\}$ are all hyperimmune. Every infinite f-thin set being an infinite subset of one of the B's, we deduce the following theorem.

Theorem 9. STS^2 *does not admit preservation of hyperimmunity.*

4 Basic Preservations of Hyperimmunity

When defining a notion, it is usually convenient to see how it relates with typical sets. There are two kinds of typicalities: genericity and randomness. Both notions admit preservation of hyperimmunity.

Theorem 10. *Fix some set Z and a countable collection of Z-hyperimmune sets $B_0, B_1, \ldots$*

1. *If G is sufficiently Cohen generic relative to Z, the B's are $G \oplus Z$-hyperimmune.*
2. *If R is sufficiently random relative to Z, the B's are $R \oplus Z$-hyperimmune.*

Note that this does not mean that the sets G and R are hyperimmune-free relative to Z. In fact, the converse holds: if G is sufficiently generic and R sufficiently random, then both are Z-hyperimmune. Some statements like the atomic model theorem (AMT), Π^0_1-genericity (Π^0_1G) and the rainbow Ramsey theorem for pairs (RRT^2_2) are direct consequences of genericity and randomness [4,11]. We can deduce from Theorem 10 that they all admit preservation of hyperimmunity.

Cohesiveness is a very useful statement in the analysis of Ramsey-type theorems as it enables to transform an arbitrary instance into a stable one [3]. A set C is *cohesive* for a sequence of sets $R_0, R_1, \ldots$ if $C \subseteq^* R_i$ or $C \subseteq^* \overline{R_i}$ for each i.

Theorem 11. COH *admits preservation of hyperimmunity.*

The proof is done by the usual construction of a cohesive set with Mathias forcing, combined with the following lemma.

Lemma 12. *For every set Z, every Z-computable Mathias condition (F, X), every $\Sigma^{0,Z}_1$ formula $\varphi(G, U)$ and every Z-hyperimmune set B, there exists an extension (E, Y) such that $X =^* Y$ and either $\varphi(G, U)$ is not essential for every set G satisfying (E, Y), or $\varphi(E, A)$ holds for some finite set $A \subseteq \overline{B}$.*

Proof. Define

$$\psi(U) = (\exists G \supseteq F)[G \subseteq F \cup X \wedge \varphi(G, U)]$$

The formula $\psi(U)$ is $\Sigma^{0,Z}_1$. By hyperimmunity of B, either $\psi(U)$ is not essential, or $\psi(A)$ holds for some finite set $A \subseteq \overline{B}$. In the first case, the condition (F, X) already satisfies the desired property. In the second case, let $A \subseteq_{\mathtt{fin}} \overline{B}$ be such that $\psi(A)$ holds. By the use property, there exists a finite set E satisfying (F, X) such that $\varphi(E, A)$ holds. Let $Y = X \setminus [0, max(E)]$. The condition (E, Y) is a valid extension. □

Weak König's lemma (WKL_0) states that every infinite, binary tree admits an infinite path.

Theorem 13. WKL_0 *admits preservation of hyperimmunity.*

Wei Wang [personal communication] observed that WKL_0 preserves hyperimmunity in a much stronger sense than COH, since cohesive sets are of hyperimmune degree [12], whereas by the hyperimmune-free basis theorem [13], WKL_0 can preserve hyperimmunities of *every* hyperimmune set simultaneously and not only countably many.

5 The Erdős-Moser Theorem and Preservation of Hyperimmunity

The Erdős-Moser theorem is a statement from graph theory which received a particular interest from reverse mathematical community as it provides, together with the ascending descending sequence principle, an alternative proof of Ramsey's theorem for pairs.

Definition 14. (Erdős-Moser theorem). *A tournament T is an irreflexive binary relation such that for all $x, y \in \omega$ with $x \neq y$, exactly one of $T(x,y)$ or $T(y,x)$ holds. A tournament T is* transitive *if the corresponding relation T is transitive in the usual sense.* EM *is the statement "Every infinite tournament T has an infinite transitive subtournament."*

The Erdős-Moser theorem was introduced in reverse mathematics by Bovykin & Weiermann [1] and then studied by Lerman, Solomon & Towsner [14] and the author [16–18]. In this section, we give a simple proof of the following theorem.

Theorem 15. EM *admits preservation of hyperimmunity.*

The proof of Theorem 15 exploits the modularity of the framework by using preservation of hyperimmunity of WKL_0. Together with the previous preservations results, this theorem is sufficient to reprove Theorem 2. We must first introduce some terminology.

Definition 16. (Minimal interval). *Let T be an infinite tournament and $a, b \in T$ be such that $T(a,b)$ holds. The* interval (a,b) *is the set of all $x \in T$ such that $T(a,x)$ and $T(x,b)$ hold. Let $F \subseteq T$ be a finite transitive subtournament of T. For $a, b \in F$ such that $T(a,b)$ holds, we say that (a,b) is a* minimal interval *of F if there is no $c \in F \cap (a,b)$, i.e., no $c \in F$ such that $T(a,c)$ and $T(c,b)$ both hold.*

Definition 17. *An* Erdős Moser condition *(EM condition) for an infinite tournament T is a Mathias condition (F, X) where*

(a) $F \cup \{x\}$ is T-transitive for each $x \in X$
(b) X is included in a minimal T-interval of F.

EM extension is Mathias extension. A set G *satisfies* an EM condition (F, X) if it is T-transitive and satisfies the Mathias extension (F, X). Basic properties of EM conditions have been stated and proven in [18].

Fix a set Z and some countable collection of Z-hyperimmune sets $B_0, B_1, \dots$ Our forcing notion is the partial order of Erdős Moser conditions (F, X) such that the B's are $X \oplus Z$-hyperimmune. Our initial condition is $(\emptyset, \omega)$. By Lemma 5.9 in [18], EM conditions are extendable, so we can force the transitive subtournament to be infinite. Therefore it suffices to prove the following lemma to deduce Theorem 15.

Lemma 18. *Fix a condition (F, X), some $i \in \omega$ and some $\Sigma_1^{0,Z}$ formula $\varphi(G, U)$. There exists an extension (E, Y) such that either $\varphi(G, U)$ is not essential for every set G satisfying (E, Y), or $\varphi(E, A)$ holds for some finite set $A \subseteq \overline{B_i}$.*

Proof. Let $\psi(U)$ be the formula "For every partition $X_0 \cup X_1 = X$, there exists some $j < 2$, a T-transitive set $G \subseteq X_j$ and a set $\tilde{A} \subseteq U$ such that $\varphi(F \cup G, \tilde{A})$ holds." By compactness, $\psi(U)$ is a $\Sigma_1^{0,X \oplus Z}$ formula. By $X \oplus Z$-hyperimmunity of B_i, we have two cases:

- Case 1: $\psi(A)$ holds for some finite set $A \subseteq \overline{B_i}$. By compactness, there exists a finite set $H \subset X$ such that for every partition $H_0 \cup H_1 = H$, there exists some $j < 2$, a T-transitive set $G \subseteq H_j$ and a set $\tilde{A} \subseteq A$ such that $\varphi(F \cup G, \tilde{A})$ holds. Given two sets U and V, we denote by $U \to_T V$ the formula $(\forall x \in U)(\forall y \in V)T(x, y)$. Each element $y \in X$ induces a partition $H_0 \cup H_1 = H$ such that $H_0 \to_T \{y\} \to_T H_1$. There exists finitely many such partitions, so by the infinite pigeonhole principle, there exists an X-computable infinite set $Y \subset X$ and a partition $H_0 \cup H_1 = H$ such that $H_0 \to_T Y \to_T H_1$. Let $j < 2$ and $G \subseteq H_j$ be the T-transitive set such that $\varphi(F \cup G, \tilde{A})$ holds for some $\tilde{A} \subseteq A \subseteq \overline{B_i}$. By Lemma 5.9 in [18], $(F \cup G, Y)$ is a valid extension.
- Case 2: $\psi(U)$ is not essential with some witness x. Then the $\Pi_1^{0,X \oplus Z}$ class $\mathscr{C}$ of sets $X_0 \oplus X_1$ such that $X_0 \cup X_1 = X$ and for every $j < 2$, every T-transitive set $G \subseteq X_j$ and every finite set $\tilde{A} > x$, the formula $\varphi(F \cup G, \tilde{A})$ does not hold is not empty. By preservation of hyperimmunity of WKL_0, there exists some partition $X_0 \oplus X_1 \in \mathscr{C}$ such that the B's are $X_0 \oplus X_1 \oplus Z$-hyperimmune. The set X_j is infinite for some $j < 2$ and the condition (F, X_i) is the desired EM extension. □

6 Thin Set Theorem and Preservation of Hyperimmunity

There exists a fundamental difference in the way SADS and STS^2 witness their failure of preservation of hyperimmunity. In the case of SADS, we construct two hyperimmune sets whereas in the case of STS^2, a countable collection of hyperimmune sets is used. This difference can be exploited to obtain further separation results.

Definition 19. (Preservation of n hyperimmunities). *A Π_2^1 statement P admits preservation of n hyperimmunities if for each set Z, each Z-hyperimmune sets $A_0, \dots, A_{n-1}$, and each P-instance $X \leq_T Z$ there exists a solution Y to X such that the A's are $Y \oplus Z$-hyperimmune.*

Theorem 6 shows that SADS does not admit preservation of 2 hyperimmunities. On the other hand, we shall see that STS^2 admits preservation of n hyperimmunities for every $n \in \omega$. Consider the following variants of the thin set theorem.

Definition 20. (Thin set theorem). *Given a function $f : [\omega]^k \to n$, an infinite set H is f-*thin *if $|f([H]^k)| \leq n-1$ (i.e. f avoids one color over H). For every $k \geq 1$ and $n \geq 2$, TS^k_n is the statement "Every function $f : [\omega]^k \to n$ has an infinite f-thin set". STS^2_n is the restriction of TS^2_n to stable colorings.*

Note that TS^2_2 is Ramsey's theorem for pairs. The following theorem is sufficient to separate TS^2 from Ramsey's theorem for pairs as $\mathsf{TS}^2 \leq_c \mathsf{TS}^2_n$ for every $n \geq 2$. The proof of preservation is rather technical and is therefore proven in appendix.

Theorem 21. *For every $n \geq 1$, TS^2_{n+1} admits preservation of n but not $n+1$ hyperimmunities.*

In the case $n = 1$, noticing that the arithmetical comprehension scheme (ACA_0) does not preserve 1 hyperimmunities as witnessed by taking any Δ^0_2 hyperimmune set, we re-obtain the separation of Ramsey's theorem for pairs from ACA_0. Hirschfeldt & Jockusch [9] asked whether TS^2_{n+1} implies TS^2_n over RCA_0. The author answered negatively in [21]. Preservation of n hyperimmunities gives the same separation.

Theorem 22. (Patey *[21]***).** *For every $n \geq 2$, let Φ be the conjunction of* COH, WKL_0, RRT^2_2, $\Pi^0_1\mathsf{G}$, EM, TS^2_{n+1}. *Over* RCA_0, *Φ does not imply any of* SADS *and* STS^2_n.

Acknowledgements. The author is thankful to Wei Wang for useful comments and discussions.

References

1. Bovykin, A., Weiermann, A.: The strength of infinitary ramseyan principles can be accessed by their densities. Ann. Pure Appl. Log. 4 (2005)
2. Cholak, P.A., Giusto, M., Hirst, J.L., Jockusch Jr, C.G.: Free sets and reverse mathematics. Reverse Math. **21**, 104–119 (2001)
3. Cholak, P.A., Jockusch, C.G., Slaman, T.A.: On the strength of Ramsey's theorem for pairs. J. Symb. Log. **66**, 1–55 (2001)
4. Csima, B.F., Mileti, J.R.: The strength of the rainbow Ramsey theorem. J. Symb. Log. **74**(04), 1310–1324 (2009)
5. Flood, S., Towsner, H.: Separating principles below WKL_0 (2014). In preparation
6. Friedman, H.M.: Fom:53:free sets and reverse math and fom:54:recursion theory and dynamics (1999). http://www.math.psu.edu/simpson/fom/
7. Friedman, H.M.: Some systems of second order arithmetic and their use. In: Proceedings of the International Congress of Mathematicians, Vancouver, vol. 1, pp. 235–242 (1974)
8. Hirschfeldt, D.R.: Slicing the truth. Lecture Notes Series, Institute for Mathematical Sciences, National University of Singapore, vol. 28 (2014)

9. Hirschfeldt, D.R., Jockusch Jr, C.G.: On notions of computability theoretic reduction between Π_2^1 principles. To appear
10. Hirschfeldt, D.R., Shore, R.A.: Combinatorial principles weaker than Ramsey's theorem for pairs. J. Symb. Log. **72**(1), 171–206 (2007)
11. Hirschfeldt, D.R., Shore, R.A., Slaman, T.A.: The atomic model theorem and type omitting. Trans. Am. Math. Soc. **361**(11), 5805–5837 (2009)
12. Jockusch, C., Stephan, F.: A cohesive set which is not high. Math. Log. Q. **39**(1), 515–530 (1993)
13. Jockusch, C.G., Soare, R.I.: Π_1^0 classes and degrees of theories. Trans. Am. Math. Soc. **173**, 33–56 (1972)
14. Lerman, M., Solomon, R., Towsner, H.: Separating principles below Ramsey's theorem for pairs. J. Math. Log. **13**(2), 1350007 (2013)
15. Patey, L.: A note on "Separating principles below Ramsey's theorem for pairs" (2013). Unpublished
16. Patey, L.: Controlling iterated jumps of solutions to combinatorial problems (2014). In preparation
17. Patey, L.: Combinatorial weaknesses of ramseyan principles (2015). In preparation
18. Patey, L.: Degrees bounding principles and universal instances in reverse mathematics (2015). Submitted
19. Patey, L.: Ramsey-type graph coloring and diagonal non-computability (2015). Submitted
20. Patey, L.: Somewhere over the rainbow Ramsey theorem for pairs (2015). Submitted
21. Patey, L.: The weakness of being cohesive, thin or free in reverse mathematics (2015). Submitted
22. Rice, B.: Thin set for pairs implies DNR. Notre Dame J. Formal Log. To appear
23. Wang, W.: The definability strength of combinatorial principles (2014)
24. Wang, W.: Some logically weak Ramseyan theorems. Adv. Math. **261**, 1–25 (2014)

Completely Regular Bishop Spaces

Iosif Petrakis(✉)

University of Munich, Munich, Germany
petrakis@math.lmu.de

Abstract. Bishop's notion of a function space, here called a Bishop space, is a constructive function-theoretic analogue to the classical set-theoretic notion of a topological space. Here we introduce the quotient, the pointwise exponential and the completely regular Bishop spaces. For the latter we present results which show their correspondence to the completely regular topological spaces, including a generalized version of the Tychonoff embedding theorem for Bishop spaces. All our proofs are within Bishop's informal system of constructive mathematics BISH.

1 Why Bishop Spaces

The theory of Bishop spaces is so far the least developed approach to constructive topology with points. Bishop introduced them in [1], where he established their connection to his notion of neighborhood spaces, a set-theoretic constructive version of a topological space, and he defined the least Bishop space over a subbase, the product of Bishop spaces and a notion of a connected Bishop space. In [2], p. 80, Bishop added some comments on them, while in [4] Bridges revived the subject, studying the morphisms between various metric spaces seen as Bishop spaces and relating Bishop spaces to apartness spaces. In [6] Ishihara related the subcategory **Fun** of the category **Bis** of Bishop spaces to the category of neighborhood spaces **Nbh**. In [8] we reported on our current development of the theory of Bishop spaces.

Our approach to topology is constructive, since we work within Bishop's informal system of constructive mathematics BISH, it is function-theoretic, since most of the notions involved are based on the concept of function, and we accept points from the beginning. Hence, the theory of Bishop spaces is an approach to constructive point-function topology, and its study is motivated by the following remarks:

- **(i)** Function-based concepts are more suitable to constructive study than set-based ones. That's why Bishop, in [2] p. 77, suggested to focus attention on Bishop spaces instead of on neighborhood spaces.
- **(ii)** Bishop's topology of functions F corresponds to the ring of real-valued continuous functions $C(X)$ on a topological space X. This allows a direct "communication" between the two theories, which does not mean though, a direct translation, due to the classical set-theoretic character of $C(X)$.
- **(iii)** The theory of Bishop spaces meets the standards of Bishop for a constructive mathematical theory: it has simple foundation and it follows the style of standard mathematics.

A. Beckmann et al. (Eds.): CiE 2015, LNCS 9136, pp. 302–312, 2015.
DOI: 10.1007/978-3-319-20028-6_31

2 Basic Definitions and Facts

If X is an inhabited set, we denote by $\mathbb{F}(X)$ the set of all functions of type $X \to \mathbb{R}$, where $\mathbb{R}$ is the set of constructive reals. A constant function in $\mathbb{F}(X)$ with value $a \in \mathbb{R}$ is denoted by $\overline{a}$, and their set by $\mathrm{Const}(X)$. In BISH a compact metric space is a complete and totally bounded metric space, and a locally compact metric space is one in which every bounded subset is included in a compact one. If X is a locally compact metric space, $\mathrm{Bic}(X)$ denotes the subset of $\mathbb{F}(X)$ of all *Bishop-continuous* functions, where $\phi \in \mathrm{Bic}(X)$, if ϕ is uniformly continuous on every bounded subset of X. Since $\mathbb{R}$ with its standard metric is locally compact, $\mathrm{Bic}(\mathbb{R})$ denotes the Bishop-continuous functions on $\mathbb{R}$.

A *Bishop space* is a pair $\mathcal{F} = (X, F)$, where X is an inhabited set and $F \subseteq \mathbb{F}(X)$ satisfies the following conditions:

(BS_1) $\mathrm{Const}(X) \subseteq F$.
(BS_2) $f \in F \to g \in F \to f + g \in F$.
(BS_3) $f \in F \to \phi \in \mathrm{Bic}(\mathbb{R}) \to \phi \circ f \in F$.
(BS_4) $f \in \mathbb{F}(X) \to \forall_{\epsilon>0}\exists_{g\in F}\forall_{x\in X}(|g(x) - f(x)| \leq \epsilon) \to f \in F$.

Bishop used the term *function space* for $\mathcal{F}$ and *topology* for F. Since the former is used in many different contexts, we prefer the term Bishop space for $\mathcal{F}$, while we use the latter, since the *topology of functions* F on X corresponds nicely to the standard *topology of opens* $\mathcal{T}$ on X. A topology F is a ring and a lattice; by BS_2 and BS_3 if $f, g \in F$, then $f \cdot g$, $f \vee g = \max\{f, g\}$, $f \wedge g = \min\{f, g\}$ and $|f| \in F$. The sets $\mathrm{Const}(X)$ and $\mathbb{F}(X)$ are topologies on X, called the *trivial* and the *discrete* topology, respectively. If F is a topology on X, then $\mathrm{Const}(X) \subseteq F \subseteq \mathbb{F}(X)$. It is straightforward that $\mathbb{F}_b(X) := \{f \in \mathbb{F}(X) \mid f \text{ is bounded}\}$ is a topology on X, and if $\mathcal{F} = (X, F)$ is a Bishop space, then $\mathcal{F}_b = (X, F_b)$ is a Bishop space, where $F_b = \mathbb{F}_b(X) \cap F$ corresponds to the ring $C^*(X)$ of the bounded elements of $C(X)$. If X is a locally compact metric space, it is easy to see that $\mathrm{Bic}(X)$ is a topology on X. The structure $\mathcal{R} = (\mathbb{R}, \mathrm{Bic}(\mathbb{R}))$ is the Bishop space of reals.

Most of the new Bishop spaces generated from old ones are defined through Bishop's inductive concept, found in [2] p. 78, of the *least topology* $\mathcal{F}(F_0)$ generated by a given inhabited *subbase* F_0 of real-valued functions on X. Conditions BS_1-BS_4, seen as inductive rules, together with the rule $f_0 \in F_0 \to f_0 \in \mathcal{F}(F_0)$ induce the following induction principle $\mathrm{Ind}_{\mathcal{F}}$ on $\mathcal{F}(F_0)$:

$$\begin{aligned}
&\forall_{f_0\in F_0}(P(f_0)) \to \\
&\forall_{a\in\mathbb{R}}(P(\overline{a})) \to \\
&\forall_{f,g\in\mathcal{F}(F_0)}(P(f) \to P(g) \to P(f+g)) \to \\
&\forall_{f\in\mathcal{F}(F_0)}\forall_{\phi\in\mathrm{Bic}(\mathbb{R})}(P(f) \to P(\phi \circ f)) \to \\
&\forall_{f\in\mathcal{F}(F_0)}(\forall_{\epsilon>0}\exists_{g\in\mathcal{F}(F_0)}(P(g) \ \wedge \ \forall_{x\in X}(|g(x) - f(x)| \leq \epsilon)) \to P(f)) \to \\
&\forall_{f\in\mathcal{F}(F_0)}(P(f)),
\end{aligned}$$

where P is any property on $\mathbb{F}(X)$. Since the identity function $\mathrm{id}_{\mathbb{R}}$ on $\mathbb{R}$ belongs to $\mathrm{Bic}(\mathbb{R})$, we get that $\mathrm{Bic}(\mathbb{R}) = \mathcal{F}(\mathrm{id}_{\mathbb{R}})$.

If $\mathcal{F} = (X, F)$ and $\mathcal{G} = (Y, G)$ are Bishop spaces, their *product* is defined as the structure $\mathcal{F} \times \mathcal{G} = (X \times Y, F \times G)$, where

$$F \times G := \mathcal{F}(\{f \circ \pi_1 \mid f \in F\} \cup \{g \circ \pi_2 \mid g \in G\}).$$

If F_0 is a subbase of F and G_0 is a subbase of G, then using $\mathrm{Ind}_{\mathcal{F}}$ we get that

$$\mathcal{F}(F_0) \times \mathcal{F}(G_0) = \mathcal{F}(\{f_0 \circ \pi_1 \mid f_0 \in F_0\} \cup \{g_0 \circ \pi_2 \mid g_0 \in G_0\}).$$

Consequently, $\mathrm{Bic}(\mathbb{R}) \times \mathrm{Bic}(\mathbb{R}) = \mathcal{F}(\{\mathrm{id}_{\mathbb{R}} \circ \pi_1\} \cup \{\mathrm{id}_{\mathbb{R}} \circ \pi_2\}) = \mathcal{F}(\pi_1, \pi_2)$. If I is a given index set and $F_{0,i} \subseteq \mathbb{F}(X, \mathbb{R})$, for every $i \in I$, we define $\bigvee_{i \in I} F_{0,i} = \mathcal{F}(\bigcup_{i \in I} F_{0,i})$. If F_i is a topology on X_i, for every $i \in I$, the product topology on $\prod_{i \in I} X_i$ is defined by $\prod_{i \in I} F_i = \bigvee_{i \in I}(F_i \circ \pi_i)$, where $F_i \circ \pi_i = \{f \circ \pi_i \mid f \in F_i\}$. As expected, $\prod_{i \in I} \mathcal{F}(F_{0,i}) = \bigvee_{i \in I}(F_{0,i} \circ \pi_i)$. If $X_i = X$, for every $i \in I$, we use the notation $\mathcal{F}^I = (X^I, F^I)$. A *Euclidean* Bishop space is a product $\mathcal{R}^I$. Simplifying our notation, $\mathrm{Bic}(\mathbb{R})^I = \bigvee_{i \in I}(\mathrm{id}_{\mathbb{R}} \circ \pi_i) = \bigvee_{i \in I} \pi_i$.

If $\mathcal{F}, \mathcal{G}$ are Bishop spaces, a *Bishop morphism*, or simply a *morphism*, from $\mathcal{F}$ to $\mathcal{G}$ is a function $h : X \to Y$ such that $\forall_{g \in G}(g \circ h \in F)$

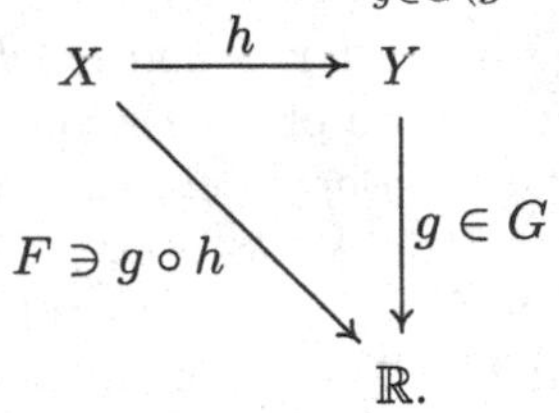

We denote the morphisms from $\mathcal{F}$ to $\mathcal{G}$ by $\mathrm{Mor}(\mathcal{F}, \mathcal{G})$. If $\mathrm{Const}(X, Y)$ denotes the constant functions from X to Y, then $\mathrm{Const}(X, Y) \subseteq \mathrm{Mor}(\mathcal{F}, \mathcal{G})$. Thus, the category **Bis** of Bishop spaces is formed with the Bishop morphisms as arrows. It is straightforward to see that $\mathcal{F} \times \mathcal{G}$ satisfies the universal property for products and that $F \times G$ is the least topology which turns the projections π_1, π_2 into morphisms. If $h \in \mathrm{Mor}(\mathcal{F}, \mathcal{G})$ is onto Y, then h is called an *epimorphism*, and we denote their set by $\mathrm{Epi}(\mathcal{F}, \mathcal{G})$. If F is a topology on X, then clearly $F = \mathrm{Mor}(\mathcal{F}, \mathcal{R})$. If G_0 is a subbase of G, then using the induction principle we get that $h \in \mathrm{Mor}(\mathcal{F}, \mathcal{G})$ iff $\forall_{g_0 \in G_0}(g_0 \circ h \in F)$, a fundamental property that we call the *lifting of morphisms*. We call a morphism h from $\mathcal{F}$ to $\mathcal{G}$ *open*, if $\forall_{f \in F} \exists_{g \in G}(f = g \circ h)$, and *strongly open*, if $\forall_{f \in F} \exists !_{g \in G}(f = g \circ h)$. Clearly, if $h \in \mathrm{Mor}(\mathcal{F}, \mathcal{G})$ such that h is 1-1 and onto Y, then $h^{-1} \in \mathrm{Mor}(\mathcal{G}, \mathcal{F})$ iff h is open. In this case h is called an *isomorphism* between $\mathcal{F}$ and $\mathcal{G}$.
Next we prove inductively the *lifting of openness*, a fundamental fact that we use here in the proof of the Theorem 4. First we need a lemma.

Lemma 1 (Well-definability lemma). *Suppose that X, Y are inhabited sets and $h : X \to Y$ is onto Y. If $f : X \to \mathbb{R}$ such that for every $\epsilon > 0$ there exists some $g : Y \to \mathbb{R}$ such that $\forall_{x \in X}(|(g \circ h)(x) - f(x)| \leq \epsilon)$, then the function $\Phi : Y \to \mathbb{R}$ defined by $\Phi(y) = \Phi(h(x)) := f(x)$, for every $y \in Y$, is well-defined i.e., $\forall_{x_1, x_2 \in X}(h(x_1) = h(x_2) \to f(x_1) = f(x_2))$.*

Proof. We fix $x_1, x_2 \in X$ such that $h(x_1) = h(x_2) = y_0$, and some $\epsilon > 0$. By our hypothesis on f there exists some $g : Y \to \mathbb{R}$ such that $\forall_{x \in X}(|(g \circ h)(x) - f(x)| \leq \frac{\epsilon}{2})$. Hence, $|g(h(x_1)) - f(x_1)| = |g(y_0) - f(x_1)| \leq \frac{\epsilon}{2}$ and $|g(h(x_2)) - f(x_2)| = |g(y_0) - f(x_2)| \leq \frac{\epsilon}{2}$. Consequently, $|f(x_1) - f(x_2)| \leq |f(x_1) - g(y_0)| + |g(y_0) - f(x_2)| \leq \frac{\epsilon}{2} + \frac{\epsilon}{2} = \epsilon$. Since ϵ is arbitrary, we get that $|f(x_1) - f(x_2)| \leq 0$, which implies that $f(x_1) = f(x_2)$.

Proposition 1 (Lifting of openness). *If $\mathcal{F} = (X, \mathcal{F}(F_0))$, $\mathcal{G} = (Y, G)$ are Bishop spaces and $h \in \mathrm{Epi}(\mathcal{F}, \mathcal{G})$, then*

$$\forall_{f_0 \in F_0} \exists_{g \in G}(f_0 = g \circ h) \to \forall_{f \in \mathcal{F}(F_0)} \exists_{g \in G}(f = g \circ h).$$

Proof. If $f = f_0 \in F_0$, then we just use our premiss. Of course, a constant function $\overline{a} : X \to \mathbb{R}$ is written as the composition $\overline{a} \circ h$, where we use the same notation for the constant function of type $Y \to \mathbb{R}$ with value a. If $f = f_1 + f_2$ such that $f_1 = g_1 \circ h$ and $f_2 = g_2 \circ h$, for some $g_1, g_2 \in G$, then $f = (g_1 + g_2) \circ h$, where $g_1 + g_2 \in G$ by BS_2. If $f = \phi \circ f'$, where $\phi \in \mathrm{Bic}(\mathbb{R})$, and there is some $g \in G$ such that $f' = g \circ h$, then $f = (\phi \circ g) \circ h$, where $\phi \circ g \in G$ by BS_3. Suppose next that $\epsilon > 0$ and $f' \in \mathcal{F}(F_0)$ such that $f' = g \circ h$, for some $g \in G$, and $\forall_{x \in X}(|f'(x) - f(x)| = |g(h(x)) - f(x)| \leq \epsilon)$. If $\Phi : Y \to \mathbb{R}$ is the function determined by the well-definability lemma such that $f = \Phi \circ h$, we get, since h is onto Y, that $\forall_{y \in Y}(|g(y) - \Phi(y)| \leq \epsilon)$. Since $\epsilon > 0$ is arbitrary, we conclude by condition BS_4 that $\Phi \in G$.

A morphism h from $\mathcal{F}$ to $\mathcal{G}$ induces the mapping $h^* : G \to F$, $g \mapsto h^*(g)$, where $h^*(g) := g \circ h$, which is a ring and a lattice homomorphism. If h is an epimorphism, then h^* is a *partial isometry* i.e., $||h^*(g)||$ exists whenever $||g||$ exists and moreover $||h^*(g)|| = ||g||$. Recall that constructively $||g|| = \sup\{|g(y)| \mid y \in Y\}$ does not always exist for some bounded element of a topology G.

The *pointwise exponential* Bishop space $\mathcal{F} \to \mathcal{G} = (\mathrm{Mor}(\mathcal{F}, \mathcal{G}), F \to G)$ corresponds to the point-open topology within the category of topological spaces **Top** and it is defined by

$$F \to G := \mathcal{F}(\{e_{x,g} \mid x \in X, g \in G\}), \quad e_{x,g} : \mathrm{Mor}(\mathcal{F}, \mathcal{G}) \to \mathbb{R}, \quad e_{x,g}(h) = g(h(x)),$$

for every $h \in \mathrm{Mor}(\mathcal{F}, \mathcal{G})$. A simple induction shows that if G_0 is a subbase of G, then $F \to \mathcal{F}(G_0) = \mathcal{F}(\{e_{x,g_0} \mid x \in X, g_0 \in G_0\})$. The *dual* Bishop space of $\mathcal{F}$ is the space $\mathcal{F}^* = (F, F^*)$, where

$$F^* := \mathcal{F}(\{\hat{x} \mid x \in X\}), \quad \hat{x} : F \to \mathbb{R}, \quad \hat{x}(f) = f(x),$$

for every $f \in F$. Clearly, $\mathcal{F}^* = \mathcal{F} \to \mathcal{R} = (\mathrm{Mor}(\mathcal{F}, \mathcal{R}), F \to \mathrm{Bic}(\mathbb{R}))$.

Although Ishihara and Palmgren constructed in [7] the quotient topological space using predicative methods, our definition of the quotient Bishop space is straightforward and permits a smooth translation of the standard classical theory of quotient topological spaces into the theory of Bishop spaces. If $\mathcal{F} = (X, F)$ is a Bishop space, Y is an inhabited set and $e : X \to Y$ is onto Y, it is straightforward to see that the set of functions G_e, defined as

$$G_e := \{g \in \mathbb{F}(Y) \mid g \circ e \in F\},$$

is a topology on Y. We call $\mathcal{G}_e = (Y, G_e)$ the *quotient Bishop space*, and G_e the *quotient* topology on Y, with respect to e. As in standard topology, the quotient topology G_e is the largest topology on Y which makes e a morphism, while if $\mathcal{H} = (Z, H)$ is a Bishop space, a function $h : Y \to Z \in \mathrm{Mor}(\mathcal{G}_e, \mathcal{H})$ iff $h \circ e \in \mathrm{Mor}(\mathcal{F}, \mathcal{H})$. The next proposition is easy to prove.

Proposition 2. *Let $\mathcal{F} = (X, F)$ be a Bishop space and $\sim$ be the equivalence relation on X defined by $x_1 \sim x_2 \leftrightarrow \forall_{f \in F}(f(x_1) = f(x_2))$. If $\pi : X \to X/\sim$ is the function $x \mapsto [x]_\sim$ and $\mathcal{F}/\sim = (X/\sim, G_\pi)$ is the quotient Bishop space, then π is a strongly open morphism from $\mathcal{F}$ to $\mathcal{F}/\sim$, and the function $\rho : F \to G_\pi$, $f \mapsto \rho(f)$, where $\rho(f)([x]_\sim) := f(x)$, is a ring and a lattice homomorphism and a partial isometry onto G_π.*

If $\mathcal{F} = (X, F)$ is a Bishop space and $A \subseteq X$, the *relative* Bishop space of $\mathcal{F}$ on A is the structure $\mathcal{F}_{|A} = (A, F_{|A})$, also called a *subspace* of $\mathcal{F}$, where

$$F_{|A} := \mathcal{F}(\{f_{|A} \mid f \in F\}).$$

If F_0 is a subbase of F, we get inductively that $F_{|A} = \mathcal{F}(\{f_{0|A} \mid f_0 \in F_0\})$. The topology $F_{|A}$ is the smallest topology G on A such that $\mathrm{id}_A \in \mathrm{Mor}(\mathcal{G}, \mathcal{F})$. If $\mathcal{G} = (Y, G)$ is a Bishop space and $e : X \to B \subseteq Y$, then $e \in \mathrm{Mor}(\mathcal{F}, \mathcal{G}) \leftrightarrow e \in \mathrm{Mor}(\mathcal{F}, \mathcal{G}_{|B})$, and if e is open as a morphism from $\mathcal{F}$ to $\mathcal{G}$, it is trivially open as a morphism from $\mathcal{F}$ to $\mathcal{G}_{|B}$. An isomorphism between $\mathcal{F}$ and a subspace of $\mathcal{G}$ is called a *topological embedding* of $\mathcal{F}$ into $\mathcal{G}$.

A topology F on X induces the *canonical* apartness relation on X, which is introduced in [4] and it is defined, for every $x_1, x_2 \in X$, by

$$x_1 \bowtie_F x_2 :\leftrightarrow \exists_{f \in F}(f(x_1) \bowtie_{\mathbb{R}} f(x_2)),$$

where $a \bowtie_{\mathbb{R}} b :\leftrightarrow a > b \vee a < b \leftrightarrow |a - b| > 0$, for every $a, b \in \mathbb{R}$. Moreover, $a \bowtie_{\mathrm{Bic}(\mathbb{R})} b \leftrightarrow a \bowtie_{\mathbb{R}} b$; if $a \bowtie_{\mathrm{Bic}(\mathbb{R})} b$, then $\phi(a) \bowtie_{\mathbb{R}} \phi(b)$, for some $\phi \in \mathrm{Bic}(\mathbb{R})$. By the obvious pointwise continuity of ϕ at a we have that if $0 < \epsilon = |\phi(b) - \phi(a)|$, $\exists_{\delta(\frac{\epsilon}{2})>0}\forall_{x \in \mathbb{R}}(|x - a| < \delta(\frac{\epsilon}{2}) \to |\phi(x) - \phi(a)| \leq \frac{\epsilon}{2})$. Hence, $\neg(|a - b| < \delta(\frac{\epsilon}{2}))$ i.e., $|a - b| \geq \delta(\frac{\epsilon}{2}) > 0$. For the converse we just use the equivalence $a \bowtie_{\mathbb{R}} b \leftrightarrow \mathrm{id}_{\mathbb{R}}(a) \bowtie_{\mathbb{R}} \mathrm{id}_{\mathbb{R}}(b)$. An apartness relation $\bowtie$ on X is called *tight*, if $\neg(x_1 \bowtie x_2) \to x_1 = x_2$, for every $x_1, x_2 \in X$. It is easy to see that if F is a topology on X, then $\bowtie_F$ is tight iff

$$\forall_{x_1, x_2 \in X}(\forall_{f \in F}(f(x_1) = f(x_2)) \to x_1 = x_2).$$

The sufficiency is mentioned in [4] and for its proof one uses the obvious fact that $\bowtie_{\mathbb{R}}$ is tight. If F_0 is a subbase of F and the restriction of every $f_0 \in F_0$ to some $A \subseteq X$ is constant, then an induction shows that the restriction of every $f \in \mathcal{F}(F_0)$ to A is constant. If $F = \mathcal{F}(F_0)$, we get that $\bowtie_F$ is tight iff

$$\forall_{x_1, x_2 \in X}(\forall_{f_0 \in F_0}(f_0(x_1) = f_0(x_2)) \to x_1 = x_2),$$

applying the previous lifting to the set $A = \{x_1, x_2\}$, where $x_1, x_2 \in X$.

3 Completely Regular Topologies of Functions

A completely regular topological space $(X, \mathcal{T})$ is one in which any pair (x, B), where B is closed and $x \notin B$, is separated by some $f \in C(X, [0,1])$. A completely regular and T_1-space satisfies classically the property

$$\forall_{x_1, x_2 \in X}(\forall_{f \in C(X)}(f(x_1) = f(x_2)) \to x_1 = x_2).$$

The importance and the "sufficiency" of the completely regular topological spaces in the theory of $C(X)$ is provided by the Stone-Čech theorem according to which, for every topological space X there exists a completely regular space ρX and a continuous mapping $\tau : X \to \rho X$ such that the induced function $g \mapsto \tau^*(g)$, where $\tau^*(g) = g \circ \tau$, is a ring isomorphism between $C(\rho X)$ and $C(X)$ (see [5] p. 41).

We call a Bishop space $\mathcal{F} = (X, F)$ *completely regular*, if its canonical apartness relation $\bowtie_F$ is tight, hence the equality of X is determined by F, which we call a *completely regular* topology. Since $\bowtie_{\mathrm{Bic}(\mathbb{R})} \leftrightarrow \bowtie_{\mathbb{R}}$ and $\bowtie_{\mathbb{R}}$ is tight, $\mathcal{R}$ is completely regular. It is immediate to see that $\mathcal{F}$ is completely regular iff $\mathcal{F}_b$ is completely regular, while if X has at least two points, then $\mathrm{Const}(X)$ is not completely regular.

In this section we prove some first fundamental results on completely regular Bishop spaces. The following version of the Stone-Čech theorem expresses the corresponding "sufficiency" of the completely regular Bishop spaces within **Bis**. Its proof is a translation of the classical one, since the quotient Bishop spaces behave as the quotient topological spaces.

Theorem 1 (Stone-Čech theorem for Bishop spaces). *For every Bishop space $\mathcal{F} = (X, F)$ there exists a completely regular Bishop space $\rho\mathcal{F} = (\rho X, \rho F)$ and a mapping $\tau : X \to \rho X \in \mathrm{Mor}(\mathcal{F}, \rho\mathcal{F})$ such that the induced mapping τ^* is a ring isomorphism between ρF and F.*

Proof. We use the equivalence relation $x_1 \sim x_2 \leftrightarrow \forall_{f \in F}(f(x_1) = f(x_2))$, for every $x_1, x_2 \in X$, and if $\tau = \pi : X \to X/{\sim}$, where $x \mapsto [x]_\sim$, we consider the quotient Bishop space $\rho\mathcal{F} = \mathcal{F}/{\sim} = (X/{\sim}, G_\pi) = (\rho X, \rho F)$. By the Proposition 2 we have that π is a morphism from $\mathcal{F}$ to $\mathcal{F}/{\sim}$ and $\rho : F \to G_\pi$ is a ring homomorphism onto G_π. We also know that $\pi^* : G_\pi \to F$ is a ring homomorphism. Since $\rho(g \circ \pi)([x]_\sim) = (g \circ \pi)(x) = g([x]_\sim)$, for every $[x]_\sim \in X/{\sim}$, we get that $\rho \circ \pi^* = \mathrm{id}_{G_\pi}$. Since $\pi^*(\rho(f)) = \rho(f) \circ \pi = f$, for every $f \in F$, we get that $\pi^* \circ \rho = \mathrm{id}_F$. Hence, π^* is a bijection (see [2] p. 17). Finally, $\rho\mathcal{F}$ is completely regular; if $\forall_{g \in \rho F}(g([x_1]_\sim) = g([x_2]_\sim))$, then $\forall_{f \in F}(f(x_1) = f(x_2))$, since $\rho(f) \circ \pi = f$ and $\rho(f) \in \rho F$, therefore $x_1 \sim x_2$ i.e., $[x_1]_\sim = [x_2]_\sim$.

Proposition 3. *Suppose that $\mathcal{F} = (X, F), \mathcal{G} = (Y, G)$ are Bishop spaces.*

(i) If $\mathcal{G}$ is isomorphic to the completely regular $\mathcal{F}$, then $\mathcal{G}$ is completely regular.
(ii) If $A \subseteq X$ and $\mathcal{F}$ is a completely regular, then $\mathcal{F}_{|A}$ is completely regular.
(iii) $\mathcal{F}$ and $\mathcal{G}$ are completely regular iff $\mathcal{F} \times \mathcal{G}$ is completely regular.
(iv) $\mathcal{F} \to \mathcal{G}$ is completely regular iff $\mathcal{G}$ is completely regular.
(v) The dual space $\mathcal{F}^*$ of $\mathcal{F}$ is completely regular.

Proof. (i) Suppose that e is an isomorphism between $\mathcal{F}$ and $\mathcal{G}$, and $y_1, y_2 \in Y$. Since e is onto Y and e^* is onto G, we get that $\forall_{g \in G}(g(y_1) = g(y_2)) \leftrightarrow \forall_{g \in G}(g(e(x_1)) = g(e(x_2))) \leftrightarrow \forall_{f \in F}(f(x_1) = f(x_2)) \rightarrow x_1 = x_2$. Hence, $y_1 = y_2$.

(ii) If $a_1, a_2 \in A$, it suffices to show that $\forall_{f \in F}(f_{|A}(a_1) = f_{|A}(a_2)) \rightarrow a_1 = a_2$. The premiss is rewritten as $\forall_{f \in F}(f(a_1) = f(a_2))$, and since F is completely regular, we conclude that $a_1 = a_2$.

(iii) The hypotheses $\forall_{f \in F}((f \circ \pi_1)(x_1, y_1) = (f \circ \pi_1)(x_2, y_2))$ and $\forall_{g \in G}((g \circ \pi_2)(x_1, y_1) = (g \circ \pi_2)(x_2, y_2))$ imply $\forall_{f \in F}(f(x_1) = f(x_2))$ and $\forall_{g \in G}(g(y_1) = g(y_2))$. For the converse we topologically embed in the obvious way $\mathcal{F}, \mathcal{G}$ into $\mathcal{F} \times \mathcal{G}$ and we use (i) and (ii).

(iv) If $\mathcal{G}$ is completely regular and $\forall_{h_1, h_2 \in \mathrm{Mor}(\mathcal{F}, \mathcal{G})} \forall_{x \in X} \forall_{g \in G}(e_{x,g}(h_1) = e_{x,g}(h_2))$ i.e., $\forall_{h_1, h_2 \in \mathrm{Mor}(\mathcal{F}, \mathcal{G})} \forall_{x \in X} \forall_{g \in G}(g(h_1(x)) = g(h_2(x)))$, then the tightness of $\bowtie_{\mathcal{G}}$ implies that $\forall_{x \in X}(h_1(x) = h_2(x))$ i.e., $h_1 = h_2$. For the converse we suppose that $\forall_{h_1, h_2 \in \mathrm{Mor}(\mathcal{F}, \mathcal{G})} \forall_{x \in X} \forall_{g \in G}(e_{x,g}(h_1) = e_{x,g}(h_2)) \rightarrow h_1 = h_2$. We fix $y_1, y_2 \in G$ and we suppose that $\forall_{g \in G}(g(y_1) = g(y_2))$. Since $\overline{y_1}, \overline{y_2} \in \mathrm{Const}(X, Y) \subseteq \mathrm{Mor}(\mathcal{F}, \mathcal{G})$, we have that $\forall_{x \in X} \forall_{g \in G}(e_{x,g}(\overline{y_1}) = g(\overline{y_1}(x)) = g(y_1) = g(y_2) = g(\overline{y_2}(x)) = e_{x,g}(\overline{y_2}))$. Hence, $\overline{y_1} = \overline{y_2}$ i.e., $y_1 = y_2$.

(v) Since $\mathcal{F}^* = \mathcal{F} \rightarrow \mathcal{R}$ and $\mathcal{R}$ is completely regular, we use (iv).

The proof of the Proposition 3(iii) works for an arbitrary product of Bishop spaces too. As in classical topology one can show that the quotient of a completely regular Bishop space need not be completely regular. If $\mathcal{F} = (X, F)$ and $\mathcal{G}_i = (Y_i, G_i)$ are Bishop spaces, for every $i \in I$, the family $(h_i)_{i \in I}$, where $h_i : X \rightarrow Y_i$, for every $i \in I$, *separates the points* of X, if $\forall_{x,y \in X}(\forall_{i \in I}(h_i(x) = h_i(y)) \rightarrow x = y)$.

Theorem 2 (Embedding lemma for Bishop spaces). *Suppose that* $\mathcal{F} = (X, F)$ *and* $\mathcal{G}_i = (Y_i, G_i)$ *are Bishop spaces and* $h_i : X \rightarrow Y_i$*, for every* i *in some index set* I*. If the family of functions* $(h_i)_{i \in I}$ *separates the points of* X*,* $h_i \in \mathrm{Mor}(\mathcal{F}, \mathcal{G}_i)$*, for every* $i \in I$*, and* $\forall_{f \in F} \exists_{i \in I} \exists_{g \in G_i}(f = g \circ h_i)$*, then the evaluation map* $e : X \rightarrow Y = \prod_{i \in I} Y_i$*, defined by* $x \mapsto (h_i(x))_{i \in I}$*, is a topological embedding of* $\mathcal{F}$ *into* $\mathcal{G} = \prod_{i \in I} \mathcal{G}_i$*.*

Proof. First we show that e is 1−1; $e(x_1) = e(x_2) \leftrightarrow (h_i(x_1))_{i \in I} = (h_i(x_2))_{i \in I} \leftrightarrow \forall_{i \in I}(h_i(x_1) = h_i(x_2)) \rightarrow x_1 = x_2$. By the lifting of morphisms we have that $e \in \mathrm{Mor}(\mathcal{F}, \mathcal{G}) \leftrightarrow \forall_{g \in G}(g \circ e \in F) \leftrightarrow \forall_{i \in I} \forall_{g \in G_i}((g \circ \pi_i) \circ e = g \circ (\pi_i \circ e) \in F) \leftrightarrow \forall_{i \in I} \forall_{g \in G_i}(g \circ h_i \in F) \leftrightarrow \forall_{i \in I}(h_i \in \mathrm{Mor}(\mathcal{F}, \mathcal{G}_i))$. Next we show that e is open i.e., $\forall_{f \in F} \exists_{g \in G}(f = g \circ e)$. If $f \in F$, by hypothesis (iii) there is some $i \in I$ and some $g \in G_i$ such that $f = g \circ h_i$. Since $g \circ \pi_i \in G = \prod_{i \in I} G_i$ we have that $(g \circ \pi_i)(e(x)) = (g \circ \pi_i)((h_i(x))_{i \in I}) = g(h_i(x)) = f(x)$, for every $x \in X$, hence $f = (g \circ \pi_i) \circ e$. Thus, e is open as a morphism from $\mathcal{F}$ to $\mathcal{G}$, therefore it is open as a morphism from $\mathcal{F}$ to $\mathcal{G}_{|e(X)}$.

According to the classical Tychonoff embedding theorem, the completely regular topological spaces are precisely those which can be embedded in a product of the closed unit interval $\mathcal{I}$. In the following characterization of the tightness of the canonical apartness relation it is $\mathcal{R}$ which has the role of $\mathcal{I}$.

Theorem 3 (Tychonoff embedding theorem for Bishop spaces). *Suppose that $\mathcal{F} = (X, F)$ is a Bishop space. Then, $\mathcal{F}$ is completely regular iff $\mathcal{F}$ is topologically embedded into the Euclidean Bishop space $\mathcal{R}^F$.*

Proof. If $\mathcal{F}$ is completely regular, using the embedding lemma we show that the mapping $e : X \to \mathbb{R}^F$, defined by $x \mapsto (f(x))_{f \in F}$, is a topological embedding of $\mathcal{F}$ into $\mathcal{R}^F$. The topology F is a family of functions of type $X \to \mathbb{R}$ that separates the points of X, since the separation condition is exactly the tightness of $\bowtie_F$. That every $f \in F$ is in $\mathrm{Mor}(\mathcal{F}, \mathcal{R})$ is already mentioned. If we fix some $f \in F$, then $f = \mathrm{id}_{\mathbb{R}} \circ f$, and since $\mathrm{id}_{\mathbb{R}} \in \mathrm{Bic}(\mathbb{R})$, the condition (iii) of the embedding lemma is satisfied. If $\mathcal{F}$ is topologically embedded into $\mathcal{R}^F$, then $\mathcal{F}$ is completely regular, since by Proposition 3 a Euclidean Bishop space is completely regular and $\mathcal{F}$ is isomorphic to a subspace of a completely regular space.

If $\mathcal{F} = (X, F)$, $\mathcal{G} = (Y, G)$ are Bishop spaces and $h \in \mathrm{Mor}(\mathcal{F}, \mathcal{G})$, then using Theorem 3 one shows, as in the classical case, the existence of a mapping $\rho h : \rho X \to \rho Y \in \mathrm{Mor}(\rho\mathcal{F}, \rho\mathcal{G})$ such that the following diagram commutes

$$\begin{array}{ccc} X & \xrightarrow{\ h\ } & Y \\ \big\downarrow{\scriptstyle \tau} & & \big\downarrow{\scriptstyle \tau'} \\ \rho X & \xrightarrow[\ \rho h\ ]{} & \rho Y, \end{array}$$

where τ, τ' are the morphisms determined by Theorem 1. Next we prove directly a generalized form of the Tychonoff embedding theorem.

Theorem 4 (General Tychonoff embedding theorem). *Suppose that $\mathcal{F} = (X, \mathcal{F}(F_0))$ is a Bishop space. Then, $\mathcal{F}$ is completely regular iff $\mathcal{F}$ is topologically embedded into the Euclidean Bishop space $\mathcal{R}^{F_0}$.*

Proof. If $\mathcal{F}$ is completely regular, we show directly that the mapping $e : X \to \mathbb{R}^{F_0}$, defined by $x \mapsto (f_0(x))_{f_0 \in F_0}$, is a topological embedding of $\mathcal{F}$ into $\mathcal{R}^{F_0}$. Since the tightness of $\bowtie_{\mathcal{F}(F_0)}$ is equivalent to $\forall_{x_1, x_2 \in X}(\forall_{f_0 \in F_0}(f_0(x_1) = f_0(x_2)) \to x_1 = x_2))$, we get that e is 1−1. Using our remark on the relative topology given with a subbase, we get that, since $\mathrm{Bic}(\mathbb{R})^{F_0} = \bigvee_{f_0 \in F_0} \pi_{f_0}$, its restriction to $e(X)$ is $(\mathrm{Bic}(\mathbb{R})^{F_0})_{|e(X)} = (\bigvee_{f_0 \in F_0} \pi_{f_0})_{|e(X)} = \bigvee_{f_0 \in F_0} (\pi_{f_0})_{|e(X)}$. By the lifting of morphisms we have that $e \in \mathrm{Mor}(\mathcal{F}, (\mathcal{R}^{F_0})_{|e(X)})$ iff $\forall_{f_0 \in F_0}((\pi_{f_0})_{|e(X)} \circ e = f_0 \in \mathcal{F}(F_0))$, which holds trivially. In order to prove that e is open it suffices by the lifting of openness on the epimorphism $e : X \to e(X)$ to show that $\forall_{f_0 \in F_0} \exists_{g \in (\mathrm{Bic}(\mathbb{R})^{F_0})_{|e(X)}}(f_0 = g \circ e)$. Since $f_0 = (\pi_{f_0})_{|e(X)} \circ e$ and $(\pi_{f_0})_{|e(X)} \in (\mathrm{Bic}(\mathbb{R})^{F_0})_{|e(X)}$, for every $f_0 \in F_0$, we are done. The converse is proved as in the proof of the Theorem 3.

If we consider $F_0 = F$, then $\mathcal{F}(F_0) = F$ and the general Tychonoff embedding theorem implies Theorem 3. If $X = \mathbb{R}^n$, then $\mathrm{Bic}(\mathbb{R})^n = \mathcal{F}(\pi_1, \ldots, \pi_n)$, and the above embedding e is $\mathrm{id}_{\mathbb{R}^n}$, since $\boldsymbol{x} \mapsto (\pi_1(\boldsymbol{x}), \ldots, \pi_n(\boldsymbol{x})) = \boldsymbol{x}$.

Proposition 4. *Suppose that $\mathcal{F} = (X, F)$ is a completely regular Bishop space, $\mathcal{G} = (Y, G)$ is Bishop space, and $\tau : X \to Y \in \mathrm{Mor}(\mathcal{F}, \mathcal{G})$. Then, τ^* is onto F iff τ is a topological embedding of $\mathcal{F}$ into $\mathcal{G}$ such that $G_{|\tau(X)} = \{g_{|\tau(X)} \mid g \in G\}$.*

Proof. We suppose that τ^* is onto F and we show first that τ is $1-1$; suppose that $\tau(x_1) = \tau(x_2)$, for some $x_1, x_2 \in X$. We have that $\forall_{f \in F}(f(x_1) = f(x_2))$, since by the onto hypothesis of τ^*, if $f \in F$, there is some $g \in G$ such that $f(x_1) = (g \circ \tau)(x_1) = g(\tau(x_1)) = g(\tau(x_2)) = (g \circ \tau)(x_2) = f(x_2)$. By the complete regularity of $\mathcal{F}$ we conclude that $x_1 = x_2$. By the lifting of morphisms we get directly that if $\tau \in \mathrm{Mor}(\mathcal{F}, \mathcal{G})$, then $\tau \in \mathrm{Mor}(\mathcal{F}, \mathcal{G}_{|\tau(X)})$. The onto hypothesis of τ^* i.e., $\forall_f \exists_{g \in G}(f = g \circ \tau)$, implies that $\forall_f \exists_{g' \in G_{|\tau(X)}}(f = g' \circ \tau)$, where $g' = g_{|\tau(X)}$. Hence, $\tau : X \to \tau(X)$ is an isomorphism between $\mathcal{F}$ and $\mathcal{G}_{|\tau(X)}$. Next we show that $\{g_{|\tau(X)} \mid g \in G\}$ is a topology on $\tau(X)$, therefore by the definition of the relative topology we get that $G_{|\tau(X)} = \{g_{|\tau(X)} \mid g \in G\}$. Clearly, $\overline{a}_{|\tau(X)} = \overline{a}$, $(g_1 + g_2)_{|\tau(X)} = g_{1|\tau(X)} + g_{2|\tau(X)}$ and $(\phi \circ g)_{|\tau(X)} = \phi \circ g_{|\tau(X)}$, where $\phi \in \mathrm{Bic}(\mathbb{R})$. Suppose that $h : \tau(X) \to \mathbb{R}$, $\epsilon > 0$ and $g \in G$ such that $\forall_{y \in \tau(X)}(|g(y) - h(y)| \leq \epsilon) \leftrightarrow \forall_{x \in X}(|g(\tau(x)) - h(\tau(x))| \leq \epsilon)$. Since $g \circ \tau \in F$ and ϵ is arbitrary, we conclude by the condition BS_4 that $h \circ \tau \in F$, hence, by our onto hypothesis of τ^*, there is some $g \in G$ such that $g \circ \tau = h \circ \tau$ i.e., $g_{|\tau(X)} = h$. For the converse we fix some $f \in F$ and we find $g \in G$ such that $f = \tau^*(g)$. Since $\tau : X \to \tau(X)$ is open, there exists some $g' \in G_{|\tau(X)}$ such that $f = g' \circ \tau$, and since $G_{|\tau(X)} = \{g_{|\tau(X)} \mid g \in G\}$, there is some $g \in G$ such that $g' = g_{|\tau(X)}$, hence $f = g_{|\tau(X)} \circ \tau = g \circ \tau = \tau^*(g)$.

As in [5] p. 155 for $C(X)$, the Proposition 4 implies the Tychonoff embedding theorem; if F is completely regular and $e : X \to \mathbb{R}^F$ is defined by $x \mapsto (f(x))_{f \in F}$, then $e^* : \bigvee_{f \in F} \pi_f \to F$ is onto F, since $e^*(\pi_f) = \pi_f \circ e = f$, therefore e is an embedding of $\mathcal{F}$ into $\mathcal{R}^F$ such that $(\bigvee_{f \in F} \pi_f)_{|e(X)} = \{g_{|e(X)} \mid g \in \bigvee_{f \in F} \pi_f\}$.

Proposition 5. *If $\mathcal{F} = (X, F)$ and $\mathcal{G} = (Y, G)$ are Bishop spaces, then:*

(i) If $e \in \mathrm{Mor}(\mathcal{F}, \mathcal{G})$, then $\hat{x} \circ e^* = \widehat{e(x)}$, for every $x \in X$, and $e^* \in \mathrm{Mor}(\mathcal{G}^*, \mathcal{F}^*)$.
(ii) The mapping $\hat{}: X \to F^*$, defined by $x \mapsto \hat{x}$, is in $\mathrm{Mor}(\mathcal{F}, \mathcal{F}^{**})$ and it is $1-1$ iff $\mathcal{F}$ is completely regular.
(iii) If $\mathcal{G}$ is completely regular, then $\forall_{e_1, e_2 \in \mathrm{Mor}(\mathcal{F}, \mathcal{G})}(e_1^* = e_2^* \to e_1 = e_2)$.

Proof. (i) By the lifting of morphisms we have that $e^* \in \mathrm{Mor}(\mathcal{G}^*, \mathcal{F}^*) \leftrightarrow \forall_{x \in X}(\hat{x} \circ e^* \in G^*)$. But $\hat{x} \circ e^* = \widehat{e(x)} \in G^*$, since $(\hat{x} \circ e^*)(g) = \hat{x}(e^*(g)) = \hat{x}(g \circ e) = (g \circ e)(x) = g(e(x)) = \widehat{e(x)}(g)$, for every $g \in G$.

(ii) Since $\mathcal{F}^{**} = (F^*, F^{**})$, where $F^{**} = \mathcal{F}(\{\hat{f} \mid f \in F\})$, $\hat{f} : F^* \to \mathbb{R}$, and $\hat{f}(\theta) = \theta(f)$, for every $\theta \in F^*$, by the lifting of morphisms we have that $\hat{} \in \mathrm{Mor}(\mathcal{F}, \mathcal{F}^{**}) \leftrightarrow \forall_{f \in F}(\hat{f} \circ \hat{} \in F)$. But $\hat{f} \circ \hat{} = f$, since $(\hat{f} \circ \hat{})(x) = \hat{f}(\hat{x}) = \hat{x}(f) = f(x)$, for every $x \in X$. Since $\hat{x} = \hat{y} \leftrightarrow \forall_{f \in F}(\hat{x}(f) = \hat{y}(f)) \leftrightarrow \forall_{f \in F}(f(x) = f(y))$, the injectivity of $\hat{}$ implies the tightness of $\bowtie_F$ and vice versa.

(iii) By (i) we have that $\hat{x} \circ e_1^* = \widehat{e_1(x)}$ and $\hat{x} \circ e_2^* = \widehat{e_2(x)}$, for every $x \in X$. Since $e_1^* = e_2^*$, we get that $\widehat{e_1(x)} = \widehat{e_2(x)}$, for every $x \in X$. By (ii) and the complete regularity of $\mathcal{G}$ we get that $e_1(x) = e_2(x)$, for every $x \in X$.

Proposition 6. *Suppose that $\mathcal{F} = (X, F)$ and $\mathcal{G} = (Y, G)$ are completely regular Bishop spaces and $\mathcal{E} : G \to F$ is an isomorphism between $\mathcal{G}^*$ and $\mathcal{F}^*$. Then, there exists an (unique) isomorphism $e : X \to Y$ between $\mathcal{F}$ and $\mathcal{G}$ such that $\mathcal{E} = e^*$ iff $\forall_{x \in X} \exists_{y \in Y}(\hat{x} \circ \mathcal{E} = \hat{y})$ and $\forall_{y \in Y} \exists_{x \in X}(\hat{y} = \hat{x} \circ \mathcal{E})$.*

Proof. The necessity follows by applying the Proposition 5(i) on e and e^{-1}. For the converse we suppose that $\forall_{x \in X} \exists_{y \in Y}(\hat{x} \circ \mathcal{E} = \hat{y})$ and we show that $\forall_{x \in X} \exists!_{y \in Y}(\hat{x} \circ \mathcal{E} = \hat{y})$; if $\hat{x} \circ \mathcal{E} = \hat{y_1} = \hat{y_2}$, then by the complete regularity of $\mathcal{G}$ we get that $y_1 = y_2$. We define $e : X \to Y$ by $x \mapsto y$, where y is the unique element of Y such that $\hat{x} \circ \mathcal{E} = \hat{y}$. Similarly, we suppose that $\forall_{y \in Y} \exists_{x \in X}(\hat{y} = \hat{x} \circ \mathcal{E})$ and we show that $\forall_{y \in Y} \exists!_{x \in X}(\hat{y} = \hat{x} \circ \mathcal{E})$; if $\hat{y} = \hat{x_1} \circ \mathcal{E} = \hat{x_2} \circ \mathcal{E}$, then $\hat{y} \circ \mathcal{E}^{-1} = \hat{x_1} = \hat{x_2}$ and by the complete regularity of $\mathcal{F}$ we conclude that $x_1 = x_2$. We define $j : Y \to X$ by $y \mapsto x$, where x is the unique element of X such that $\hat{x} \circ \mathcal{E} = \hat{y}$. Next we show that $j = e^{-1}$, or equivalently that $e \circ j = \mathrm{id}_Y$ and $j \circ e = \mathrm{id}_X$; for the first equality we have that $\hat{y} = \widehat{j(y)} \circ \mathcal{E}$ and also that $\widehat{j(y)} \circ \mathcal{E} = \widehat{e(j(y))}$, which implies that $\hat{y} = \widehat{e(j(y))}$. By the complete regularity of $\mathcal{G}$ we get that $y = e(j(y))$. For the second equality we have that $\hat{x} \circ \mathcal{E} = \widehat{e(x)}$ and $\widehat{e(x)} = \widehat{j(e(x))} \circ \mathcal{E}$, which implies that $\hat{x} \circ \mathcal{E} = \widehat{j(e(x))} \circ \mathcal{E}$, and consequently $\hat{x} = \widehat{j(e(x))}$. By the complete regularity of $\mathcal{F}$ we get that $x = j(e(x))$. Hence, e is a bijection. Next we show that $\mathcal{E}(g) = g \circ e$, for every $g \in G$; since the first part of our hypothesis can be written as $\forall_{x \in X}(\hat{x} \circ \mathcal{E} = \widehat{e(x)})$, we get that $(\hat{x} \circ \mathcal{E})(g) = \widehat{e(x)}(g) \leftrightarrow \mathcal{E}(g)(x) = g(e(x)) \leftrightarrow \mathcal{E}(g)(x) = (g \circ e)(x)$, for every $g \in G$ and $x \in X$. Since $g \circ e = \mathcal{E}(g) \in F$, for every $g \in G$, we conclude that $e \in \mathrm{Mor}(\mathcal{F}, \mathcal{G})$, while if $f \in F$, since $\mathcal{E}$ is onto F, there exists some $g \in G$ such that $\mathcal{E}(g) = g \circ e = f$ i.e., e is open.

A formalization of the previous proof requires Myhill's axiom of nonchoice, which is considered compatible with BISH (see [3], p. 75).

References

1. Bishop, E.: Foundations of Constructive Analysis. McGraw-Hill, New York (1967)
2. Bishop, E., Bridges, D.: Constructive Analysis. Grundlehren der Mathematischen Wissenschaften 279. Springer, Heidelberg (1985)
3. Bridges, D., Reeves, S.: Constructive mathematics in theory and programming practice. Philosophia Mathe. **3**, 65–104 (1999)
4. Bridges, D.S.: Reflections on function spaces. Ann. Pure Appl. Logic **163**, 101–110 (2012)
5. Gillman, L., Jerison, M.: Rings of Continuous Functions. Van Nostrand, Princeton (1960)

6. Ishihara, H.: Relating bishop's function spaces to neighborhood spaces. Ann. Pure Appl. Logic **164**, 482–490 (2013)
7. Ishihara, H., Palmgren, E.: Quotient topologies in constructive set theory and type theory. Ann. Pure Appl. Logic **141**, 257–265 (2006)
8. Petrakis, I.: Bishop spaces: constructive point-function topology. In: Mathematisches Forschungsinstitut Oberwolfach Report No. 52/2014, Mathematical Logic: Proof Theory, Constructive Mathematics, pp. 26–27. doi:10.4171/OWR/2014/52

Computing Equality-Free String Factorisations

Markus L. Schmid(✉)

Fachbereich IV – Abteilung Informatikwissenschaften,
Universität Trier, 54286 Trier, Germany
MSchmid@uni-trier.de

Abstract. A factorisation of a string is equality-free if each two factors are different; its size is the number of factors and its width is the maximum length of any factor. To decide, for a string w and a number m, whether w has an equality-free factorisation with a size of at least (or a width of at most) m are NP-complete problems. We further investigate the complexity of these problems and also study the converse problems of computing a factorisation that is to a large extent not equality-free, i.e., a factorisation of size at least (or width at most) m such that the total number of different factors does not exceed a given bound k.

Keywords: String factorisations · NP-hard string problems · FPT

1 Introduction

Many classical hard string problems can be defined in terms of factorisations of strings that satisfy certain properties. For example, the well-known problem of computing the shortest common superstring of given strings $w_1, \ldots, w_k$ (see, e. g., [1]) asks whether there exists a short string x that, for every i, $1 \leq i \leq k$, has a factorisation $u_i \cdot w_i \cdot v_i$. Since a string w is a subsequence of a string u if u has a factorisation $v_1 \cdot v_2 \cdots v_k$ and w has a factorisation $v_{i_1} \cdot v_{i_2} \cdots v_{i_n}$ with $1 \leq i_1 < \ldots < i_n \leq k$, the famous LONGEST COMMON SUBSEQUENCE and SHORTEST COMMON SUPERSEQUENCE problems can as well be described in terms of factorisations. Another example of a string problem that has recently attracted much attention is the problem to decide for two words x and y and a given k whether they have factorisations $u_1 \cdot u_2 \cdots u_k$ and $v_1 \cdot v_2 \cdots v_k$, respectively, such that $(u_1, \ldots, u_k)$ is a permutation of $(v_1, \ldots, v_k)$, i. e., the MINIMUM COMMON STRING PARTITION problem. See [1] for a survey on the multivariate analysis of NP-hard string problems.

In this paper we are concerned with so-called equality-free factorisations, recently introduced in [2,3]. A factorisation $u_1 \cdot u_2 \cdots u_k$ is *equality-free* if every factor is distinct, i. e., $|\{u_1, u_2, \ldots, u_k\}| = k$. In [2,3], Condon et al. investigate the problem of deciding whether a given string w has an equality-free factorisation of *width* at most m, where the width is the maximum length of any factor.[1]

[1] This problem is also mentioned in [1]; furthermore, in [6], the hardness of computing an equality-free factorisation with only palindromes as factors is investigated.

A. Beckmann et al. (Eds.): CiE 2015, LNCS 9136, pp. 313–323, 2015.
DOI: 10.1007/978-3-319-20028-6_32

A motivation for this problem comes from gene synthesis. Since it is only possible to produce short pieces of DNA (so-called *oligo fragments*) artificially, longer DNA sequences are usually obtained by a self-assembly of many oligos into the desired DNA sequence; thus, the task is to find the right oligos for successful self-assembly. Computing equality-free factorisations with bounded width is an abstraction of this problem: the width bound represents the necessity for short oligos and the equality-freeness models the condition that each two oligos must not be too similar in order to not hybridise with each other (see [2,3] for more details). This problem is NP-complete, even if the width bound is 2 or the alphabet is binary (see [3]). We revisit this problem and show that it is fixed-parameter tractable if both the width bound and the alphabet size are parameters.

If instead of a small width, we are looking for an equality-free factorisation with a large *size*, i. e., a large number of factors, then we obtain a different NP-complete problem (see [4]). This variant is motivated by injective pattern matching with variables (which is identical to the special case of solving word equations, where the left side of the equation does not contain variables and different variables must be replaced by different words), see [4] for more details. We show that computing equality-free factorisations with large size is fixed-parameter tractable if parameterised by the size bound. However, the question whether the problem remains hard for fixed alphabets is still open.

We also consider the converse of computing equality-free factorsations, i. e., computing factorisations that are to a large extent not equality-free (or *repetitive*). Our measure of repetitiveness is the number of different factors in the factorisation. If this number is small (in comparison to the size or width of the factorisation), then many factors are repeated. This yields an interesting combinatorial question in its own right: how many different words are needed in order to cover a given word? Furthermore, it is motivated by data compression, since a factorisation with many repeated factors can be used in order to compress a word, e. g., by using a dictionary of the different factors. We can show that deciding on whether a word w has a factorisation of width at most m and with at most k different factors is NP-complete, even if $m = 2$. On the other hand, if k or the alphabet size is a constant, then the problem can be solved in polynomial time. In contrast to this, if m is a lower bound on the size of the factorisation, then the problem can be solved in polynomial time if either m, k or the alphabet size is a constant, but it is open, whether the problem is NP-complete in general.

As a tool for proving some of our main results, we also investigate the problem of deciding whether a given word w has an equality-free factorisation with only factors from a given finite set F of words. It turns out that this problem is NP-complete even for binary alphabets. However, it is in FPT if $|F|$ is a parameter and in P if we drop the equality-freeness condition.

Due to space constraints, not all results are formally proven.

2 Basic Definitions

Let $\mathbb{N} = \{1, 2, 3, \ldots\}$. By $|A|$, we denote the cardinality of a set A. Let Σ be a finite alphabet of *symbols*. A *word* or *string* (over Σ) is a sequence of symbols from Σ. For any word w over Σ, $|w|$ denotes the length of w and ε denotes the *empty word*, i. e., $|\varepsilon| = 0$. The symbol Σ^+ denotes the set of all non-empty words over Σ and $\Sigma^* = \Sigma^+ \cup \{\varepsilon\}$. For the *concatenation* of two words w_1, w_2 we write $w_1 \cdot w_2$ or simply $w_1 w_2$. For every symbol $a \in \Sigma$, by $|w|_a$ we denote the number of occurrences of symbol a in w. We say that a word $v \in \Sigma^*$ is a *factor* of a word $w \in \Sigma^*$ if there are $u_1, u_2 \in \Sigma^*$ such that $w = u_1 v u_2$. If $u_1 = \varepsilon$ or $u_2 = \varepsilon$, then v is a *prefix* (or a *suffix*, respectively) of w. For every i, $1 \leq i \leq |w|$, by $w[1..n]$ we denote the prefix of w with length n. As a convention, in this work every set of words is always a finite set.

By the term *trie*, we refer to the well-known ordered tree data structure for representing sets of words.

Factorisations. For any word $w \in \Sigma^+$, a *factorisation of* w is a tuple $p = (u_1, u_2, \ldots, u_k) \in (\Sigma^+)^k$, $k \in \mathbb{N}$, with $w = u_1 u_2 \ldots u_k$. Every word u_i, $1 \leq i \leq k$, is called a *factor (of p)* or simply *p-factor*. For the sake of readability, we sometimes represent a factorisation $(u_1, u_2, \ldots, u_k)$ in the form $u_1 \centerdot u_2 \centerdot \ldots \centerdot u_k$.

Let $p = (u_1, u_2, \ldots, u_k)$ be an arbitrary factorisation. We define the following parameters: $\mathsf{sf}(p) = \{u_1, u_2, \ldots, u_k\}$ (the *set of factors*), $\mathsf{s}(p) = k$ (the *size*), $\mathsf{c}(p) = |\mathsf{sf}(p)|$ (the *cardinality*) and $\mathsf{w}(p) = \max\{|u_i| \mid 1 \leq i \leq k\}$ (the *width*). A factorisation p is *equality-free* if $\mathsf{s}(p) = \mathsf{c}(p)$.

Problems. We now define the different problems to be investigated in this work.

Equality-Free Factor Cover (EFFC)
Instance: A word w and a set F of words represented as a trie.
Question: Does there exist an equality-free factorisation p of w with $\mathsf{sf}(p) \subseteq F$?

Maximum Equality-Free Factorisation Size (MaxEFF-s)
Instance: A word w and a number m, $1 \leq m \leq |w|$.
Question: Does there exist an equality-free factorisation p of w with $\mathsf{s}(p) \geq m$?

Maximum Repetitive Factorisation Size (MaxRF-s)
Instance: A word w, numbers m, $1 \leq m \leq |w|$, and k, $1 \leq k \leq |w|$.
Question: Does there exist a factorisation p of w with $\mathsf{s}(p) \geq m$ and $\mathsf{c}(p) \leq k$?

The problems MaxEFF-s and MaxRF-s where m is interpreted as an upper bound on the width instead of a lower bound on the size are denoted by MinEFF-w and MinRF-w. In the remainder of the paper, the symbol m is reserved as the bound on the size or width (depending on the problem under consideration) of the factorisation and k as the bound on the cardinality of the factorisation, respectively. For any problem K from above and any fixed alphabet Σ, K_Σ denotes the problem K, where the input word is over Σ.

We shall now illustrate these definitions with an example.

Example 1. Let p = aab . ba . cba . aab . ba . aab be a factorisation. We note that $\mathsf{sf}(p) = \{$aab, ba, cba$\}$, $\mathsf{s}(p) = 6$, $\mathsf{c}(p) = 3$ and $\mathsf{w}(p) = 3$. The factorisation p is not equality-free. Furthermore, (abbcbaabbc, 6) $\in$ MaxEFF-s (witnessed by a.bb.c.ba.ab.bc), whereas (abbcbaabbc, 7) $\notin$ MaxEFF-s. On the other hand, ((aabbcc)^2, m) $\in$ MinEFF-w if and only if $m \geq 2$, whereas ((aabbcc)^3, m) $\in$ MinEFF-w if and only if $m \geq 3$.

Parameterised Complexity. We consider decision problems as languages over some alphabet Γ. A *parameterisation (of Γ)* is a polynomial time computable mapping $\kappa : \Gamma^* \rightarrow \mathbb{N}$ and a *parameterised problem* is a pair (Q, κ), where Q is a problem (over Γ) and κ is a parameterisation of Γ. We usually define κ implicitly by describing which part of the input is the parameter. A parameterised problem (Q, κ) is *fixed-parameter tractable* if there is an *fpt-algorithm* for it, i. e., an algorithm that solves Q on input x in time $\mathcal{O}(f(\kappa(x)) \times p(|x|))$ for recursive f and polynomial p. The class of fixed-parameter tractable problems is denoted by FPT. Note that if a parameterised problem becomes NP-hard if the parameter is set to a constant, then it is not in FPT unless P = NP. For detailed explanations on parameterised complexity, the reader is referred to [5].

3 Main Results

We begin this section with some preliminary observations that are necessary for proving some of the main results.

If in an equality-free factorisation p, we join one of the longest factors with one of its neighbours, then the resulting factorisation is still equality-free and has size $\mathsf{s}(p) - 1$.

Observation 1. A word w has an equality-free factorisation p with $\mathsf{s}(p) \geq m$, $m \in \mathbb{N}$, if and only if it has an equality-free factorisation p' with $\mathsf{s}(p') = m$.

We can check whether a factorisation p of a word w is equality-free in time $\mathcal{O}(|w|)$ by inserting all factors in a trie and checking for each factor if it is already contained in the trie. If a set F of words is given as a trie, then we can check in a similar way whether or not $\mathsf{sf}(p) \subseteq F$ in time $\mathcal{O}(|w|)$.

Observation 2. Let $w \in \Sigma^+$ and let p be a factorisation of w. It can be decided in time $\mathcal{O}(|w|)$ whether or not p is equality-free.

Observation 3. Let $w \in \Sigma^+$ and let F be a set of words over Σ represented as a trie. It can be decided in time $\mathcal{O}(|w|)$ whether or not $\mathsf{sf}(p) \subseteq F$.

The following result is straightforward, but it nevertheless contributes to our understanding of the complexity of the considered problems.

Proposition 1. *The problems* EFFC, MAXEFF-s, MINEFF-w, MAXRF-s *and* MINRF-w *are in* FPT *with respect to parameter* $|w|$.

3.1 The Problem EFFC

As mentioned in the introduction, the problem EFFC is not our main concern. However, its investigation, as we shall see, yields some valuable insights with respect to equality-free factorisations and we also obtain an algorithm that shall be used later in order to prove tractability results with respect to the problems MAXRF-s and MINRF-w.

We first show that EFFC is NP-complete, even for fixed binary alphabets.

Theorem 1. *Let Σ be an alphabet with $|\Sigma| = 2$. Then* EFFC_Σ *is* NP*-complete.*

Proof. Since we can guess a factorisation p and check in polynomial time whether it is equality-free and $\mathsf{sf} \subseteq F$, EFFC_Σ is in NP. Let (w, m) be an instance of MINEFF-w_Σ and let F be the set of all factors of w of length at most m. We note that $|F| \leq \sum_{i=1}^{m} |w| - (i-1) \leq m \times |w|$ and F can be constructed in time $\mathcal{O}(|F| \times m)$. The word w has an equality-free factorisation p with $\mathsf{w}(p) \leq m$ if and only it has an equality-free factorisation p' with $\mathsf{sf}(p') \subseteq F$. Since MINEFF-w_Σ is NP-complete (see [3]), EFFC_Σ is NP-complete as well. □

In addition to the alphabet size, the cardinality of the given set F of factors is another natural parameter and we can show that the hardness is not preserved if we bound $|F|$ by a constant (in contrast to bounding $|\Sigma|$ (Theorem 1)).

Theorem 2. *The Problem* EFFC *can be solved in time $\mathcal{O}(|w|^{|F|+1})$.*

Theorem 2 is obtained by an analysis of a brute-force algorithm, coupled with the observation that if an equality-free factorisation p satisfies $\mathsf{sf}(p) \subseteq F$, then its size must be bounded by $|F|$. As we shall see next, a more sophisticated approach, which relies on encoding the different factors of F as single symbols and factorisations as words over these symbols, yields an fpt-algorithm for EFFC with respect to the parameter $|F|$ (note that this does not make the algorithm of Theorem 2 obsolete, since for large $|F|$ it might still be faster than the fpt-algorithm).

Theorem 3. *The Problem* EFFC *can be solved in time $\mathcal{O}(|w| \times (2^{|F|} - 1) \times |F|!)$.*

Proof. Let $w \in \Sigma^*$ and $F = \{u_1, u_2, \ldots, u_\ell\}$ be an instance of EFFC. Furthermore, let $\Gamma = \{1, 2, \ldots, \ell\}$, let $h : \Gamma^* \to \Sigma^*$ be a morphism defined by $h(i) = u_i$, $i \in \Gamma$, and let $S = \{v \in \Gamma^+ \mid |v|_i \leq 1, i \in \Gamma\}$. There exists a word $v = j_1 j_2 \ldots j_m \in S$ with $h(v) = w$ if and only if $p = (u_{j_1}, u_{j_2}, \ldots, u_{j_m})$ is an equality-free factorisation of w with $\mathsf{sf}(p) \subseteq F$. Therefore, we can solve EFFC by checking for each word $v \in S$ whether or not $h(v) = w$, which can be done in time $\mathcal{O}(|w| \times |S|)$. We conclude the prove by observing that $S = \bigcup_{\Gamma' \subseteq \Gamma, \Gamma' \neq \emptyset} S_{\Gamma'}$, where $S_{\Gamma'} = \{v \in \Gamma'^+ \mid |v|_i = 1, i \in \Gamma'\}$. Since $|S_{\Gamma'}| = |\Gamma'|!$ and the sets $S_{\Gamma'}$ are pairwise disjoint, we have $|S| = \sum_{\Gamma' \subseteq \Gamma, \Gamma' \neq \emptyset} |\Gamma'|! \leq \sum_{i=1}^{2^\ell - 1} \ell! = (2^\ell - 1) \times \ell!$. Since $\ell = |F|$, we obtain a total running time of $\mathcal{O}(|w| \times (2^{|F|} - 1) \times |F|!)$. □

Next, we investigate the impact of the equality-freeness condition itself, i. e., we consider the problem FC, which is identical to EFFC with the only difference that the factorisation p of w with $\mathsf{sf}(p) \subseteq F$ does not need to be equality-free. This problem is similar to the problem EXACT BLOCK COVER (recently investigated by Jiang et al. in [8]), which differs from FC only in that instead of a set we are given a sequence of factors and every factor of the sequence has to be used exactly once (in particular, this coincides with the variant of MINIMUM COMMON STRING PARTITION where the partition of one of the two strings is already fixed). While EXACT BLOCK COVER is NP-complete (see [8]), FC can be solved in polynomial time by dynamic programming. This demonstrates that it is really the equality-freeness condition that makes EFFC hard and, in addition, we obtain a useful tool to devise algorithms for solving variants of the problems MAXRF-s and MINRF-w later on in Sect. 3.3.

Theorem 4. *The problem* FC *can be solved in time* $\mathcal{O}(|F| \times |w|^2)$.

Proof. We define a dynamic programming algorithm. Let w be a word and $F = \{u_1, u_2, \ldots, u_\ell\}$. For every n, m, $1 \leq m \leq n \leq |w|$, let $T[n, m] = 1$ if there exists a factorisation p of size m of $w[1..n]$ with $\mathsf{sf}(p) \subseteq F$ and $T[n, m] = 0$ otherwise. Obviously, (w, F) is a positive instance of FC if and only if $T[|w|, m] = 1$ for some m, $1 \leq m \leq |w|$. We can now solve FC on instance (w, F) by computing all the $T[n, m]$, $1 \leq m \leq n \leq |w|$, in the following way.

In time $\mathcal{O}(|w| \times |F|)$, we first construct a table S with $|w|$ rows and ℓ columns with $S[n, i] = 0$, $1 \leq n \leq |w|$, $1 \leq i \leq \ell$. Then, by using the Knuth-Morris-Pratt algorithm [9], for every i, $1 \leq i \leq \ell$, we set $S[n, i] = 1$ if u_i is a suffix of $w[1..n]$. Since the Knuth-Morris-Pratt algorithm has running time $\mathcal{O}(|w| + |u_i|)$, building up this table can be done in time $\sum_{i=1}^{\ell}(|w| + |u_i|) \leq \sum_{i=1}^{\ell} 2|w| = \mathcal{O}(|F| \times |w|)$. Then, for every n, m, $1 \leq m \leq n \leq |w|$, we initialise $T[n, m] = 0$, which requires time $\mathcal{O}(|w|^2)$, and, for every i, $1 \leq i \leq \ell$, we set $T[|u_i|, 1] = 1$ if $S[|u_i|, i] = 1$, which requires time $\mathcal{O}(|F|)$. We note that, for every n, m with $2 \leq m \leq n \leq |w|$, $T[n, m] = 1$ if and only if there exists a word $u_i \in F$ that is a suffix of $w[1..n]$ (i.e., $S[n, i] = 1$) with $T[n - |u_i|, m - 1] = 1$. Thus, for every n, m, $2 \leq m \leq n \leq |w|$, we can compute $T[n, m]$ in time $\mathcal{O}(|F|)$, provided that all $T[n', m - 1]$, $n' < n$, have already been computed, which is satisfied if we iterate over m, $2 \leq m \leq |w|$, in an outer loop and over n, $m \leq n \leq |w|$, in an inner loop. Hence, all the elements $T[n, m]$, $1 \leq m \leq n \leq |w|$, are computed in time $\mathcal{O}(|F| \times |w|^2)$. □

3.2 The Problems MaxEFF-s and MinEFE-w

In this section, we investigate the problems MINEFF-w and MAXEFF-s. Their NP-completeness is established in [2,4], respectively, but in [3] it is additionally shown that MINEFF-w remains NP-complete even if the bound on the width is 2 or the alphabet is fixed and binary. In particular, this means that, unless P = NP, MINEFF-w is not in FPT with respect to parameter m or $|\Sigma|$. However, if we let both m and $|\Sigma|$ be parameters at the same time, then MINEFF-w is fixed-parameter tractable:

Theorem 5. MINEFF-w$_\Sigma$ *can be solved in time* $\mathcal{O}(m^{m^2 \times |\Sigma|^m + 2} \times |\Sigma|^m)$.

Proof. Let (w, m) be an instance of MINEFF-w$_\Sigma$. For every ℓ, $1 \leq \ell \leq m$, there are $|\Sigma|^\ell$ words of length ℓ and therefore $\sum_{\ell=1}^{m} |\Sigma|^\ell \leq m \times |\Sigma|^m$ words of length at most m. Consequently, if a factorisation p satisfies $\mathsf{w}(p) \leq m$, then $\mathsf{s}(p) \leq m \times |\Sigma|^m$. Furthermore, if for an equality-free factorisation p of w we have $\mathsf{s}(p) \leq m \times |\Sigma|^m$ and $\mathsf{w}(p) \leq m$, then $|w| \leq m^2 \times |\Sigma|^m$. Hence, if $|w| > m^2 \times |\Sigma|^m$ (which can be checked in time $\mathcal{O}(m^{m^2 \times |\Sigma|^m + 2} \times |\Sigma|^m)$), then there is no equality-free factorisation p of w with $\mathsf{w}(p) \leq m$. If, on the other hand,

$|w| \leq m^2 \times |\Sigma|^m$, then we can enumerate all factorisations of w that have a width of at most m and check for each such factorisation whether or not it is equality-free in time $\mathcal{O}(|w|) = \mathcal{O}(m^2 \times |\Sigma|^m)$ (see Observation 2). Since there are at most $m^{|w|} \leq m^{m^2 \times |\Sigma|^m}$ such factorisations, the statement of the theorem follows. □

For the problem MAXEFF-s, i. e., deciding on the existence of an equality-free factorisation with a size of at least m (instead of a width of at most m), we encounter a slightly different situation. First of all, it is still an open problem whether MAXEFF-s remains NP-complete if the alphabet is fixed:

Open Problem 1. *Let Σ be an alphabet. Is* MAXEFF-s_Σ NP-*complete?*

From an intuitive point of view, for the problem MINEFF-w, the bound on the width can conveniently be exploited in order to design gadgets for encoding an NP-hard problem (see [3] and also the proof of Theorem 12). A lower bound on the size seems to provide fewer possibilities for controlling the structure of the factorisation, which makes it difficult to express another NP-complete problem by MAXEFF-s (especially if we have only a constant number of symbols at our disposal). On the other hand, a constant alphabet does not seem to help in order to find an equality-free factorisation with a size of at least m in polynomial time.

However, if we consider m as a constant, then the problem is not NP-complete anymore; in fact, it is even fixed-parameter tractable with respect to m:

Theorem 6. *The problem* MAXEFF-s *can be solved in time* $\mathcal{O}((\frac{m^2+m}{2} - 1)^m)$.

Proof. Let (w, m) be an instance of MAXEFF-s. If $|w| > \Sigma_{i=1}^m i = \frac{m^2+m}{2}$, which can be checked in time $\mathcal{O}((\frac{m^2+m}{2} - 1)^m)$, then the factorisation $(u_1, u_2, \dots, u_m)$ of w with $|u_i| = i$, $1 \leq i \leq m-1$, and $|u_m| = |w| - |u_1 u_2 \dots u_{m-1}|$ is equality-free, since each two factors have a different length. If, on the other hand, $|w| \leq \frac{m^2+m}{2} - 1$, then we can enumerate all factorisations of size m of w in time $\mathcal{O}(|w|^{m-1})$ and, by Observation 2, check in time $\mathcal{O}(|w|)$ for each such factorisation whether or not it is equality-free. Since w has an equality-free factorisation of size at least m if and only if it has an equality-free factorisation of size exactly m (see Observation 1), this solves the problem MAXEFF-s in time $\mathcal{O}(|w|^m) = \mathcal{O}((\frac{m^2+m}{2} - 1)^m)$. □

3.3 The Problems MaxRF-s and MinRF-w

In this section, we investigate the problem of finding a factorisation of a word w with as few different factors as possible. Since (w) is always a solution, we also impose an upper bound on the width of the factorisation or a lower bound on its size. In a sense, a factorisation p of this kind is to a large extent repetitive, since if k is much smaller than $\mathsf{s}(p)$ or $\frac{|w|}{\mathsf{w}(p)}$, then many factors must be repeated.

We shall see that if k or $|\Sigma|$ are constants, then both MAXRF-s and MINRF-w can be solved in polynomial time. If, on the other hand, m is a constant, then MAXRF-s can be solved in polynomial time as well, whereas MINRF-w is NP-complete even for $m = 2$. Unfortunately, we are not able to answer whether MAXRF-s is NP-complete in general.

We now first investigate the problem MAXRF-s.

Theorem 7. *The problem* MAXRF-s *can be solved in time* $\mathcal{O}(k^2 \times |w|^{2k+3})$.

Proof. Let (w, m, k) be an instance of MAXRF-s with $m \leq |w|$ and $k \leq |w|$ (otherwise, it would be a negative instance). Let $F_w = \{u \mid u \text{ is a factor of } w\}$. For every $F \subseteq F_w$ with $|F| \leq k$, we run the algorithm defined in the proof of Theorem 4 on input (w, F). If $T[|w|, \ell] = 1$, for an ℓ, $m \leq \ell \leq |w|$, then there is a factorisation p of w with $\mathsf{s}(p) \geq m$ and $\mathsf{sf}(p) \subseteq F$; since $|F| \leq k$, this implies $\mathsf{c}(p) \leq k$. To carry out this procedure, we have to enumerate all subsets $F \subseteq F_w$ with $|F| \leq k$. Since $|F_w| \leq |w|^2$, for every ℓ, $1 \leq \ell \leq k$, there at most $|F_w|^\ell \leq |w|^{2\ell}$ subsets $F \subseteq F_w$ with $|F| = \ell$. Thus, there are $\sum_{i=1}^{k} |w|^{2i} \leq k \times |w|^{2k}$ subsets to investigate. For each subset F, we run the algorithm of the proof of Theorem 4 in time $\mathcal{O}(|F| \times |w|^2)$, and check for every ℓ, $m \leq \ell \leq |w|$, whether or not $T[|w|, \ell] = 1$, which requires time $\mathcal{O}(|w|)$. Hence, the total running time of this procedure is $\mathcal{O}(k \times |w|^{2k} \times k \times |w|^3) = \mathcal{O}(k^2 \times |w|^{2k+3})$. □

From Theorem 7 we can conclude with moderate effort that MAXRF-s can also be solved in time that is exponential only in m or $|\Sigma|$. To this end, we observe that if, for an instance (w, m, k) of MAXRF-s, we have $k \geq |\Sigma|$, then splitting w in only factors of length 1 yields a factorisation p with $\mathsf{c}(p) \leq |\Sigma| \leq k$ and $\mathsf{s}(p) = |w| \geq m$, and if $k \geq m$, then any factorisation p of w of size m satisfies $\mathsf{c}(p) \leq m \leq k$ and $\mathsf{s}(p) \geq m$. If, on the other hand, k is bounded by $|\Sigma|$ or m, then the procedure used in the proof of Theorem 7 has a running time that is exponential only in $|\Sigma|$ or m, respectively, which yields the following results:

Theorem 8. *Let Σ be an alphabet. Then the problem* MAXRF-s_Σ *can be solved in time* $\mathcal{O}(|\Sigma|^2 \times |w|^{2|\Sigma|+1})$.

Theorem 9. *The problem* MAXRF-s *can be solved in time* $\mathcal{O}(m^2 \times |w|^{2m+1})$.

The probably most interesting question, which, unfortunately, is still open is whether the general version of MAXRF-s can also be solved in polynomial time.

Open Problem 2. *Is* MAXRF-s NP-*complete?*

We now turn to the problem MINRF-w. In an analogous way as done in the proof of Theorem 7, we can show that MINRF-w can be solved in time exponential only in k, too. The only difference is that instead of running the algorithm of Theorem 4 for every subset of the set of all factors of w, it is sufficient to only consider all subsets of the set of all factors of w that have a length of at most m.

Theorem 10. MINRF-w *can be solved in time* $\mathcal{O}(k^2 \times m^k \times |w|^{k+3})$.

In a similar way as Theorem 8 follows from Theorem 7, i.e., by bounding k in terms of $|\Sigma|$, we can conclude from Theorem 10 the next result.

Theorem 11. *Let Σ be an alphabet. Then the problem* MINRF-w$_\Sigma$ *can be solved in time* $\mathcal{O}(|\Sigma|^2 \times m^{(|\Sigma|-1)} \times |w|^{|\Sigma|+2})$.

While for problem MAXRF-s it was also possible to bound k in terms of m, for MINRF-w, we can only observe that (w, m, k) must be a positive instance if $k \geq \lceil \frac{|w|}{m} \rceil$, but in case $k < \lceil \frac{|w|}{m} \rceil$, the algorithm of the proof of Theorem 10 has a running time exponential in $|w|$ and it does not seem possible to solely bound k in terms of m. We now justify this intuition by showing that MINRF-w is NP-complete, even if $m = 2$. First, we recall the hitting set problem (see [7]):

HITTING SET (HS)
Instance: $U = \{x_1, \dots, x_\ell\}$, $S_1, \dots, S_n \subseteq U$ and $q \in \mathbb{N}$.
Question: Does there exist $T \subseteq U$ with $|T| \leq q$ and $T \cap S_i \neq \emptyset$, $1 \leq i \leq n$?

We now give a reduction from HS to MINRF-w with $m = 2$. Let $(U, S_1, \dots, S_n, q)$ be an instance of HS. We assume that, for every i, j, $1 \leq i < j \leq n$, $|S_i| = |S_j| = r$ (note that HS reduces to the variant where all sets S_i have the same cardinality r by adding $r - |S_i|$ new elements to every S_i). For the sake of concreteness, we assume $S_i = \{y_{i,1}, y_{i,2}, \dots, y_{i,r}\}$, $1 \leq i \leq n$. We define an alphabet $Sigma = U \cup \{\$_{i,j} \mid 1 \leq i \leq n, 1 \leq j \leq r-1\} \cup \{¢\}$ and a word $w = ¢¢\, v_1\, ¢\, v_2\, ¢ \dots ¢\, v_n\, ¢$, where, for every i, $1 \leq i \leq n$, $v_i = y_{i,1}\$_{i,1}y_{i,2}\$_{i,2} \dots \$_{i,r-1}y_{i,r}$. The following lemma states that this transformation from an HS instance to a word over Σ is in fact a reduction from HS to MINRF-w.

Lemma 1. *There exists a set $T \subseteq U$ with $|T| \leq q$ and $T \cap S_i \neq \emptyset$, $1 \leq i \leq n$, if and only if w has a factorisation p with $\mathsf{w}(p) \leq 2$ and $\mathsf{c}(p) \leq n(r-1) + q + 1$.*

Proof. We start with the *only if* direction and assume that there exists a set $T \subseteq U$ with $|T| \leq q$ and $T \cap S_i \neq \emptyset$, $1 \leq i \leq n$. We now construct a factorisation p of w with the desired properties. We let every single occurrence of ¢ be a factor of p; thus, it only remains to split every v_i, $1 \leq i \leq n$, into factors of size at most 2, which is done as follows. For every i, $1 \leq i \leq n$, let j_i, $1 \leq j_i \leq r$, be arbitrarily chosen such that $y_{i,j_i} \in T$ (since $T \cap S_i \neq \emptyset$, $1 \leq i \leq n$, such j_i exist). Then, for every i, $1 \leq i \leq n$, we factorise v_i into

$$y_{i,1}\$_{i,1} \cdot y_{i,2}\$_{i,2} \cdot \dots \cdot y_{i,j_i-1}\$_{i,j_i-1} \cdot y_{i,j_i} \cdot \$_{i,j_i}y_{i,j_i+1} \cdot \dots \cdot \$_{i,r-1}y_{i,r}.$$

Obviously, this results in a factorisation p of w with $\mathsf{w}(p) \leq 2$. Furthermore, $\mathsf{sf}(p)$ contains the factor ¢, at most $|T|$ factors x with $x \in T$ and, for every i, $1 \leq i \leq n$, j, $1 \leq j \leq r-1$, a distinct factor of length 2 that contains the symbol $\$_{i,j}$ (the distinctness of these factors follows from the fact that each symbol $\$_{i,j}$ has only one occurrence in w). This implies that $\mathsf{c}(p) \leq 1 + |T| + n(r-1) \leq 1 + q + n(r-1)$, which concludes the *only if* direction of the proof.

In order to prove the *if* direction, we assume that there exists a factorisation p of w with $\mathsf{w}(p) \leq 2$ and $\mathsf{c}(p) \leq 1 + q + n(r-1)$. We now modify p step by step such that every modification maintains $\mathsf{w}(p) \leq 2$ and $\mathsf{c}(p) \leq 1 + q + n(r-1)$. Since w starts with ¢¢ and $\mathsf{w}(p) \leq 2$, we can conclude that ¢ or ¢¢ is a factor of p. If ¢¢ is a factor of p, then we can split it into ¢ • ¢ without increasing $\mathsf{c}(p)$, since the factor ¢¢ is then not a factor of p anymore and we get at most ¢ as a new factor. Every factor of p that contains the symbol ¢ is either the factor ¢ or of the form x¢ or ¢x for some $x \in U$. If, for an $x \in U$, we split all occurrences of factor x¢ in p into x • ¢, then we may produce the new factor x (recall that ¢ is already a factor), but we also necessarily lose x¢ as a factor; thus, $\mathsf{c}(p)$ does not increase. If we apply this modification with respect to all $x \in U$ and all factors x¢ and ¢x, then we obtain a factorisation in which every single occurrence of the symbol ¢ in w is also a factor of p, $\mathsf{w}(p) \leq 2$ and $\mathsf{c}(p) \leq 1+q+n(r-1)$. For every i, $1 \leq i \leq n$, $|v_i|$ is odd, which implies that the factorisation of v_i (according to p) must contain a factor of length 1 and, by the structure of v_i, this factor must be of the form $x \in U$. This particularly implies that for the set T of all elements of U that occur as a factor in p, we must have $T \cap S_i \neq \emptyset$, $1 \leq i \leq n$. Now $\mathsf{sf}(p)$ contains ¢, all $n(r-1)$ factors containing a symbol $\$_{i,j}$, $1 \leq i \leq n$, $1 \leq j \leq r-1$, and the factors in T. Thus, $\mathsf{c}(p) = 1 + n(r-1) + |T| \leq 1 + n(r-1) + q$, which implies $|T| \leq q$. □

We note that the MinRF-w instance $(w, 2, n(r-1) + q + 1)$ can be constructed from the HS instance $(U, S_1, \ldots, S_n, q)$ in polynomial time and that MinRF-w is in NP (we can guess and verify a factorisation). Hence, from the NP-completeness of HS (see [7]) and Lemma 1, we can conclude the following:

Theorem 12. *The problem* MinRF-w *is* NP*-complete even if* $m \leq 2$.

References

1. Bulteau, L., Hüffner, F., Komusiewicz, C., Niedermeier, R.: Multivariate algorithmics for NP-hard string problems. EATCS Bull. **114**, 31–73 (2014)
2. Condon, A., Maňuch, J., Thachuk, C.: Complexity of a collision-aware string partition problem and its relation to oligo design for gene synthesis. In: Hu, X., Wang, J. (eds.) COCOON 2008. LNCS, vol. 5092, pp. 265–275. Springer, Heidelberg (2008)
3. Condon, A., Maňuch, J., Thachuk, C.: The complexity of string partitioning. In: Kärkkäinen, J., Stoye, J. (eds.) CPM 2012. LNCS, vol. 7354, pp. 159–172. Springer, Heidelberg (2012)
4. Fernau, H., Manea, F., Mercaş, R., Schmid, M.L.: Pattern matching with variables: fast algorithms and new hardness results. In: Leibniz International Proceedings in Informatics (LIPIcs), Proceedings 32nd Symposium on Theoretical Aspects of Computer Science, STACS 2015, vol. 30, pp. 302–315 (2015)
5. Flum, J., Grohe, M.: Parameterized Complexity Theory. Springer, New York (2006)
6. Gagie, T., Inenaga, S., Karkkainen, J., Kempa, D., Piatkowski, M., Puglisi, S.J., Sugimoto, S.: Diverse palindromic factorization is NP-complete. Technical report 1503.04045 (2015). http://arxiv.org/abs/1503.04045
7. Garey, M.R., Johnson, D.S.: Computers and Intractability. W.H. Freeman and Company, San Francisco (1979)

8. Jiang, H., Su, B., Xiao, M., Xu, Y., Zhong, F., Zhu, B.: On the exact block cover problem. In: Gu, Q., Hell, P., Yang, B. (eds.) AAIM 2014. LNCS, vol. 8546, pp. 13–22. Springer, Heidelberg (2014)
9. Knuth, D.E., Morris, J.H., Pratt, V.R.: Fast pattern matching in strings. Commun. ACM **6**(2), 323–350 (1977)

Towards the Effective Descriptive Set Theory

Victor Selivanov(✉)

A.P. Ershov Institute of Informatics Systems SB RAS, Novosibirsk, Russia
vseliv@iis.nsk.su

Abstract. We prove effective versions of some classical results about measurable functions and derive from this extensions of the Suslin-Kleene theorem, and of the effective Hausdorff theorem for the computable Polish spaces (this was established in [2] with a different proof) and for the computable ω-continuous domains (this answers an open question from [2]).

Keywords: Weakly computable cb_0-space · Computable Polish space · Computable ω-continuous domain · Effective hierarchy · Suslin-Kleene theorem · Effective Hausdorff theorem

1 Introduction

Classical descriptive set theory (DST) [11] deals with hierarchies of sets, functions and equivalence relations in Polish spaces. Theoretical Computer Science, in particular Computable Analysis [23], motivated an extension of the classical DST to non-Hausdorff spaces; a noticeable progress was achieved for the ω-continuous domains and quasi-Polish spaces [4,20].

Theoretical Computer Science and Computable Analysis especially need an effective DST for some effective versions of the mentioned classes of topological spaces. A lot of useful work in this direction was done within Classical Computability Theory but only for the discrete space ω, the Baire space $\mathcal{N}$, and some of their relatives [17,22]. For a systematic work to develop the effective DST for effective Polish spaces see e.g. [8,13,15]. There was also some work on the effective DST for effective domains and approximation spaces [2,20,21].

In this paper we try to make a next step towards the "right" version of effective DST beyond effective Polish spaces. The task seems non-trivial since even the recent search for the "right" effective versions of topological spaces for Computable Analysis resulted in proliferation of different notions of effective spaces of which it is quite hard to choose really useful ones.

We start in the next section with fixing some classes of effective spaces which have good effective DST-properties. Section 3 recalls definitions of effective versions of the classical hierarchies. In Sect. 4 we prove effective versions of some

V. Selivanov—Supported by a Marie Curie International Research Staff Exchange Scheme Fellowship within the 7th European Community Framework Programme, by DFG Mercator Programme at the University of Würzburg, and by RFBR project 13-01-00015a.

A. Beckmann et al. (Eds.): CiE 2015, LNCS 9136, pp. 324–333, 2015.
DOI: 10.1007/978-3-319-20028-6_33

classical DST-results about measurable functions. In Sects. 5 and 6 we derive from this an extension of the Suslin-Kleene theorem [15] and the effective Hausdorff theorem for computable metric spaces (this is earlier established in [2] with a different proof) and for the computable ω-continuous domains (this answers Open Problem 5.15 from [2]).

2 Classes of Effective Spaces

All topological spaces considered here are assumed to be countably based and satisfying the T_0-separation axiom (cb$_0$-spaces, for short). By *weakly computable cb$_0$-space* we mean a pair (X,τ) where X is a non-empty set of points and $\tau:\omega\to P(X)$ is a numbering of a base of a T_0-topology in X such that for some computable functions f,g we have $\tau_x\cap\tau_y=\tau_{f(x,y)}$ and $\bigcup\tau[W_x]=\tau_{g(x)}$ (where $\{W_n\}$ is the standard numbering of c.e. sets [17]). Note that, for a function $h:X\to Y$ and a set $A\subseteq X$, $h[A]$ denotes the image $\{h(a)\mid a\in A\}$. Abusing notation, we often abbreviate (X,τ) to X. Informally, $\tau[\omega]$ is a collection of open sets (called effectively open sets) which is rich enough to have the usual closure properties of open sets effectively.

As morphisms between weakly computable cb$_0$-spaces (X,ξ) and (Y,η) we use *effectively continuous functions*, i.e. functions $f:X\to Y$ such that the numbering $\lambda n.f^{-1}(\eta_n)$ is reducible to ξ (in particular, $f^{-1}(A)\in\xi[\omega]$ whenever $A\in\eta[\omega]$). Recall that *numbering* is any function with domain ω, the *reducibility relation* $\mu\leq\nu$ on numberings means that $\mu=\nu\circ f$ for some computable function f on ω, and numberings μ,ν are *equivalent* (in symbols, $\mu\equiv\nu$) iff $\mu\leq\nu$ and $\nu\leq\mu$. By *effective homeomorphism* between X,Y we mean a bijection f between X and Y such that both f and f^{-1} are effectively continuous.

If (X,τ) is a weakly computable cb$_0$-space then any non-empty subset Y of X has the induced structure τ_Y of weakly computable cb$_0$-space defined by $\tau_Y(n)=Y\cap\tau(n)$. For weakly computable cb$_0$-spaces (X,ξ) and (Y,η), by a *weakly computable embedding of X into Y* we mean an injection $f:X\to Y$ such that $\lambda n.f[\xi_n]\equiv\eta_{f[X]}$. Obviously, such an f is an effective homeomorphism between (X,ξ) and $(f[X],\eta_{f[X]})$.

By *weakly computable cb$_0$-base structure* we mean a pair (X,β) where X is a non-empty set of points and $\beta:\omega\to P(X)$ is a numbering of a base of a T_0-topology in X such that there is a c.e. sequence $\{A_{ij}\}$ with $\beta_i\cap\beta_j=\bigcup\beta[A_{ij}]$ for all $i,j\geq 0$. It is easy to check that any weakly computable cb$_0$-space is a weakly computable cb$_0$-base structure, and any cb$_0$-base structure (X,β) induces the weakly computable cb$_0$-space (X,β^*) where $\beta^*(n)=\bigcup\beta[W_n]$. We say that a weakly computable cb$_0$-base structure (X,β) *induces a weakly computable cb$_0$-space* (X,τ) if $\tau\equiv\beta^*$.

By *c.e. cb$_0$-space* we mean a weakly computable cb$_0$-space (X,τ) such that the predicate $\tau_n\neq\emptyset$ is c.e. The notion of c.e. cb$_0$-base structure is obtained by a similar strengthening of the notion of weakly computable cb$_0$-base structure. Note that if (X,β) is a c.e. cb$_0$-base structure then (X,β^*) is a c.e. cb$_0$-space. Similar spaces were introduced and studied in [9,12,21] under different names. For such spaces one can naturally define the notions of a computable function and show that the computable functions coincide with the effectively continuous ones.

Any computable metric space (X, d, ν) [23] gives rise to a c.e. cb_0-base structure (X, β) were $\beta_{\langle m,n\rangle} = B(\nu_m, \varkappa_m)$ is the basic open ball with center ν_m and radius $\varkappa_m$ ($\varkappa$ is a computable numbering of the rationals). Note that the recursively presented metric spaces from [15] are computable metric spaces but not vice versa; nevertheless, they give rise to the same class of Polish spaces [6]. By *computable Polish space* we mean a c.e. cb_0-space (X, τ) induced by a computable complete metric space (X, d, ν), i.e. $\tau \equiv \beta^*$. Most of the popular Polish spaces are computable.

By *computable ω-continuous domain* [1] we mean a pair $(X, \{b_n\})$ where X is an ω-continuous domain and $\{b_n\}$ is a numbering of a (domain) base in X modulo which the approximation relation $\ll$ is c.e. Any computable ω-continuous domain $(X, \{b_n\})$ gives rise to a c.e. cb_0-base structure (X, β) where $\beta_n = \{x \mid b_n \ll x\}$. Most of the popular ω-continuous domains are computable.

As we will see below, both computable Polish spaces and computable ω-continuous domains have some attractive effective DST-properties. In contrast, arbitrary c.e. cb_0-spaces seem too general to admit a reasonable effective DST. Thus, it makes sense to look for a subclass of c.e. cb_0-spaces with good effective DST that contains both computable Polish spaces and computable ω-continuous domains. A similar problem in classical DST was resolved by M. de Brecht [4] suggested the important notion of a quasi-Polish space, so it makes sense to search for a natural effective version of quasi-Polish spaces. A reasonable candidate was suggested in [2]. A *convergent approximation space* is a triple $(X, \mathcal{B}, \ll)$ consisting of a T_0-space X and a binary relation $\ll$ on a basis $\mathcal{B}$ such that for all $U, V, T \in \mathcal{B}$: $U \ll V$ implies $V \subseteq U$, $U \subseteq T$ and $U \ll V$ imply $T \ll V$, for any $x \in U$ there is $W \in \mathcal{B}$ with $x \in W \gg U$, any sequence $U_0 \ll U_1 \ll \cdots$ is a neighborhood basis of some point. By *effective convergent approximation space* we mean a triple $(X, \beta, \ll)$, $\beta : \omega \to P(X)$, where $(X, \beta[\omega], \ll)$ is a convergent approximation space such that the relation $\beta_m \ll \beta_n$ is c.e. and any β_n is non-empty. In particular, (X, β) is a c.e. cb_0-base structure. We immediately obtain the following effectivization of Proposition 3.5 in [2].

Proposition 1. *Computable Polish spaces and computable ω-continuous domains can be naturally considered as effective convergent approximation spaces.*

Proof. If X is computable Polish, choose a compatible computable metric space (X, d, ν), and let β be the numbering of basic open balls. For such balls U, V, let $U \ll V$ iff the closure of V is contained in U and $diam(V) \leq diam(U)/2$. Then $(X, \beta \ll)$ is an effective convergent approximation space.

If $X$is a computable ω-continuous domain, choose a domain basis $\{b_n\}$ such that "$b_m \ll b_n$" is c.e. and all basic open sets $\beta_n := \{x \mid b_n \ll x\}$ are non-empty. For such basic opens U, V, let $U \ll V$ iff $V = \{x \mid c \ll x\}$ for some $c \in U$. Then $(X, \beta, \ll)$ is an effective convergent approximation space. □

Note that if (X, ξ) and (Y, η) are weakly computable cb$_0$-base structures then so is also $(X \times Y, \xi \times \eta)$ where $(\xi \times \eta)_{\langle m,n \rangle} = \xi_m \times \eta_n$, and the product topology on $X \times Y$ is induced by the basis $(\xi \times \eta)[\omega]$. One easily checks that all classes of effective spaces defined in this section are closed under this cartesian product.

3 Effective Hierarchies in Weakly Computable Cb$_0$-Spaces

In any weakly computable cb$_0$-space (X, τ) one can naturally define [21] effective versions of the classical hierarchies of DST [11,15,22] denoting, as usual, levels of the effective hierarchies in the same manner as levels of the corresponding classical hierarchies, using the lightface letters Σ, Π, Δ instead of the boldface $\mathbf{\Sigma}, \mathbf{\Pi}, \mathbf{\Delta}$ used for the classical hierarchies. We are not completely precise here, in the precise version we have to define levels of the hierarchies together with their "canonical" numberings (an alternative apparently equivalent approach to the effective Borel hierarchy uses the so called effective Borel codes [14]), [15].

First let us sketch the definition of the effective Borel hierarchy. *Finite effective Borel hierarchy* in (X, τ) is the sequence $\{\Sigma^0_n(X)\}_{n<\omega}$ defined as follows: $\Sigma^0_0(X) = \{\emptyset\}$; $\Sigma^0_1(X)$ is the class of effective open sets equipped with the numbering τ; $\Sigma^0_2(X)$ is the class of sets $\bigcup \beta[W_x]$, $x \geq 0$, where $\beta : \omega \to P(X)$ is the numbering of finite Boolean combinations of effective open sets induced by τ and a Gödel numbering of Boolean terms; $\Sigma^0_n(X)$ $(n \geq 3)$ is the class of sets $\bigcup \gamma[W_x]$, $x \geq 0$, where γ is the numbering of $\Pi^0_{n-1}(X)$ induced by the numbering of $\Sigma^0_{n-1}(X)$ existing by induction.

The transfinite extension of $\{\Sigma^0_n(X)\}_{n<\omega}$ is also defined in a natural way. In place of ω_1 in classical DST one has to take the first non-computable ordinal ω_1^{CK}. In fact, to obtain reasonable effectivity properties one should enumerate levels $\Sigma^0_{(a)}$ of the transfinite hierarchy not by computable ordinals $\alpha < \omega_1^{CK}$ but rather by their names $|a|_O = \alpha$ in the Kleene notation system $(O; <_O)$ $(a \mapsto |a|_O$ is a surjection from $O \subseteq \omega$ onto $\omega_1^{CK})$, see [17]. Levels of the transfinite version are defined in the same way as for the finite levels, using the effective induction along the well-founded set $(O; <_O)$. In this way we obtain the *effective Borel hierarchy* $\{\Sigma^0_{(a)}(X)\}_{a \in O}$ *in* $X = (X, \tau)$. This hierarchy is extensional, i.e. $\Sigma^0_{(a)}(X) = \Sigma^0_{(b)}(X)$ whenever $|a|_O = |b|_O$, so it may be sometimes easier denoted as $\{\Sigma^0_\alpha(X)\}_{\alpha < \omega_1^{CK}}$.

For every ordinal α, define the operation D_α sending sequences of sets $\{A_\beta\}_{\beta<\alpha}$ to sets by

$$D_\alpha(\{A_\beta\}_{\beta<\alpha}) = \bigcup \{A_\beta \setminus \bigcup_{\gamma<\beta} A_\gamma \mid \beta < \alpha,\, r(\beta) \neq r(\alpha)\}$$

where $r : \alpha \to \{0, 1\}$ is the parity function distinguishing even and odd ordinals. For any ordinal α and class of sets $\mathcal{C}$, let $D_\alpha(\mathcal{C})$ be the class of sets $D_\alpha(\{A_\beta\}_{\beta<\alpha})$, where $A_\beta \in \mathcal{C}$ for all $\beta < \alpha$.

The *effective Hausdorff hierarchy* $\{\Sigma^{-1,\alpha}_{(a)}(X)\}_{a\in O}$ *over* $\Sigma^0_\alpha(X)$ is defined as follows: $\Sigma^{-1,\alpha}_{(a)}(X)$ is the class of sets of the form $D_{|a|}(\{A_b\}_{b<_O a})$, where $\{A_b\}_{b<_O a}$ ranges over the $\Sigma^0_\alpha(X)$-computable sequences (naturally identified with sequences $\{A_\beta\}_{\beta<|a|}$). For $\alpha = 1$, we abbreviate $\Sigma^{-1,\alpha}_{(a)}(X)$ to $\Sigma^{-1}_{(a)}(X)$. WARNING: the effective Hausdorff hierarchy is not extensional.

The *effective Luzin hierarchy* is the family of pointclasses $\{\Sigma^1_n\}_{n<\omega}$ defined by induction as follows: $\Sigma^1_0(X) = \Sigma^0_2(X)$, $\Sigma^1_{n+1}(X) = \{pr_X(A) \mid A \in \Pi^1_n(\mathcal{N} \times X)\}$ where $pr_X(A)$ is the projection of A along the $\mathcal{N}$-axis. In this way we obtain the sequence $\{\Sigma^1_n(X)\}$ of pointclasses in any weakly computable cb$_0$-space $X = (X, \tau)$.

The introduced hierarchies have many properties well known in particular cases from effective DST [15,17]: the natural inclusions of levels of any given hierarchy, the mutual inclusions between levels of different hierarchies, the closure of any level under certain set-theoretic operations and under preimages of effectively continuous functions. For a future reference, we only give an example of such a property related to subspaces.

Proposition 2. *Let* (X, τ) *be a weakly computable cb*$_0$*-space,* (Y, τ_Y) *a subspace, and* Γ *a level of an introduced hierarchy. Then* $\Gamma(Y) = \{Y \cap A \mid A \in \Gamma(X)\}$.

We also give the following effective version of a result in [4].

Proposition 3. *For any weakly computable cb*$_0$*-space* (X, τ)*, the equality relation* $=_X$ *on* X *is in* $\Pi^0_2(X \times X)$.

Proof. For any $x, y \in X$ we have: $x \neq y$ iff there is some n such that $x \in \tau_n \not\ni y$ or $y \in \tau_n \not\ni x$, i.e. $\neq_X$ coincides with $\bigcup_n(((\tau_n \times X)\setminus(X \times \tau_n)) \cup ((X \times \tau_n)\setminus(\tau_n \times X)))$. By the definition of $X \times X$, $\neq_X$ is in $\Sigma^0_2(X \times X)$. □

4 On the Effective Descriptive Theory of Functions

For levels Γ, E of the effective Borel hierarchy and for any weakly computable cb$_0$-spaces X, Y, let $\Gamma E(X, Y)$ (resp. $\Gamma E[X, Y]$) denote the class of functions $f : X \to Y$ such that $f^{-1}(B) \in \Gamma(X)$ for each $B \in E(Y)$ effectively in B, (resp. $f[A] \in E(Y)$ for each $A \in \Gamma(X)$ effectively in A). In the case $\Gamma = E$ we abbreviate $\Gamma E(X, Y)$ to $\Gamma(X, Y)$ and $\Gamma E[X, Y]$ to $\Gamma[X, Y]$.

The introduced notions are effective versions of the corresponding notions from [16] and include some notions already considered in Computable Analysis (see e.g. [3,23]. In particular, $\Sigma^0_1(X, Y)$ is the class of effectively continuous functions, $\Sigma^0_1[X, Y]$ is the class of effectively open functions, $\Sigma^0_2\Sigma^0_1(X, Y)$ is the class of effectively Σ^0_2-measurable (or effective Baire class 1) functions.

Our first result is an effective version of the classical fact that any Polish space is a continuous open image of the Baire space [22, Theorem 1.3.7] (for Polish spaces a closely related fact was announced in [7]).

Theorem 1. *Let X be a computable Polish space or a computable ω-continuous domain. Then there exist functions $f : \mathcal{N} \to X$ and $s : X \to \mathcal{N}$ such that $f \circ s = id_X$, $f \in \Sigma^0_1(\mathcal{N}, X) \cap \Sigma^0_1[\mathcal{N}, X]$, and $s \in \Sigma^0_2\Sigma^0_1(X, \mathcal{N}) \cap \Pi^0_2[X, \mathcal{N}]$.*

Proof. Let $(X, \beta, \ll)$ be the effective convergent approximation space for X from the proof of Proposition 1. Since the relation "$\beta_m \ll \beta_n$" is c.e., there is a computable function $g : \omega^+ \to \omega$ such that $g(n) = n$ for each $n < \omega$ and $\{g(\sigma n) \mid n < \omega\} = \{m \mid \beta_{g(\sigma)} \ll \beta_m\}$ for each $\sigma \in \omega^+$. In particular, for the sets $U_\sigma := \beta_{g(\sigma)}$ we have $X = \bigcup_n U_n$, $U_\sigma = \bigcup_n U_{\sigma n}$, and $U_\sigma \ll U_{\sigma n}$.

For any $p \in \mathcal{N}$, let $f(p) \in X$ be the unique element with the neighborhood base $\{U_{p[n+1]}\}$ [2] where $p[m] := (p(0) \cdots p(m-1))$. Note that if X is computable Polish then $f(p) = lim_n x_n$ (where x_n is the center of the ball $U_{p[n+1]}$), and if X is a computable ω-continuous domain then $f(p) = sup\{b_{g(p)[n+1]} \mid n < \omega\}$ (obviously, $b_{g(p[1])} \ll b_{g(p[2])} \ll \cdots$). Therefore, $f : \mathcal{N} \to X$ is computable, hence $f \in \Sigma^0_1(\mathcal{N}, X)$.

For any $x \in X$, define $p = s(x) \in \mathcal{N}$ as follows. If X is computable Polish then (by induction on i) $p(i) := \mu n(x \in U_{p[i]n})$, and if X is a computable ω-continuous domain then $p(i) := \mu n(x \in U_{p[i]n} \wedge \forall j < i(b_j \ll x \to b_j \in U_{p[i]n}))$. Then we clearly have $f \circ s = id_X$, in particular f is surjective (in the "Polish case" this is obvious while in the "ω-continuous case" the second conjunction summand guarantees that $x = sup\{b_{g(s(x))[n+1]} \mid n < \omega\} = f(s(x))$). The same argument shows that $f[\sigma \cdot \mathcal{N}] = U_\sigma$ for each $\sigma \in \omega^+$, hence $f \in \Sigma^0_1[\mathcal{N}, X]$.

Furthermore, $s^{-1}(\sigma \cdot \mathcal{N})$ is a finite Boolean combination of basic open sets in X that may be explicitly written for any given $\sigma \in \omega^+$, namely $s^{-1}(\sigma \cdot \mathcal{N}) = U_\sigma \setminus A$ where A is the union of U_ρ for all ρ lexicographically less than σ. It follows that $s \in \Sigma^0_2\Sigma^0_1(X, \mathcal{N})$.

Obviously, $s[X] = \{p \in \mathcal{N} \mid s(f(p)) = p\}$, hence $s[X] = \{p \mid \forall i(p(i) = \mu n(f(p) \in U_{p[i]n}))\}$ in the "Polish case" and

$$s[X] = \{p \mid \forall i(p(i) = \mu n(f(p) \in U_{p[i]n} \wedge \forall j < i(b_j \ll f(p) \to b_j \in U_{p[i]n}))\}$$

in the "ω-continuous case". Since f is computable and $\{U_\sigma\}_{\sigma \in \omega^+}$ is Σ^0_1-computable, $s[X] \in \Pi^0_2(\mathcal{N})$. For any $A \subseteq X$, we have $s[A] = \{p \in s[X] \mid f(p) \in A\} = s[X] \cap f^{-1}(A)$. If now $A \in \Pi^0_2(X)$ then $f^{-1}(A) \in \Pi^0_2(\mathcal{N})$, hence $s[A] \in \Pi^0_2(\mathcal{N})$. It follows that $s \in \Pi^0_2[X, \mathcal{N}]$. □

For weakly computable cb$_0$-spaces X and Y, we say that X is an *effective retract* of Y iff there exist effectively continuous functions $s : X \to Y$ (called a section) and $r : Y \to X$ (called a retraction) such that $r \circ s = id_X$. We will use the following

Proposition 4. *Let (X, ξ) and (Y, η) be weakly computable cb$_0$-spaces and Γ a non-zero level of the effective Borel hierarchy.*

(1) *If $f : X \to Y$ is a weakly computable embedding with $f[X] \in \Gamma(Y)$ then $f \in \Gamma[X, Y]$.*
(2) *If X is an effective retract of Y via a section-retraction pair (s, r) then $s \in \Pi^0_2[X, Y]$.*

Proof. 1. Let $A \in \Gamma(X)$. Since (X, ξ) is effectively homeomorphic to the subspace $(f[X], \eta_{f[X]})$, $f[A] \in \Gamma(f[X])$. By Proposition 2, $f[A] = f[X] \cap S$ for some $S \in \Gamma(Y)$. Since Γ is closed under intersection, $f[A] \in \Gamma(Y)$.

2. First we check that s is a weakly computable embedding, i.e. $\lambda n.s[\xi_n] \equiv \eta_{s[X]} = \lambda n.s[X] \cap \eta_n$. Since $s[\xi_n] = s[X] \cap f^{-1}(\xi_n)$, $\lambda n.s[\xi_n] \leq \eta_{s[X]}$ via a computable function g satisfying $f^{-1}(\xi_n) = \eta_{g(n)}$. Since $s[s^{-1}(\eta_n)] = s[X] \cap \eta_n$, $\eta_{s[X]} \leq \lambda n.s[\xi_n]$ via a computable function h satisfying $s^{-1}(\eta_n) = \xi_{h(n)}$, s is really an effective embedding. Since $s[X] = \{y \in Y \mid sr(y) = y\}$, $s[X] \in \Pi^0_2(Y)$ by Proposition 3. It remains to use item 1. □

Our second result is an effective version of the classical fact that any perfect Polish space contains a homeomorphic copy of the Cantor (or Baire) space (see e.g. [11]):

Theorem 2. *Let X be a perfect computable Polish space, or a computable reflective ω-algebraic domain, or a computable 2-reflective ω-algebraic domain. Then there exists a weakly computable embedding $g : \mathcal{C} \to X$ such that $g \in \Sigma^0_1(\mathcal{C}, X) \cap \Pi^0_2[\mathcal{C}, X]$. The same holds with $\mathcal{N}$ in place of $\mathcal{C}$.*

Proof. In the "Polish case", there is a basic open ball B which is perfect in the subspace topology. Clearly, there is a computable sequence $\{B_\sigma\}_{\sigma \in 2^*}$ of basic open balls such that $B_\emptyset = B$, $B_{\sigma 0} \cap B_{\sigma 1} = \emptyset$, the closure $\bar{B}_{\sigma i}$ of $B_{\sigma i}$ is contained in B_σ, and $diam(B_{\sigma i}) \leq diam(B_\sigma)/2$.

For any $p \in \mathcal{C}$, let $g(p)$ be the unique element of $\bigcap_n B_{p[n]}$. Then $g : \mathcal{C} \to X$ is a computable topological embedding, hence $g \in \Sigma^0_1(\mathcal{C}, X)$. Since

$$g[\mathcal{C}] = \bar{B}_\emptyset \cap (\bar{B}_0 \cup \bar{B}_1) \cap (\bar{B}_{00} \cup \bar{B}_{01} \cup \bar{B}_{10} \cup \bar{B}_{11}) \cap \cdots$$

and $\bar{B}_\sigma \in \Pi^0_1(X)$ uniformly in σ, $g[\mathcal{C}] \in \Pi^0_1(X)$. By Proposition 4, $g \in \Pi^0_1[\mathcal{C}, X]$, and also $g \in \Pi^0_2[\mathcal{C}, X]$. Since there is a weakly computable embedding $h : \mathcal{N} \to \mathcal{C}$ with $h[\mathcal{N}] \in \Pi^0_2(\mathcal{C})$), $h \in \Pi^0_2[\mathcal{N}, \mathcal{C}]$ by Proposition 4. Thus, $g \circ h \in \Pi^0_2[\mathcal{N}, X]$.

In the "reflective case" we use the result in [19,20] that the domain $\omega^{\leq\omega}$ is an effective retract of X, let $s : \omega^{\leq\omega} \to X$ be the corresponding effectively continuous section. By Proposition 4, $s \in \Pi^0_2[\omega^{\leq\omega}, X]$. The inclusion $i : \mathcal{N} \to \omega^{\leq\omega}$ is a weakly computable embedding such that $i[\mathcal{N}] \in \Pi^0_2(\omega^{\leq\omega})$, hence $s \circ i \in \Pi^0_2[\mathcal{N}, X]$. Since $\mathcal{C}$ is an effective retract of $\mathcal{N}$, the assertion also holds for the Cantor space.

For the "2-reflective case" the proof is almost the same as for the "reflective case". We use the result in [19,20] that the domain $\omega^{\leq\omega\top}$ obtained from $\omega^{\leq\omega}$ by joining a top element $\top$, is an effective retract of X. The inclusion $i : \mathcal{N} \to \omega^{\leq\omega\top}$ is again a weakly computable embedding with $i[\mathcal{N}] \in \Pi^0_2[\omega^{\leq\omega\top}]$. □

We call weakly computable cb$_0$-spaces X, Y *Γ-isomorphic* iff there is a Γ-isomorphism between them, i.e. a bijection $f : X \to Y$ from $\Gamma(X, Y) \cap \Gamma[X, Y]$. Below we use this notion for the pointclass $\Gamma = \Delta^0_{<\omega} := \bigcup_{n<\omega} \Sigma^0_n$.

Our third result is an effective version of the classical fact that any two uncountable Polish spaces are Borel isomorphic (see e.g. [11]):

Theorem 3. *Let X be a perfect computable Polish space, or a computable reflective ω-algebraic domain, or a computable 2-reflective ω-algebraic domain. Then X is $\Delta^0_{<\omega}$-isomorphic to $\mathcal{N}$.*

Proof. By Theorem 1, there is an injection $s : X \to \mathcal{N}$ such that $s \in \Sigma^0_2\Sigma^0_1(X, \mathcal{N}) \cap \Pi^0_2[X, \mathcal{N}]$. By Theorem 2, there is an injection $g : \mathcal{N} \to X$ such that $g \in \Sigma^0_1(\mathcal{N}, X) \cap \Pi^0_2[\mathcal{N}, X]$. Let h be the bijection between $\mathcal{N}$ and X obtained from g, s by the standard Schröder-Bernstein back-and-fourth argument. One easily checks that h is a desired $\Delta^0_{<\omega}$-isomorphism. □

5 The Suslin-Kleene Theorem

We say that a weakly computable cb$_0$-space X *satisfies the Suslin-Kleene theorem* iff $\bigcup\{\Sigma^0_\alpha(X) \mid \alpha < \omega_1^{CK}\} = \Delta^1_1(X)$ (since the inclusion from left to right holds for any X, the condition is equivalent to $\bigcup\{\Sigma^0_\alpha(X) \mid \alpha < \omega_1^{CK}\} \supseteq \Delta^1_1(X)$). Which weakly computable cb$_0$-spaces satisfy the Suslin-Kleene theorem? According to classical results of Kleene [17], $\omega, \mathcal{N}$ are among these spaces. The next theorem extends this to many natural spaces but first we establish the following:

Proposition 5. *Let (X, ξ) and (Y, η) be weakly computable cb$_0$-spaces, $f : X \to Y$ a function in $\Delta^0_{<\omega}(X, Y)$ and Γ an infinite level of the effective Borel hierarchy or a non-zero level of the effective Luzin hierarchy. Then $A \in \Gamma(Y)$ implies $f^{-1}(A) \in \Gamma(X)$ effectively w.r.t. the canonical numberings*

Proof. The case $\Gamma \in \{\Sigma^0_\alpha, \Pi^0_\alpha \mid \omega \leq \alpha < \omega_1^{CK}\}$ is checked by a straightforward induction on $\alpha \geq \omega$. The case $\Gamma \in \{\Sigma^1_n, \Pi^1_n \mid 1 \leq n < \omega\}$ is checked by a straightforward induction on n, so we give some details only for $\Gamma = \Sigma^1_1$. Let $A \in \Sigma^1_1(Y)$, then $A = pr_Y(B)$ for some $B \in \Pi^0_2(\mathcal{N} \times Y)$, then $f^{-1}(A) = pr_X g^{-1}(B)$ for some $g \in \Delta^0_{<\omega}(\mathcal{N} \times X, \mathcal{N} \times Y)$. Since $g^{-1}(B) \in \Delta^0_{<\omega}(\mathcal{N} \times X) \subseteq \Sigma^1_1(\mathcal{N} \times X)$ and Σ^1_1 is closed under the projection along $\mathcal{N}$-axis, $f^{-1}(A) \in \Sigma^1_1(X)$. □

Theorem 4. *Let X be a perfect computable Polish space, or a computable reflective ω-algebraic domain, or a computable 2-reflective ω-algebraic domain. Then X satisfies the Suslin-Kleene theorem.*

Proof. By Theorem 3, there is a $\Delta^0_{<\omega}$-isomorphism $h : \mathcal{N} \to X$. Let $A \in \Delta^1_1(X)$. By Proposition 5, $h^{-1}(A) \in \Delta^1_1(\mathcal{N})$, hence $h^{-1}(A) \in \Sigma^0_\alpha(\mathcal{N})$ for some infinite computable ordinal α. By Proposition 5, $A \in \Sigma^0_\alpha(X)$. □

As it follows from Theorem 1.2.(9) in [15], the Suslin-Kleene Theorem holds in any computable Polish space.

6 The Effective Hausdorff Theorem

We say that a weakly computable cb$_0$-space X *satisfies the effective Hausdorff theorem* iff $\Delta^0_2(X) = \bigcup\{\Sigma^{-1}_{(a)}(X) \mid a \in O\}$. Since the inclusion from right to left holds for any X, the equality is equivalent to the converse inclusion. Here we investigate which weakly computable cb$_0$-spaces satisfy the effective Hausdorff theorem. We need the following easy fact.

Proposition 6. *Let X, Y be weakly computable cb_0-spaces, $f : X \to Y$ an effectively continuous, effectively open surjection, and X satisfy the effective Hausdorff theorem. Then Y satisfies the effective Hausdorff theorem.*

Proof. We have to show the inclusion $\Delta_2^0(Y) \subseteq \bigcup\{\Sigma_{(a)}^{-1}(Y) \mid a \in O\}$. Let $A \in \Delta_2^0(Y)$. Since f is effectively continuous, $f^{-1}(A) \in \Delta_2^0(X)$, hence $f^{-1}(A) \in \Sigma_{(a)}^{-1}(X)$ for some $a \in O$, i.e. $f^{-1}(A) = \bigcup_{b<_O a}\{B_b \setminus \bigcup_{c<_O b} B_c \mid r(b) \neq r(a)\}$ for some $\Sigma_1^0(X)$-computable sequence $\{B_b\}_{b<_O a}$. Since f is effectively open, the sequence $\{A_b\}_{b<_O a}$, $A_b := f[B_b]$, is $\Sigma_1^0(Y)$-computable, so it suffices to check that $A = \bigcup_{b<_O a}\{A_b \setminus \bigcup_{c<_O b} A_c \mid r(b) \neq r(a)\}$. Let $y \in A$, and choose $x \in f^{-1}(y)$. Then $x \in f^{-1}(A)$, hence $x \in \bigcup_{b<_O a} B_b$, hence $y \in \bigcup_{b<_O a} A_b$. Choose the smallest (w.r.t. $<_O$) $b <_O a$ with $y \in A_b$, then it remains to check that $r(b) \neq r(a)$. Suppose toward a contradiction that $r(b) = r(a)$. Since $A_b = f[B_b]$, $y = f(x')$ for some $x' \in B_b$. Since $x' \in f^{-1}(y) \subseteq f^{-1}(A)$, $x' \in B_d$ for some $d <_O b$. Then $y \in f(B_d) = A_d$, contradicting the choice of b.

Conversely, let $y \in Y \setminus A$. If $y \notin \bigcup_{b<_O a} A_b$ we are done. Otherwise, the argument of the previous paragraph applies. □

Since $\mathcal{N}$ satisfies the effective Hausdorff theorem by [18] and any computable Polish space (as well as any computable ω-continuous domain) is an effectively continuous and effectively open image of $\mathcal{N}$ by Theorem 1, the next result is an immediate corollary of Proposition 6.

Theorem 5. *Let X be a computable Polish space or a computable ω-continuous domain. Then X satisfies the effective Hausdorff theorem.*

In [5] the last result was proved for the space ω, in [18] the fact was established for the Baire space, in [10] it was obtained for the finite-dimensional Euclidean spaces, and for the computable Polish spaces the result was established in [2]. Our proof here is different from and shorter than the proof in [2]. The case of ω-continuous domain was left open in [2].

7 Conclusion

The effective DST is still in its early stage, in particular there are many open questions related to this paper. E.g., the "right" computable version of quasi-Polish space is still not completely clear. Also, the status of the effective Hausdorff-Kuratowski theorem seems to be widely open.

Acknowledgement. I thank the anonymous referees for useful comments and bibliographical hints.

References

1. Abramsky, S., Jung, A.: Domain theory. In: Abramsky, S., Gabbay, D.M., Maibaum, T.S.E. (eds.) Handbook of Logic in Computer Science, vol. 3, pp. 1–168. Clarendon Press, Oxford (1994)

2. Becher, V., Grigorieff, S.: Borel and Hausdorff hierarchies in topological spaces of Choquet games and their effectivization. Math. Struct. Comput. Sci. Available on CJO (2014). doi:10.1017/S096012951300025X
3. Brattka, V.: Effective Borel measurability and reducibility of functions. Math. Logic Q. **51**(1), 19–44 (2005)
4. de Brecht, M.: Quasi-Polish spaces. Ann. Pure Appl. Logic **164**, 356–381 (2013)
5. Ershov, Y.L.: On a hierarchy of sets 1,2,3 (in Russian). Algebra i Logika **7**(4), 15–47 (1968)
6. Gregoriades, V., Kispeter, T., Pauly, A.: A comparison of concepts from computable analysis and effective descriptive set theory. Mathematical structures in computer science (submitted to). arxiv.org/pdf/1403.7997
7. Gregoriades V.: Effective refinements of classical theorems in descriptive set theory (informal presentation). Talks at Conference "Computability and Complexity in Analysis" in Nancy (2013). http://cca-net.de/cca2013/slides/
8. Gregoriades, V.: Classes of polish spaces under effective Borel isomorphism. Mem. Amer. Math. Soc. (to appear). www.mathematik.tu-darmastadt.de/gregoriages/papers/eff.webpage.pdf
9. Grubba, T., Schröder, M., Weihrauch, K.: Computable metrization. Math. Logic Q. **53**, 381–395 (2007)
10. Hemmerling, A.: The Hausdorff-Ershov hierarchy in Euclidean spaces. Arch. Math. Logic **45**, 323–350 (2006)
11. Kechris, A.S.: Classical Descriptive Set Theory. Graduate Texts in Mathematics, vol. 156. Springer, New York (1995)
12. Korovina, M.V., Kudinov, O.V.: Basic principles of Σ-definability and abstract computability. Schriften zur Theoretischen Informatik, Bericht Nr 08–01, Univeversität Siegen (2008)
13. Louveau, A.: Recursivity and compactness. In: Müller, G.H., Scott, D.S. (eds.) Higher Set Theory. Lecture Notes in Mathematics, vol. 669, pp. 303–337. Springer, Heidelberg (1978)
14. Louveau, A.: A separation theorem for Σ^1_1-sets. Trans. Amer. Math. Soc. **260**(2), 363–378 (1980)
15. Moschovakis, Y.N.: Descriptive Set Theory. North Holland, Amsterdam (2009)
16. Motto Ros, L., Schlicht, P., Selivanov, V.: Wadge-like reducibilities on arbitrary quasi-Polish spaces. Mathematical structures in computer science (to appear). arXiv:1304.1239 [cs.LO]
17. Rogers Jr, H.: Theory of Recursive Functions and Effective Computability. McGraw-Hill, New York (1967)
18. Selivanov, V.L.: Wadge degrees of ω-languages of deterministic Turing machines. Theor. Inf. Appl. **37**, 67–83 (2003)
19. Selivanov, V.L.: Variations on the Wadge reducibility. Siberian Adv. Math. **15**, 44–80 (2005)
20. Selivanov, V.L.: Towards a descriptive set theory for domain-like structures. Theor. Comput. Sci. **365**(3), 258–282 (2006)
21. Selivanov, V.L.: On the difference hierarchy in countably based T_0-spaces. Electron. Notes Theor. Comput. Sci. **221**, 257–269 (2008)
22. Gao, S.: Invariant Descriptive Set Theory. CRC Press, New York (2009)
23. Weihrauch, K.: Computable Analysis. Springer, Berlin (2000)

On Computability of Navier-Stokes' Equation

Shu Ming Sun[1], Ning Zhong[2], and Martin Ziegler[3](✉)

[1] Virginia Tech, Blacksburg, USA
sun@math.vt.edu
[2] University of Cincinnati, Cincinnati, USA
zhongn@ucmail.uc.edu
[3] TU Darmstadt, Darmstadt, Germany
ziegler@mathematik.tu-darmstadt.de
http://m.zie.de

Abstract. We approach the question of whether the Navier-Stokes Equation admits recursive solutions in the sense of Weihrauch's Type-2 Theory of Effectivity: A suitable encoding ("representation") is carefully constructed for the space of solenoidal vector fields in the L_q sense over the d-dimensional open unit cube with zero boundary condition. This is shown to render both the Helmholtz projection and the semigroup generated by the Stokes operator uniformly computable in the case $q = 2$.

1 Introduction

The (physical) Church-Turing Hypothesis [10] postulates that every physical phenomenon or effect can, at least in principle, be simulated by a sufficiently powerful digital computer up to arbitrarily prescribable precision. Its validity had been challenged, though, in the rigorous framework of Recursive Analysis: there is a computable $\mathcal{C}^1$ initial condition to the Wave Equation that leads to an incomputable solution [6,8]. The controversy was later resolved by demonstrating that, in the both physically [1,18] and mathematically more appropriate Sobolev space settings, the solution is computable uniformly in the initial data [13]. Recall that functions f in a Sobolev space are not defined pointwise but by local averages in the L_q sense [1] (e.g. $q = 2$ corresponding to energy) with derivatives understood in the distributional sense. This led to a series of investigations on the computability of linear and nonlinear partial differential equations [14–16].

The (incompressible) Navier-Stokes Equation

$$\partial_t \boldsymbol{u} - \triangle \boldsymbol{u} + (\boldsymbol{u} \cdot \nabla)\boldsymbol{u} + \nabla P = \boldsymbol{f}, \quad \nabla \cdot \ \boldsymbol{u} = 0, \qquad \boldsymbol{u}(0) = \boldsymbol{u}_0 \tag{1}$$

describes the motion of a viscous incompressible fluid filling a rigid box $\overline{\Omega}$. The vector field $\boldsymbol{u} = \boldsymbol{u}(x,t) = \left(u^{(1)}, u^{(2)}, \ldots, u^{(d)}\right)$ represents the velocity of the

The first author is partially supported by the National Science Foundation under grant No. DMS-1210979; the last author acknowledges support from German Research Foundation (DFG) under grant Zi 1009/4-1 and from IRTG 1529.

[1] We use $q \in [1, \infty]$ to denote the norm index, P for the pressure field, p for polynomials, $\mathcal{P}$ for sets of (tuples of) the latter, and $\mathbb{P}$ for the Helmholtz Projection.

A. Beckmann et al. (Eds.): CiE 2015, LNCS 9136, pp. 334–342, 2015.
DOI: 10.1007/978-3-319-20028-6_34

fluid and $P = P(\boldsymbol{x}, t)$ is the scalar pressure with gradient ∇P; $\nabla \cdot \boldsymbol{u}$ denotes componentwise divergence; $\boldsymbol{u} \cdot \nabla$ means, in Cartesian coordinates, $u^{(1)}\partial_x + u^{(2)}\partial_y + \ldots$; and the function $\boldsymbol{u}_0 = \boldsymbol{u}_0(\boldsymbol{x})$ describes the initial velocity and $\boldsymbol{f}$ a given external force. Equation (1) thus constitutes a system of $d+1$ partial differential equations for $d+1$ functions.

The question of global existence and smoothness of its solutions is one of the Millennium Prize Problems posted by the Clay Mathematics Institute at the beginning of the 21st century; cmp. [17]. Local strong existence in time has been established, though, over various spatial L_q settings [3]. Numerical solution methods are abundant, often based on pointwise (or even uniform, rather than L_q) approximation and struggling with computational artefacts [5]. In fact, the very last of seven open problems listed in the addendum to [7] asks for a "recursion theoretic study of [...] the Navier-Stokes equation". Moreover, it has been suggested [9] that a hydrodynamical system could in principle be incomputable in the sense that it allows simulation of universal Turing computation and thus 'solves' the Halting problem. Indeed, recent progress towards (a negative answer to) the Millennium Problem [11] proceeds by simulating a computational process in the vorticity dynamics to construct a blowup in finite time for a PDE similar to (1).

1.1 Overview

Using the rigorous framework of Recursive Analysis we approach the problem of computing, given an initial condition $\boldsymbol{u}_0$ and inhomogenity $\boldsymbol{f}$, the strong solution of (1) in the space $X_2(\Omega)$ of square-integrable vector fields $\boldsymbol{u}$ on the open unit cube $\Omega := (-1;1)^d$ with zero boundary condition $\boldsymbol{u}\big|_{\partial\Omega} \equiv 0$ which are divergence-free (aka solenoidal) in the sense that $\triangle \cdot \boldsymbol{u} \equiv 0$ holds. We follow a common strategy used in classical existence proofs over spaces $L_q(\Omega)$, cmp [2]:

(i) Eliminate the pressure P by (Helmholtz) projecting the equation onto the space $X_q(\Omega)$ of solenoidal (i.e. divergence-free) solutions;
(ii) solve the associated linear equation for the Stokes operator $\mathbb{A} := -\mathbb{P}\triangle$ using semigroup methods and spectral estimates;
(iii) extend (ii) to incorporate the nonlinearity, e.g. using an iterative approximation/Banach fixed point argument;
(iv) recover the pressure by solving a Poisson Problem.

The present work effectivizes steps (i) and (ii): We introduce a natural representation (in the sense of Weihrauch's TTE) of the space $X_q(\Omega)$ of solenoidal L_q functions on $\Omega = (-1;1)^d$ with zero boundary conditions; and derive a representation for all (total but not necessarily bounded) linear operators on this space. We employ spectral analysis in order to establish that the Stokes operator gives rise to a computable semigroup; hence showing that the solution to the Stokes Dirichlet problem is uniformly computable. We also show that the Helmholtz Projection $\mathbb{P} : \big(L_2(\Omega)\big)^2 \to X_2(\Omega)$ is computable.

2 Representing Divergence-Free Boundary-Free L_q Functions on Ω

Assume that Ω is the d-dimensional open cube $(-1;1)^d$, $d=2,3$. Let $X_q(\Omega)$ (or X_q if the context is clear) be the closure of the set $\{\boldsymbol{u} \in C_0^\infty(\Omega)^d : \nabla \cdot \boldsymbol{u} = 0\}$ in L_q-norm, $\mathcal{P}$ the set of all polynomials of d real variables with rational coefficients, and $\mathcal{P}^0_{\text{div}}$ the set of all d–tuples of polynomials in $\mathcal{P}$ which are divergence-free in Ω and vanish on $\partial\Omega$. Let us call a vector field $\boldsymbol{f}$ satisfying $\nabla \cdot \boldsymbol{f} = 0$ on Ω and $\boldsymbol{f} = 0$ on $\partial\Omega$ *divergence-free and boundary-free.* A vector-valued function $\boldsymbol{p}$ is called a polynomial of degree N if each of its components is a polynomial of degree less than or equal to N with respect to each variable and at least one component is a polynomial of degree N.

Proposition 1. *(a) $\mathcal{P}^0_{div}$ is dense in X_q (in L_q-norm).*
(b) A polynomial $\boldsymbol{p} = (p_1, p_2) = (\sum_{i,j=0}^N a^1_{i,j} x^i y^j, \sum_{i,j=0}^N a^2_{i,j} x^i y^j)$ belongs to $\mathcal{P}^0_{div}$ iff its coefficients satisfy the following systems of equations:

$$\begin{aligned} (i+1)a^1_{i+1,j} + (j+1)a^2_{i,j+1} &= 0, \quad 0 \le i,j \le N-1; \\ (i+1)a^1_{i+1,N} &= 0, \quad 0 \le i \le N-1; \\ (j+1)a^2_{N,j+1} &= 0, \quad 0 \le j \le N-1; \end{aligned} \tag{2}$$

and for all $0 \le i,j \le N$,

$$\sum_{i=0}^N a^1_{i,j} = \sum_{i=0}^N a^2_{i,j} = \sum_{i=0}^N (-1)^i a^1_{i,j} = \sum_{i=0}^N (-1)^i a^2_{i,j} = 0; \tag{3}$$

$$\sum_{j=0}^N a^1_{i,j} = \sum_{j=0}^N a^2_{i,j} = \sum_{j=0}^N (-1)^j a^1_{i,j} = \sum_{j=0}^N (-1)^j a^2_{i,j} = 0. \tag{4}$$

One may attempt to use $\mathcal{P}^0_{\text{div}}$ as a set of names for coding the elements in the space $X_q(\Omega)$. However the closure of $\mathcal{P}^0_{\text{div}}$ in L_q-norm contains $X_q(\Omega)$ as a proper subspace - already in the one-dimensional case, $1 - x^{2n}$ is boundary-free on $(-1;1)$ but its L_q–limit is not - $\mathcal{P}^0_{\text{div}}$ is therefore "too big"for representing $X_q(\Omega)$. We need to "trim" polynomials in $\mathcal{P}^0_{\text{div}}$ so that for any such modified convergent sequence its limit also belongs to $X_q(\Omega)$. Recall that $X_q(\Omega)$ is the closure in L_q-norm of all divergence-free C^∞ functions with compact supports in Ω. Thus, if we can modify the polynomials in $\mathcal{P}^0_{\text{div}}$ to meet the conditions that each modified polynomial is divergence-free on Ω with a compact support contained in Ω and the set of all modified polynomials is still dense in X_q, then these modified polynomials can be used as Cauchy codes for representing $X_q(\Omega)$.

Now for the details. For readability, we only show the modifying process in the case where $\Omega = (-1,1)^2$. The same argument applies to the three dimensional case. Let $\Omega_k = (-1+2^{-k}, 1-2^{-k})^2$, $k \in \mathbb{N}$, and $\Gamma_k = \partial\Omega_k$ be the boundary of Ω_k. We first use a scaling argument to "trim"the supports of the polynomials in $\mathcal{P}^0_{\text{div}}$ to compact sets contained in Ω. For each $k \ge 1$, each $\boldsymbol{p} = (p_1, p_2) \in \mathcal{P}^0_{\text{div}}$, and $j = 1$ or 2, define

$$G_{j,k,p}(x,y) = \begin{cases} p_j(\frac{x}{1-2^{-k}}, \frac{y}{1-2^{-k}}), -1+2^{-k} \leq x, y \leq 1-2^{-k} \\ 0, \qquad \text{otherwise} \end{cases} \tag{5}$$

Then $G_{j,k,p}$ and $\boldsymbol{G}_{k,p}(x,y) = \big(G_{1,k,p}(x,y), G_{2,k,p}(x,y)\big)$ have the following properties:

(a) $\boldsymbol{G}_{k,p}$ has compact support $\bar{\Omega}_k$ contained in Ω.
(b) $G_{j,k,p}$ is a polynomial of rational coefficients in Ω_k.
(c) $\boldsymbol{G}_{k,p}$ is continuous on $[-1,1]^2$.
(d) Since $\boldsymbol{p}$ vanishes on the boundary of Ω, it follows from (5) that $\boldsymbol{G}_{k,p}(x,y)$ vanishes in the exterior region of Ω_k, including its boundary $\partial\Omega_k$.
(e) $\boldsymbol{G}_{k,p}$ is divergence-free in Ω_k following the calculation below:

For $(x,y) \in \Omega_k$, we have that $(\frac{x}{1-2^{-k}}, \frac{y}{1-2^{-k}}) \in \Omega$ and

$$\begin{aligned} \frac{\partial G_{1,k,p}}{\partial x}(x,y) + \frac{\partial G_{2,k,p}}{\partial y}(x,y) &= \frac{1}{1-2^{-k}}\frac{\partial p_1}{\partial u}(u,v) + \frac{1}{1-2^{-k}}\frac{\partial p_2}{\partial v}(u,v) \\ &= \frac{1}{1-2^{-k}}\left[\frac{\partial p_1}{\partial u}(u,v) + \frac{\partial p_2}{\partial v}(u,v)\right] \quad = \quad 0 \end{aligned}$$

where $u = \frac{x}{1-2^{-k}}$ and $v = \frac{y}{1-2^{-k}}$. We note that the last equality holds because $\boldsymbol{p}$ is divergence-free in Ω.

Definition 2 *For each $k \in \mathbb{N}$, let $\mathcal{P}^{0,k}_{div} = \{\boldsymbol{G}_{k,p} \,|\, \boldsymbol{p} \in \mathcal{P}^0_{div}\}$.*

From the discussion above it follows that every function in $\mathcal{P}^{0,k}_{\text{div}}$ is a divergence-free polynomial of rational coefficients on Ω_k that vanishes in $[-1,1]^2 \setminus \Omega_k$ and is continuous on $[-1,1]^2$. However, although the functions in $\mathcal{P}^{0,k}_{\text{div}}$ are continuous on $[-1,1]^2$ and C^∞ in Ω_k, they can be non-differentiable on Ω. To use these functions as names for coding elements in $X_q(\Omega)$, they need to be smoothed along the boundary $\partial\Omega_k$ so that they become C^∞ on Ω. A standard technique for smoothing a function is to convolute it with a C^∞ function. We use this technique to modify functions in $\mathcal{P}^{0,k}_{\text{div}}$. Let

$$\gamma(x) = \begin{cases} \alpha e^{-\frac{1}{1-\|x\|^2}} \text{ if } \|x\| = \max\{|x_1|, |x_2|\} < 1, \\ 0 \qquad \text{otherwise,} \end{cases} \tag{6}$$

where α is a constant such that the integral $\int_{\mathbb{R}^2} \gamma(x)\,dx = 1$. The constant α is computable, since integration on continuous functions is computable [12]. Let $\gamma_k(x) = 2^{2k}\gamma(2^k x)$. Then, for all $k \in \mathbb{N}$, γ_k is a C^∞ function having for support the closed square $[-2^{-k}, 2^{-k}]^2$ and $\int_{\mathbb{R}^2} \gamma_k(x)\,dx = 1$. Let $\tilde{\mathcal{P}}_k = \{\gamma_n * \boldsymbol{p} \,|\, \boldsymbol{p} \in \mathcal{P}^{0,k}_{\text{div}}, n \geq k+1\}$, where $\gamma_n * \boldsymbol{p}$ is the convolution of γ_n and $\boldsymbol{p}$ defined componentwise as follows:

$$(\gamma_n * p)(x,y) = \int_{-1}^{1}\int_{-1}^{1} \gamma_n(x-s, y-t)\,p(s,t)\,ds\,dt$$

It is easy to see that the support of $\gamma_n * \boldsymbol{p}$ is contained in the closed square $[-1+2^{-(k+1)}, 1-2^{-(k+1)}]^2$. It is also known classically that $\gamma_n * \boldsymbol{p}$ is a $\mathcal{C}^\infty$ function. Since γ_n is a computable function and the integration is computable, the map $(n, \boldsymbol{p}) \mapsto \gamma_n * \boldsymbol{p}$ is computable.

Lemma 3 *Every function in* $\tilde{\mathcal{P}}_k = \{\gamma_n * \boldsymbol{p} \mid \boldsymbol{p} \in \mathcal{P}^{0,k}_{div},\ n \geq k+1\}$ *is divergence-free in* Ω, $k \in \mathbb{N}$.

Lemma 4 *The set* $\tilde{\mathcal{P}} = \bigcup_{k=1}^{\infty} \tilde{\mathcal{P}}_k$ *is dense in* $X_q(\Omega)$.

From Lemmas 3 and 4 it follows that $\tilde{\mathcal{P}}$ is a countable set that is dense in $X_q(\Omega)$ (in L_q-norm) and every function in $\tilde{\mathcal{P}}$ is C^∞, divergence-free in Ω, and having a compact support contained in Ω; in other words, $\tilde{\mathcal{P}} \subset \{\boldsymbol{u} \in C_0^\infty(\Omega)^d : \nabla \cdot \boldsymbol{u} = 0\}$, $d = 2$ or 3. Thus, $X_q(\Omega) =$ the closure of $\tilde{\mathcal{P}}$ in L_q-norm. This fact indicates that $\tilde{\mathcal{P}}$ is qualified to serve as codes for representing $X_q(\Omega)$. But, in order to get a "computable name" for the Stokes operator (the definition will be given shortly), we need to expand $\tilde{\mathcal{P}}$ as follows. For each $n \in \mathbb{N}$, since γ_n is computable and C^∞, the Laplace $\triangle\gamma_n$ is also computable and C^∞. We also note that, following from Lemma 3 and $(\triangle\gamma_n) * p = \triangle(\gamma_n * p)$, $(\triangle\gamma_n) * p$ is divergence-free. Let $\mathfrak{P} = \{\gamma_n * \boldsymbol{p},\ \triangle\gamma_n * \boldsymbol{p} : \boldsymbol{p} \in \mathcal{P}^{0,k}_{\mathrm{div}}, n \geq k+1\}$. Then $\mathfrak{P}$ is countable and is dense in X_q. It is this set $\mathfrak{P}$ which we use as a set of "codes." We now introduce the desired representation for $X_q(\Omega)$.

Since the function $\phi : \bigcup_{N=0}^{\infty} \mathbb{Q}^{(N+1)^2} \times \mathbb{Q}^{(N+1)^2} \to \{0,1\}$, where

$$\phi\big((r_{i,j})_{0\leq i,j\leq N}, (s_{i,j})_{0\leq i,j\leq N}\big) = \begin{cases} 1 \text{ if (2), (3), and (4) are satisfied} \\ \quad \text{(with } r_{i,j} = a^1_{i,j} \text{ and } s_{i,j} = a^2_{i,j}) \\ 0 \text{ otherwise} \end{cases}$$

is computable, there is a total computable function on $\mathbb{N}$ that enumerates $\mathcal{P}^0_{\mathrm{div}}$. Then it follows from the definition of $\tilde{\mathcal{P}}$ that there is a total computable function $\alpha : \mathbb{N} \to \mathfrak{P}$ that enumerate $\mathfrak{P}$; thus $(X_q, d, \mathfrak{P}, \alpha)$ is a computable metric space with $d(\boldsymbol{f}, \boldsymbol{g}) = \|\boldsymbol{f} - \boldsymbol{g}\|_{L_q}$. Let $\delta_{X_q} : \mathbb{N}^\omega \to X_q$ be the standard Cauchy representation of X_q; that is, every function $\boldsymbol{w} \in X_q$ is encoded by a sequence $(\boldsymbol{w}_k)_k$ in $\mathfrak{P}$ such that $\|\boldsymbol{w} - \boldsymbol{w}_k\|_{L_q} \leq 2^{-k}$. The sequence $(\boldsymbol{w}_k)_k$ is called a δ_{X_q}-name of $\boldsymbol{w}$.

2.1 Representing Closed Operators

Since a Cauchy representation is admissible [12, Definition 3.2.7], a mapping $\mathbb{A} : X_q \to X_q$ is continuous if and only if it is $(\delta_{X_q}, \delta_{X_q})$-continuous. Therefore, the standard representation $[\delta_{X_q} \to \delta_{X_q}] :\subset \mathbb{N}^\omega \to C(X_q, X_q)$ constitutes a representation of the set of all continuous total functions from X_q to X_q and, in particular, of the linear continuous operators.

In the following, we will mainly consider closed *un*bounded linear operators defined on X_q with $q = 2$. Let $\mathcal{A}$ be the set of all linear operators, bounded or unbounded, on X_q with non-empty closed graphs. Since $\mathcal{A} \not\subseteq C(X_q, X_q)$, $\mathcal{A}$ cannot be represented by $[\delta_{X_q} \to \delta_{X_q}]$. Instead, we use the representation δ_G of $\mathcal{A}$ defined in [16, p.516] as follows.

Definition 5 $\delta_G(p) = \mathbb{A} :\iff$ *there are* p_i, q_i *in the domain of* δ_{X_q} *such that* $p = \langle p_0, q_0, p_1, q_1, \ldots \rangle$ *and* $\{(\delta_{X_q}(p_i), \delta_{X_q}(q_i)) \mid i \in \mathbb{N}\}$ *is dense in* graph($\mathbb{A}$).

Thus a δ_G-name of $\mathbb{A} \in \mathcal{A}$ lists a set dense in the graph of $\mathbb{A}$.

3 Computability of the Stokes Semigroup

Let $\mathbb{A} = -\mathbb{P}\triangle$ be the Stokes operator (see, for example, [2, §2]) with the domain

$$D(\mathbb{A})_q = X_q \cap \{\boldsymbol{u} \in H_q^2(\Omega); \boldsymbol{u} = 0 \text{ on } \partial\Omega\}$$

where $\mathbb{P} : (L_q(\Omega))^2 \to X_q$ is the Helmholtz projection and $H_q^2(\Omega)$ the Sobolev space of twice weakly (in space) differentiable functions. It is known from the classical study that $\mathbb{A}$ is a closed linear operator whose domain $D(\mathbb{A})_q$ is dense in X_q. When $q = 2$, $\mathbb{A}$ is a positive self-adjoint linear operator on X_2; thus $\mathbb{A}^* = \mathbb{A}$ and $0 < \langle \mathbb{A}\boldsymbol{u}, \boldsymbol{u}\rangle := \int_\Omega \mathbb{A}\boldsymbol{u}(\boldsymbol{x}) \cdot \boldsymbol{u}(\boldsymbol{x})\, d\boldsymbol{x}$ for $\boldsymbol{u} \in D(\mathbb{A})_2$, $\boldsymbol{u} \not\equiv 0$. Hence $-\mathbb{A}$ is the infinitesimal generator of an analytic semigroup. We note that $\mathfrak{P} \subset D(\mathbb{A})_q$. Moreover, for any $\boldsymbol{p} \in \mathcal{P}_{\text{div}}^{0,k}$ and $n \geq k+1$, it follows from Lemma 3 that $\triangle\gamma_n * \boldsymbol{p}$ is divergence-free with a compact support contained in Ω and, consequently, $\triangle\gamma_n * \boldsymbol{p} \in X_q$; thus $-\mathbb{A}(\gamma_n * \boldsymbol{p}) = \mathbb{P}\triangle(\gamma_n * \boldsymbol{p}) = \mathbb{P}(\triangle\gamma_n * \boldsymbol{p}) = \triangle\gamma_n * \boldsymbol{p}$. It then follows from Definition 5 that $\{(\gamma_n * \boldsymbol{p}, \triangle\gamma_n * \boldsymbol{p}) \mid \boldsymbol{p} \in \mathcal{P}_{\text{div}}^{0k}, n \geq k+1\}$ is a computable δ_G-name of $-\mathbb{A}$.

Using the Stokes operator $\mathbb{A}$ the Eq. (1) can be rewritten as an evolution equation in X_q:

$$\partial_t \boldsymbol{u} + \mathbb{A}\boldsymbol{u} + \mathbb{B}\boldsymbol{u} = \mathbb{P}\boldsymbol{f} \quad (t > 0), \qquad \boldsymbol{u}(0) = \boldsymbol{a} \in X_q \tag{7}$$

where $\mathbb{B}\boldsymbol{u} = \mathbb{P}(\boldsymbol{u} \cdot \nabla)\boldsymbol{u}$ and, since $\boldsymbol{u} \in X_q$, $\mathbb{P}\boldsymbol{u} = \boldsymbol{u}$.

For the linear homogeneous part of (7)

$$\partial_t \boldsymbol{v} + \mathbb{A}\boldsymbol{v} = 0, \qquad \boldsymbol{v}(0) = \boldsymbol{a} \tag{8}$$

its solution is $\boldsymbol{v}(t) = e^{-\mathbb{A}t}\boldsymbol{a}$, where $e^{-\mathbb{A}t}$ is the analytic semigroup generated by the infinitesimal generator $-\mathbb{A}$. It follows from [16, THEOREM 5.4.2] that $\boldsymbol{v}$ can be computed by a Turing algorithm from any data type $(-\mathbb{A}, \delta, M)$, where $0 < \delta < \pi/2$ satisfying $\rho(-\mathbb{A}) \supset \Sigma_\delta = \{\lambda \in \mathbb{C} : |\arg\lambda| < \frac{\pi}{2} + \delta\} \cup \{0\}$ and $M > 0$ satisfying $|(\lambda + \mathbb{A})^{-1}| \leq M/|\lambda|$ for all $\lambda \in \Sigma_\delta$ and $\lambda \neq 0$, $\rho(-\mathbb{A})$ is the resolvent of $-\mathbb{A}$. Thus to show that v can be computed from the data defining the initial-boundary value problem (8), it suffices to show that δ and M are computable from $\mathbb{A}$.

Proposition 6 *For the linear homogenous Eq. (8), the solution operator* $S : X_2 \to C(\mathbb{R}; X_2)$, $\boldsymbol{a} \mapsto e^{-\mathbb{A}t}\boldsymbol{a}$ *is* $(\delta_{X_2}, [\rho \to \delta_{X_2}])$*-computable.*

Proof Recall that for the Stokes operator $-\mathbb{A}$, $\{(\gamma_n * \boldsymbol{p}, \triangle\gamma_n * \boldsymbol{p}) : \boldsymbol{p} \in \mathcal{P}_{\text{div}}^{0,k}, n \geq k+1\}$ is a δ_G-name of $-\mathbb{A}$. We show that for any given $0 < \delta < \pi/2$ satisfying $\rho(-\mathbb{A}) \supset \Sigma_\delta$, we have

$$|(\lambda + \mathbb{A})^{-1}| \leq \frac{1}{|\lambda| \sin\delta} \tag{9}$$

for all $\lambda \in \Sigma_\delta = \{\lambda \in \mathbb{C} \mid |\arg \lambda| < \frac{\pi}{2} + \delta\} \cup \{0\}$, $\lambda \neq 0$, where $|(\lambda + \mathbb{A})^{-1}|$ denotes the operator norm of $(\lambda + \mathbb{A})^{-1}$. Since the spectrum of the Stokes operator $\mathbb{A}$ lies on the positive real line, we can choose $\delta = \pi/4$; thus $M = 1/\sin(\pi/4) = \sqrt{2}$. Then it follows from (9) and [16] that the solution $\boldsymbol{v(t)} = e^{-\mathbb{A}t}\, \boldsymbol{a}$ of the linear homogenous Eq. (8) is computable from $(-\mathbb{A}, \pi/4, \sqrt{2})$.

It remains to show the claim (9). For any $\lambda \in \Sigma(\delta)$, write $\lambda = |\lambda|(\cos\theta + i\sin\theta)$, where $|\theta| < \pi - \delta$. Then for $\boldsymbol{u} \in D(\mathbb{A})_2$,

$$\begin{aligned} \left|\langle(\lambda + \mathbb{A})\boldsymbol{u},\, \boldsymbol{u}\rangle\right|^2 &= \left||\lambda|\cos\theta\,\|\boldsymbol{u}\|_2^2 + \langle\mathbb{A}\boldsymbol{u},\, \boldsymbol{u}\rangle + i|\lambda|\sin\theta\,\|\boldsymbol{u}\|_2^2\right|^2 = \\ &= |\lambda|^2\cos^2\theta\|\boldsymbol{u}\|_2^4 + 2|\lambda|\cos\theta\,\langle\mathbb{A}\boldsymbol{u},\, \boldsymbol{u}\rangle + \langle\mathbb{A}\,\boldsymbol{u},\, \boldsymbol{u}\rangle^2 + |\lambda|^2\sin^2\theta\,\|\boldsymbol{u}\|_2^4 \end{aligned}$$

where $\|\boldsymbol{u}\|_2 = \sqrt{\langle\boldsymbol{u}, \boldsymbol{u}\rangle}$ denotes the L_2-norm of $\boldsymbol{u}$. Now if $-\pi/2 < \theta < \pi/2$, then $\cos\theta > 0$ and we have $|\langle(\lambda + \mathbb{A})\boldsymbol{u},\, \boldsymbol{u}\rangle| > |\lambda|\|\boldsymbol{u}\|_2^2 > |\lambda|\sin\delta\,\|\boldsymbol{u}\|_2^2$; recall that $\langle\mathbb{A}\boldsymbol{u},\, \boldsymbol{u}\rangle > 0$. If $\pi/2 \le \theta < \pi - \delta$ or $\pi + \delta < \theta \le 3\pi/2$, then we have $|\langle(\lambda + \mathbb{A})\boldsymbol{u},\, \boldsymbol{u}\rangle| = \left||\lambda|\cos\theta\,\|\boldsymbol{u}\|_2^2 + \langle\mathbb{A}\ \boldsymbol{u},\, \boldsymbol{u}\rangle + i|\lambda|\sin\theta\,\|\boldsymbol{u}\|_2^2\right| \ge |\lambda|\,|\sin\theta|\,\|\boldsymbol{u}\|_2^2 > |\lambda|\sin\delta\,\|\boldsymbol{u}\|_2^2$. Thus for any $\lambda \in \Sigma_\delta$ with $\lambda \neq 0$, we have

$$|\langle(\lambda + \mathbb{A})\boldsymbol{u}, \boldsymbol{u}\rangle| > |\lambda|\sin\delta\,\|\boldsymbol{u}\|_2^2 \tag{10}$$

Note that, by the Schwartz-Cauchy inequality, $|\langle(\lambda + \mathbb{A})\boldsymbol{u},\, \boldsymbol{u}\rangle| \le \|(\lambda + \mathbb{A})\boldsymbol{u}\|_2\,\|\boldsymbol{u}\|_2$. Thus it follows from (10) that $\|(\lambda + \mathbb{A})\boldsymbol{u}\|_2\|\boldsymbol{u}\|_2 \ge |\lambda|\sin\delta\|\boldsymbol{u}\|_2^2$. Let us set $\boldsymbol{f} = (\lambda + \mathbb{A})\boldsymbol{u}$. Then $\boldsymbol{u} = (\lambda + \mathbb{A})^{-1}\boldsymbol{f}$ and $\|\boldsymbol{f}\|_2\|(\lambda + \mathbb{A})^{-1}\boldsymbol{f}\|_2 = \|(\lambda + \mathbb{A})\boldsymbol{u}\|_2\|\boldsymbol{u}\|_2 \ge |\lambda|\sin\delta\,\|(\lambda + \mathbb{A})^{-1}\boldsymbol{f}\|_2^2$, which implies that $\|(\lambda + \mathbb{A})^{-1}\boldsymbol{f}\|_2 \le \frac{1}{|\lambda|\sin\delta}\|\boldsymbol{f}\|_2$. □

4 Extension to the Nonlinear Problem and Helmholtz Projection

Next we consider the nonlinear problem

$$\partial_t\boldsymbol{u} + \mathbb{A}\boldsymbol{u} + \mathbb{B}\boldsymbol{u} = 0, \quad \boldsymbol{u}(0) = \boldsymbol{a} \tag{11}$$

where $\mathbb{A} = -\mathbb{P}\triangle$ is the Stokes operator and $\mathbb{B}\boldsymbol{u} = \mathbb{P}(\boldsymbol{u}\cdot\nabla)\boldsymbol{u}$. In its integral form the problem (11) can be rewritten as follows

$$\boldsymbol{u}(t) = e^{-t\mathbb{A}}\boldsymbol{a} - \int_0^t e^{-(t-s)\mathbb{A}}\mathbb{B}\boldsymbol{u}(s)\,ds, \quad t > 0 \tag{12}$$

Classically, the existence of the solution of the problem (12) is proved by showing that the sequence of approximate solutions constructed by the following iteration scheme

$$\boldsymbol{u}_0(t) = e^{-t\mathbb{A}}\boldsymbol{a}, \quad \boldsymbol{u}_{n+1}(t) = \boldsymbol{u}_0(t) - \int_0^t e^{-(t-s)\mathbb{A}}\mathbb{B}\boldsymbol{u}_n(s)\,ds, \quad n \ge 0 \tag{13}$$

is contracting near $t = 0$; thus the sequence converges to a unique limit. Since the limit satisfies the integral Eq. (12), it is the solution of the initial value problem (11) near $t = 0$.

Thus, as an important step towards a rigorous algorithm computing the solution of Eq. (12) on input $\boldsymbol{a}$, we establish

Proposition 7. *The Helmholtz projection* $\mathbb{P} : (L_2(\Omega))^2 \to X_2$ *is* $(\delta_{L_2}, \delta_{X_2})$-*computable.*

Proof (Sketch). It follows from [4, pp.40] that for each $\boldsymbol{u} = \begin{pmatrix} u^{(1)} \\ u^{(2)} \end{pmatrix} = (u^{(1)}, u^{(2)})^T$ with $(u^{(1)}, u^{(2)}) \in (L_2(\Omega))^2$

$$\mathbb{P}\boldsymbol{u} = \begin{pmatrix} -\partial_y \varphi \\ \partial_x \varphi \end{pmatrix} = (-\partial_y \varphi, \partial_x \varphi)^T \tag{14}$$

where the scalar function φ is the solution of the boundary value problem:

$$\triangle \varphi = \partial_x u^{(2)} - \partial_y u^{(1)} \text{ in } \Omega, \qquad \varphi = 0 \text{ on } \partial\Omega \tag{15}$$

To show that the projection is $(\delta_{L_2}, \delta_{X_2})$-computable, it suffices to show that a δ_{X_2}-name of $\mathbb{P}\boldsymbol{u}$ can be computed from any given δ_{L_2}-name of $\boldsymbol{u}, \boldsymbol{u} \in (L_2(\Omega))^2$. □

5 Conclusion and Perspectives

We have, in the rigorous sense of Recursive Analysis, established effectivizations of two of the four steps of common classical existence proofs of strong in time and spatially L_2–solutions to the Navier-Stokes Equation on the open unit cube with zero boundary condition: computability of the Helmholtz projection and of the analytic semigroup generated by the Stokes operator.

Future endeavors will cover the remaining two steps, thus proving local computability of strong solutions; and extend to more general domains and boundary conditions.

References

1. Beggs, E., Costa, J.F., Tucker, J.V.: Axiomatising physical experiments as Oracles to algorithms. Philos. Trans. R. Soc. Math. Phys. Eng. Sci. **370**, 3359–3384 (2012)
2. Giga, Y.: Time and spatial analyticity of solutions of the Navier-Stokes equations. Comm. Partial Differ. Equ. **8**, 929–948 (1983)
3. Giga, Y., Miyakawa, T.: Solutions in L^r of the Navier-Stokes initial value problem. Arch. Ration. Mech. Anal. 89(3) 5.VIII, 267–281 (1985)
4. Girault, V., Raviart, P.-A.: Finite Element Methods for Navier-Stokes Equations. Series in Computational Mathematics, vol. 5. Springer, Heidelberg (1986)
5. Patel, M.K., Markatos, N.C., Cross, M.: A critical evaluation of seven discretization schemes for convection-diffusion equations. Int. J. Numer. Meth. Fluids **5**(3), 225–244 (1985)
6. Pour-El, M.B., Richards, J.I.: The wave equation with computable initial data such that its unique solution is not computable. Adv. Math. **39**(4), 215–239 (1981)
7. Pour-El, M.B., Richards, J.I.: Computability in Analysis and Physics. Springer, Heidelberg (1989)

8. Pour-El, M.B., Zhong, N.: The wave equation with computable initial data whose unique solution is nowhere computable. Math. Logic Q. **43**(4), 499–509 (1997)
9. W.D. Smith: On the uncomputability of hydrodynamics, NEC preprint (2003)
10. Soare, R.I.: Computability and recursion. Bull. Symbolic Logic **2**, 284–321 (1996)
11. Tao, T.: Finite time blowup for an averaged three-dimensional Navier-Stokes equation. arXiv:1402.0290 (submitted); http://terrytao.wordpress.com/2014/02/04/
12. Weihrauch, K.: Computable Analysis: An Introduction. Texts in Theoretical Computer Science. An EATCS Series. Springer, Heidelberg (2000)
13. Weihrauch, K., Zhong, N.: Is wave propagation computable or can wave computers beat the turing machine? Proc. Lond. Math. Soc. **85**(2), 312–332 (2002)
14. Weihrauch, K., Zhong, N.: Computing the solution of the Korteweg-de Vries equation with arbitrary precision on turing machines. Theor. Comput. Sci. **332**, 337–366 (2005)
15. Weihrauch, K., Zhong, N.: Computing schrödinger propagators on type-2 turing machines. J. Complex. **22**(6), 918–935 (2006)
16. Weihrauch, K., Zhong, N.: Computable analysis of the abstract Cauchy problem in Banach spaces and its applications I. Math. Logic Q. **53**, 511–531 (2007)
17. Wiegner, M.: The Navier-Stokes equations - a neverending challenge? Jahresbericht der Deutschen Mathematiker Vereinigung (DMV) **101**(1), 1–25 (1999). http://dml.math.uni-bielefeld.de/JB_DMV/JB_DMV_101_1.pdf
18. Ziegler, M.: Physically-relativized church-turing hypotheses: physical foundations of computing and complexity theory of computational physics. Appl. Math. Comput. **215**(4), 1431–1447 (2009)

Kalmár and Péter: Undecidability as a Consequence of Incompleteness

Máté Szabó(✉)

Department of Philosophy, Carnegie Mellon University,
161 Baker Hall, Pittsburgh, PA 15213, USA
mszabo@andrew.cmu.edu

Abstract. László Kalmár and Péter Rózsa "proved that the existence of (...) undecidable problems follows from Gödel's Theorem on relatively undecidable problems" ([6], p. vii). Unfortunately, the only available document of their joint work is Kalmár's sketch of the proof in his [3]. In the following, I assemble a paper from Kalmár's manuscripts on this issue.

Keywords: History of computing · Undecidability · Incompleteness · László Kalmár · Péter Rózsa

1 Introduction

László Kalmár gave a talk, entitled *On Unsolvable Mathematical Problems* [3] at *The Tenth International Congress of Philosophy* in Amsterdam in 1948. In it he sketched a proof to show that "Church's theorem[1] is a consequence, or even a particular case, of the (generally formulated) Gödel–Rosser [incompleteness] theorem." (p. 758) Before turning to his sketch of the proof Kalmár refers to Péter Rózsa's work:

> In a paper which has not been published yet, Miss Péter has shown that Church's theorem is a consequence of Gödel's theorem, applied to a formal system due to Hilbert and Bernays, in which all calculations of values of effectively calculable functions can be carried out. Miss Péter's proof led me to the following remark. (pp. 757–758.)

This remark is actually Kalmár's sketch. Péter's paper that was mentioned by Kalmár was never published.

Péter also mentions this result in the *Preface to the English Edition* of her book, *Playing with Infinity*, written in 1960:

I would like to thank Wilfried Sieg, Jeremy Avigad, Zalán Gyenis, Réka Bence, and the anonymous referees for their insightful comments, and the cordial assistance of the employees of the Klebelsberg Library at the University of Szeged.

[1] That is, the undecidability of predicate logic.

A. Beckmann et al. (Eds.): CiE 2015, LNCS 9136, pp. 343–352, 2015.
DOI: 10.1007/978-3-319-20028-6_35

> The reader should remember that the book mirrors my methods of thinking as they were in 1943;[2] I have hardly altered anything in it. Only the end has been altered substantially. Since then, László Kalmár and I have proved that the existence of (...) undecidable problems follows from Gödel's Theorem on relatively undecidable problems (...). ([6], p. vii).

The details of the proof were not published in this book, as it was written for a wider audience.

Thus, unfortunately, the only publicly available document of their joint work is Kalmár's 3 page conference summary [3]. However, the Kalmár Nachlass at the Klebelsberg Library at the University of Szeged contains multiple manuscripts on this topic. Out of the boxes dedicated to published papers, which also contain preprints and alternate versions of his published papers:

- Folder 23 contains a 12 page (cca. A4 or letter size) typed version of Kalmár's [3] in English.

Among the regular boxes which contain manuscripts, notes, sketches, etc.:

- Folder 156 contains a proof on 6 typed pages (cca. A4 or letter size) in Hungarian, but is labeled as if it contained some notes for a talk at the Institute Henri Poincarè, about Gödel's results in French. The notes are undated but they are most likely from the 1960s. Kalmár visited Paris twice in 1963. If the manuscript was found among those papers it suggests that it was written in that year.
- Folder 213 contains a 10 page (cca. A5 or half of letter size) handwritten document in English, entitled *Two more abstract forms of Gödel's theorem.* The manuscript was prepared for a talk Kalmár gave at the Polish Mathematical Society on 7 October 1966. Although this manuscript does not contain the proof, I will use it, as it is almost identical to the first 4 pages of Folder 156, but was written in English.
- Folder 221 contains 26 handwritten pages (cca. A5 or half of letter size) in English, that constitutes two versions of a document, entitled *On the application of Gödel's theorem to a proof of Church's Theorem.* The manuscripts were prepared for talks Kalmár gave at the University of Oxford, the University of Leicester, and the London School of Economics on 12, 16, and 18 of May 1967, respectively.
- Folder 347 contains a 19 page (cca. A5 or half of letter size) handwritten document in English, entitled *On the relation between Gödel's and Church's theorems.* László Kalmár and Rózsa Péter are listed as the authors. The manuscript is unfinished and incomplete at some places but was clearly intended for publication. Again, the manuscript is undated, but it can be found among the boxes which are labeled as "Most likely from the 1970s."

In the following, I have assembled a paper from the above manuscripts. It begins with the introductory paragraph of Folder 347. It is followed, to preserve the authenticity of Kalmár's work,[3] by the whole content of Folder 213. The

[2] The year of the completion of the first Hungarian edition.
[3] Even in cases where the wording is unusual.

remaining part is my translation of Folder 156 where I unified the notation with that of Folder 213. While translating Folder 156 I took the similar parts of Folder 347 into account and used Kalmár's wording.[4] The structures and the typesetting of the original manuscripts are preserved. In the assembled paper the footnotes either point to smaller differences between the manuscripts or quote different versions to make the arguments more comprehensible. 'Hungarian version' will always refer to Folder 156 while 'English version' will always refer to Folder 213.

Before presenting the details of the actual paper, I want to convey to the reader the broad idea behind the proof. Péter's work led Kalmár to the following observation:

> In the conditions under which Gödel's (or Rosser's) theorems hold for a formal system S, nowhere is postulated that the proofs in S are real proofs of some standard form, based on generally accepted ways of inference. ([3], p. 758).

Based on this observation, abstract forms of Gödel's Theorem are formulated with as few restrictions as possible on proofs of formal systems. The theorem is formulated in such a broad way that the application of any algorithm, for solving a fixed problem set, can be considered as a proof. Thus, the broadly formulated Gödel Theorem allows us to infer that, for any algorithm, either it does not give an answer for a problem from the set or it does not give a correct answer. I hope this will help the reader in following the paper.

In the assembled paper, after the introduction of Folder 347, points 1–4 of Folder 213 contain the definitions and terminology for what Kalmár calls "weak arithmetics" and in point 5 the abstract form of Gödel's Theorem is asserted and proved for these arithmetics. Points 6–7 present the notion of "strong arithmetics" in a similar way and, once again, point 8 contains the abstract form of Gödel's Theorem for those arithmetics. It is followed by the last part of Folder 156 in which Church's Theorem is shown to be "a consequence, or even a particular case" of Gödel's Theorem.

2 Kalmár and Péter: Undecidability as a Consequence of Incompleteness

From Folder 347:[5]

This paper contains a proof of the existence of arithmetical problems unsolvable by any method of a certain kind [as a "consequence, or even a particular case" of

[4] Although Folder 221 also contains the result and is entirely in English, but its presentation is significantly different from that of Folder 156. I chose to use Folder 213, written in English, supplemented by my translation of the last one and a half pages of Folder 156, as the latter is easier to follow.

[5] The title was given by me. I believe it is more indicative of the content than other titles of the manuscripts, such as: *On the relation between Gödel's and Church's theorems*, and *On the application of Gödel's theorem to a proof of Church's Theorem.*

Gödel's theorem].[6] The proof in question takes its origin from discussions made in words and in correspondence since 1942 about Gödel's and Church's theorems. The authors of the present paper played the principal part in these discussions; however, also Mr. J. Surányi,[7] Th. Vargha,[8] T. Szele,[9] P. Bernays, G. Alexits,[10] St. Fenyő[11] and others took part in them and the authors are very much indebted for their valuable suggestions.[12]

From Folder 213:

1. Definition. Deductive theory:[13] $\theta = < F, P, \omega >$, where
(i) F, P are [arbitrary][14] sets;
(ii) $\omega \in F^P$.

Terminology: (i) F is called the set of formulae (of θ);
(ii) P is called the set of proofs (in θ);
(iii) for $p \in P$, $f = \omega(p) \in F$ is called the conclusion of the proof p;
and p is called a proof of the formula f;
(iv) f is called a thesis[15] (of θ) (notation: $\vdash f$), iff $\exists p\,(p \in P \wedge \omega(p) = f)$.

2. Gödel's theorem applies for deductive theories in which at least a part of arithmetic can be formulated; such deductive theories will be called arithmetics. They are not particular deductive theories (as, e.g., Abelian groups are particular groups) but rather deductive theories supplied with an additional structure (as, e.g., ordered groups are groups supplied with order structure).

We shall consider two forms of Gödel's theorem:
(i) a more general one for so-called weak arithmetics,
(ii) and a more familiar one for so-called strong arithmetics.

[6] The parenthetical comment was added by me. The quote is from ([3], p. 758).

[7] János Surányi (1918–2006), mathematician working in the fields of logic and algebra.

[8] Tamás Varga (1919–1987), educationist, expert in the education of mathematics.

[9] Tibor Szele (1918–1955), mathematician working in the field of algebra.

[10] György Alexits (1899–1978), mathematician working in the field of analysis.

[11] István Fenyő (1917–1987), mathematician working in the field of analysis.

[12] The first page of Folder 221 lists some additional information on the development of their results: "Idea of proof of Church's theorem based on a general form of Gödel's theorem: Rózsa Péter, letter to Bernays, 1945." Sadly, the letters from Péter to Bernays which have been written around this time and are preserved in the library of ETH Zürich do not mention this result. About the development towards abstract forms of Gödel's Theorem, Kalmár cites Chauvin's [1] which led to [4] and [5], his early, abstract formulations of Gödel's Theorem in French. Finally, he quotes Henkin's review [2] of [4] and [5]: "The abstractions are not fruitful because no new particular cases which are of importance in themselves are brought by these means within the scope of Gödel's theorems." (p. 230).

[13] It is called "Formal system" in Hungarian. Thus, in this paper "deductive theory" and "formal system" are interchangeable.

[14] The parenthetical comment was added by me. The word 'arbitrary' occurs only in the Hungarian version.

[15] In manuscripts written in English, Kalmár uses "thesis" instead of "theorem." I conjecture that he uses "thesis" because of the unrestricted character of proofs in θ.

3. Definition. Weak arithmetic: $A =< F, P, \omega, T, \psi, \iota, \bar{\iota} >$, where
(i) $< F, P, \omega >= \theta$ is a deductive theory (hence the notions 'formula', 'proof', 'conclusion', 'thesis' can be regarded as already defined for weak arithmetics);
(ii) T is a set;
(iii) $\psi \in N^{T \times N}$ with $N = \{0, 1, 2, \cdots\}$;
(iv) $\iota \in F^{T \times N}$ and $\bar{\iota} \in F^{T \times N}$.

Terminology: (i) T is called the set of terms [not necessarily of θ; terms are used to represent arithmetical functions of one argument];[16]
(ii) for $t \in T$, $k \in N$, $n = \psi(t, k) \in N$ is called the value of the function represented by t for the argument k; and the function $\varphi_t = \hat{k}\psi(t, k) \quad (= \lambda k.\psi(t, k))$ defined by $\varphi_t(k) = \psi(t, k)$ for $k \in N$ is called the function represented by t; and a function $\varphi \in N^N$ is called representable (in A) iff $\exists t(t \in T \wedge \varphi_t = \varphi)$, i.e.

$$\exists t(t \in T \wedge \forall k(k \in N \rightarrow \psi(t, k) = \varphi(k)));$$

(iii) for $t \in T$, $n \in N$, $\quad f = \iota(t, n) \in F$ is called the identical inequality of φ_t to n (or the formula expressing that φ_t does not take the value n for any argument); and $g = \bar{\iota}(t, n) \in F$ is called the negated identical inequality of φ_t to n (or the formula expressing that φ_t does take the value n for some argument).

4. Conditions for Gödel's theorem.
1.1, 1.2, $\cdots$: "expressivity condition" (requiring that some reasonings can be expressed in the arithmetic in question; the conditions requiring that some propositions can be expressed in it are included in the definitions);
2.1, 2.2: "regularity (or recursivity) conditions";
3.1, 3.2: "consistency conditions".

For the first form of Gödel's theorem we need one expressivity and two regularity conditions but no consistency condition.

1.1. $\forall t_1 t_2 n_1 n_2 (t_1 \in T \wedge t_2 \in T \wedge n_1 \in N \wedge n_2 \in N \wedge \iota(t_1, n_1) = \iota(t_2, n_2) \rightarrow \rightarrow n_1 = n_2)$,

i.e., n is uniquely determined by $\iota(t, n)$ (It would be quite reasonable to require that also t is uniquely determined by $\iota(t, n)$; however, such a condition is not needed.)[17]

[16] The parenthetical comment was added by me. In Folder 347 Kalmár mentions some instances of classes of functions that the terms T can range over: (1) general recursive functions; (2) primitive recursive functions; (3) elementary functions.

[17] Here $\iota(t_1, n_1) = \iota(t_2, n_2)$ states the identity of two formulas. In the Hungarian version the same property is required for $\bar{\iota}$ as well. Furthermore, the Hungarian version contains only the informal assertion of this property, that is, n is uniquely determined by $\iota(t, n)$.

2.1. The sets T and P are enumerable.

Terminology: Let be τ a fixed one-to-one mapping of T into $N' = N - \{0\} = \{1, 2, \cdots\}$ and π a fixed one-to-one mapping of P into N'. Then

(i) $n = \tau(t)$ is called the Gödel number of the term t; and if there is a t (and therefore only one) with $n = \tau(t)$, this t is denoted also by t_n; otherwise, t_n is not defined;

(ii) $k = \pi(p)$ is called the Gödel number of the proof p; and if there is a p (and hence only one) with $k = \pi(p)$, this p is denoted also by p_k; otherwise, p_k is not defined;

(iii) $f \in F$ is called a diagonal formula iff $\exists tn(t \in T \wedge n \in N \wedge f = \iota(t, n) \wedge \tau(t) = n)$; i.e. the diagonal formulae are the formulae of the form $\iota(t_n, n)$ with t_n defined;

(iv) for a diagonal formula $f = \iota(t, n)$, the number n (uniquely determined owing to 1.1) is called the index of f;

(v) $p \in P$ is called a diagonal proof iff $\omega(p)$ is a diagonal formula; and in this case its index is called also the index of p;

(vi) the arithmetical function γ defined then:

$$\gamma(k) = \begin{cases} \text{the index of } p_k \text{ if } p_k \text{ is defined and it is a diagonal proof;} \\ 0 \text{ if } p_k \text{ is either not defined or it is defined but it is not a diagonal proof} \end{cases}$$ [18]

is called the Gödel function of A (belonging to the mappings τ and π).

2.2. For suitable mappings τ and π, the Gödel function γ is representable.[19]

5. First form of Gödel's theorem. For every weak arithmetic A for which the conditions 1.1, 2.1 and 2.2 are satisfied, $\exists tn(t \in T \wedge n \in N \wedge (1^0 \vee 2^0 \vee 3^0))$ where

$1^0 \vdash \iota(t, n) \wedge \exists k(k \in N \wedge \varphi_t(k) = n)$,

$2^0 \vdash \bar{\iota}(t, n) \wedge \forall k(k \in N \rightarrow \varphi_t(k) \neq n)$,

$3^0 \neg \vdash \iota(t, n) \wedge \neg \vdash \bar{\iota}(t, n)$;[20]

---

[18] p_k is defined and is a diagonal proof means that $\omega(p_k)$ is a diagonal formula, thus $\omega(p_k) = \iota(t_n, n)$ for some $t \in T$, $n \in N$ where t_n is defined. The index of p_k is the index of $\omega(p_k)$, thus n.

[19] Here the Hungarian version says that for suitable mappings γ is an elementary function (and thus representable) and adds the following remark: "From now on let τ and π be such fixed mappings [that γ is an elementary function]. In theory it is easy but technically it is tedious to show that 2.2 holds for those formal (axiomatic) systems that are used in mathematics." (Folder 156, p. 3).

[20] $\neg \vdash$ should be understood as $\nvdash$.

i.e. either $\iota(t,n)$ is a thesis but what it expresses is false for some argument k, φ_t takes the value n; or $\bar{\iota}(t,n)$ is a thesis but what it expresses is false, for φ_t does not take the value n for any argument k; or neither $\iota(t,n)$, nor $\bar{\iota}(t,n)$ is a thesis.

(Meta-)Proof. Owing to 2.2, we have (for suitable mappings τ and π) $\exists t(t \in T \wedge \gamma = \varphi_t)$. Be $n = \tau(t)$. Then we have $n \in N'$ (hence $n \neq 0$) and $t = t_n$. Now,

(i) if we have $\vdash \iota(t,n)$, be p one of the proofs of $\iota(t,n)$ and $k = \pi(p)$. Then $p = p_k$ is a diagonal proof with index n, for we have $\omega(p_k) = \iota(t,n) = \iota(t_n,n)$. Hence, by the definition of the function γ, we have

$$\varphi_t(k) = \gamma(k) = n,$$

hence we have case 1^0.

(ii) if we have $\neg \vdash \iota(t,n)$[21] then for any $k \in N$,
either p_k is not defined,
or p_k is defined but not a diagonal proof,
or p_k is defined and it is a diagonal proof, i.e.
the proof of some diagonal formula $\iota(t_m,m) \neq \iota(t_n,n)$, hence $m \neq n$.

Hence, by the definition of the function γ, we have
either $\varphi_t(k) = \gamma(k) = 0 \neq n$,
or $\varphi_t(k) = \gamma(k) = m \neq n$;
thus, we have $\varphi_t(k) \neq n$ for any $k \in N$.

Hence, if $\vdash \bar{\iota}(t,n)$, we have case 2^0,
and if $\neg \vdash \bar{\iota}(t,n)$, we have case 3^0, qu.e.d.

6. Definition. Strong arithmetic: $A =< F, P, \omega, T, \psi, \iota, \bar{\iota}, \varepsilon, \bar{\varepsilon} >$, where
(i) $A =< F, P, \omega, T, \psi, \iota, \bar{\iota} >$ is a weak arithmetic
(hence the notions 'formula', 'proof', 'conclusion', 'thesis', 'term', 'value of the function represented by a term for an argument $k \in N'$', 'function represented by a term', 'representable function', 'identical inequality', 'negated identical inequality' can be regarded as already defined for strong arithmetics too, and provided conditions 1.1, 2.1 and 2.2 to be satisfied, first form of Gödel's theorem holds);
(ii) $\varepsilon \in F^{T \times N \times N}$ and $\bar{\varepsilon} \in F^{T \times N \times N}$.

Terminology. For $t \in T$, $k \in N$, $n \in N$, $f = \varepsilon(t,k,n) \in F$ is called the equation of $\varphi_t(k)$ to n (or the formula expressing that φ_t takes the value n for the argument k); and $g = \bar{\varepsilon}(t,k,n) \in F$ is called the inequality of $\varphi_t(k)$ to n (or the formula expressing that φ_t does not take the value n for the argument k).

[21] $\neg \vdash$ should also here be understood as $\nvdash$. Thus, in (ii) it is assumed $\iota(t,n)$ is not provable, and depending on whether $\bar{\iota}(t,n)$ is provable or not, it leads to 2^0 or 3^0, respectively. See the last two lines of the proof.

7. Additional conditions for the second form of Gödel's theorem (three more expressivity as well as two consistency conditions):

1.2. $\forall tkn(t \in T \wedge k \in N \wedge n \in N \wedge \varphi_t(k) = n \rightarrow \vdash \varepsilon(t,k,n))$;
1.3. $\forall tkln(t \in T \wedge k \in N \wedge l \in N \wedge n \in N \wedge \vdash \varepsilon(t,k,l) \wedge l \neq n \rightarrow$
$\rightarrow \vdash \overline{\varepsilon}(t,k,n))$;
1.4. $\forall tkn(t \in T \wedge k \in N \wedge n \in N \wedge \vdash \iota(t,n) \rightarrow \vdash \overline{\varepsilon}(t,k,n))$;
3.1. $\forall tkn(t \in T \wedge k \in N \wedge n \in N \rightarrow \neg(\varepsilon(t,k,n) \wedge \overline{\varepsilon}(t,k,n))$;
3.2. $\forall tn(t \in T \wedge n \in N \rightarrow \neg(\vdash \overline{\iota}(t,n) \wedge \forall k(k \in N \rightarrow \vdash \overline{\varepsilon}(t,k,n))))$.

(1.2 is a particular case of the condition to the effect that numerical facts can be proved; 1.3 is another particular case of the same condition (for numerically true inequalities); 1.4 is a particular case of the dictum de omni; 3.1 is a particular case of the consistency and 3.2 a particular case of the ω-consistency of A.)

8. Second form of Gödel's theorem. For every strong arithmetic A for which the conditions 1.1, 1.2, 1.3, 1.4, 2.1, 2.2, 3.1, and 3.2 are satisfied, $\exists tn(t \in T \wedge n \in N \wedge 3^0)$, i.e. for which neither $\iota(t,n)$ nor $\overline{\iota}(t,n)$ is a thesis.

(Meta-)Proof. Owing to 1.1, 2.1 and 2.2 as well as first form of Gödel's theorem, $\exists tn(t \in T \wedge n \in N \wedge (1^0 \vee 2^0 \vee 3^0))$. Hence, it satisfies to prove that if also 1.2, 1.3, 1.4, 3.1 and 3.2 are satisfied, neither case 1^0 nor case 2^0 can arise.

Now, in case 1^0, we should have on the one hand $\vdash \iota(t,n)$, hence, owing to 1.4, $\forall k \vdash \overline{\varepsilon}(t,k,n)$; on the other hand, for a suitable $k \in N$, $\varphi_t(k) = n$, hence, owing to 1.2, $\vdash \varepsilon(t,k,n)$, which is impossible if 3.1 is satisfied.

On the other hand, in case 2^0, we should have, on the one hand, $\vdash \overline{\iota}(t,n)$, on the other hand, for any $k \in N$, $\varphi_t(k) = n$; hence, if we put $\varphi_t(0) = l_0$, $\varphi_t(1) = l_1, \cdots$, we have $l_k \neq n$ for any $k \in N$. Now, owing to 1.2 we have $\vdash \varepsilon(t,k,l_k)$, hence, owing to 1.3, $\vdash \overline{\varepsilon}(t,k,n)$ for any $k \in N$ which is impossible if 3.2 is satisfied.

Hence, only 3^0 can be the case, qu.e.d.

From Folder 156:

Church's theorem as a consequence of the first form of Gödel's theorem.

Consider the following problem set (with the parameter (e,i) where $e \in E$ [the set of elementary expressions containing at most one free variable][22] and $i \in N$):

In case of which $e \in E$ and $i \in N$ does the φ_e elementary function take the value i somewhere?

[22] The parenthetical comment was added by me. As the Hungarian version deals only with elementary functions, E is used everywhere instead of T. Hopefully, this will not lead to any confusion.

An algorithm for solving this problem set is a method[23] which is applicable for every value of (e, i) as the "input parameter" (maybe even multiple ways) and it always (for every input parameter and for every application) gives an answer, which is either "it does take the value" or "it does not take the value". An algorithm is good if all applications of it lead to true results.[24]

Let $\mathbf{A}$[25] be such an algorithm and let us assign the following deductive theory $< F, P, \omega >$ to it:

F is the set of ordered triples in the form $\langle e, i, 0\rangle$ and $\langle e, i, 1\rangle$ where e runs through the set E and i runs through N. ($F = E \times N \times \{0, 1\}$.)

P is the set of all applications of $\mathbf{A}$ (for any input parameter (e, i)).

ω: if the application a ($\in P$) of $\mathbf{A}$ on input parameter (e, i) gives the answer "it does take the value" then let $\omega(a) = \langle e, i, 1\rangle$, if it gives the answer "it does not take the value" then let $\omega(a) = \langle e, i, 0\rangle$.

We call the algorithm $\mathbf{A}$ regular if $< F, P, \omega >$ satisfies the conditions 2.1 and 2.2. It can be shown (in theory it is easy but technically it is tedious) that Markov's normal algorithms and partially recursive algorithms are regular, for instance.[26]

---

23 The word "method" (instead of procedure) comes from Folder 347. Although it is unusual wording I tried to incorporate as much of Kalmár's original manuscripts in English as I could. In the following pages the words "application", "good" and "regular" algorithm are also taken from Folder 347.

24 Here, once again, the wording comes from Folder 347. At a similar point of that folder Kalmár remarks that "instead of defining formally" what an algorithm or method is, "we formulate the facts needed in the sequel, concerning (...) their applications and results." (p. 14).

25 In the Hungarian version Kalmár uses A instead of $\mathbf{A}$ because previously it was not used to denote deductive theories. This distinction between the two A-s almost disappears since in the next step a deductive theory is assigned to the algorithm.

26 In the English version 2.1 requires T and P to be enumerable. In the Hungarian version it requires only P to be enumerable. It does not lead to any problems, since the set E which is used here instead of T is clearly enumerable and thus satisfies the enumerability requirement. Thus, 2.1 is obviously satisfied.

2.2 is a "regularity condition" which requires that the relationship between the terms T and the applications (proofs) P is regular in some sense. In particular, here, it requires that there are such Gödel numberings of the elementary functions and of the applications of the algorithm that the Gödel function γ is an elementary function as well (and thus it is representable). In footnote 19 I already quoted Kalmár, claiming that 2.2 can be satisfied.

If instead of elementary functions some other class of arithmetical functions is used, e.g. the class of general recursive functions, then the task is to show that the corresponding Gödel function is a member of the class in question, and hence it is representable.

The last comment about Markov's normal algorithms and partially recursive algorithms being regular claims that the usual notions of algorithms or procedures all fall under the above open ended informal definition.

Church's Theorem:
There is no algorithm that is good (always leads to true results), regular, and solves the above problem set.

Proof. The above $< F, P, \omega >$ deductive theory satisfies 1.1 if $\iota(e,i) = \langle e,i,0\rangle$ and $\bar{\iota}(e,i) = \langle e,i,1\rangle$. Let **A** be a regular algorithm; then 2.1 and 2.2 are satisfied. Thus, the first form of Gödel's theorem holds for $< F, P, \omega >$, hence there exists $e \in E$ and $i \in N$ such that

(1) either $\langle e,i,0\rangle$ is provable, that is, an a ($\in P$) application of **A** gives the answer on the input parameter (e,i), i.e. for the question whether φ_e takes the value i or not, that "it does not take the value", however the answer is not correct because at the same time, according to 1^0, there exists an $n \in N$ that $\varphi_e(n) = i$;

(2) or $\langle e,i,1\rangle$ is provable, that is, an a ($\in P$) application of **A** gives the answer for the question above that "it does take the value", however the answer is not correct because at the same time (according to 2^0)

$$\varphi_e(0) \neq i, \varphi_e(1) \neq i, \varphi_e(2) \neq i, ...;$$

(3) or neither $\langle e,i,0\rangle$ nor $\langle e,i,1\rangle$ is provable, that is, the algorithm **A** (in contradiction with the definition of the notion of algorithm) does not answer the question above.

References

1. Chauvin, A.: Structures logiques. CR Hebdomadaires Séances Acad. Sci. **228**, 1085–1087 (1949)
2. Henkin, L.: Review of Kalmár's [4] and [5]. J. Symbolic Logic **15**(3), 230 (1950)
3. Kalmár, L.: On unsolvable mathematical problems. In: Beth, E.W., Pos, J., Hollak, J. (eds.) Proceedings of the Tenth International Congress of Philosophy (Amsterdam), pp. 756–758. North-Holland, Amsterdam (1949)
4. Kalmár, L.: Une forme du théorème de Gödel sous des hypothèses minimales. CR Hebdomadaires Séances Acad. Sci. **229**, 963–965 (1949)
5. Kalmár, L.: Quelques formes générales du théorème de Gödel. CR Hebdomadaires Séances Acad. Sci. **229**, 1047–1049 (1949)
6. Péter, R.: Playing with Infinity. Translated by Zoltan Dienes. Dover, New York (1976)

How to Compare Buchholz-Style Ordinal Notation Systems with Gordeev-Style Notation Systems

Jeroen Van der Meeren(✉) and Andreas Weiermann

Department of Mathematics, Ghent University, Krijgslaan 281, 9000, Gent, Belgium
{jeroen.vandermeeren,andreas.weiermann}@ugent.be

Abstract. By a syntactical construction we define an order-preserving mapping of Gordeev's ordinal notation system $PRJ(P)$ into Buchholz's ordinal notation system $OT(P)$ where P represents a limit ordinal. Since Gordeev already showed that $OT(P)$ can be considered as a subsystem of $PRJ(P)$, we obtain a direct proof of the equality of the order types of both systems. We expect that our result will contribute to the general program of determining the maximal order types of those well-quasi-orders which are provided by gap-embeddability relations considered by Friedman [10], Gordeev [5,7] and Kriz [8].

Keywords: Ordinal notation systems · Embeddings · Buchholz's notation system · Gordeev's notation system · Order type · Well-partial-orderings

1 Introduction

Over the last decades ordinal notation systems played a central role in determining the proof-theoretic strength of various formal systems for arithmetic and set theory. Proof-theoretic ordinals have in the meantime successfully been calculated for the standard systems of reverse mathematics and even for theories as strong as $(\Pi^1_2\text{-}\mathsf{CA}_0)^-$ although the details for the latter, even after polishing, remain notoriously complicated (e.g., see [9]).

An important but still widely open problem in this area of proof theory is Feferman's *natural well-ordering problem* [4]. It is a conceptual question about when a representation of a well-ordering is considered *natural*. In this article we are not aiming at answering this problem, but we think that it is worthwhile to single out natural properties of existing ordinal notation systems since they might be typical for notation systems in general.

An important facet of this investigation is to study the role of associated well-quasi-orders or well-partial-orders. These orders recently regained considerable interest due to a general formula (suggested by the second author) which predicts

J. Van der Meeren—The first author wants to thank his funding organization Fellowship of the Research Foundation - Flanders (FWO).

A. Beckmann et al. (Eds.): CiE 2015, LNCS 9136, pp. 353–362, 2015.
DOI: 10.1007/978-3-319-20028-6_36

the maximal order type of a given well-partial-order which is defined in terms of finite trees [11,12]. Recent investigations show that this formula is able to handle Friedman-style gap-embeddability for finite labels and a remaining challenge is to verify it for Friedman [10], Gordeev [5,7] and Kriz-style [8] embeddability relations for possibly transfinite ordinal labels.

It turns out that the maximal order types of the associated well-partial-orders can be represented by different natural ordinal notations systems and the question is whether there exists a natural embedding between such representations. In this article we carry out this investigation for the systems $PRJ(P)$ defined by Gordeev [6] and $OT(P)$ defined by Buchholz [2]. Gordeev [6] already showed that $OT(P)$ can be considered as a subsystem of $PRJ(P)$. Other natural embeddings between different systems have already been defined. E.g., Buchholz [1] presented an embedding between systems based on the $\overline{\theta}$-functions and ordinal diagrams. Using the embedding of this article one gets a complete picture by composing the embeddings under consideration.

Gordeev's ordinal notation system [6] is particularly appealing because of its nice behavior with respect to iterated collapsing: $D_\mu D_\nu \alpha = D_{\min\{\mu,\nu\}}\alpha$, where equality means here that the normal forms of the terms are the same. By way of contrast, the treatment of terms of the form $D_\mu D_\nu \alpha$ requires very delicate considerations and Pohlers called this the *worst understood part of local predicativity* [3].

We have not succeeded in defining the D_μ of the Gordeev system by natural functions on the ordinals. Instead, we use, for defining our embedding, a peculiar syntactical operation on terms from Buchholz's system. This definition is different from corresponding definitions for ordinal diagrams and it seems that the nature of Gordeev's system differs considerably from ordinal diagrams. In our perspective Gordeev's approach to notation systems is largely independent of the approaches studied in other sources.

2 Preliminaries

Given a limit ordinal $(P, <)$. In this section, we introduce the notation systems of Buchholz and Gordeev. For $\alpha > 0$, let Ω_α be the αth uncountable cardinal and define Ω_0 as 1. So $\Omega_1 = \aleph_1$ and $\Omega_\omega = \aleph_\omega$.

2.1 Buchholz's Ordinal Notation System

Buchholz's ordinal notation system is based on the ψ_ν-functions [2]. For the sake of completeness, we give the definition of these functions. In [2], the ψ_ν are introduced for $\nu \leq \omega$, but we can generalize them to $\nu \in P$, where $(P, <)$ is a limit ordinal (see [6]).

Definition 1. *The definitions of $C_\nu(\alpha)$ and $\psi_\nu\alpha$ proceeds by transfinite recursion on α simultaneously for all $\nu \in P$. Assume that we have $C_\nu(\xi)$ and $\psi_\nu\xi$ for all $\xi < \alpha$. Define $C_\nu(\alpha)$ as the least set X of ordinals such that*

1. $\Omega_\nu \subseteq X$,
2. $\forall \beta, \gamma \in X(\beta + \gamma \in X)$,
3. $(\forall \xi \in X \cap \alpha)(\forall \mu \in P)(\psi_\mu \xi \in X)$.

Define $\psi_\nu \alpha$ *as* $\min\{\gamma : \gamma \notin C_\nu(\alpha)\}$.

For more information on Buchholz's ψ, we refer to [2]. We state some basic facts of these collapsing functions without mentioning the proofs.

Lemma 1. *For ordinals* $\nu \in P$ *and* α, β,

1. $\psi_\nu 0 = \Omega_\nu$.
2. $\Omega_\nu \leq \psi_\nu \alpha < \Omega_{\nu+1}$.
3. $\mu < \nu$ *yields* $\psi_\mu \alpha < \psi_\nu \beta$.
4. $\alpha \leq \beta$ *yields* $\psi_\nu \alpha \leq \psi_\nu \beta$.
5. $\psi_\nu \alpha$ *is additively closed.*

Based on these functions, one can define the following notation system.

Definition 2. *Let* $T(P)$ *be the least set of terms satisfying*

- $0 \in T(P)$.
- *If* $\alpha \in T(P)$ *and* $\nu \in P$, *then* $D_\nu \alpha \in T(P)$. *We call* $D_\nu \alpha$ *a principal term.*
- *If* $\alpha_0, \ldots, \alpha_n \in T(P)$ *are principal terms and* $n \geq 1$, *then* $\alpha_0 \oplus \cdots \oplus \alpha_n \in T(P)$.

$\alpha_{\pi(0)} \oplus \cdots \oplus \alpha_{\pi(n)}$ is seen as the same term as $\alpha_0 \oplus \cdots \oplus \alpha_n$ for every permutation π of $\{0, \ldots, n\}$. Therefore, $\oplus$ can be interpreted as the natural commutative sum between ordinals.

Definition 3. *Assume* $\alpha, \beta \in T(P)$. *Then* $\alpha < \beta$ *is valid if one of the conditions is satisfied.*

- $\alpha = 0$ *and* $\beta \neq 0$.
- $\alpha = D_\mu \alpha'$, $\beta = D_\nu \beta'$ *and* $\mu < \nu$ *or* $\mu = \nu$ *and* $\alpha' < \beta'$.
- $\alpha = \alpha_0 \oplus \cdots \oplus \alpha_n$, $\beta = \beta_0 \oplus \cdots \oplus \beta_m$ $(n + m \geq 1)$ *with* $\alpha_0 \geq \cdots \geq \alpha_n$, $\beta_0 \geq \cdots \geq \beta_m$ *and*
  - *either* $n < m$ *and* $\forall i \leq n(\alpha_i = \beta_i)$,
  - *or* $\exists i \leq \min\{n, m\}(\alpha_i < \beta_i$ *and* $\forall j < i(\alpha_j = \beta_j))$.

It is easy to see that $D_0 0$ is the successor of 0 in $(T(P), <)$. Furthermore, $D_\nu \alpha$ is additively closed. We want to define a subset of $T(P)$ which corresponds to terms which are in normal form. For this purpose, we introduce $G_\mu \alpha$, a set of coefficients.

Definition 4

- $G_\mu 0 := \emptyset$.
- $G_\mu(\alpha_0 \oplus \cdots \oplus \alpha_n) := G_\mu(\alpha_0) \cup \cdots \cup G_\mu(\alpha_n)$.
- $G_\mu(D_\nu \alpha) := \begin{cases} \emptyset & if\, \nu < \mu, \\ G_\mu(\alpha) \cup \{\alpha\} & if\, \nu \geq \mu. \end{cases}$

Definition 5. *$OT(P)$ is a subset of $T(P)$ which only uses terms in a specific normal form.*

- $0 \in OT(P)$,
- *If $\alpha_0, \dots, \alpha_n$ are principal terms in $OT(P)$, $n \geq 1$ and $\alpha_0 \geq \dots \geq \alpha_n$, then $\alpha_0 \oplus \dots \oplus \alpha_n \in OT(P)$.*
- *If $\alpha \in OT(P)$, $\nu \in P$ and $G_\nu(\alpha) < \alpha$, then $D_\nu \alpha \in OT(P)$.*

The elements of $OT(P)$ are called ordinal terms.

One can prove that $OT(P)$ is a linear ordering. In general, the interpretation of $D_\nu \alpha$ in the ordinal numbers is $\psi_\nu \alpha$. For clarity reasons (Gordeev's system also uses the notation D_ν), we write ψ_ν instead of D_ν from now on. It is also worth to mention that the normal non-commutative sum $\alpha + \beta$ between ordinals can be well-defined on $OT(P)$.

2.2 Gordeev's Ordinal Notation System

Gordeev's ordinal notation system can be found in [6]. We remark that in [6] the Ω_ν are defined in a different way: there the Ω_ν are equal to our $\Omega_{\nu+1}$. The main difference between Gordeev's system and Buchholz's system is that $D_\mu D_\nu \alpha = D_{\min\{\mu,\nu\}} \alpha$ holds in Gordeev's system, whereas this is in general not the case in the system of Buchholz. Equality here means the identity of normal forms.

Definition 6. *Let $T'(P)$ be the least set of terms satisfying*

- *$0 \in T'(P)$ and $\Omega_\nu \in T'(P)$ if $\nu \in P$ and $\nu > 0$.*
- *If $\alpha \in T'(P)$ and $\nu \in P$, then $D_\nu \alpha \in T'(P)$. We call $D_\nu \alpha$ and Ω_ν principal terms.*
- *If $\alpha_0, \dots, \alpha_n \in T'(P)$ are principal terms and $n \geq 1$, then $\alpha = \alpha_0 \oplus \dots \oplus \alpha_n \in T'(P)$.*

We define the normal form α^* of $\alpha \in T'(P)$ and the ordering $<$ on $T'(P)$ simultaneously as follows.

Definition 7. *Assume that $\alpha, \beta \in T'(P)$.*

- $0^* := 0$ *and* $\Omega_\nu^* := \Omega_\nu$.
- $(\alpha_0 \oplus \dots \oplus \alpha_n)^* := (\alpha_{\pi(0)})^* \oplus \dots \oplus (\alpha_{\pi(n)})^*$, *where π is a permutation on $\{0, \dots, n\}$ such that $(\alpha_{\pi(0)})^* \geq \dots \geq (\alpha_{\pi(n)})^*$.*
- $(D_\nu \alpha)^* := \begin{cases} \alpha^* & \text{if } \alpha < \Omega_{\nu+1}, \\ D_\nu(\beta) & \text{if } \alpha \geq \Omega_{\nu+1} \text{ and } \alpha^* = D_\mu \beta \text{ with } \mu > \nu, \\ D_\nu(\alpha^*) & \text{otherwise.} \end{cases}$
- *$\alpha < \beta$ is valid if one of the following holds*
  - *$\alpha = 0$ and $\beta \neq 0$.*
  - *$\alpha^* = \alpha_0 \oplus \dots \oplus \alpha_n$ and $\beta^* = \beta_0 \oplus \dots \oplus \beta_m$ and one of the following is valid.*

1. $n < m$ *and* $\forall i \leq n(\alpha_i = \beta_i)$,
2. $\exists i \leq \min\{n, m\}(\alpha_i < \beta_i$ *and* $\forall j < i(\alpha_j = \beta_j))$.

- $\alpha^* = \Omega_\mu$ *and* $\beta^* = \Omega_\nu$ *with* $\mu < \nu$.
- $\alpha^* = D_\mu\alpha'$ *and* $\beta^* = \Omega_\nu$ *with* $\mu < \nu$.
- $\alpha^* = \Omega_\mu$ *and* $\beta^* = D_\nu\beta'$ *with* $\mu \leq \nu$.
- $\alpha^* = D_\mu\alpha'$ *and* $\beta^* = D_\nu\beta'$ *with* $\mu < \nu$.
- $\alpha^* = D_\mu\Omega_\sigma$ *and* $\beta^* = D_\mu\Omega_\delta$ *with* $\sigma < \delta$.
- $\alpha^* = D_\mu(\alpha_1 \oplus \cdots \oplus \alpha_n)$ *and* $\beta^* = D_\mu(\beta_1 \oplus \cdots \oplus \beta_m)$, *where all* α_i, β_i *are principal and one of the following holds*
  1. $\alpha_1 \oplus \cdots \oplus \alpha_n < \beta_1 \oplus \cdots \oplus \beta_m$ *and* $\forall i \leq n(D_\mu(\alpha_i) < \beta)$,
  2. $m > 1$ *and* $\exists j \leq m(\alpha \leq D_\mu(\beta_j))$.

We remark that these definitions yield $\alpha < \beta \Leftrightarrow \alpha^* < \beta^*$ for $\alpha, \beta \in T'(P)$. Furthermore, if $\alpha \in T'(P)$, then $t = D_\nu D_\mu s$ can never be a *subterm* of α^*: if $\beta \in T'(P)$, then $(D_\mu D_\nu \beta)^* = (D_{\min\{\mu,\nu\}}\beta)^*$. Additionally, $\alpha + \beta$ is also here well-defined. Let $r(\alpha)$ be the least $\nu \in P$ such that $\alpha < \Omega_{\nu+1}$.

Gordeev's ordinal notation system $PRJ(P)$ is defined as $(T'(P), <)$, where the unique normal form of every term α is α^*. Following [6], one can approximate the Veblen hierarchy $\varphi\alpha\beta$ for $\alpha, \beta \in T'(P)$ as follows in $T'(P)$. First define ω^α as $D_0(\Omega_1)$ if $\alpha^* = 0$ and let ω^α be α if $\alpha^* = \Omega_\gamma$ for a certain γ. Otherwise, define ω^α as $D_{r(\alpha)}(\Omega_{r(\alpha)+1} + \alpha)$. Now, for $\alpha, \beta \in T'(P)$, define $\varphi\alpha\beta$ as ω^β if $\alpha^* = 0$. Otherwise, define $\varphi\alpha\beta$ as $D_{r(\beta)}(D_\gamma(\Omega_{\gamma+1} + \alpha) + \beta)$, where $\gamma = \max\{r(\alpha), r(\beta)\}$.

3 Connection Between the Two Notation Systems

Both ordinal notation systems have D_μ in their constructions. In Buchholz's approach, we will write ψ_μ instead of D_μ for clarity reasons. In section four of [6] it is shown that $OT(P)$ can be seen as a proper subsystem of $PRJ(P)$, although it has the same order type. Gordeev's indirect proof of the equality of the order types is based on underlying well-partial-orders. More specifically, Gordeev showed that $\alpha^\#$ is an order-preserving embedding from $OT(P)$ to $PRJ(P)$, where $\alpha^\#$ is defined by replacing every subterm $\psi_\mu\beta$ by $D_\mu(\Omega_{\mu+1} + \beta^\#)$. In this section, we will show that $PRJ(P)$ can be seen as a subsystem of $OT(P)$ using a direct embedding. Hence, the equality of order types follows directly. First, we define inductively a subsystem G of $OT(P)$. We remark that if we talk about Ω_ν in the context of $OT(P)$, we actually mean the ordinal $\psi_\nu 0$.

Definition 8. *For* $\alpha = \psi_{\mu_1}\alpha_1 \oplus \cdots \oplus \psi_{\mu_n}\alpha_n \in OT(P)$, *define* $S\alpha$ *as the ordinal* $\max\{\mu_1, \ldots, \mu_n\}$. *If this sum is in normal form, the maximum is equal to* μ_1. *Define* $S0$ *as* 0.

Definition 9

- $0 \in G$ *and if* $\nu \in P$, *then* $\Omega_\nu = \psi_\nu 0 \in G$.
- *If* $\alpha_0, \ldots, \alpha_n$ *are principal terms in* G , $n \geq 1$ *and* $\alpha_0 \geq \cdots \geq \alpha_n$, *then* $\alpha_0 + \cdots + \alpha_n \in G$.

- *If* $\nu \in P$ *and* $\alpha \in G$*, then* $\psi_\nu(g^*_\nu\alpha + \omega^\alpha) \in G$*, where we define* ω^α *as* $\psi_{S\alpha}(g^*_{S\alpha}\alpha + \alpha)$*. Here* $+$ *is non-commutative, hence* $g^*_\nu\alpha + \omega^\alpha$ *can be equal to* ω^α*.*
- *If* $\nu \in P$*, then* $g^*_\nu(0) = 0$*.*
- *If* $\nu, \mu \in P$*, then* $g^*_\nu(\Omega_\mu) = 0$*.*
- *If* $\nu \in P$*, then* $g^*_\nu(\alpha_0 \oplus \cdots \oplus \alpha_n) = \max\{g^*_\nu(\alpha_0), \ldots, g^*_\mu(\alpha_n)\}$*.*
- *If* $\nu, \mu \in P$*, then* $g^*_\nu\psi_\mu(g^*_\mu\alpha + \omega^\alpha) = \begin{cases} 0 & \text{if } \nu > \mu, \\ g^*_\nu\alpha + \omega^\alpha & \text{if } \nu \leq \mu. \end{cases}$

Ordinals of the form $\psi_\nu\xi$ where ξ is not zero or not of the form $g^*_\nu\alpha + \omega^\alpha$ with $\alpha \in G$, do not appear in G. This syntactical definition of G is very crucial and different from the ones you can find in the literature. In G, the ordinal 1 is equal to $\Omega_0 = \psi_0 0$ and the ordinal ω is equal to $\psi_0(g^*_0 0 + \omega^0)$, because $\omega^0 = \psi_{S0}(g^*_{S0}0 + 0) = \psi_0 0 = 1$.

We remark that ω^α in $OT(P)$ does *not* represent the usual exponentiation function as it is always strictly bigger than α (assertion 5 of the next lemma). The main needed property about ω^α is that it is additively closed. Assertion 7 of the next lemma shows that G is indeed a subset of $OT(P)$.

Lemma 2. *Define* $g_\mu\alpha$ *as* $\max G_\mu\alpha$*. For all* $\alpha, \beta, \gamma, \delta$ *in* G *and* μ, ν *in* P *we have*

1. $g_\mu\alpha \leq g^*_\mu\alpha$.
2. $\mu \leq \nu \Rightarrow g^*_\mu g^*_\nu\alpha \leq g^*_\mu\alpha$.
3. $\mu \leq \nu \Rightarrow g_\mu g^*_\nu\alpha \leq g^*_\mu\alpha$.
4. $\gamma, \delta < \omega^\alpha \Rightarrow \gamma + \delta < \omega^\alpha$.
5. $\alpha < \omega^\alpha$.
6. $\alpha < \beta \Rightarrow \omega^\alpha < \omega^\beta$.
7. $g_\nu(g^*_\nu\alpha + \omega^\alpha) < g^*_\nu\alpha + \omega^\alpha$.

Proof. Assertion 1–3 can be proven by induction on α.

Assertion 4 is trivial because $\psi_{S\alpha}(g^*_{S\alpha}\alpha + \alpha)$ is additively closed in $OT(P)$.

If $\alpha = 0$, then the assertion 5 is trivial. Assume $\alpha = \psi_{\mu_1}\alpha_1 + \cdots + \psi_{\mu_n}\alpha_n$. Then $\alpha < \omega^\alpha$ is true if $\psi_{\mu_1}\alpha_1 < \omega^\alpha = \psi_{S\alpha}(g^*_{S\alpha}\alpha + \alpha) = \psi_{\mu_1}(g^*_{\mu_1}\alpha + \alpha)$. This is the case if $\alpha_1 < g^*_{\mu_1}\alpha + \alpha$. If $\alpha_1 = 0$, this is trivial. Assume $\alpha_1 \neq 0$. $g^*_{\mu_1}\alpha + \alpha = \max\{g^*_{\mu_1}(\psi_{\mu_1}\alpha_1), \ldots, g^*_{\mu_1}(\psi_{\mu_n}\alpha_n)\} + \alpha > g^*_{\mu_1}(\psi_{\mu_1}\alpha_1)$. We know that $\alpha_1 = g^*_{\mu_1}\xi + \omega^\xi$ (because $\psi_{\mu_1}\alpha_1 \in G$), hence $g^*_{\mu_1}(\psi_{\mu_1}\alpha_1) = g^*_{\mu_1}\xi + \omega^\xi = \alpha_1$.

Assertion 6 can be proven by induction on the sum of the lengths of construction of α and β.

Assertion 7 follows from the previous ones. If $\nu > S\alpha$, then $g_\nu(g^*_\nu\alpha + \omega^\alpha) = \max\{g_\nu(g^*_\nu\alpha), g_\nu(\omega^\alpha)\} = g_\nu(g^*_\nu\alpha) \leq g^*_\nu\alpha$, hence we are done. Assume $\nu \leq S\alpha$. Then $g^*_{S\alpha}\alpha \leq g^*_\nu\alpha$. Therefore, $g_\nu(g^*_\nu\alpha + \omega^\alpha) = \max\{g_\nu(g^*_\nu\alpha), g_\nu(\omega^\alpha)\} \leq \max\{g^*_\nu\alpha, g_\nu(\psi_{S\alpha}(g^*_{S\alpha}\alpha + \alpha))\} = \max\{g^*_\nu\alpha, g_\nu(g^*_{S\alpha}\alpha + \alpha), g^*_{S\alpha}\alpha + \alpha\} = \max\{g^*_\nu\alpha, g_\nu(\alpha), g^*_{S\alpha}\alpha + \alpha\} < g^*_\nu\alpha + \omega^\alpha$. □

Now, for every $\nu \in P$, we define a function $\widetilde{D}_\nu : G \to G$. This will be the translation of D_ν in $PRJ(P)$. The definition of $\widetilde{D}_\nu$ is rather syntactical and it is the crucial idea of the embedding.

Definition 10. *For every $\nu \in P$, define $\widetilde{D}_\nu : G \to G$ as follows. If $\alpha < \Omega_{\nu+1}$ let $\widetilde{D}_\nu(\alpha)$ be α. Now, for $\alpha \geq \Omega_{\nu+1}$, define $\widetilde{D}_\nu(\alpha)$ as follows:*

- *If $\mu \geq \nu + 1$, then $\widetilde{D}_\nu(\Omega_\mu) := \psi_\nu(g^*_\nu \Omega_\mu + \omega^{\Omega_\mu})$.*
- *$\widetilde{D}_\nu(\beta_0 + \cdots + \beta_n) := \psi_\nu(g^*_\nu(\beta_0 + \cdots + \beta_n) + \omega^{\beta_0 + \cdots + \beta_n})$ if $n \geq 1$.*
- *If $\mu \geq \nu + 1$, then $\widetilde{D}_\nu(\psi_\mu(g^*_\mu \beta + \omega^\beta)) := \psi_\nu(g^*_\nu \beta + \omega^\beta)$.*

Note that $\widetilde{D}_\nu \alpha < \Omega_{\nu+1}$.

Definition 11. *Define $o : PRJ(P) \to G \subseteq OT(P)$ in the following recursive way. If $\alpha \in T'(P)$ is not in normal form, then $o(\alpha) := o(\alpha^*)$. Assume $\alpha = \alpha^*$, then define $o(\alpha)$ as follows.*

- *$o(0) :- 0$.*
- *$o(\Omega_\nu) := \Omega_\nu$.*
- *$o(\alpha_0 + \cdots + \alpha_n) := o(\alpha_0) + \cdots + o(\alpha_n)$.*
- *$o(D_\nu \alpha) := \widetilde{D}_\nu(o(\alpha))$.*

If $\alpha \in PRJ(P)$ is in normal form, then α is either 0, Ω_ν or $\alpha = D_{\mu_1}\alpha_1 + \cdots + D_{\mu_n}\alpha_n$, where α_i is also in normal form and α_i is different from $D_\nu \gamma$ for all γ. Therefore, $\alpha_i \geq \Omega_{\mu_i+1}$ and it is equal to Ω_ν for a certain ν or it is a sum. Hence, $o(D_{\mu_1}\alpha_1) = \widetilde{D}_{\mu_1} o(\alpha_1) = \psi_{\mu_1}(g^*_{\mu_1}(o(\alpha_1)) + \omega^{o(\alpha_1)})$. Note that $o(1) = o(D_0 \Omega_1) = \widetilde{D}_0 \Omega_1 = \psi_0(\omega^{\Omega_1}) = \psi_0(\psi_1(\Omega_1))$.

Lemma 3. *Assume $\alpha_1, \ldots, \alpha_n, D_\gamma(\alpha_1 + \cdots + \alpha_n), D_\gamma \beta$ ($n \geq 2$) are elements of $PRJ(P)$ in normal from, α_i are principal terms, $o(\alpha_1 + \cdots + \alpha_n) < o(\beta)$ and $o(D_\gamma(\alpha_i)) < o(D_\gamma(\beta))$. Then $o(D_\gamma(\alpha_1 + \cdots + \alpha_n)) < o(D_\gamma \beta)$.*

Proof. It is important to notice that the $D_\gamma(\alpha_i)$'s are not necessarily in normal form. $o(D_\gamma(\alpha_1 + \cdots + \alpha_n)) < o(D_\gamma \beta)$ is the same as saying

$$\psi_\gamma(g^*_\gamma(o(\alpha_1) + \cdots + o(\alpha_n)) + \omega^{o(\alpha_1 + \cdots + \alpha_n)}) < \psi_\gamma(g^*_\gamma(o(\beta)) + \omega^{o(\beta)}),$$

which is equivalent with

$$g^*_\gamma(o(\alpha_1) + \cdots + o(\alpha_n)) + \omega^{o(\alpha_1 + \cdots + \alpha_n)} < g^*_\gamma(o(\beta)) + \omega^{o(\beta)}.$$

Because $o(\alpha_1 + \cdots + \alpha_n) < o(\beta)$, we only have to prove that $g^*_\gamma(o(\alpha_1) + \cdots + o(\alpha_n)) < g^*_\gamma(o(\beta)) + \omega^{o(\beta)}$ (this is because $\omega^{o(\beta)}$ is additively closed). $g^*_\gamma(o(\alpha_1) + \cdots + o(\alpha_n)) = g^*_\gamma(o(\alpha_i))$ for a certain i. Because α_i is principal, we know that $\alpha_i = \Omega_\nu$ or $\alpha_i = D_\nu \delta$ for certain ν, δ. In the former case $g^*_\gamma(o(\alpha_i)) = 0$, hence $g^*_\gamma(o(\alpha_1) + \cdots + o(\alpha_n)) < g^*_\gamma(o(\beta)) + \omega^{o(\beta)}$ follows trivially. Assume that we are in the latter case. Then $g^*_\gamma(o(\alpha_i)) = g^*_\gamma(\psi_\nu(g^*_\nu(o(\delta)) + \omega^{o(\delta)}))$. If $\nu < \gamma$, again $g^*_\gamma(o(\alpha_i)) = 0$ and we are done. Assume $\nu \geq \gamma$. Then $g^*_\gamma(o(\alpha_i)) = g^*_\gamma(o(\delta)) + \omega^{o(\delta)}$. Now, $o(D_\gamma(\alpha_i)) = o(D_\gamma D_\nu \delta) = o(D_\gamma \delta) < o(D_\gamma(\beta))$, hence $\psi_\gamma(g^*_\gamma(o(\delta)) + \omega^{o(\delta)}) < \psi_\gamma(g^*_\gamma(o(\beta)) + \omega^{o(\beta)})$. This implies

$$g^*_\gamma(o(\delta)) + \omega^{o(\delta)} < g^*_\gamma(o(\beta)) + \omega^{o(\beta)},$$

so $g^*_\gamma(o(\alpha_i)) < g^*_\gamma(o(\beta)) + \omega^{o(\beta)}$. □

Lemma 4. *Assume $\beta_1, \ldots, \beta_n, \beta = D_\gamma(\beta_1 + \cdots + \beta_n)$ $(n \geq 2)$ are in normal form. Furthermore, suppose they are principal terms. Then $o(D_\gamma \beta_i) < o(\beta)$ for all i.*

Proof. It is worth to mention that $D_\gamma(\beta_i)$ is not necessarily in normal form. We see that $o(\beta) = \psi_\gamma(g^*_\gamma(o(\beta_1 + \cdots + \beta_n)) + \omega^{o(\beta_1 + \cdots + \beta_n)})$. Because β_i is in normal form, we obtain that $\beta_i = D_{\gamma_i}\beta'_i$ or $\beta_i = \Omega_{\gamma_i}$.

Assume that $\beta_i = \Omega_{\gamma_i}$. Then $o(D_\gamma \beta_i) = \widetilde{D}_\gamma(\Omega_{\gamma_i})$. If $\gamma_i < \gamma + 1$, then $o(D_\gamma \beta_i) = \Omega_{\gamma_i} = \psi_{\gamma_i} 0 \leq \psi_\gamma 0 < o(\beta)$. If $\gamma_i \geq \gamma + 1$, then $o(D_\gamma \beta_i) = \psi_\gamma(g^*_\gamma \Omega_{\gamma_i} + \omega^{\Omega_{\gamma_i}}) = \psi_\gamma(\omega^{\Omega_{\gamma_i}}) \leq \psi_\gamma(\omega^{o(\beta_1 + \cdots + \beta_n)}) \leq \psi_\gamma(g^*_\gamma(o(\beta_1 + \cdots + \beta_n)) + \omega^{o(\beta_1 + \cdots + \beta_n)}) = o(\beta)$.

Assume that $\beta_i = D_{\gamma_i}\beta'_i$. If $\gamma_i < \gamma$, then $o(D_\gamma \beta_i) = o(D_\gamma D_{\gamma_i}\beta'_i) = o(D_{\gamma_i}\beta'_i) = \psi_{\gamma_i}(g^*_{\gamma_i}(o(\beta'_i)) + \omega^{o(\beta'_i)}) < \psi_\gamma(g^*_\gamma(o(\beta_1 + \cdots + \beta_n)) + \omega^{o(\beta_1 + \cdots + \beta_n)}) = o(\beta)$. If $\gamma_i \geq \gamma$, then $o(D_\gamma \beta_i) = o(D_\gamma D_{\gamma_i}\beta'_i) = o(D_\gamma \beta'_i) = \psi_\gamma(g^*_\gamma(o(\beta'_i)) + \omega^{o(\beta'_i)})$. Therefore, $o(D_\gamma \beta_i) < o(\beta)$ is valid iff

$$g^*_\gamma(o(\beta'_i)) + \omega^{o(\beta'_i)} < g^*_\gamma(o(\beta_1 + \cdots + \beta_n)) + \omega^{o(\beta_1 + \cdots + \beta_n)}.$$

This is true because $g^*_\gamma(o(\beta_1 + \cdots + \beta_n)) \geq g^*_\gamma(o(\beta_i)) = g^*_\gamma(\widetilde{D}_{\gamma_i}(o(\beta'_i))) = g^*_\gamma(\psi_{\gamma_i}(g^*_{\gamma_i}(o(\beta'_i)) + \omega^{o(\beta'_i)})) = g^*_\gamma(o(\beta'_i)) + \omega^{o(\beta'_i)}$. □

Theorem 1. *o is an order preserving embedding from $PRJ(P)$ to $OT(P)$.*

Proof. We prove that $\alpha < \beta$ implies $o(\alpha) < o(\beta)$ by induction on the sum of the lengths of constructions of α and β. If α or β are not in normal form, then we can easily conclude the assertion from the induction hypothesis. Assume from now on that α and β are in normal form. If $\beta = 0$, then the assertion is trivial. Suppose that $\beta > 0$.

Case (i) $\beta = \Omega_\delta$ with $\delta \in P$.
If $\alpha \neq D_\gamma \alpha'$, then $o(\alpha) < o(\beta)$ easily follows from $\alpha < \beta$. Assume $\alpha = D_\gamma \alpha'$. Then $\gamma < \delta$, hence $o(\alpha) = \widetilde{D}_\gamma o(\alpha') < \Omega_{\gamma+1} \leq \Omega_\delta = o(\Omega_\delta)$.

Case (ii) $\beta = \beta_0 + \cdots + \beta_n$ with $n \geq 1$.
If $\alpha \neq D_\gamma \alpha'$, then $o(\alpha) < o(\beta)$ easily follows from $\alpha < \beta$. Assume $\alpha = D_\gamma \alpha'$. Then $\beta_0 = \alpha$ or $\alpha < \beta_0$. Hence, $\alpha \leq \beta_0$. So, $o(\alpha) \leq o(\beta_0) < o(\beta)$.

Case (iii) $\beta = D_\delta \beta'$.
Because β is in normal form, we obtain $\beta' \geq \Omega_{\delta+1}$. If $\alpha = 0$, then $o(\alpha) < o(\beta)$ follows easily.

Subcase 1: $\alpha = \alpha_0 + \cdots + \alpha_n$.
$\alpha < \beta$ yields $\alpha_n \leq \cdots \leq \alpha_0 < \beta$. Therefore, $o(\alpha) < o(\beta)$.

Subcase 2: $\alpha = \Omega_\gamma$.
$\alpha < \beta$ implies $\gamma \leq \delta$. We know $o(\alpha) = \Omega_\gamma$. Furthermore, $\beta' \geq \Omega_{\delta+1}$ yields $\Omega_{\delta+1} = o(\Omega_{\delta+1}) \leq o(\beta')$. So, $o(\beta) = \widetilde{D}_\delta(o(\beta')) = \psi_\delta(\xi) > \Omega_\delta \geq \Omega_\gamma = o(\alpha)$. We conclude that $o(\alpha) < o(\beta)$.

Subcase 3: $\alpha = D_\gamma \alpha'$.

Because both α and β are in normal form, we obtain that $\alpha' \geq \Omega_{\gamma+1}$ and $\beta' \geq \Omega_{\delta+1}$. Hence, $o(\alpha) = \psi_\gamma \xi$ and $o(\beta) = \psi_\delta \zeta$ for certain ξ and ζ. $\alpha < \beta$ yields either $\gamma < \delta$ (hence $o(\alpha) < o(\beta)$) or $\gamma = \delta$ and one of the following holds

1. $\alpha' = \Omega_\nu$, $\beta' = \Omega_\mu$ and $\nu < \mu$.
2. $\alpha' = \alpha_1 + \cdots + \alpha_n$, $\beta' = \beta_1 + \cdots + \beta_m$ and one of the following is valid
   (a) $\alpha' < \beta'$ and $\forall i \leq n(D_\gamma(\alpha_i) < \beta = D_\gamma(\beta'))$.
   (b) $m > 1$ and $\exists j \leq m(\alpha \leq D_\gamma(\beta_j))$.

If we are in case 1, $o(\alpha) < o(\beta)$ easily follows.

Assume that we are in case 2(a). It is not possible that both $n = m = 1$ in this case, otherwise we would be in case 1 (α_i, β_i are principal). Assume $n > 1$. Then $\alpha = D_\gamma(\alpha_1 + \cdots + \alpha_n)$, $\beta = D_\gamma \beta'$, $\alpha_1, \ldots, \alpha_n$ are in normal form and $\alpha_1 + \cdots + \alpha_n < \beta'$. Additionally, the induction hypothesis yields $o(D_\gamma(\alpha_i)) < o(\beta) = o(D_\gamma \beta')$ and $o(\alpha_1 + \cdots + \alpha_n) < o(\beta')$. Lemma 3 then implies $o(\alpha) = o(D_\gamma(\alpha_1 + \cdots + \alpha_n)) < o(D_\gamma \beta') = o(\beta)$.

Assume now that we are in case 2(a) and $n = 1$. The normal form of α yields $\alpha = D_\gamma \alpha_1$ and α_1 is equal to Ω_ν for $\nu \geq \gamma + 1$. Then $\alpha' = \alpha_1 < \beta'$ yields that there is a certain β_j bigger than Ω_ν, hence $\Omega_\nu \leq o(\beta_j) < o(\beta')$. Now,

$$o(\alpha) = \psi_\gamma(g^*_\gamma \Omega_\nu + \omega^{\Omega_\nu}) = \psi_\gamma(\omega^{\Omega_\nu}) < \psi_\gamma(g^*_\gamma(o(\beta')) + \omega^{o(\beta')}) = o(\beta)$$

holds iff $\omega^{\Omega_\nu} < g^*_\gamma(o(\beta')) + \omega^{o(\beta')}$. The latter is true because $\Omega_\nu < o(\beta')$.

If we are in case 2(b), the induction hypothesis yields $o(\alpha) \leq o(D_\gamma \beta_j)$. From Lemma 4 we also obtain $o(D_\gamma \beta_j) < o(\beta)$. Hence $o(\alpha) < o(\beta)$. □

The main theorem implies that Gordeev's $PRJ(P)$ can be interpreted as a subsystem G of Buchholz's $OT(P)$. Hence, the order type of $PRJ(P)$ is less than or equal to the order type of $OT(P)$. In section four of [6] it is shown that $OT(P)$ can be seen as a subsystem of $PRJ(P)$, from which we can conclude that they have exactly the same order type. It is also worth the mention that the composition of $o(\cdot)$ and $\cdot^\#$ from the beginning Sect. 3 is not the identity:

$$o((\psi_\nu \beta)^\#) = \widetilde{D}_\nu(\Omega_{\nu+1} + o(\beta^\#)) = \psi_\nu(g^*_\nu(o(\beta^\#)) + \omega^{\Omega_{\nu+1} + o(\beta^\#)}),$$

which is in general not equal to $\psi_\nu \beta$. To explain our results more intuitively, we present the image under o of the approximation of (the first level of) Veblen's hierarchy $\varphi 0 \beta$ in $PRJ(P)$ (see Subsect. 2.2) without mentioning the proofs: $o(\varphi 0 \beta)$ is equal to $o(1) = \psi_0(\psi_1(\Omega_1))$ if $\beta^* = 0$. If $\beta = \Omega_\gamma$, then $o(\varphi 0 \beta)$ is Ω_γ. Otherwise,

$$o(\varphi 0 \beta) = \psi_{r(\beta)}(g^*_{r(\beta)}(\beta) + \psi_{r(\beta)+1}(g^*_{r(\beta)+1}(\beta) + \Omega_{r(\beta)+1} + \beta)).$$

In a similar way, one can also calculate $o(\varphi \alpha \beta)$ for arbitrary $\alpha, \beta \in T'(P)$.

With regard to this topic, some open questions remain. E.g. what is the order type of Gordeev's system if we only have $+1$ and not the general $+$ operator? Is it ε_0 if $P = \omega$? Furthermore, what happens if P is a well-partial-order? Is the maximal order type of $OT(P)$ then equal to the order type of $OT(o(P))$, where $o(P)$ is the maximal order type of the well-partial-order P?

References

1. Buchholz, W.: Normalfunktionen und konstruktive Systeme von Ordinalzahlen. In: $\models$ ISILC Proof Theory Symposion (Proceedings of the International Summer Institute and Logic Colloquium, Kiel, 1974). Lecture Notes in Mathematics, vol. 500, pp. 4–25. Springer, Berlin (1975)
2. Buchholz, W.: A new system of proof-theoretic ordinal functions. Ann. Pure Appl. Log. **32**(3), 195–207 (1986)
3. Buchholz, W., Feferman, S., Pohlers, W., Sieg, W.: Iterated Inductive Definitions and Subsystems of Analysis: Recent Proof-theoretical Studies. Lecture Notes in Mathematics, vol. 897. Springer, Berlin (1981)
4. Crossley, J.N., Kister, J.B.: Natural well-orderings. Arch. Math. Logik Grundlag. **26**(1–2), 57–76 (1986/1987)
5. Gordeev, L.: Generalizations of the one-dimensional version of the Kruskal-Friedman theorems. J. Symb. Log. **54**(1), 100–121 (1989)
6. Gordeev, L.: Systems of iterated projective ordinal notations and combinatorial statements about binary labeled trees. Arch. Math. Log. **29**(1), 29–46 (1989)
7. Gordeev, L.: Generalizations of the Kruskal-Friedman theorems. J. Symb. Log. **55**(1), 157–181 (1990)
8. Kříž, I.: Well-quasiordering finite trees with gap-condition. Proof of Harvey Friedman's conjecture. Ann. of Math. (2) **130**(1), 215–226 (1989)
9. Rathjen, M.: An ordinal analysis of parameter free Π^1_2-comprehension. Arch. Math. Log. **44**(3), 263–362 (2005)
10. Simpson, S.G.: Nonprovability of certain combinatorial properties of finite trees. In: Harrington, L.A., Morley, M.D., Ščedrov, A., Simpson, S.G. (eds.) Harvey Friedman's Research on the Foundations of Mathematics. Studies in Logic and the Foundations of Mathematics, vol. 117, pp. 87–117. North-Holland, Amsterdam (1985)
11. Van der Meeren, J., Rathjen, M., Weiermann, A.: Well-partial-orderings and the big Veblen number. Arch. Math. Log. **54**(1–2), 193–230 (2015)
12. Weiermann, A.: A computation of the maximal order type of the term ordering on finite multisets. In: Ambos-Spies, K., Löwe, B., Merkle, W. (eds.) CiE 2009. LNCS, vol. 5635, pp. 488–498. Springer, Heidelberg (2009)

Author Index

Printed in the United States
By Bookmasters